Quick Reference Guide to *SuperCalc® 4*

<E> signifies <Enter> or <Return>   TM Trade MarK of COMPUTER ASSOCIATES INTERNATIONAL, INC.

Commands

/ Arrange —
- Column letter — <E> for entire col.; ascending sort ; no adjust — row or block range — Ascend — Adjust — Go — row (secondary) — Ascend
- Row number — col. or block range — Descend — No adj. — Options — col. (secondary) — Descend
- <E> for entire row.; ascending sort ; no adjust

/ Blank —
- range --- <E>
- <E> for current cell
- * graph range

/ Copy -
- from range ---, to upper/left cell of destination range --- — <E> adjust — No adjust
- from * graph number (1-9), to graph number (1-9) <E> — , options — Ask for adjust / Values only / + - * /

/ Delete —
- Row — row range (to delete one or more rows of data) --- <E>
- Column — column range (to delete one or more columns of data) --- <E> — Left (data to right of deletion moves left)
- Block — block range --- <E> — Up (data below deletion moves up)
- File — filename (to delete a file from disk) <E> or F3 for Directory

/ Edit —
- any cell --- <E>
- <E> for current cell

/ Format —
- Global level — Accept selected formats — Text aligned - Left, Right, Center
- Column level - column range --- — Integer for no decimals — * for asterisk linear display
- Row level ---- row range to 254 or Remaining — General (num. with best fit) — User-defined format - (1-8)
- Entry level - any range --- — Exponential numbers only — Hide values
- Define table (User-defined formats. $n,nnn; — $ for two decimal places — Default settings (G,R,TL,9)
  (neg. #); O=blank; %; dec. places; scaling — Right numeric justification — Width of columns - (0-127) column width
  — Left numeric justification

/ Global —
- Optimum Spreadsheet Conditions menu — memory usage (max. speed or spreadsheet data-space)
- Keep (default menus, & drive — spreadsheet boundary (up to 255 col. by 9999 rows)
- Graphics menus — screen control
- Row, Column, or Dependency order of calc. — Quit Global
- Iteration control — Colors menu (graph component colors)
- Manual or Automatic recalculation — Fonts menu (graph type styles)
- + or - sets on/off toggles state — Layout menu (graph or sheet size & graph position)
- Formula display (on/off) — Options menu (graph appearance & device settings)
- * for test entry (on/off) — Device Selection menu (select graphics printer or plotter)
- Labels for range names (on/off) — Quit — for Delta = 0.01
- Protect to edit protected cells (on/off) — Quit Global — Delta — cell containing Delta---<E>
- Border display (on/off) — Solve (SuperCalc4 controls iter's) — Range to converge to Delta---<E>
- Next to auto-advance cursor (on/off) — Fixed (1-99) fixed max. number of iter's <E>
- Tab to skip blank & protect cells (on/off)

/ Insert —
- Row of empty cells — row range (one or more empty rows) ---<E>
- Column of empty cells — column range (one or more empty columns) ---<E> — Right (dislocated data moves right)
- Block of empty cells — block range (an empty rectangle) ---<E> — Down (dislocated data moves down)

/ Load —
- filename, — All (contents & values)
- F3 for Directory — Values only
  — Consolidate
  — Part from range ---, to upper/left cell --- — <E> adjust — No adjust
  — Names (loads range names from file specified) — , options — Ask for adjust / Values only / + - * /
  — Graphs from graph range ---, to fist graph number (1-9)

/ Move —
- Block from block range ---, to range ---
- Row from row range ---, to row number --- (will be top row if move is up; bot. row if move is down)
- Column from col. range ---, to col. letter --- (will be left col. if move is left; right col. if move is right)

/ Name —
- Create range name ---, range --- <E> — Left - asgns names from rightmost cells of range ---<E>
- Delete range name --- <E> — Up - asgns names from bottommost cells of range ---<E>
- Labels — Right - asgns names from leftmost cells of range ---<E>
- Zap — No cancels command, Yes removes all Range Names — Down - asgns names from topmost cells of range ---<E>

/ Output —
- Printer — Range (for output) — Quit (exit Opt.) or All (resets all Opt. to defaults)
- File — filename, — Go (to output the file) — Report (Formatted, Contents)
  — F3 for Directory — Console (to preview on-screen) — Layout (Page-length, Width, Margins)
  — Line (sends line-feed) — Paper (Wait, Auto page, Double space, Line-feed
  — Change name — Page (sends form-feec) — Borders (State, Character) only)
  — Backup — Options (menu) — Titles (Auto, Manual, None)
  — Overwrite — Zap (to reset options) — Copies (number) <E>
  — Align (resets top of page) — Headers or Footers (1 to 4)
  — Quit (exit Output) — Setup (setup string for printer)

/ Protect — range --- <E> or for current cell

/ Quit —
- Yes to exit from SuperCalc4 (does not save current work)
- No to cancel this command — filename
- To quit & re-load SC4 or load another program — F3 for Directory

/ Save —
- filename, — Change name — All (contents & values)
- F3 for directory — Backup — Values only — All (contents & values) - range --- <E>
  — Overwrite — Part — Values only - range --- <E>

/ Title —
- Horizontal lock Both
- Vertical lock Clear

/ Unprotect — range --- <E> or for current cell

/ View —
- Show current graph on-screen
- Data — Variable range --- <E> — Quit Options — Axis scale labels
  — SPACE to skip — Format — Time-Labels — format
  — - to clear — Variable-Labels — options
- Graph-Type — Pie Bar Stacked-Bar — Point-Labels —
  — Line X-Y Area Hi-Lo — % pie segment labels — Quit
- Time-Labels — range --- <E> — Explosion — All pie segments — Width
  — - to clear — Quit Headings — None — Default
- Variable-Labels — variable range --- <E> — Main - cell --- <E> — Segments - (1-8)
- Point-Labels — SPACE to skip — Sub - cell --- <E> — One Variable — variable A-J <E>
  — - to clear — X-axis - cell --- <E> — Pie-Mode — All Variables — element 1-9999 <E>
- Headings — Y-axis - cell --- <E> — X-axis — cell --- <E>
- Options — Scaling — Y-axis — SPACE for Auto-Scaling
- ? for current graph description summary
- (1-9) graph #

/ Window —
- Horizontal or Vertical split Synchronize split-wise scroll
- Clear to right or below split Unsynchronize split-wise scroll

/ Zap —
- No or Yes to delete current spreadsheet; retains settings of Global menus, Output Setup, & Directory
- Contents, same as Yes, but also retains User-defined format, default disk drive

// Data mgt. —
- Quit Data mgt.
- Input — range --- <E> — ↓ next record — Yes to accept
- Criterion — - to clear — ↑ previous reccrd — No to reject
- Output — → next field — → next field
- Find — ← previous field — ← previous field
- Extract records specified — <E> to cancel — ← to cancel
- Select records or veto
- Remain at current location

// Export —
- 1-2-3, SC3 — Change — DIF ONLY — All
- XDIF, DIF — filename, — Backup — Colwise — Values (1-2-3, SC3, XDIF only)
- CSV — F3 for Directory — Overwrite — Rowwise — Part or range --- <E>

// Import —
- 1-2-3, CSV — DIF ONLY — All or Values only (1-2-3, XDIF)
- XDIF, NUMBERS — filename, — Colwise — Consolidate (1-2-3, XDIF, DIF, CSV only)
- DIF, TEXT — F3 for Directory — Rowwise — Part from range ---, options (1-2-3, XDIF, DIF, CSV only)

// Macro —
- Learn range --- <E>, (Alt-F4 toggles LEARN mode; Alt-F6 toggles DIRECT mode) — All, Macros only, Labels & Macros,
- Read or Write - filename, range --- — Comments & Macros

INTRINSIC FUNCTIONS AND OPERATORS

Logical Functions (Partial List)
- AND(cond_a,condb)
- FALSE
- IF(cond,c/v1,c/v2)
- NOT(cond)
- OR(cond_a,condb)
- TRUE

Financial Functions
- FV(pmt,int,per)
- IRR((guess,)range)
- NPV(disc,range)
- PMT(prin,int,per)
- PV(pmt,int,per)

Mathematical Functions
- ABS(c/v)
- ACOS(c/v)
- ASIN(c/v)
- COS(c/v)
- EXP(c/v)
- EXP(c/v)
- INT(c/v)
- LN(c/v)
- LOG(c/v)
- MOD(c/v1,c/v2)
- PI
- RAN
- ROUND(c/v,places)
- SIN(c/v)
- SQRT(c/v)
- TAN(c/v)

Special Functions
- ERROR
- HLOOKUP(c/v,range,offset)
- LOOKUP(c/v,range)
- NA
- VLOOKUP(c/v,range,offset)

Statistical Functions
- AV(range)
- COUNT(range)
- MAX(range)
- MIN(range)
- STD(range)
- SUM(range)
- VAR(range)

+	Plus
-	Minus
*	Multiply
/	Divide
^ (or **)	Raise to a power
%	Percent
=	Equal
<>	Not equal to
<	Less than
<=	Less than or equal
>	Greater than
>=	Greater than or equal

c/v - a Cell reference or Value
cond - algebraic Condition(e.g. A5>5)
disc - Discount rate
guess - a Guess
int - Interest rate
per - Period or term
places - number of Places after decimal
pmt - Payment
prn - Principal
range - a Range of cells (e.g. B3.C6)

FUNCTION KEY ACTION

Crtl-Break Cancels Command Line
- F1 Help
- F2 Edit Current Cell Contents
- F3 Displays Named Range DIR
- F4 Converts to Abs Cell Ref
- F5 GoTo Command (also =)
- F6 Change Window (also ;)
- F7 Calculate Sheet
- F8 Resume Macro
- F9 Graph on Plotter
- F10 Graph on Screen

PREFIXES FOR CELL ENTRIES

Letters and most symbols begin text mode
" begins forced text mode
' begins repeating entry
Numbers, functions, cell references, +, -, ., begin formulas

RANGES CELL COLUMN ROW

SINGLE CELL
  D5 is col D row 5
BLOCK OF CELLS
  Two opposite Corner Cel's
  D5.F48 or F48.D5
FULL COLUMN: F
MULTI-COLUMNS: B.G
PARTIAL COLUMN: C3.C10
FULL ROW: 39
MULTI-ROWS: 45.83
PARTIAL ROW: G25.M25

Quick Reference Guide to SuperCalc³ RELEASE 2

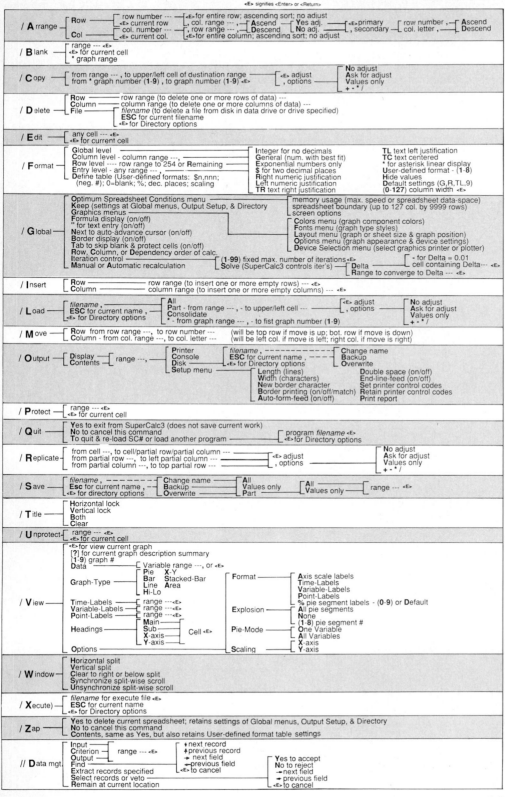

<E> signifies <Enter> or <Return>

/ Arrange
- Row
  - row number --- <E> for entire row; ascending sort; no adjust
  - <E> current row
  - , col. range ---, ⌐ Ascend ⌐ Yes adj. ⌐ , <E> primary ⌐ row number , ⌐ Ascend
  - Descend — No adj. — , secondary — col. letter , — Descend
- Col
  - col. number --- , row range ---,
  - <E> current col.
  - <E> for entire column; ascending sort; no adjust

/ Blank
- range --- <E>
- <E> for current cell
- * graph range

/ Copy
- from range ---, to upper/left cell of destination range ⌐ ⌐ <E> adjust ⌐ No adjust
- from * graph number (1-9), to graph number (1-9) <E> — , options — Ask for adjust / Values only / + - * /

/ Delete
- Row — row range (to delete one or more rows of data) ---
- Column — column range (to delete one or more columns of data) ---
- File — filename (to delete a file from disk in data drive or drive specified) — ESC for current filename — <E> for Directory options

/ Edit
- any cell --- <E>
- <E> for current cell

/ Format
- Global level
- Column level - column range ---,
- Row level ---- row range to 254 or Remaining
- Entry level - any range ---,
- Define table (User-defined formats: $n,nnn; (neg. #); 0=blank; %; dec. places; scaling
- Integer for no decimals
- General (num. with best fit)
- Exponential numbers only
- $ for two decimal places
- Right numeric justification
- Left numeric justification
- TR text right justification
- TL text left justification
- TC text centered
- * for asterisk linear display
- User-defined format - (1-8)
- Hide values
- Default settings (G,R,TL,9)
- (0-127) column width <E>

/ Global
- Optimum Spreadsheet Conditions menu
- Keep (settings at Global menus, Output Setup, & Directory)
- Graphics menus
- Formula display (on/off)
- " for text entry (on/off)
- Next to auto-advance cursor (on/off)
- Border display (on/off)
- Tab to skip blank & protect cells (on/off)
- Row, Column, or Dependency order of calc.
- Iteration control
- Manual or Automatic recalculation
  - memory usage (max. speed or spreadsheet data-space)
  - spreadsheet boundary (up to 127 col. by 9999 rows)
  - screen colors
    - Colors menu (graph component colors)
    - Fonts menu (graph type styles)
    - Layout menu (graph or sheet size & graph position)
    - Options menu (graph appearance & device settings)
    - Device Selection menu (select graphics printer or plotter)
  - (1-99) fixed max. number of iterations<E>
  - Solve (SuperCalc3 controls iter's)
    - Delta ⌐ - for Delta = 0.01 — cell containing Delta--- <E>
    - Range to converge to Delta --- <E>

/ Insert
- Row — row range (to insert one or more empty rows) --- <E>
- Column — column range (to insert one or more empty columns) --- <E>

/ Load
- filename ,
- ESC for current name
- <E> for Directory options
  - All
  - Part - from range ---, to upper/left cell --- ⌐ <E> adjust ⌐ No adjust
  - Consolidate — , options — Ask for adjust / Values only / + - * /
  - * - from graph range ---, to fist graph number (1-9)

/ Move
- Row from row range ---, to row number --- (will be top row if move is up; bot. row if move is down)
- Column - from col. range ---, to col. letter --- (will be left col. if move is left; right col. if move is right)

/ Output
- Display
- Contents ⌐ range ---,
  - Printer
  - Console
  - Disk — filename — ⌐ Change name — ⌐ All
  - Setup menu — ESC for current name , — Backup — Values only — ⌐ All
    - <E> for Directory options — Overwrite — Part — Values only — ⌐ range --- <E>
  - Length (lines)
  - Width (characters)
  - New border character
  - Border printing (on/off/match)
  - Auto-form-feed (on/off)
  - Double space (on/off)
  - End-line-feed (on/off)
  - Set printer control codes
  - Retain printer control codes
  - Print report

/ Protect
- range --- <E>
- <E> for current cell

/ Quit
- Yes to exit from SuperCalc3 (does not save current work)
- No to cancel this command
- To quit & re-load SC# or load another program — program filename <E> — <E>for Directory options

/ Replicate
- from cell ---, to cell/partial row/partial column ---
- from partial row ---, to left partial column --- ⌐ <E> adjust ⌐ No adjust
- from partial column ---, to top partial row --- — , options — Ask for adjust / Values only / + - * /

/ Save
- filename ,
- Esc for current name, —
- <E> for directory options
  - Change name — All
  - Backup — Values only — ⌐ All
  - Overwrite — Part — Values only — ⌐ range --- <E>

/ Title
- Horizontal lock
- Vertical lock
- Both
- Clear

/ Unprotect
- range --- <E>
- <E> for current cell

/ View
- <E> for view current graph
- [?] for current graph description summary
- (1-9) graph #
- Data ⌐ Variable range ---, or <E>
- Graph-Type — Pie / X-Y / Bar / Stacked-Bar / Line / Area / Hi-Lo
  - Format — Axis scale labels / Time-Labels / Variable-Labels / Point-Labels / % pie segment labels - (0-9) or Default
- Time-Labels ⌐ range ---<E>
- Variable-Labels ⌐ range ---<E>
- Point-Labels — range ---<E>
  - Explosion — All pie segments / None / (1-8) pie segment #
- Headings — Main / Sub — Cell <E>
  - X-axis / Y-axis
  - Pie-Mode — One Variable / All Variables
- Options
  - Scaling — X-axis / Y-axis

/ Window
- Horizontal split
- Vertical split
- Clear to right or below split
- Synchronize split-wise scroll
- Unsynchronize split-wise scroll

/ Xecute)
- filename for execute file <E>
- ESC for current name
- <E> for Directory options

/ Zap
- Yes to delete current spreadsheet; retains settings of Global menus, Output Setup, & Directory
- No to cancel this command
- Contents, same as Yes, but also retains User-defined format table settings

// Data mgt.
- Input
- Criterion — range --- <E>
- Output
- Find ⌐ ↑ next record / ↓ previous record / → next field / ← previous field / <E> to cancel
- Extract records specified
- Select records or veto ⌐ Yes to accept / No to reject / → next field / ← previous field / <E> to cancel
- Remain at current location

INTRINSIC FUNCTIONS AND OPERATORS

Logical Functions
- AND(cond$_a$,cond$_b$)
- FALSE
- IF(cond,c/v$_1$,c/v$_2$)
- NOT(cond)
- OR(cond$_a$,cond$_b$)
- TRUE

Financial Functions
- FV(pmt,int,per)
- IRR([guess,]range)
- NPV(disc,range)
- PMT(prin,int,per)
- PV(pmt,int,per)

Mathematical Functions
- ABS(c/v)
- ACOS(c/v)
- ASIN(c/v)
- ATAN(c/v)
- COS(c/v)
- EXP(c/v)
- INT(c/v)
- LN(c/v)
- LOG(c/v)
- MOD(c/v$_1$,c/v$_2$)
- PI
- RAN
- ROUND(c/v,places)
- SIN(c/v)
- SQRT(c/v)
- TAN(c/v)

Special Functions
- ERROR
- LOOKUP(c/v,range)
- NA

Statistical Functions
- AV(range)
- COUNT(range)
- MAX(range)
- MIN(range)
- SUM(range)

Operators
- +   Plus
- -   Minus
- *   Multiply
- /   Divide
- ^ (or **) Raise to power
- %   Percent
- =   Equal to
- <> Not equal to
- <   Less than
- <= Less than or equal
- >   Greater than
- >= Greater than or equal

- c/v - a Cell reference or Value
- cond - some algebraic Condition (e.g., A5>5.)
- disc - Discount rate
- guess - a Guess
- int - Interest rate
- per - Period or term
- places - number of Places after the decimal
- pmt - Payment
- prin - Principal
- range - a Range of cells (e.g., B3,C6)

FUNCTION KEY ACTION
- F1 or ?   Help
- F2 or Ctrl Z   Clears/Aborts
- F9 or Ctrl Y   Graph on Plotter
- F10 or Ctrl T   Graph on Screen

PREFIXES FOR CELL ENTRIES
Letters and most symbols begin text mode
" begins forced text mode
' begins repeating entry
Numbers, functions, cell references, +, -, ., cell addresses begin formulas

RANGES CELL, COLUMN, ROW

SINGLE CELL
D5 is column D row 5

BLOCK OF CELLS
Two Opposite Corner Cells
D5.F48
H1.A17

FULL COLUMN
F

MULTI-COLUMNS
B.G

PARTIAL COLUMN
C3.C10

FULL ROW
39

MULTI-ROWS
45.83

PARTIAL ROW
G25.M25

CAREERS
in Engineering
and Technology

Wind power's potential value is in its ability to substitute for the use of oil, coal, or nuclear energy. The most promising form of wind machine is the large, horizontal-axis wind turbine with propeller-type rotor blades.

4TH EDITION

CAREERS
in Engineering
and Technology

GEORGE C. BEAKLEY

DONOVAN L. EVANS

DELOSS H. BOWERS

Arizona State University

Macmillan Publishing Company
NEW YORK
Collier Macmillan Publishers
LONDON

Macmillan Publishing Company
866 Third Avenue, New York, New York   10022

Collier Macmillan Canada, Inc.

Library of Congress Cataloging-in-Publication Data

Beakley, George C.
   Careers in engineering and technology.

   Includes bibliographical references and index.
   1. Engineering—Vocational guidance.   2. Technology—
Vocational guidance.   I. Evans, Donovan L.   II. Bowers,
Deloss H.   III. Title.
TA157.B39   1987   620'.0023   86-31172
ISBN   0-02-307620-8

Printing: 1 2 3 4 5 6 7 8   Year: 7 8 9 0 1 2 3 4 5 6

Preface

This 4th edition charts some previously unexplored territory containing approximately 75 percent new content and 25 percent updated and revised material. Chapters 1, 4, and 5 are updated versions of material from the previous editions, aimed at explaining the make-up of the technical team (scientists, engineers, technologists, technicians, and craftsmen) and the array of career fields and work opportunities that are available. The authors are especially appreciative of the work of Jack Stadmiller in preparing textual materials used in Chapter 1, and to Kelly Stadmiller for her work in preparing textual materials used in Chapter 3. Keith Roe collaborated with the authors in preparing textual materials in Chapter 5.

The purpose of Chapter 3, Study Habits to Maximize Success, is to bring about a friendly transition for the student from the high school environment to that which is typical of engineering and technology curricula.

The authors believe that the student must gain an early awareness of the full responsibilities of the engineer and technologist to society. Previous editions have addressed this point, but in the brief span of time that has elapsed since their first introduction, society's problems seem only to have become more critical. Chapters 2 and 6 place increased emphasis on these responsibilities. Also, throughout the text, a special effort has been made to impart to the student a professional attitude and personal concern to produce designs compatible with the occupational health and safety of everyone who might be affected.

Those who create, artists-designers-architects-engineers, have always found it expedient to communicate to others the essence of their ideas through drawings, sketches, diagrams, or other graphic means. Until the latter part of the eighteenth century, most design drawings were made freehand. For example, Leonardo da Vinci's middle sixteenth century drawings depicting his creative ideas are still easy to comprehend today. Beginning about 1820, engineers in this country began to be taught projection drawing, based on the French system that was first developed by Gaspard Monge (1746–1818). Projection drawings are most useful in the manufacturing process, particularly with regard to the production of mechanical components. As a result, courses in engineering drawing and descriptive geometry have been standard components of most engineering curricula until the past few years.

However, traditional engineering drawing methods stressing orthographic projection no longer serve as a common medium of graphic communication in many engineering disciplines such as microelectronics, advanced chemical processes, nuclear power, and computing systems. The methods of freehand drawing used by da Vinci are still effective today in enabling engineers from all disciplines to clearly communicate the essence of their creative thought. Chapter 7, Freehand Drawing and Visualization, has been developed to enable students (by class instruction or by self-study) to enhance their spatial visualization skills and acquire a freehand drawing proficiency that will be universally useful to engineers of all disciplines for the visual communication of ideas and design concepts. Martha Heier, who has extensive experience in teaching freehand drawing skills to university students, prepared most of the textual material for the chapter, including the student assignments.

A recent nationwide study by the Education Testing Service has shown that during the last two decades, the competence of high school students in spatial visualization has diminished at an alarming rate. This is an essential skill for engineers, architects, and medical doctors. We believe that since drawing is "seeing," the practice of drawing stimulates the ability to visualize, something that the more creative persons in our society appear to do naturally. Exercising the visualization skills of engineers can only improve their performance in the creative and communication aspects of their work.

The revolution in microprocessor development in the last few years has dramatically changed and enhanced the engineer's and technologist's reliance on the computer. All indications point to this reliance accelerating in the future, with mainframe computing power being built into desktop units. And, low cost portable computers smaller than a normal textbook, yet containing an electronic spreadsheet and general equation solver (with graphics), are quickly reaching costs low enough to make their use widespread.

We believe that freshman engineers and technologists can and should be required to become familiar with both the hardware and software that are an intimate part of this revolution. As a result, we have put special emphasis on the computer in this latest edition. Chapters 8, 9, and 10 are devoted to the introduction of these marvelous and useful tools. Chapter 15 is then devoted to design problems that lend themselves to solution by the computer.

In Chapter 8, we present an introduction to both hardware and software. We believe that both programming languages (BASIC, FORTRAN, C, Pascal, for example) and applications programs are useful and necessary for engineers and technologists. However, since programming languages are usually presented in separate courses in most curricula, we do not attempt to cover their use in this text. We do, however, cover two types of applications programs: Computer-Aided Drafting (CAD) programs, and electronic spreadsheets.

The rapid acceptance of the microcomputer in business and industry has created many changes. One of the most dramatic has been in the area of design and drafting. CAD has been placed within reach of even the smallest companies, and its rapid penetration of this market is changing the nature of documentation drawing. Not only are the tools of traditional drafting becoming obsolete, but the nature of the skills needed to produce documentation drawing are also changing.

In Chapter 9, we give an introduction to modern design documentation methods with CAD through a unique side-by-side presentation of both AutoCAD[1] and VersaCAD[2] programs for microcomputers. Both of these are recognized to be powerful production-oriented CAD programs. Special insets called *Topics* are used to assist the student in mastering and reviewing the subject material. Students find this new technology tremendously exciting and, in our experience, have been able to assimilate it quite rapidly. The chapter is written for all students, including those who have not had a traditional drafting course. The material emphasizes the ability to make engineering drawings using any CAD system, and provides examples using two popular microcomputer based drawing systems, AutoCAD and VersaCAD.

In Chapter 10, we present an introduction to electronic spreadsheets since we are convinced that these revolutionary inventions are extremely useful to the engineer. We present material for two of the most widely used spreadsheets, SuperCal4[3] and Lotus 1-2-3[4]. *Topics* are again used to deliver a side-by-side presentation of these two programs.

In building toward a design experience for the student reader, we devote Chapters 11, 12, and 13 to some useful and often necessary rudiments of engineering. This includes precision and accuracy of numbers, statistics, dimensions, units, and engineering economy. The textual material in Chapter 13 was prepared originally for the text *Engineering,* 5/e. Dr. John Keats was the primary author of this chapter. We have found that engineering economy, at least to the extent of present worth analysis, can be understood and used by freshmen engineering students. In fact, this modeling tool can be understood and exercised by freshmen students without the semesters of study involved in most other engineering technical analysis tools.

Chapters 14 and 15 are intended to introduce freshmen engineering students to engineering design through the practice of the profession rather than by merely explaining what engineers and technologists do. Chapter 14, Engineering Design, stresses the principles of design which we believe need emphasizing at the freshman level. It clearly delineates between engineering design, the process, and an engineering DESIGN, the product of the process. We have come to believe that the more detailed aspects of the design process and of the DESIGN (which are not treated directly in this text) are more properly relegated to upper level design courses in a student's major. In this text, we strive only to have the student understand the essence of engineering design.

Chapter 15 contains a collection of design projects that adequately allow the student *to experience* engineering design. These are class tested projects which involve and promote an understanding of the design process and which all involve the use of the computer. Spreadsheets are excellent computer tools to aid in the solution of these problems. Equation solvers such as TK!Solver[5] may also be used to good advantage.

The projects in Chapter 15 are of the "paper study" variety—feasibility or conceptual studies. Because of the computer, freshmen students no longer are restricted to doing only inventive type projects in which the purpose is to build some artifact. The emphasis can now be placed on establishing the plans and specifications for an artifact, rather than construction of it. Students still must use their creative abilities, but they experience the tie that must be made between their ideas and the necessary engineering modeling and the iterations typical of the design process.

George C. Beakley
Donovan L. Evans
Deloss H. Bowers

[1] tm Autodesk Inc.
[2] tm T & W Systems, Inc.
[3] tm Computer Associates, Inc.
[4] tm Lotus Development, Inc.
[5] tm Technical Systems, Inc.

Contents

Acknowledgments

Frontispiece	Southern California Edison Company
Figure 1-3	General Electric Forum
Figure 1-4	Matheson Gas Producers
Figure 1-5	National Aeronautics and Space Administration
Figure 1-7	General Electric Research and Development Center
Figure 2-1	Western Electric
Figure 2-2	United States Navy
Figure 2-3	Exxon Chemical Company, USA
Figure 2-4	Uniroyal, Inc.
Figure 2-5	Bethlehem Steel Corporation
Figure 2-6	H. Armstrong Roberts Photo
Figure 2-7	Stennett Heaton Photo, Courtesy Neil A. Maclean Co., Inc.
Figure 2-9	Floyd A. Craig, Christian Life Commission, Southern Baptist Convention
Figure 2-10	Planned Parenthood—World Population
Figure 2-12	Planned Parenthood—World Population
Figure 2-13	Ministers Life and Casualty Union
Figure 2-14	Monsanto Company
Figure 2-15	Ambassador College
Figure 2-16	Adolph Coors Company
Figure 2-18	Ford Motor Company
Figure 2-19	Citizens for Clean Air
Figure 2-21	Southcoast Air Quality Management Division
Figure 2-22a	Torit Corporation
Figure 2-22b	*Arizona Republic*
Figure 2-23	George Kranse; Presbyterian Ministers Fund Life Insurance
Figure 2-24	Planned Parenthood—World Population
Figure 2-25	United States—Department of Agriculture
Figure 2-26	Floyd A. Craig, Christian Life Commission, Southern Baptist Convention
Figure 2-27	The Carborundum Company
Figure 2-28	Shell Oil Company
Figure 2-29	Phil Stitt, *Arizona Architect*
Figure 2-30	Floyd A. Craig, Christian Life Commission, Southern Baptist Convention
Figure 2-32	Planned Parenthood—World Population
Figure 2-34	Xerox Corporation
Figure 2-35	Hewlett-Packard
Figure 2-36	Hewlett-Packard
Figure 2-37	*Kaiser News*
Figure 2-38	Southern California Edison Company
Figure 2-39	Pacific Gas and Electric Company
Figure 2-40	Southern California Edison Company
Figure 2-41	Humble Oil and Refining Company
Figure 2-43	Los Angeles County Air Pollution Control District
Figure 2-44	Shell Oil Company
Figure 2-45	American Telephone and Telegraph Company
Figure 2-46	*Arizona Republic*
Figure 2-47	Peter Kiewit Son's Company
Figure 2-48	Institute of Traffic Engineers
Figure 2-49	American Iron & Steel Institute
Figure 2-50	American Express Company
Figure 2-51	Polaroid Corporation
Figure 2-52	General Telephone & Electronics
Figure 2-55	Floyd A. Craig, Christian Life Commission, Southern Baptist Convention
Figure 2-56	Modified after drawing of Ron Thomas, *Kaiser News*
Figure 2-58	The Carborundum Company
Figure 2-59	General Electric Company
Figure 3-4	*Engineering Graphics*
Figure 4-1	McDonald Douglas
Figure 4-3	International Harvester Company
Figure 4-4	Aluminum Company of America
Figure 4-5	General Motors Research Laboratories
Figure 4-7	Glass Container Manufacturers Institute
Figure 4-8	American Institute of Plant Engineers
Figure 4-9	Bethlehem Steel Corporation
Figure 4-10	*Engineering Times*
Figure 4-11	General Electric Research and Development Center
Figure 4-12	General Electric Research and Development Center
Figure 4-13	RCA Electronic Corporation
Figure 4-14	Cessna Aircraft Company
Figure 4-15	Information Handling Services
Figure 4-16	Phillips Petroleum Company
Figure 4-17	Renault USA, Inc.
Figure 4-18	Union Carbide Corporation
Figure 4-19	Cities Service Company
Figure 4-20	Phelps Dodge Corporation
Figure 4-21	Dr. John McKlveen
Figure 4-22	Southern California Edison Company
Figure 4-23	Atlantic Richfield Company
Figure 4-24	Exxon Company, USA
Figure 5-1	AT&T Bell Laboratories
Figure 5-2	Dr. J.E. Cermack, Director, Fluid Dynamics and Diffusion Laboratory, Colorado State University
Figure 5-3	Fafnir Bearing Division of Textron
Figure 5-6	Information Handling Services
Figure 5-7	Information Handling Services
Figure 5-8	Naval Undersea Center
Figure 5-9	*Industrial Engineering*
Figure 5-12	Bethlehem Steel Corporation
Figure 5-13	American Institute of Plant Engineers
Figure 5-14	NCR Corporation
Figure 5-15	International Paper Company
Figure 5-16	Arkansas Best Corporation
Figure 5-17	Union Carbide Corporation
Figure 5-18	Fisher Scientific Company
Figure 6-1	Ernst and Ernst
Figure 6-2	General Dynamics
Figure 6-4	Federal Highways Administration
Figure 6-8	Modine Manufacturing Company
Figure 6-9	American Youth Magazine
Figure 6-12	*Industrial Research*
Figure 6-13	American Lung Association

Chapter 1

The Technical Team

We live in a world that through the ages has been greatly influenced by technological change. The history of man[1] is an account of how his creative designs have been used to affect political, economic, and social development. Sometimes these designs have developed from the pressures of need from evolving civilizations [1-1]. At other times abilities to produce and meet needs have led the way for civilizations to advance. In every age timely inventions and scientific breakthroughs have been largely the result of individual efforts. Archimedes (287–212 B.C.) [1-2], Johann Gutenberg (1398–1468), Leonardo da Vinci (1452–1519), Galileo (1564–1642), Robert Boyle (1627–1691), Robert Hooke (1635–1703), Isaac Newton (1642–1727), James Watt (1736–1819), Thomas Alva Edison (1847–1931) [1-3], Nikola Tesla (1856–1943), Lee DeForest (1973–1961), Orville and Wilbur Wright (1867–1948), and Albert Einstein (1879–1955) [1-4] are but a few of those creative individuals who have led the way for us today to have an improved quality of life. Some were artisans and craftsmen. Some were engineers and scientists. Others, had they been living today, might have been called technologists.

In centuries past, eras of history were often referred to by special names, such as the Stone Age, the Industrial Age, the Agrarian Age, or the Middle (or Dark) Ages. The past 50 years has seen a number of these designations come and go. For example, we have all become

---

[1] In discussing activities of mankind—the human species—the authors use the generic word *man* to represent both male and female.

[1-1] History is an account of the ever-increasing complexity of man's inventions.

Hoe

Plow

Wheel and Axle

Animal-Powered Wheel

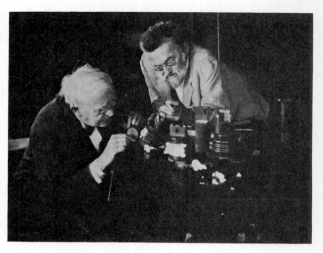

[1-3] Nations the world over acknowledge their indebtedness to Thomas Alva Edison, inventor, and Charles Proteus Steinmetz, electrical engineer, for their significant inventions relating to electricity.

[1-2] The Archimedes water screw is still used today to lift water.

I know this world is ruled by Infinite Intelligence. It required Infinite Intelligence to create it and requires Infinite Intelligence to keep it on its course. Everything that surrounds us—everything that exists— proves that there are Infinite Laws behind it. There can be no denying this fact. It is mathematical in its precision.

Thomas Alva Edison, 1847–1931

No man really becomes a fool until he stops asking questions.

Charles P. Steinmetz, 1865–1923

[1-4] Albert Einstein, father of the nuclear age, emphasized the importance of simplicity in all matters. He believed that harmony will result if people act in accordance with principles founded on consciously clear thinking and experience.

One day, about noon, going towards my boat, I was exceedingly surprised with the print of a man's naked foot on the shore.

Daniel Defoe, 1660–1731
Robinson Crusoe

[1-5] Man's safe exploration of the moon will long stand as one of his greatest engineering achievements.

accustomed to talking about the Atomic Age, the Space Age [1-5], the Computer Age, the Age of High Technology, and more recently the age of the Information Society.[2]

For this country, undoubtedly one of the most significant happenings has been the specialization that has occurred within the educational system. In Aristotle's time (250 B.C.), for example, it was possible for an individual to be learned and conversant with most of the scientific and mathematical knowledge of that era. A person could focus his attention on a problem or idea and, using most of the scientific principles that were then known, effect a workable design. In more recent times it has been estimated that the great abundance of engineering and scientific knowledge has been doubling every decade. It is literally impossible for any one person to have at his or her grasp even a small fragment of the known principles of engineering and science. Within the past half century major projects of innovation have been brought into use because of the employment of a concept known as the *technical team* [1-6]. Since World War II there has been a movement toward improved productivity, standardization, and higher-quality products. Automation and the use of industrial robots have placed even greater emphasis on the use of the team concept.

The technical team idea is used in many countries today and is not exclusive to America. It involves dividing complex engineering or scientific tasks into five specialized occupational subgroups. The

[2]The *Information Society* is a term that has come into use to describe the environment where knowledge (not land, money, machines, or armies) is all powerful. Those who possess, sell, deal in, and control information (such as governments, large corporations, banks, etc.) are the ones who wield great power.

CRAFTSMAN

TECHNICIAN

TECHNOLOGIST

ENGINEER

SCIENTIST

[1-6] Technical team occupational spectrum.

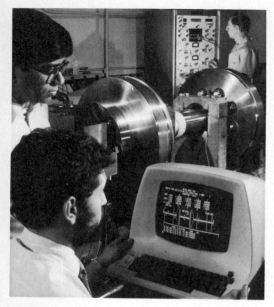

people who choose to work in each of the subgroups are highly skilled and educated specifically for that task. Together they play a major role in increasing the world trade competitiveness of our nation. The team members who make up these five groups are:

Craftsmen

Technicians

Technologists

Engineers

Scientists

In this country the educational system is especially geared to developing a person's expertise in one of these categories. The categories form a spectrum of work that feeds American industry [1-7].

As indicated in the spectrum, the boundaries separating the occupational groups are not well defined. For this reason it is quite often the case that one person will be educated as a scientist (such as a physicist) but may often function on the job as an engineer. A person educated as an engineer might function effectively on occasion in the role of a scientist, or even as a technologist. One of the purposes of this chapter is to clarify the normal roles of each member of the technical team. This is especially important because the education and training required for each of these five occupational categories is uniquely different. The work functions of the people in each category are so dissimilar that it is unlikely that a person educated to function in one category (e.g., as a scientist) would be happy and effective working in another category (e.g., as a technician).

Since the education and training required for one to be successful is substantially different for each of the career categories, it is important that we consider the requirements of each.

THE CRAFTSMAN

An acquired personal skill is the hallmark of the craftsman, who takes great pride in transforming materials from their raw state to finished products. Historically, the craftsman came on the scene before any of the other members of the technical team. Prior to the industrial revolution, the master craftsman held a position of considerable esteem in the community. Usually, he operated his own shop, determined the designs, set standards, and directed the apprentice workmen and other artisans who worked for him. The period of apprenticeship was usually about seven years, and individual skills were mastered by constant repetition. Paul Revere, an outstanding master craftsman of the colonial era, was skilled in the casting and working of metal, particularly silver and copper.

Since the products were usually sold, the average craftsman was sensitive to the style, marketability, and serviceability required by his customers. He was, therefore, accustomed to working within definite limits. In due time, however, the urgent need for craftsmen skills diminished and the role of the craftsman changed. Machines were developed to do many of the routine tasks, and industrial workers

Occupational description

Craftsman—one who performs with skill and dexterity tasks which are often repetitive and require manual use of tools or equipment. Many of these skills are learned and perfected through practice and experience. Rarely supervises others.

found themselves to be more captives of the machines than the reverse. Also, as factories became more dominant a worker no longer was personally responsible for making a product from start to finish. He now specialized in a single task—one of a series of operations—repeating this task over and over to help achieve the desired production quota. When this happened, the pride of accomplishment so necessary to motivate the craftsman to excel in producing his finished product disappeared. Because of this trend, many individuals who aspired to be true craftsmen moved into other areas of work.

Today, crafts continue to work in a variety of technologically related areas. They provide the same intense feeling of pride and satisfaction for the individual as did the domestic-type craft work of several decades ago. Electricians, carpenters, plumbers, welders, tool and die makers, pattern makers, and precision machinists are typically employed in today's modern term "craftsman." Today a high school education is usually needed for entry into these specialty areas. In most cases this is followed by an intensified period of skill training that may vary from 6 to 18 months, followed sometimes by years of practice, for one to become highly skilled.

THE TECHNICIAN

Science and technology have become more and more complex, requiring tremendous support systems to perform many of the detailed functions of operating and maintaining equipment. Design drawings must be prepared and data must be collected. The technician is a key element in all of these activities. He or she is concerned with the "how" to do things, rather than the "why," or "if" they should be done.

The engineering technician works with equipment, primarily assembling and testing the component parts of designs that are designed by others. In this respect the technician is an experimentalist—an Edison-type thinker rather than an Einstein- or Steinmetz-type thinker. His preference is given to assembly, repair, or to making improvements in technical equipment by learning its characteristics rather than by studying the scientific or engineering basis for its original design.

As a semiprofessional, the technician possesses many of the skills of a craftsman and frequently is personally able to effect required physical changes in engineering hardware. Because most work is concerned directly with such equipment, it is often possible to suggest answers to difficult problems that have not been thought of previously.

The technician typically works in a laboratory, out in the field on a construction job, or he might troubleshoot problems on a production line, rather than, for example, work at a desk. There are, however, technicians who do their work in drafting rooms or offices under the direction of engineers. Technicians are frequently found in research laboratories, where they provide effective service in repair-

[3] Daniel M. Hull and Leno S. Pedrotti, "Challenges and Changes in Engineering Technology," *Engineering Education*, May 1986, p. 729.
[4] Ibid.

Occupational Description

Technician—performs routine equipment checks and maintenance. Carries out plans and designs of engineers. Sets up scientific experiments. Seldom supervises others.

One type of equipment that will remain in the technological environment is the computer. It will no doubt change and become more complex, but it will not go away. Technicians will continue to use the computer in three major ways: (1) as an information-management tool to acquire, store, manipulate and present data; (2) as a control device to receive information from sensors about the condition of some system or process, to compare these conditions to what is desired, to decide what changes are needed and to send out change orders to control the devices; and (3) as an instrument to produce layouts of buildings, devices and circuits and to test the efficiency, feasibility, strength, integrity and utility of these items before they are built. *For the foreseeable future, technicians will need to be able to communicate with computers via the keyboard/CRT, to understand fundamental programming principles, and to interface control signals and input/output devices to computers.*[3]

Defining the Technician[4]

Tasks

The technician of the future may be expected to:

1. Perform tests of mechanical, optical, hydraulic, pneumatic, electrical, thermal, and electronic/digital components or systems and prepare appropriate technical reports covering the tests.
2. Obtain, select, compile and use technical information from computer-controlled measuring, recording and display instruments.
3. Use computers to analyze and interpret information.

4. Prepare or interpret engineering drawings and sketches and write reports, working procedures and detailed specifications of equipment.
5. Design, help develop or modify products, techniques and applications in laboratory and industrial settings.
6. Plan, supervise or assist in the installation and inspection of complex scientific apparatus, computer equipment and control systems.
7. Operate, maintain and repair apparatus and equipment incorporating computer-controlled systems.
8. Advise, plan and estimate costs as field representatives of manufacturers or distributors of technical apparatus, equipment, services and products.
9. Apply knowledge of science and mathematics while providing direct technical assistance to physical scientists or engineers engaged in scientific research, experimentation and design.

Needed Skills and Abilities

To perform the tasks listed above, technicians must also have the following special abilities and skills:

1. The ability to apply and use the basic principles, concepts and laws of physics and technology.
2. A facility with mathematics, including the ability to use algebra, trigonometry and analytic geometry as problem-solving tools. An understanding of higher mathematics—including computer language and some calculus—may be required.
3. The ability to analyze, troubleshoot and repair systems that are composed of subsystems in three or more of the following areas: electronic, electrical, mechanical, thermal, fluidic and optical.
4. A facility in the use of materials, processes, apparatus, procedures, equipment, methods, and techniques commonly used in the technology.
5. A knowledge of the field of specialization with an understanding of current engineering applications and industrial processes in the field.
6. A facility in the use of computers for information management, equipment and process control, and design.
7. Communication skills that include the ability to record, analyze, interpret, synthesize and transmit facts and ideas with objectivity—orally, graphically and in writing.

What Are the Current and Future High-Tech Areas?

To develop an all-inclusive list of high-tech areas for technician education and training probably is not possible, but current offerings should include:

- Computers/electronics
- Telecommunications/electronics
- Robotics/automated manufacturing
- Laser/electro-optics
- Biomedical equipment
- Instrumentation and control

Some new technologies already can be seen on the horizon. One of these is computer-aided drafting and design (CADD) or computer-aided design and computer-aided manufacturing (CAD/CAM).[5]

ing equipment, setting up experiments, and accumulating scientific data. They are very important in this role because it is often the case that scientists themselves are not too adept or interested in carrying out tasks of this type.

In construction work, technicians also play key roles. They are needed to accomplish tasks in surveying, to make estimates of material and labor costs, to be responsible occasionally for the coordination of skilled labor and the work of subcontractors, and for the delivery of materials to the job sites.

A substantial number of engineering technicians are employed in the manufacturing and electronics industry and by public utilities. They may carry out standard calculations, serve as technical salesmen, make estimates of costs, or assist in preparing service manuals, such as for electronic equipment and plant operation and maintenance. They install and maintain, make checks on, and frequently modify electrical and mechanical equipment. As a group they are important problem-solving individuals whose interests are directed more to the practical than to the theoretical. Technicians learn fundamental scientific theory and master the mathematical topics most useful in analyzing and solving problems. The majority of their studies, however, are oriented more to the practical than to the theoretical.

The technician's education typically requires two full years of collegiate-level study. Generally, this work is taken in a technical institute or community college and leads to an associate degree in technology. In many instances this school work is transferable to a senior academic institution and in some cases may be applied as credit toward a bachelor of engineering technology program. In most cases, however, these courses would not be applicable toward a degree in engineering or science.

THE TECHNOLOGIST

The technologist assists the engineer, and many people today employed as technologists in industry can more appropriately be described as engineering technologists. Engineering technology is that part of the engineering field which requires the application of scientific and engineering knowledge and methods combined with technical skills to support engineering activities.

The technologist differs from the craftsman and technician in increased knowledge of scientific and engineering theory and methods.

The engineering technologist is the most recent of the technological team to appear, having emerged in the last two decades. Technologists work in the occupational domain between the craftsman-technician and the engineer. Their areas of interest and education are typically less theoretical and mathematically based than those of the engineer. Technologists are employed in such diverse areas as construction, operation, maintenance, and production. Also, as a group they are more likely to be identified as design organizer-producers rather than design innovators.

[5] Ibid.

Ordinarily, the engineering technologist will concentrate his or her activities on the development of various design components of systems that have been designed and developed by engineers. They are valuable in research for liaison work and possess the ability to assume a variety of technical supervisory and management roles in industry.

Engineering technologists are usually graduates of baccalaureate-degree programs in engineering technology.

Occupational Description

Technologist—applies engineering principles for industrial production, construction, and operation. Works with engineering design components. Occasionally supervises others.

Fifty percent of the equipment, procedures and processes in today's fast-paced industries will be changed in three to four years.[6]

THE ENGINEER

The engineer is a problem solver—a creative designer. Using mathematics as a tool and science (particularly the physical sciences) as a factual or theoretical basis, he or she is interested in the useful application of scientific principles to create functional devices, systems, and processes which enhance the well-being of the human race.

An engineer is a person who does not enjoy the status quo. He or she is always trying to improve a design or concept. Two functions are performed in this role: first, recognition of problems or conditions that need improvement; and second, the use of one's reservoir of knowledge and skill in innovative thought to produce one or more acceptable solutions.

Thus, as a personal characteristic the engineer should possess a persistently inquiring and creative mind—yet one that is receptive to producing realistic innovative action. This must include a fundamental understanding of the laws of nature and of mathematics, as well as being able to recognize and interpret their application to real-life situations. Although these characteristics are most often associated with those who provide our nation with technological leadership in commerce and industrial production, they are also highly desired qualities to attain for those who want to follow careers in management, medicine, and law.

The baccalaureate degree in engineering is required for entry into industry, and the master's degree (a total of five years of college) is preferred for many types of work.

Occupational Description

Engineer—identifies and solves problems. An innovator in applying principles of science to produce economically feasible designs. Frequently supervises others.

THE SCIENTIST

The scientist is a searcher for truth, and is the most theoretical of the team members. He or she seeks, through observation and experimentation, to understand the physical universe. This includes the development of hypotheses, principles, and methods whereby knowledge can be advanced. The scientist is interested in what occurs and why, as far as increasing mankind's understanding of the world we live in.

The primary objectives of the scientist are (1) to expand existing fields of knowledge, (2) to correlate observations and experimental

---

[6] Ibid.

Occupational Description

Scientist—searches for new knowledge concerning the nature of man and the universe. Infrequently involved in supervisory work.

data into a formulation of laws, (3) to learn new theories and to explore their meanings, and (4) in general, to broaden the horizons of science. The scientist is typically a theoretician who is concerned about *why* natural phenomena occur. As suggested above, the scientist's primary objective is to expand the world's reservoir of knowledge, and he or she finds little reason to pause to search for practical uses for any newfound truths that may be discovered. There are, of course, *applied scientists* who work to find specific uses for new knowledge, such as devising a new instrument or synthesizing an improved medicine. For this reason it is often difficult for the average person to tell the difference between an applied scientist and an engineer. In terms of schooling, the person educated as a scientist will generally have earned a doctorate (at least seven years of college) in some field of natural science.

Table 1-1 summarizes representative job positions of the five team members in various engineering fields.

We must recognize that as individuals we do not always fit neatly into one specific area of the technical occupational spectrum. Rather, our interests and aptitudes may span one or two of the work areas—or even across the entire spectrum. If so, good! However, educational programs leading to each of the various career choices are usually quite different in their composition.

In a general sense, Figure [1-8] indicates how such factors as degree of aptitude for manipulating mathematical expressions and the degree of satisfaction that is derived from manual artistry will influence the probability of lasting interest in these various career fields.

This text will concentrate more attention on the career opportunities for engineering technologists and engineers. Since the engineering technologist is the newest of the career fields making up the technical team, more attention will be directed specifically to this career field here.

WHO IS THE TECHNOLOGIST?

Now that we have a general picture of the members of the technical team, their functions, and relationships to each other, we will focus

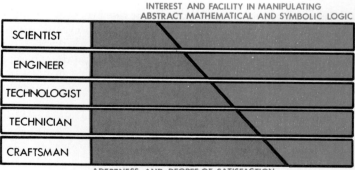

[1-8] Occupational preferences.

TABLE 1-1 REPRESENTATIVE TECHNICAL TEAM POSITIONS FOR VARIOUS ENGINEERING FIELDS

Engineering field	Craftsman	Technician	Technologist	Engineer	Scientist
Aerospace	Wind-tunnel builder	Helicopter production foreman	Wind-tunnel operator	Power plant engineer	Magneto-hydro-dynamist
Agricultural	Farm machinery mechanic	Food technician	Land surveyor	Equipment engineer	Agronomist
Architectural	Model builder	Architectural drafter	Delineator	Construction engineer	Acoustic scientist
Biomedical	Biomedical instrument repair technician	Testing laboratory technician	Laboratory technologist	Product development engineer	Biologist
Chemical	Glass blower	Refinery technician	Chemical analyst	Chemical process engineer	Physical chemist
Civil	Heavy equipment operator	Photogrammetist	Hydrologist	Sanitary engineer	Organic polymer chemist
Computer	Computer operator	Computer programmer	Software designer	Hardware design engineer	Mathematician
Construction	Carpenter	Inspector	Estimator	Superintendent	Geologist
Electrical	Transmission linesman	Electronic technician	Customer representative	Systems/project engineer	Solid-state physicist
Industrial	Tool and die maker	Time-and-motion engineer	Production supervisor	Director of manufacturing	Information scientist
Mechanical	Machinist	Robotics technician	Laboratory test supervisor	Design engineer	Physicist
Metallurgical	Welder	Metallographer	Materials evaluator	Materials development engineer	Physical metallurgist
Mining	Hoist operator	Process control technician	Extractive metallurgist	Mill superintendent	Chemical metallurgist
Nuclear	Reactor mechanic	Power plant operator	Radiation specialist	Safety engineer	Radiation health specialist
Petroleum	Driller	Compressed gases tester	Drill hole logger	Site engineer	Geophysicist
Educational and experience requirements	Special trade school, 1 year or less Experience 4 to 6 years Continuing education	Community college minimum Experience 2 to 4 years Continuing education	Baccalaureate degree minimum Experience 1 to 3 years Continuing education	Baccalaureate degree minimum, master's and doctorate degrees desirable Continuing education	Doctorate minimum, postdoctorate desirable Continuing education

our attention on the engineering technologist. What typically describes the aptitudes and interests of those who enter this field of work, and what are the educational requirements to become an engineering technologist?

The technologist is primarily one who is interested in the operation or function of engineering designs and devices. In general, the technologist's work parallels (but at a different level) the work of engineers or scientists. The engineering technologist's work is less theoretical in nature and is usually directed toward the applications of known concepts. Unlike the scientist, who is concerned with theory and the discovery of new truths, the technologist is interested in the equipment or means which translate the scientist's findings and the engineer's designs into working systems. The technologist has many interests that are similar to those of the engineer, but he or she may not be as interested in problem solving or as scientifically informed, mathematically adept, or cost-conscious as the engineer.

Engineering technologists perform work in virtually every phase of engineering and science. They may work in research, development, design, model testing, experimentation, and data processing. They may also assist in production planning, conducting quality control tests, and directing time-and-motion studies. They could also serve as technical maintenance specialists, technical writers, or sales representatives. Under the direction of engineers or scientists, the engineering technologist must be able to carry out proven operational techniques and procedures.

In order to fit into the technical team, the engineering technologist must have some understanding of scientific and mathematical theory, as well as having specialized training in some specific branch of engineering technology. Some mathematical ability is needed to analyze and solve engineering problems and certain additional instruction and training is necessary to provide skill and competence in the operation of complex equipment and instrumentation. It is also necessary to understand the fundamentals of computer programming and to assist in data analysis and problem solving.

In addition to being expected to work without close engineering supervision, the technologist will frequently be required to supervise the work of other technologists, technicians, and craftsmen. In this capacity, it is mandatory that he or she be able to communicate effectively ideas and instructions to others, both orally and in the written word. Therefore, it is essential that technologists have a good vocabulary and be familiar with the basic rules of grammar and be able to use the tools effectively.

ACADEMIC PREPARATION

One of the most important things to know about engineering and engineering technology is that they *are* different. It is generally not possible for one who has been educated as an engineering technologist to become an engineer. In the same way most registered nurses do not learn to become medical doctors. The second most important thing to understand is that it is the educational path that one follows that determines whether he or she will become an engineer or an engineering technologist. One should be careful not to confuse career designation with the work of engineers and engineering technologists. Quite often their work overlaps and it is common in industry for engineers and engineering technologists to work together, and on occasion, perform the same tasks. This concept may be illustrated by reviewing the corresponding analogy to medicine. If you are quite ill and it has been determined that you need an injection of a particular type of medicine, you are not concerned with who gives you the medicine. The nurse has special training and expertise to accomplish the task. On the other hand, the medical doctor attending you may prefer to give the injection. However, regardless of who administers the medicine, you never get confused about which role the doctor and nurse play. They attained their respective designations by pursuing quite different, prescribed, educational paths. Different educational paths differentiate the engineer from the engineering technologist.

From Table 1-1 we see that the minimum educational requirement of both career designations is the same—a baccalaureate degree in a selected specialization. Of course, there are a variety of specialty fields in each. Some representative fields are shown in Table 1-1. All of the quality programs in both engineering and engineering technology conform to the same basic structure whose minimums have been specified by the Accreditation Board of Engineering and Technology.[6] Let us examine these curricula criteria.

ENGINEERING PROGRAMS

The course work for the four years must include at least:

1. One year equivalent of a combination of mathematics (above algebra and trigonometry) and basic sciences.
2. One-half year equivalent of humanities and social sciences.
3. One-year equivalent of the engineering sciences (the theory of engineering).
4. One-half year equivalent of engineering design (the application of engineering theory to the solution of practical problems).

ENGINEERING TECHNOLOGY PROGRAMS

The course work for the four years must include at least:

1. Three-fourths year equivalent of a combination of mathematics and basic sciences.
2. Three-fourths year equivalent of humanities, social sciences, and written and oral communication.
3. One and one-half year equivalent of engineering technology courses, with special emphasis on practice oriented subjects.

---

[6] Accreditation Board for Engineering and Technology, United Engineering Center, 345 East 47th Street, New York, NY 10017.

Some general conclusions can be drawn from the above. Engineering curricula are more intensely mathematical and scientifically based than engineering technology curricula. Engineering technology curricula are more practice oriented than engineering curricula and thus are more laboratory intensive. Both types of curricula are intensely computer based. Table 1-2 is a comparative summary of these two fields of study.

Most American industries employ substantially more technicians and engineering technologists than engineers. The ratios vary from industry to industry, but on the average it takes three or four technologists or technicians to support one engineer. Again, this is somewhat analogous to the health care system, where in a hospital it takes about four nurses to support one medical doctor.

Because the fields of specialization and job functions are the same, or similar, for both engineering and engineering technologists, Chapters 4, 5, and 6 are written so that the content is applicable to both career fields. In most cases in these chapters a reference to "engineering" can be applied equally to the "engineering technologist."

TABLE 1-2 COMPARISON OF NEW ENGINEERING AND FOUR-YEAR ENGINEERING TECHNOLOGY GRADUATES[7]

Comparison Factor	New Engineering Graduate	New Four-Year Engineering Technology Graduate
Technical interest	Education of engineering graduate is relatively broad. The engineer has an analytical, creative mind that is challenged by open-ended technical problems.	Education of technology graduate is relatively specialized. The technologist has an applications orientation that is challenged by specific technical problems.
Technical capability	Engineers use basic knowledge of energy, forces, materials, and their physical and chemical behavior to develop products and services beneficial to humankind.	Technologist utilizes a knowledge of technical sciences and applied physical sciences to produce services beneficial to humankind.
Technical practices	Engineers develop new procedures to advance the state of the art.	Technologists apply established procedures to utilize the state of the art.
Typical beginning job aspirations	The BSE entering industry would typically aspire to an entry-level position in conceptual design, systems engineering, or product research and development.	The BSET entering industry would typically aspire to an entry-level position in product design, product development, technical operations, or technical service and sales.
Adaptability to current industrial practices	Upon graduation, an engineer typically requires a period of "internship" since the engineering program stresses basic fundamentals.	Upon graduation, a technologist may be ready to begin technical assignments immediately since the technology academic program stresses relatively current industrial practices and design procedures.
Average starting salary	Similar, with engineering salaries generally being slightly higher than engineering technology salaries.	

[7] Adapted from *Career Choice: Engineering or Engineering Technology* (Milwaukee, Wis.: Milwaukee School of Engineering, 1986).

PROBLEMS

1-1. Interview an engineer and write a 500-word essay concerning his or her work.

1-2. Survey the job opportunities for engineers, scientists, and technicians. Discuss the differences in opportunity and salary.

1-3. Discuss the role of the engineer in government.

1-4. Frequently technical personnel in industry are given the title "engineer" in lieu of other benefits. Discuss the difficulties that arise as a result of this practice.

1-5. Write an essay on the differences between the work of the engineer, the scientist, and the technician.

1-6. Write an essay on the differences between the education of the engineer, the scientist, the technologist, and the technician.

1-7. Interview an engineering technician and write a 500-word essay concerning her work.

1-8. Discuss the role of the engineering technician in the aircraft industry.

1-9. Investigate the opportunities for employment of electronic technicians. Write a 500-word essay concerning your findings.

1-10. Investigate the differences in educational requirements of the engineer and the technician. Discuss your conclusions.

1-11. Classify the following items as to the most probable assignment in the occupational spectrum:

A. Detail drawing of a small metal mounting bracket.

B. Assisting an engineer in determining the pH of a solution.

C. Boring a hole in an aluminum casting to fit a close tolerance pin.

D. Determining the behavior of flow of a viscous fluid through a pipe elbow.

E. Determining the percentage of carbon in a series of steel specimens to be used in fabricating cutter bits.

F. Designing a device to permit the measurement of the temperature of molten zinc at a location approximately 150 feet from a vat.

G. Preparation of a laboratory report concerning the results of a series of tests on an assortment of prospective heat-curing bonding adhesives.

H. Preparing a work schedule for assigning manpower for a two-day test of a small gasoline engine.

I. Determining the effects of adding ammonia to the intake air of a gas turbine engine.

J. Fabrication of 26 identical transistorized circuits using printed circuit boards.

K. Calculation of the area of an irregular tract of land from a surveyor's field notes.

L. Preparation of a proposal to study the effect of sunlight on anodized and unanodized aluminum surfaces.

M. Design and fabrication of a device to indicate the rate of rainfall at a location several hundred feet from the sensing apparatus.

N. Interpretation of the results of a test on a punch-tape-controlled milling machine.

O. Preparing a computer program to determine the location of the center of gravity of an airplane from measured weight data.

1-12. A large office building is to be constructed, and tests of the load-bearing capacity of the underlying soil are to be made. Outline at least one way each member of the technical team (scientists, engineers, technologists, technicians, and craftsmen) is involved in the work of determining suitability of the soil for supporting a building.

Chapter 2

Modern Challenges for Engineers and Technologists

[2-1] Where does the firefly get its light?

There is another design that is far better. It is the design that nature has provided. . . . It is pointless to superimpose an abstract, man-made design on a region as though the canvas were blank. It isn't. Somebody has been there already. Thousands of years of rain and wind and tides have laid down a design. Here is our form and order. It is inherent in the land itself—in the pattern of the soil, the slopes, the woods—above all, in the patterns of streams and rivers.

William H. Whyte
The Last Landscape

[2-2] In the beginning God created the heavens and the earth.

Genesis 1:1
The Holy Bible

Our earth is a magnificent planet, unlike others that we know about. The more we learn about the earth, the more wonderful it becomes. As we move about on its surface, we recognize that we are actually living in two worlds at the same time. One is a *natural world,* the other, a *man-made* world. The former, a creation of God, is manifested by certain laws of nature, the latter by designs of man—engineering designs. These designs are the things that have been made by man since his habitation of the earth—his creations, inventions, tools, and products—which have been the means of his survival on this earth.

The natural world existed long before man first made his appearance on the earth, and it will continue to exist, with or without man. We learn about the laws of nature through a study of the various sciences. If we use these laws wisely, we can make our lives safe, comfortable, and productive. However, when we defy these laws, or ignore them, we invariably suffer unpleasant and sometimes deadly consequences.

Biologists have named us *Homo sapiens*—Man the Thinker. All that man has created from and in the world is a product of his thoughts . . . his mind. Thought without action is a mere pastime of philosophers. So man might more appropriately be called *Homo faber*—Man the Skillful, or perhaps Man the Maker. We might even describe him as Man the Tool User, or Man the Fabricator. The authors prefer to use the term *Man*[1] *the Engineer.*

[1] In discussing activities of mankind—the human species—the author uses the generic word *man* to represent both male and female.

WE ARE ALL ENGINEERS

The statement above may seem farfetched to you at first. But, consider, if you will, your world as you first discovered it as a child. Like other children, when you first became aware of the world outside yourself, you were curious. You wanted to know about things. You asked many questions of older persons. This attribute of curiosity is also one of the strongest characteristics of engineers. Next, as a child indulging in one of the activities that children like best, you played "make believe"—and you built or made things. You constructed your designs out of such things as big and little boxes, building blocks, Tinker Toys, Lincoln Logs, or perhaps Erector Sets. When you didn't have manufactured toys, you "pretended," using sticks, rocks, dirt, sand on the beach, or even an available puddle of water. You splashed it, diverted it, dammed it, or maybe changed its flow in as many ways as you could imagine.

In other words, you were a builder of play houses, tree houses, forts, castles, mountains (of dirt, snow, or sand), cars, rafts and boats, and other childhood fantasies. As you grew older, your building projects became more sophisticated, as did your ability to repair, modify, and invent more complex devices or ways of doing things. Actually, many of these actions on your part were elementary forms of engineering design.

Most individuals do not choose to focus their interests, ambitions, and education to become occupational practicing engineers. However, for most people their innate urge to create never completely leaves them. Most continue to find other ways to make creative contributions. They may paint a landscape, write a poem, compose a piece of music, devise a new accounting system, create a new hair style, develop a new computer program . . . or engage themselves in a host of other mind-stimulating design activities.

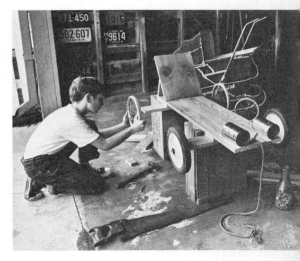

[2-3] Creative effort is a function of the mind, not the age of the individual.

Imagination is more important than knowledge.
Albert Einstein
On Science, 1931

My toys were all tools—they still are.
Henry Ford, 1863–1947

[2-4] The system.

TODAY'S ENGINEERING ENVIRONMENT

Chapter 1 considered man's history in innovation and engineering design. In every age timely inventions and scientific breakthroughs have been largely the result of individual efforts. We found that man has always been able to meet the challenges placed before him. Today's engineers must lead our society in being equally successful. This chapter explores some of the conditions and challenges that confront today's engineers. Some important questions will be asked such as: How can we be sure that our natural environment continues to provide an enhanced quality of life for all people in every part of the earth? And . . . what is my own responsibility?

The earth is a constantly changing system. Until man arrived on the scene the changes that took place over the geological ages were the result of natural processes. However, through time, as man's numbers have grown, he has exerted his own influence on the system. During the past century his population increased at such an alarming rate that his "artificial" changes have taken precedence over the slower natural changes of the system.

[2-5] We must have a good understanding of the natural order of the earth.

The act of creation and the act of appreciation of beauty are not in essence, distinguishable.
H. E. Huntley
The Divine Proportion, 1970

Engineering is the art of directing the great sources of power in nature for the use and convenience of man.
Thomas Tredgold, 1783–1829

It [the Universe] is an infinite sphere whose centre is everywhere, its circumference nowhere.
Blaise Pascal, 1623–1662
Pensées

Beauty is in the eyes of the beholder.
Margaret Wolfe Hungerford, 1855–1897.
Molly Brown

The chess-board is the world, the pieces are the phenomena of the universe, the rules of the game are what we call the laws of Nature. The player on the other side is hidden from us. We know that his play is always fair, just, and patient. But also we know, to our cost, that he never overlooks a mistake or makes the smallest allowance for ignorance.
Thomas H. Huxley, 1825–1895
A Liberal Education

Perhaps more than the members of any other profession, engineers throughout history have made possible the incremental changes in our way of life—some desirable, some undesirable. However, too often the engineer has not been involved in the political, religious, economic, demographic, and military decisions that have determined the ultimate use and effect of his creations—his designs.

The system—the earth—has undergone many artificial but significant changes in the last 100 years, many of which are detrimental to maintaining a desirable quality of life and too many of which are irreversible in their influence. Once man questioned, "Am I my brother's keeper?" Now the answer is no longer one of uncertainty; it is, "Yes, I am!" To survive on this earth we *must* be concerned about our neighbor's well-being. For example, when insecticides can be plowed into farmlands in the midwest and a few months later their traces found in animals in such remote locations as the Antarctic, man can no longer ignore the ultimate consequences of his actions—his designs. By analyzing particulate matter in the upper atmosphere, scientists at the Mauna Loa Observatory in Hawaii can tell with great accuracy when spring plowing starts in northern China.[2] The situation is becoming critical and people in all parts of the world *must* begin to show a special concern for the welfare of their fellowmen and of the plants and animals that share this planet with man.

The engineer of today must be more aware of the effects that his designs could have on the total system, on the lives of his fellowmen, both now and in the future, and on his own well-being. In this regard his designs must always reflect a consciousness for the safety and health of all people. He is responsible also to warn those in government and business of the consequences and possible misuse of his designs.

Before we can realistically evaluate the effects of our designs, we must first have a good understanding of the natural order of the system in which we and our fellowmen live. Since the whole realm of the physical and life sciences is devoted to gaining such understanding, no in-depth attempt will be made in this book to address that part of the task. Rather, the discussion here will be devoted to an exploration of the natural constraints of the system and particularly to the major challenges facing the engineer today—especially with regard to restoring the system to a state of change more nearly in harmony with its natural rate of change. However, lasting solutions to complex problems of this magnitude are possible *only* if the general populace demands and supports a national priority sufficient to supply engineers with resources adequate to sustain their designs.

THE BIOSPHERE

The earth and its inhabitants form a complex system of constantly changing interrelationships. From the beginning of time until a few thousand years ago, the laws of nature programmed the actions of all

[2] Josef R. Parrington et al., "Asian Dust: Seasonal Transport to the Hawaiian Islands," *Science*, April 8, 1983.

living things in relation to each other and to their environments. From the beginning, however, change was ever present. For example, as glaciers retreated, new forests grew to reclaim the land, ocean levels and coastlines changed, the winding courses of rivers altered, lakes appeared, and fish and animals migrated, as appropriate, to inhabit the new environments. Changes in climate and/or topography always brought consequential changes in the distribution and ecological relationships of all living things—plants and animals alike. Almost invariably these changes brought about competitive relationships between the existing and migrating species. In this way certain competing forms were forced to adapt to new roles, while others became extinct. As a result of this continual change in the ecological balance of the earth over eons of time, there currently exist some 1,300,000 different kinds of plants and animals that make their homes in rather specific locations. Only a few, such as the cockroach, housefly, body louse, and house mouse, have been successful in invading a diversity of environments—this because (willingly and unwillingly) they followed man in his travels. Presumably, even these would be confined to specific regions if man did not exist on the earth.

Primitive man was concerned with every facet of his environment and he had to be acutely aware of many of the existing ecological interrelationships. For example, he made it his business to know those places most commonly frequented by animals that he considered to be good to eat or whose skins or pelts were valued for clothing. He distinguished between the trees, plants, and herbs, and he knew which would provide him sustenance. Although by today's standards of education men of earlier civilizations might have been classified as "unlearned," they certainly were not ignorant. The Eskimo, for example, knew long ago that his sled dogs were susceptible to the diseases of the wild arctic foxes, and the Masai of east Africa have been aware for centuries that malaria is caused by mosquito bites.[3]

From century to century man has continued to add to his store of knowledge and understanding of nature. In so doing he has advanced progressively from a crude nomadic civilization in which he used what he could find useful to him in nature, to one sustained by domestication and agriculture in which he induced nature to produce more of the things that he wanted, and currently to one in which he is endeavoring to use the technologies of his own design to control the forces of nature.

For thousands of years after man first inhabited the earth, populations were relatively small and, because man was mobile, the cumulative effect of his existence on his environment was negligible. If, perchance, he did violence to a locality (e.g., caused an entire forest to be burned), it was a relatively simple matter for him to move to another area and to allow time and nature to heal the wound. Because of the expanded world population, this alternative is no longer available. Now he affects not only his immediate environment but that of the whole earth.

[2-6] There are too few beautiful areas to be found today that have been spared the ravages of man.

No man is an island, entire of itself.
John Donne, 1573–1631
Devotions

[2-7] And when the span of man has run its course sitting upon the ruins of his civilization will be a cockroach, calmly preening himself In the rays of the setting sun.

---

[3] Peter Farb, *Ecology* (New York: Time, Inc., 1963), p. 164.

I have long believed that our Nation has a God-given responsibility to preserve and protect our natural resource heritage. Our physical health, our social happiness, and our economic well-being will be sustained only to the extent that we act as thoughtful stewards of our abundant natural resources.

Ronald Reagan
Message to Congress,
July 11, 1984

The first law of ecology is that everything is related to everything else.

Barry Commoner, 1917–

The air, the water, and the ground are free gifts to man and no one has the power to portion them out in parcels. Man must drink and breathe and walk and therefore each man has a right to his share of each.

James Fenimore Cooper
The Prairie, 1827

Ecology—the study of plants and animals in relation to their natural environment.

Harper's Encyclopedia of Science, 1967

The health of nations is more important than the wealth of nations.

Will Durant, 1885–1981

We must and will be sensitive to the delicate balance of our ecosystems, the preservation of endangered species, and the protection of our wilderness lands. We must and will be aware of the need for conservation, conscious of the irreversible harm we can do to our natural heritage, and determined to avoid the waste of our resources and the destruction of the ecological systems on which these precious resources are based.

Ronald Reagan
The White House,
July 11, 1984

Edible—good to eat, and wholesome to digest, as a worm to a toad, a toad to a snake, a snake to a pig, a pig to a man, and a man to a worm.

Ambrose Bierce
The Devil's Dictionary

Man's existence is a part of, not independent of, nature—and specifically it is most concerned with that part of nature that is closest to the surface of the earth known as the *biosphere*. This is the wafer-thin skin of air, water, and soil comprising only a thousandth of the planet's diameter and measuring less than 8 miles thick.[4]

It might be said to be analogous to the skin of an apple. However, this relatively narrow space encompasses the entire fabric of life as we know it—from virus to field mouse, man, and whale. Most life forms live within a domain extending from ½ mile below the surface of the ocean to 2 miles above the earth's surface, although very few creatures can live in the deep ocean or above a 20,000-foot altitude. A number of processes of nature provide a biosphere with a delicate balance of characteristics that are necessary to sustain life. The life cycles of all living things, both fauna and flora, are interdependent and inextricably interwoven to form a delicately balanced *ecological system* that is as yet not completely understood by man. We do know, however, that not only is every organism affected by the environment of the "world" in which it lives, but it also has some effect on this environment.[5] The energy necessary to operate this system comes almost entirely from the sun and is utilized primarily through the processes of photosynthesis and heat. These processes are cyclic and are often referred to as the *chain of life cycle* [2-8]. Elements in the soil are combined with carbon dioxide (by photosynthesis) to produce plants. In turn, they serve as the energy basis for other life forms. The entire cycle is activated by energy from the sun. In accordance with the second law of thermodynamics, there is a loss of energy at each conversion in the cycle. Any changes that man exerts on any part of the system will affect its tenuous balance and cause internal adjustment of either its individual organisms, its environment, or both. The extent and magnitude of the modifications that man has exerted on this ecological system have increased immeasurably within the past few years, particularly as a consequence of his rapid population growth. Certain of these modifications are of particular concern to today's engineer.

Even during the period of the emergence of agriculture and domestication of animals, man began to alter the ecological balance of his environment. Eventually, some species of both plants and animals became extinct, while the growth of others was stimulated artificially. All too often man has not been aware of the extent of the consequences of his actions and, more particularly, of the irreversibility of the alterations and imbalances that he may have caused in nature's system. His concerns have more often been directed toward *subduing* the earth than to *replenishing* it. Over the period of a few thousand years nature's law of "survival of the fittest" was gradually replaced by man's law of "survival of the most desirable." From man's short-term point of view, this change represented a significant

---

[4] R. C. Cook, ed., "The Thin Slice of Life," *Population Bulletin*, Vol. 24, No. 5 (1968), p. 101.

[5] Marston Bates, "The Human EcoSystem," in National Academy of Sciences–National Research Council, *Resources and Man* (San Francisco: W. H. Freeman, 1969), p. 25.

ENERGY LOST

BASIC ELEMENTS IN THE SOIL + CO$_2$

THE FOOD CYCLE

GREEN PLANTS

ENERGY LOST

ENERGY LOST

SOLAR RADIATION   ENERGY GAIN

SUN

SOLAR RADIATION
ENERGY GAIN

HERBIVORES

OMNIVORES

ENERGY LOST

ENERGY LOST

CARNIVORES

[2-8] The chain of life cycle.

And God blessed them, and God said unto them, be fruitful, and multiply, and *replenish* the earth, and *subdue* it: and have dominion over the fish of the sea, and over the fowl of the air, and over every living thing that moveth upon the earth.

Genesis 1:28
The Holy Bible

"And on the Seventh Day"

In the end,
There was Earth, and it was with form and beauty.
And Man dwelt upon the lands of the Earth, the
    meadows and trees, and he said,
"Let us build our dwellings in this place of beauty."
And he built cities and covered the Earth with
    concrete and steel.
And the meadows were gone.
And Man said, "It is good."
On the second day, Man looked upon the waters of
    the Earth.
And Man said, "Let us put our wastes in the waters
    that the dirt will be washed away."
And Man did.
And the waters became polluted and foul in smell.
And Man said, "It is good."
On the third day, Man looked upon the forests of
    the Earth and saw they were beautiful.
And Man said, "Let us cut the timber for our homes
    and grind the wood for our use."
And Man did.
And the lands became barren and the trees were
    gone.
And Man said, "It is good."
On the fourth day, Man saw that animals were in
    abundance and ran in the fields and played in
    the sun.
And Man said, "Let us cage these animals for our
    amusement and kill them for our sport."
And Man did.
And there were no more animals on the face of the
    Earth.
And Man said, "It is good."
On the fifth day, Man breathed the air of the Earth.
And Man said, "Let us dispose of our wastes into
    the air for the winds shall blow them away."
And Man did.
And the air became heavy with dust and all living
    things choked and burned.
And Man said, "It is good."
On the sixth day, Man saw himself and seeing the
    many languages and tongues, he feared and
    hated.
And Man said, "Let us build great machines and
    destroy these lest they destroy us."
And Man built great machines and the Earth was
    fired with the rage of great wars.
And Man said, "It is good."
On the seventh day, Man rested from his labors
    and the Earth was still, for Man no longer dwelt
    upon the Earth.
And it was good.

New Mexico State Land Office

advantage to him. Whereas once he was forced to gather fruit and nuts and to hunt animals to provide food and clothing for himself and his family—a life filled with uncertainty, at best—now he could simplify his food-gathering processes by increasing the yield of such crops as wheat, corn, rice, and potatoes. In addition, certain animals, such as cows, sheep, goats, and horses, were protected from their natural enemies and, in some instances, their predators were completely annihilated. Such eradication seemed to serve man's immediate interests, but it also eliminated nature's way of maintaining an ecological balance. Man, in turn, was also affected by these changes. At one time only the strongest of his species survived, and the availability or absence of natural food kept his population in balance with the surroundings. With domestication these factors have become less of a problem and "survival of the fittest" no longer governs his increase in numbers. In general, today both strong and weak live and

[2-9] Man's continued misuse of the world's resources can only lead to a depletion of the essentials for life.

"THAT'S NOT A PLANET — IT'S AN INCUBATOR."

[2-10]

procreate. Because of this condition the world's population growth has begun to mount steadily and *alarmingly* . . . because of the manifold problems that accompany large populations and for which solutions are still to be found.

It many respects the young engineer of today lives in a world that is vastly different from the one known to his or her grandfather or great grandfather. Without question we enjoy a standard of living unsurpassed in the history of mankind; yet, in spite of the significant agricultural and technological advances made in this generation, over half the world's population still lives in perpetual hunger. Famine, disease, and war continue to run rampant throughout portions of the earth, and wastes from our own technology continue to mount steadily. Nevertheless, the world's population continues to grow and further compounds these problems.

THE POPULATION EXPLOSION: A RACE TO GLOBAL FAMINE

Until recently the growth of the human race was governed by the laws of nature in a manner similar to the laws of nature controlling the growth of all other living things. However, as man's culture changed from nomadic to agrarian to technological, he began to alter nature's population controls significantly. Through control of disease and pestilence his average life span has been extended by a factor of three. His ability to supply his family consistently with food and clothing has also been improved immeasurably. Combinations of these two factors have caused his population to increase in geometric progression: 2-4-8-16-32-64-128, and so on.[6] Initially it took hundreds of thousands of years for a significant change to occur in the world population. However, as the population numbers became larger, and particularly in more recent times as man's life span began to lengthen as a result of his gaining some control over starvation, disease, and violence, the results of geometric growth began to have a profound effect. Whereas it has taken an estimated 2 million years for the world population to reach slightly over 3 billion persons, it will take only 30 years to add the next 3 billion *if present growth rates remain unchanged*. The significance of this problem is dramatically illustrated by Figure [2-11].

As in the case of man's altering his ecological environment, he has not always been wise enough to anticipate all the various effects of his changes. In the case of his own propagation, he has managed to introduce "death control," but birth rates have continued to climb, particularly in the underdeveloped countries. It is estimated that the average annual increase in world population in 1650 was only 0.3 percent.[7] In 1900 this annual growth rate had increased to 0.9 percent; in 1930, 1.0 percent; in 1960, 2.0 percent; but in 1983

---

[6]T. R. Malthus, *An Essay on the Principle of Population As It Affects the Future Improvements of Mankind*, 1798. Facsimile reprint in 1926 for J. Johnson, Macmillan, London.

[7]J. M. Jones, *Does Overpopulation Mean Poverty?* (Washington, D.C.: Center for International Economic Growth, 1962), p. 13.

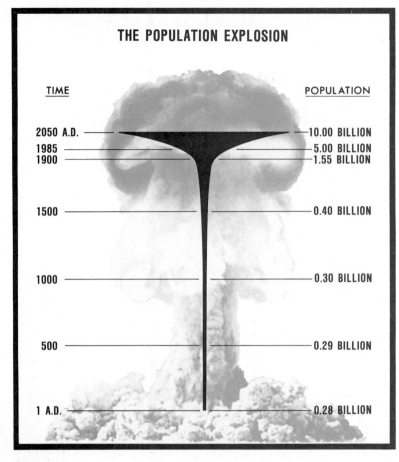

THE POPULATION EXPLOSION

TIME	POPULATION
2050 A.D.	10.00 BILLION
1985	5.00 BILLION
1900	1.55 BILLION
1500	0.40 BILLION
1000	0.30 BILLION
500	0.29 BILLION
1 A.D.	0.28 BILLION

[2-11]

By the close of this century the world may have to feed as many as 2 billion additional people. Most of them will be born in developing countries, especially in marginal lands ill-suited for food production.

Donald L. Plucknett and Nigel J. H. Smith
Science, July 16, 1982, p. 215

The arithmetic of population growth is awesome, and sobering: The earth gains 150 new persons per minute, 9,100 per hour, 219,100 per day and 79.6 million per year and does so mostly in the nations least able to cope with the burden of added people to feed and clothe, and minds to nourish.

John Collins
U.S. News & World Report, July 23, 1984

had decreased slightly to 1.7 percent. This is approximately double the United States annual growth rate. Unless the world rates decline significantly, which does not now seem likely, a worldwide crisis is fast approaching. Unfortunately, 92 percent of the increase in the world's population between 1986 and 2000 is predicted to occur in the less-developed countries.

The earth's land area, only 10 percent of which appears to be arable, is fixed and unexpandable, and a shortage of food and water is already an accepted fact of life in many countries. Although conditions in some slum areas of the United States are very bad, they bear little resemblance to many areas of the world where people grovel in filth and live little better than animals. Unfortunately, Paul Ehrlich's description of a visit to India could just as well have referred to a similar visit to a multitude of other countries.[8]

> I have understood the population explosion intellectually for a long time. I came to understand it emotionally one stinking hot night in Delhi a couple of years ago. My wife and daughter and I were

[8] P. R. Ehrlich, *The Population Bomb* (New York: Ballantine Books, 1968), p. 15.

[2-12] Traditionally, one person signifies loneliness; two persons—companionship; three persons—a crowd. In more recent times a new concept has been added: multitudes signify pollution and loneliness.

Very few Americans, picking and choosing among the piles of white bread in a supermarket, have ever appreciated the social standing of white bread elsewhere in the world. To be able to afford white bread is a dream that awaits fulfillment for billions of the world's population. To afford it signifies that one enjoys all the comforts of life.

Isabel Cary Lundberg
Harper's Magazine

[2-13]

returning to our hotel in an ancient taxi. The seats were hopping with fleas. The only functional gear was third. As we crawled through the city, we entered a crowded slum area. The temperature was well over 100°, and the air was a haze of dust and smoke. The streets seemed alive with people. People eating, people washing, people sleeping. People visiting, arguing, and screaming. People thrusting their hands through the taxi window, begging. People defecating and urinating. People clinging to buses. People herding animals. People, people, people, people. As we moved slowly through the mob, hand horn squawking, the dust, noise, heat, and cooking fires gave the scene a hellish aspect.

Over one-fourth of the human beings now living on the earth are starving, and another one-third are ill fed. The underdeveloped countries of the world are incapable of producing enough food to feed their populations. For example, the growth in Africa's food supply compares favorably with that for the world as a whole. However, its increase in human numbers is far more rapid. The continent has the world's fastest population growth, as well as widespread soil erosion and desertification. The food production in Africa has fallen 11 percent since 1970.[9] This deficiency is 20 million tons of food each year and will grow to a staggering 100 million tons of food per year by 1990. For these people to be fed an adequate diet, the current world food production would have to double by 1990, *which appears to be an impossible task.* The quantity of food available is not the only problem; it must also be of the proper quality. Again, in central Africa every other baby born dies before the age of 5 *even*

---

[9] U.S. Department of Agriculture, Economic Research Service, *World Indices of Agricultural and Food Production, 1950–82* (Washington, D.C.: USDA, 1983).

[2-14] One-half of the people in the world had food to eat this morning. . .   [2-15] . . . the other one-half went hungry.

though food is generally plentiful. They die from a disease known as *kwashiorkor,* which is caused by a lack of sufficient protein in the diet. The magnitude of the problem continues to grow as the world population increases. More population means more famine. It also means more crowding, more disease, less sanitation, more waste and garbage, more pollution of air, water, and land, and ultimately . . . the untimely death of millions.

In the United States, just as in the underdeveloped countries of the world, the population has tended to migrate to cities and some of the cities have grown to monstrous size, often called *megalopolises.* For example, if current world trends continue, Mexico City will be the world's largest city by the year 2000, bulging with 25 to 30 million people.[10] In this country, these migrations have been brought about by the widespread use of mechanized agriculture and the impoverishment of the soil. With these migrations special problems have arisen. No longer is a city dweller self-sufficient in his ability to provide sustenance, shelter, and security for his family. Rather, his food, water, fuel, and power must all be brought to him by others, and his wastes of every kind must be taken away. Most frequently his work is located many miles from his home, and his reliance on a transportation system becomes critical. He finds himself vulnerable to every kind of public emergency, and the psychological pressures of city life often lead to mental illness or escape into the use of alcohol and drugs. The incidence of crime increases, and his clustering invites rapid spread of disease and pestilence. In general, cities are enormous consumers of electrical and chemical energy and producers of staggering amounts of wastes and pollution. Today 70 percent of the people in our country live on 1 percent of the land, and the exodus from the countryside has diminished only slightly. For the past decade, there have been regional shifts in population from the north central and northeastern states to the South and the West.

What are the implications for the engineer of these national and international sociological crises? The engineer is particularly affected because he or she is an essential participant among those whose creative efforts should be directed *to improving* man's physical and economic lot. First, we must learn all we can about the extent and causes of the technological problems that have resulted, and then direct the necessary energies and abilities to solve them. In general, we must recognize our responsibility to restore the equilibrium to the ecological system of nature in those cases in which it has become unbalanced. This requires that we be cognizant of the manifold effects of our designs *prior to their implementation.* All engineering designs are subject to failure under both predictable and unpredictable conditions. Some such failures could have catastrophic effects with regard to the health and safety of people, while others are of lesser import. It is our responsibility to design redundant engineering control systems for those failure situations that could be hazardous (such as in the design of processing systems and in the control of hazardous

[2-16] Wheat is one of the world's most important food crops because each grain contains a relatively large amount of useful protein. Even though engineering designs and advances in agricultural science have made possible a substantially increased production, the disparity between food supply and world population continues to increase. The problem is not merely one of increasing the production of food products. The economic, social, political, and engineering aspects of the logistics of distribution must also receive attention.

U.S. POPULATION GROWTH

1900 - 76 MILLION

1940 - 132 MILLION

1980 - 231 MILLION

2000 - 280 MILLION

SOURCE: U.S. BUREAU OF THE CENSUS

[2-17]

---

[10] United Nations, Department of International Economics and Social Affairs, "Estimates and Projections of Urban, Rural, and City Populations (1950–2025: The 1980 Assessment)," *ST-ESA-R-45* (New York: UN, 1982), Table 8, p. 61.

[2-18] Go west, young man.
*John Babsone Lane Soule, 1815–1891
Article in the Terre Haute, Indiana Express, 1981*

[2-19] Chronic exposure to lead in food or air can cause anemia, convulsions, kidney damage, and brain damage. In the United States, approximately 675,000 children between six months and five years old have high blood-lead levels.
*Joseph L. Annest et al.
Blood-Lead Levels for Persons 6 Months–
74 Years of Age, United States,
1976–1980, Department of Health and Human
Services*

emissions from processes and operations. The task is not an easy one, and the challenge is great, but the consequences are too severe to be disregarded.

Just as you learn that the physical laws of nature *do* govern the universe, so also must you be aware that nature's laws governing the procession and diversity of life on this planet are equally valid and unyielding. In the remainder of this chapter we consider the severity and complexity of several problems that confront today's society and, more particularly, the engineer's social and humanitarian responsibility for their solution.

OUR POLLUTED PLANET

In the last few years the average U.S. citizen has become aware that our "spaceship earth" is undergoing many severe and detrimental ecological changes, which may take hundreds of years to repair. Some genetic changes may be irreversible. Pathological effects may be delayed in an individual for many years (such as the contraction of lung cancer 20 years after exposure to asbestos); also, insidious, genetic mutations (from ionizing radiation or mutagenic chemicals) that become manifest several generations later. Unfortunately, man is not always able to distinguish effects that are of a temporary nature from those with long-term consequences. Frequently, man's most damaging actions to the environment are either of an incremental or visually indistinguishable nature, and for this reason he participates willingly in them. In some measure man's reactions are dulled by the slowness of deterioration. This is somewhat analogous to the actions of a frog that will die rather than jump out, when placed in a bucket of water that is being *slowly* heated. In constrast, if the frog is pitched into a bucket of boiling water, he will immediately jump out and thereby avoid severe injury. The engineer in particular must learn to understand such cause-and-effect relationships so that his designs will not become detrimental to the orderly and natural development of all life. In the past two decades significant progress has been made in improving the working environment. The landmark Occupational Safety and Health Act (OSHA) established as national policy that American workers should be provided with a safe and healthful place of employment. The Clean Air Act of 1970 and 1972 amendments to the Federal Water Pollution Control Act (since renamed the Clean Water Act) focused primarily on cleaning up "conventional" pollutants—smoke and sulfur oxides in the air, oxygen-depleting discharges to the water, and solid wastes on the land.

THE AIR ENVIRONMENT

The atmosphere, which makes up the largest fraction of the biosphere, is a dynamic system that absorbs continuously a wide range of solids, liquids, and gases from natural and man-made sources. These substances often travel through the air, disperse, and react with one another and with other substances. Eventually, most of these constituents find their way into a depository, such as the ocean, or to a receptor, such as man. A few, such as helium, escape from the biosphere. Others, such as carbon dioxide, may enter the atmosphere faster than they can be absorbed and thus gradually accumulate in the air.

Clean, dry air contains 78.09 percent nitrogen by volume and 20.94 percent oxygen. The remaining 0.97 percent is composed of carbon dioxide, helium, argon, krypton, and xenon, as well as very small amounts of some other inorganic and organic gases whose amounts vary with time and place in the atmosphere. Varying amounts of contaminants continuously enter the atmosphere from natural and man-made processes that exist on the earth. That portion of these substances which interacts with the environment to cause toxicity, disease, aesthetic distress, physiological effects, or environmental decay has been labeled by man as a pollutant. Environmental contaminants may be in the form of a chemical (or biological) agent or physical agent, such as noise, electromagnetic radiation, and so on. In general, the actions of people are the primary cause of pollution and, as population increases, the attendant pollution problems

The sky is the daily bread of the eyes.
Ralph Waldo Emerson
Journal, May 25, 1843

Hell is a city much like London—a populous and smoky city.
Percy Bysshe Shelley
Peter Bell the Third, 1819

I durst not laugh, for fear of opening my lips and receiving the bad air.
Shakespeare
Julius Caesar

[2-20] We can't agree on what to call Noah's new invention. He wants to call it *fire* . . . I want to call it *pollution*.

[2-21] Air pollution can make your eyes water and your throat burn. It can cause dizziness, blurred vision, coughing, chest discomfort and impaired breathing. During episodes of heavy air pollution, scores of people come to hospital emergency rooms with serious breathing problems, and premature deaths from heart and lung diseases jump dramatically. Furthermore, many scientists are convinced that air pollution contributes to three major types of chronic disease that kill millions of people annually—heart disease, lung disease and cancer.

Besides these health effects, dirty air has other impacts. It injures crops, flowers, shrubs and forests; it corrodes and dirties buildings, statues, fabrics and metals. When air pollution emissions return to the earth in the form of acid rain, populations of fish and other organisms in sensitive lakes can be decimated. Air pollution can impair visibility on the highway and in national parks. And, it may even be altering the earth's climate.

Deborah A. Sheiman, 1981
LWVEF Environmental Quality Department

increase at the same geometric rate. This is not a newly recognized relationship, however. The first significant change in man's effect on nature came with his deliberate making of a fire. *No other creature on earth starts fires.* (The author concedes that the infamous Chicago fire of 1871 was blamed on Mrs. O'Leary's cow!) Prehistoric man built a fire in his cave home for cooking, heating, and to provide light for his family. Although the smoke was sometimes annoying, no real problem was perceived to exist with regard to pollution of the air environment. However, when his friends or neighbors visited him and also built fires in the same cave, even prehistoric man recognized that he then had an *air-pollution problem.* People in some nineteenth-century cities with their hundreds of thousands of smoldering soft-coal grates coughed amid a thicker and deadlier smog than any modern city can concoct. Today the natural terrain that surrounds large cities is recognized as having a significant bearing on the air-pollution problem. However, this is not an altogether new concept either. Historians tell us that the present Los Angeles area, which in recent years has become a national symbol of comparison for excessive smog levels,[11] was known as the "Valley of Smokes" when the Spaniards first arrived.[12] In recent years air pollution has become a problem of world concern.

In the United States the most common air pollutants are carbon monoxide, sulfur oxides, hydrocarbons, ozones, nitrogen oxides, and suspended particulates. Their primary sources are motor vehicles, industry, electrical power plants, space heating, and refuse disposal, with approximately 60 percent of the bulk being contributed by motor vehicles and 17 percent by industry. It seems probable that by the year 2000 America's streets will contain 20 percent more automobiles than the current 125 million.

In 1969, every man, woman, and child in the United States was producing an average of 1400 pounds per year of air pollutants. The problem was one of serious proportions. The National Air Quality Standards Act of 1970, which specified that motor vehicle exhaust emissions should be reduced by 90 percent by January 1, 1975, has provided the impetus for a sincere and necessary national commitment. In the 15 years since that time, over $300 billion has been spent for air pollution control. Industry supplied approximately 60 percent of these funds, the federal government 30 percent, and the consumer public 10 percent. The results have been a 50 percent decrease in measurable outdoor toxic air contaminants. But by focusing on outside air, we have missed many serious problems created by pollutants indoors, often referred to as occupational environmental contamination. Since on an average, American workers spend about one-fourth of their total life (about one-half of their waking hours) in the workplace, it is of utmost importance that this environment be free of contamination.

---

[11] The term *smog* was coined originally to describe a combination of smoke and fog, such as was common in London when coal was widely used for generating power and heating homes. More recently it has come to mean the accumulation of photochemical reaction products that result largely from the action of the radiant energy of the sun on the emissions of internal combustion engines (automobile exhaust).

[12] H. C. Wohlers, *Air Pollution—The Problem, the Source, and the Effects* (Philadelphia: Drexel Institute of Technology, 1969), p. 1.

It has been found that the significantly increasing volume of particulate matter entering the atmosphere scatters the incoming sunlight. This reduces the amount of energy that reaches the earth and lowers its temperature. The decreasing mean global temperature of recent years has been attributed to the rising concentrations of airborne particles in the atmosphere.[13] A counteracting phenomenon, commonly referred to as the *greenhouse effect*, is caused by the presence of water vapor in the atmosphere, and to a lesser extent carbon dioxide and ozone, which combine to act in a manner similar to the glass in a greenhouse. Light from the sun arrives as short-wavelength radiation (visible and ultraviolet) and passes through it to heat the earth, but the relatively long-wavelength infrared radiation (heat radiation) that is emitted by the earth is absorbed by the carbon dioxide and water vapor—thereby providing an abnormal and additional heating effect to the earth. Although carbon dioxide occurs naturally as a constituent of the atmosphere and is not normally classified as an air pollutant, man generates an abnormally large amount of it in those combustion processes that utilize coal, oil, and natural gas. Some have estimated that if the carbon dioxide content in the atmosphere continues to increase at the present rate, the mean global temperature could rise by almost 4°C in the next 40 to 50 years. This might become a matter of great importance, because small temperature increases could cause a partial melting of the ice caps of the earth (causing flooding of coastal land, towns, and cities) with consequential and devastating effects to man.[14] A recent study by the National Academy of Sciences concludes that the impact of increasing CO_2 may not be as serious as others fear, although there are many factors that still need further investigation.[15]

Air pollution can cause death, impair health, reduce visibility, bring about vast economic losses, and contribute to the general deterioration of both our cities and countryside. Even though significant improvements have been made in the last two decades to reduce pollution levels, the overall situation has continued to deteriorate because in many cases the composition of the pollutants has changed also. Today, many pollutants that are released are geneotoxic and will cause genetic damage that will affect future generations. The problem must be viewed as one of protecting the health (genetic or otherwise) of the entire biosystem. It is therefore a matter of grave importance that engineers of all disciplines consciously incorporate in their designs sufficient constraints, controls, and safeguards to ensure that they do not contribute to the pollution of the atmosphere. In addition, they must apply their ingenuity and problem-solving abilities to eliminating air pollution where it exists and restoring the natural environment. The preferable course of action is for the engineer to produce designs that do not contribute to pollution in any form.

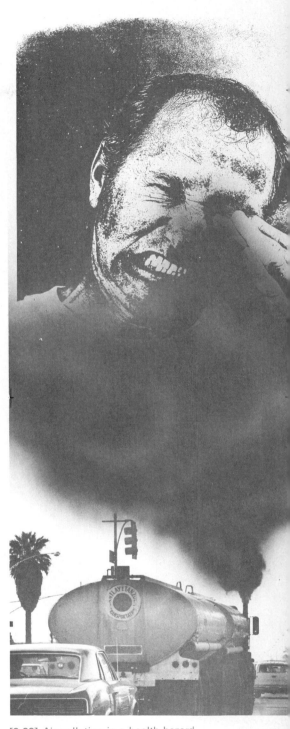

[2-22] Air pollution is a health hazard.

---

[13] R. E. Newell, "The Global Circulation of Atmospheric Pollutants," *Scientific American,* January 1971, p. 40.

[14] The Conservation Foundation, *State of the Environment,* (Washington, D.C.: CF, 1984), p. 103.

[15] U.S. Environmental Protection Agency, Office of Policy and Resource Management, Strategic Studies Staff, "Can We Delay a Greenhouse Warming?" (Washington, D.C.: EPA, September 1983).

[2-23] Oil and grease discharges are responsible for polluting many of the rivers and lakes of America.

THE QUEST FOR WATER QUALITY_____

Water is the most abundant compound to be found on the face of the earth and, next to air, it is the most essential resource for man's survival. The per capita daily water withdrawal in the United States is over 1000 gallons per day, and this demand continues to grow. Early man was most concerned with the quality (purity) of his drinking water, and even he was aware that certain waters were contaminated and could cause illness or death. In addition, modern man has found that he must be concerned also with the quantity of the water available for his use. An abundant supply of relatively pure water is no longer available in most areas. Today, water pollution, *the presence of toxic or noxious substances or heat in natural water sources,* is considered to be one of the most pressing social and economic issues of our time. Unlike the nation's relatively consistent and successful improvements in air quality in the last decade, success in cleaning up surface waters has not been as impressive. Some streams and rivers have improved and the "dying" Great Lakes, Erie and Ontario, are reviving, with fewer algae blooms and growing fish populations.[16] Unfortunately, many streams and lakes have been degraded or have shown no significant change since the early 1970s.

The processes of nature have long made use of the miraculous ability of rivers and lakes to "purify themselves." After pollutants find their way into a water body they are subject to dilution or settling, the action of the sun, and to being consumed by beneficial bacteria. The difficulty arises when man disturbs the equilibrium of the ecosystem by dumping large amounts of his wastes into a particular water body, thereby intensifying the demand for purification. In time the body of water cannot meet the demand, organic debris accumulates, anaerobic areas develop, fish die, and putrification is

---
[16] The Conservation Foundation, *State of the Environment* (Washington, D.C.: CF, 1982), p. 97.

[2-24] Pollution does not necessarily need to accompany poverty, but it most frequently does.

the result. This process also occurs in nature, but it may take many thousands of years to complete the natural processes of deterioration. Man can alter nature's time scale appreciably.

Many city water-treatment plants merely remove the particulate matter and disinfect the available water with chlorine to kill bacteria, since they were not originally designed to remove pesticides, herbicides, and other organic and inorganic chemicals that may be present.[17] Industrial liquid wastes frequently carry large quantities of dissolved material which should be recycled. Silver salts from photographic processes are but one example. This problem has become acute in a number of areas in recent months as hundreds of new contaminants have been discovered in deep wells, streams, and lakes: bacteria and viruses, chemicals from solid-state electronics processes, detergents, municipal sewage, acid from mine drainage, pesticides and weed killers, radioactive substances, phosphorus from fertilizers, trace amounts of metals and drugs, and other organic and inorganic chemicals. As the population continues to increase, the burden assumed by the engineer to design more comprehensive wastewater treatment plants also mounts. Indeed, the well-being of entire communities may depend on the engineer's design abilities, because it is now a recognized condition of population increase that "everyone cannot live upstream."

The presence of radioactive wastes and excess heat are relatively new types of pollution to water bodies, but they are of no less importance for the engineer to take into account in designing. All radioactive materials are biologically injurious. Therefore, radioactive substances that are normally emitted by nuclear power plants are suspected of finding their way into the ecological food chain, where they could cause serious problems. For this reason all radioactive wastes should be isolated from the biological environment during

[17] Gene Bylinsky, "The Limited War on Water Pollution," *The Environment* (New York: Harper & Row, 1970), p. 20.

Modern man . . . has asbestos in his lungs, DDT in his fat, and strontium 90 in his bones.
Today's Health, *April 1970*

Every citizen has a personal stake in the quality of the nation's waters. We need water that is safe to drink, safe to swim in, habitable for aquatic life, free of nuisance conditions, and usable for agriculture and industry. Health, jobs, and the quality of our lives are thus affected in many ways by water quality.
Environmental Quality—1977
The Eighth Annual Report of the Council on Environmental Quality

Soil conservation must be given a higher priority in the future. This will mean not only obvious measures such as terracing and tree planting, but also less conventional approaches including a much greater emphasis on organic fertilizers (including biological nitrogen fixation), minimum tillage techniques, and, in the tropics, farming methods that preserve negative cover and protect the soil from baking by sun and battering by rain, such as use of dead or living mulches and agro-forestry.
Paul Harrison
Ambio, *1984, p. 167*

[2-25] Soil erosion is a major source of water pollution in agricultural areas.

[2-26] The character of a nation is revealed by examining its garbage.

Sooner or later closed-cycle manufacturing, materials recycling, energy recovery from wastes, and the return to the land of fiber and nutrients will certainly occur.

Ariel Parkinson
Environment, *December 1983*

the "life" of the isotopes (as much as 600 years or longer in some cases), and this must be incorporated in the design.

The heat problem arises because electrical power generating plants require great quantities of water for cooling. Although the heated discharge water from a nuclear power plant is approximately the same temperature as that from a fossil-fuel power plant, the quantity of water must be increased by approximately 40 percent. The warmer water absorbs less oxygen from the atmosphere, and this decelerates the normal rate of decomposition of organic matter. This abnormal heat also unbalances the life cycles of fish, which, being cold blooded, cannot regulate their body temperatures correspondingly. If the number of large power plants increases, this problem could loom larger than ever.

Engineering designs of the future must take into account all these factors to ensure for all the nation's inhabitants a water supply that is both sufficient in quantity and unpolluted in quality. This is not only the engineer's challenge, but his responsibility as well.

SOLID-WASTE DISPOSAL AND RECYCLING MATERIALS

We are living in a most unusual time—a time when possibly the most valuable tangible asset that a person could own would be a "bottomless hole." Never before in history have so many people had so much garbage, refuse, trash, and other wastes to dispose of, and at the same time never before has there been such a shortage of "dumping space." The proliferation of refuse, however, is only partly attributable to the population explosion. A substantial portion of the blame must be assumed by an affluent society that is careless with its increasing purchasing power and that has demonstrated a decided distaste for secondhand articles.

The most popular method of solid-waste disposal has long been "removal from the immediate premises." For centuries man has been aware of the health hazards that accompany the accumulations of garbage. Historians have recorded that a sign at the city limits of ancient Rome warned all persons to transport their refuse outside the city or risk being fined. Also, it has been recognized that in the Middle Ages the custom of dumping garbage in the streets was largely responsible for the proliferation of disease-carrying rats, flies, and insects that made their homes in piles of refuse.

On a per capita basis the United States has established an unenviable record with regard to the generation of solid waste, refuse, and garbage:

1920: 2.75 pounds per person per day.

1970: 5.5 pounds per person per day.

1985: 10.0 pounds per person per day.

Industries in this country generate about 2500 pounds of hazardous waste per capita annually. From New York to Los Angeles, cities throughout the nation are rapidly depleting their disposal space, and there is considerable concern that too little attention has been given to what is fast becoming one of man's most distressing problems—solid-waste disposal. Why is trash becoming such a problem? The answer seems to lie partially in man's changing value system. Just a generation or two ago, thrift and economy were considered to be important tenets of American life and few items with any inherent value were discarded. Today, we live in a "throwaway society." More and more containers of all types are being made of inert plastic or glass, or nondegradable aluminum, and everything from furniture to clothing is being made from disposable paper products, which are sold by advertising that challenges purchasers to "discard when disenchanted." The truth is that most "consumers" in America are becoming "users and discarders," but this fact is not always recognized.

Unfortunately, most current waste-disposal practices make no attempt to recover any of the potential values that are in solid wastes. The problem is not just one of wasting materials. We also throw away energy when we discard products. Only 4 percent as much energy is required to recycle aluminum as to produce it originally from bauxite. To recycle copper would save 90 percent and to produce steel from scrap would save almost 50 percent. The methods of disposing of refuse in most common use today are dumping and/or burning in the open, sanitary land fills, burial in abandoned mines, dumping at sea, and grinding in disposal systems followed by flushing in sewers. Some edible waste, such as garbage, is fed to hogs.

Waste is a human concept. In nature, nothing is wasted, for everything is part of a continuous cycle [2-8]. Even the death of an animal provides nutrients that will eventually be reincorporated in the chain of life. For example, a newspaper has been "consumed"

[2-27] The world will little note nor long remember what we say here; but it can never forget what they did here. . . .

Abraham Lincoln
Address at Gettysburg, *1863*

[2-29] I think that I shall never see
     A billboard lovely as a tree
     Indeed, unless the billboards fall
     I'll never see a tree at all.

Ogden Nash
Song of the Open Road, *1945*

[2-28] Pollution is a personal matter.

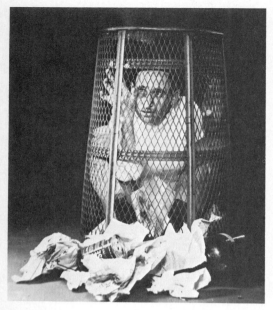

[2-30] As the earth becomes more crowded, there is no longer an "away." One person's trash basket is another's living space. (National Academy of Sciences)

when its purchaser has finished reading it. But one person's waste newspapers are another person's recyclable fibers or cellulose insulation.[18] Also, a few cities have begun to convert their municipal garbage into energy. Landfill gases are created through bacterial decomposition of organic waste, and consist of approximately 55 percent methane (CH_4), 45 percent CO_2, and trace amounts of other gases [2-31]. At present less than 1 percent of the solid waste generated in the United States each year is used to generate energy. Several Scandinavian countries convert up to 40 percent of their garbage into useful energy.[19]

Since almost all engineering designs will eventually be discarded due to wear or obsolescence, it is imperative that consideration for disposal be given to each design *at the time that it is first produced*. In addition, the well-being of society as we know it appears to depend in some measure on the creative design abilities of the engineer to devise new processes of recycling wastes by either changing the physical form of wastes or the manner of their disposal, or both. Such designs must be accomplished within the constraints of economic considerations and without augmenting man's other pollution problems: air, water, and sound. Basically, the solution of waste disposal is a matter of attitude, ingenuity, and economics—all areas in which the engineer can make significant contributions.

In addition to reducing the cost of increasing solid-waste disposal, recycling of discardables will save energy and expensive raw

[18] Denis Hayes, *Repairs, Reuse, Recycling—First Steps Toward a Sustainable Society* (Washington, D.C.: Worldwatch Institute, 1978), p. 6.

[19] *Two Energy Futures* (Washington D.C.: American Petroleum Institute, 1980), p. 108.

At least two-thirds of the material resources that we now waste could be reused without important changes in our life-styles. With products designed for durability and for ease of recycling, the waste streams of the industrial world could be reduced to small trickles. And with an intelligent materials policy, the portion of our resources that is irretrievably dissipated could eventually be reduced to almost zero.
Dennis Hayes, Repairs, Reuse, Recycling—First Steps Toward a Sustainable Society *(Washington, D.C.: Worldwatch Institute, September 1978)*

Feedlots now produce more organic waste than the total sewage from all U.S. municipalities.
Barry Commoner
The Closing Circle, *1971*

The paper packaging for McDonald's first eight billion hamburgers used up 890 square miles of forest.
Bruce Hannon
Letters to the Editor,
Not Man Apart,
September 1972

[2-31] The recovery of methane from garbage has become a profitable practice.

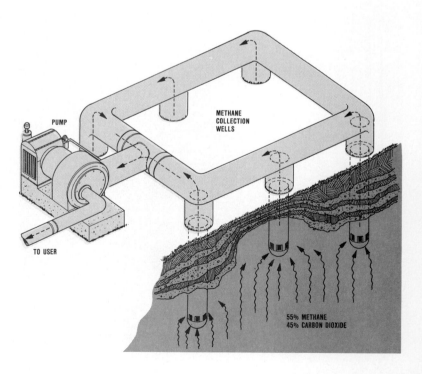

PUMP

METHANE
COLLECTION
WELLS

TO USER

55% METHANE
45% CARBON DIOXIDE

materials, as well as protecting the environment. It has been estimated that throwing away an aluminum beverage container wastes an amount of energy equivalent to that which would be wasted by pouring out such a container half-filled with gasoline. Failing to recycle a weekday edition of the *Washington Post* or the *Los Angeles Times* wastes just about as much.[20]

Some progress in recycling aluminum has been made in the past two decades, growing from 16 percent to 32 percent of the total production.[21] This represents 54 percent of all aluminum beverage cans used each year. Paper consumption has more than doubled during this period, but less than one-fourth is recycled paper. The recovery rate of this country's annual steel consumption is a mere 35 percent.

The placement of debris in space (nose cones, pieces of disintegrated satellites, nuts, and bolts) is becoming a matter of concern to NASA. It is estimated that over 15,000 man-made objects are circling the earth in space, with at least 5000 of these being tracked continuously. Perhaps it is time to consider the design of a space debris scavenger satellite.[22]

THE RISING CRESCENDO OF UNWANTED SOUND

A silent world is not only undesirable but impossible to achieve. Man's very nature is psychologically sensitive to the many sounds that come to his ears. For example, he is pleased to hear the gurgle and murmur of a brook or the soothing whispering wind as it filters through overhead pine trees, but his blood is likely to chill if he recognizes the whirring buzz of a rattlesnake or hears the sudden screech of an automobile tire as it slides on pavement. He may thrill to the sharp bugle of a far-off hunting horn, but his thoughts often tend to lapse into dreams of inaccessible places as a distant train whistle penetrates the night.[23] Yes, sounds have an important bearing on man's sense of well-being. Although the average city dweller's ears continue to alert him to impending dangers, their sensitivity is far less acute than that of people who live in less densely populated areas. It is said that, even today, aborigines living in the stillness of isolated African villages can easily hear each other talking in low conversational tones at distances as great as 100 yards, and that their hearing acuity diminishes little with age.[24] Even as man's technology has brought hundreds of thousands of desirable and satisfying inno-

[2-32] Adequate housing is a concern for people everywhere.

Woe unto them that join house to house that lay field to field, till there be no place, that they may be placed alone in the midst of the earth!

Isaiah 5:7–8
The Holy Bible

Noise has only recently been recognized as a serious environmental pollutant. In comparison to efforts to control air and water pollution, the control of noise pollution is still in its infancy.
Ronald M. Buege, Director of Environmental Health
Journal of Environmental Health, *Vol. 45, No. 5*

---

[20]William V. Chandler, *State of the World 1984*, Worldwatch Institute (New York: W. W. Norton, 1984), p. 95.

[21]Aluminum Association, *Aluminum Statistical Review for 1981*, Washington, D.C.

[22]Peter T. White, "The Fascinating World of Trash," *National Geographic Magazine*, April 1983, p. 456.

[23]*Noise: Sound Without Value* (Washington, D.C.: Committee on Environmental Quality of the Federal Council for Science and Technology, 1968), p. 1.

[24]The Editors of *Fortune*, *The Environment* (New York: Harper & Row, 1970), p. 136.

Noise is the only pollutant that some people actually want. Truckdrivers want the meanest sounding rig. A company tried to sell a quiet vacuum cleaner years ago, but few wanted them. They would never sell because they were too quiet.
Edward DiPolmere, Association of Noise Control Officials
Quoted in U.S. News & World Report, *July 16, 1984*

Not many sounds in life, and I include all urban and all rural sounds, exceed in interest a knock at the door.
Charles Lamb
Valentine's Day, 1823

It will be generally admitted that Beethoven's Fifth Symphony is the most sublime noise that has ever penetrated into the ear of man.
Edward Morgan Forster, 1879–1970
Howards End

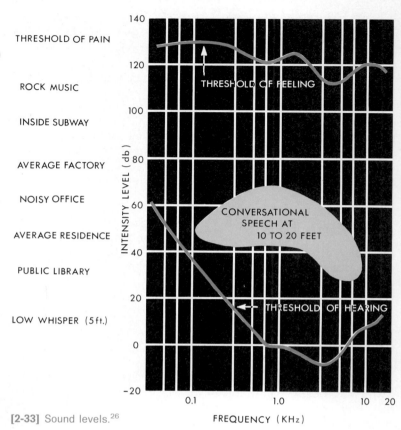

[2-33] Sound levels.[26]

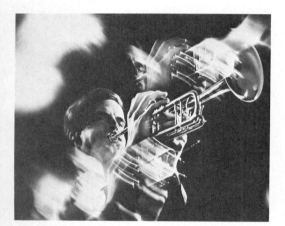

[2-34] Man is affected psychologically by the sounds that he hears.

[2-35]

vations, it has also provided the means for the retrogression of his sense of hearing—for deafness caused by a deterioration of the microscopic hair cells that assist in transmitting sound from the ear to the brain. It has been found that prolonged exposure to intense sound levels will produce permanent hearing loss, and it matters not that such levels may be considered to be pleasing. Some people purport to enjoy "rock" music concerts at sound levels exceeding 110 decibels[25] [2-33]. Today noise-induced hearing loss looms as one of America's major health hazards.

Noise is generally considered to be any annoying or unwanted sound. Noise (like sound) has two discernible effects on man. One causes a deterioration of his sensitivity of hearing, the other affects his psychological state of mind. The adverse effects of noise have long been recognized as a form of environmental pollution. Julius Caesar was so annoyed by noise that he banned chariot driving at night, and, prior to 1865, studies in England reported substantial hearing losses among blacksmiths, boilermakers, and railroad men.[27]

---

[25] A *decibel* (abbreviated dB) is a measure of sound intensity or pressure change on the ear.

[26] Adapted from W. E. Woodson and D. W. Conover, *Human Engineering Guide for Equipment Designers* (Berkeley: University of California Press, 1964), pp. 4–10.

[27] Aram Glorig, in *Noise as a Public Health Hazard*, W. D. Ward and J. E. Fricke, eds. (Washington, D.C.: The American Speech and Hearing Association, 1969), p. 105.

TABLE 2-1   PERMISSIBLE NOISE EXPOSURES[a]

Duration per Day (hr)	Sound Level (dBA Slow Response)
8	90
6	92
4	95
3	97
2	100
$1\frac{1}{2}$	102
1	105
$\frac{1}{2}$	110
$\frac{1}{4}$ or less	115

[a]Noise measured on the A scale of a sound level meter is given in dB(A) units. The A scale measures noise processed by the human ear. It gives greatest weight to noise in the octave band centered on 2 kHz. Noise of low frequency receives less weight.
Source: OSHA-29CFR 1910.956.

He who sleeps in continual noise is wakened by silence.

> *William Dean Howells, 1837–1920*
> Pordenone

Men trust their ears less than their eyes.

> *Herodotus*
> *ca. 485–ca. 425 B.C.*

The crescendo of noise—whether it comes from truck or jackhammer, siren or airplane—is more than an irritating nuisance. It intrudes on privacy, shatters serenity and can inflict pain. We dare not be complacent about this ever mounting volume of noise. In the years ahead, it can bring even more discomfort—and worse—to the lives of people.

> *Lyndon Baines Johnson*

Unnecessary noise is the most cruel absence of care which can be inflicted either on sick or well.

> *Florence Nightingale*
> *1820–1910*

Nature has given man one tongue and two ears, that we may hear twice as much as we speak.

> *Epictetus*
> Fragments, A.D. 90

Psychologists have found that music does things to you whether you like it or not. Fast tempos invariably raise your pulse, respiration, and blood pressure; slow music lowers them.

> *Doron K. Antrim*

It has only been in the last 50 years or so, however, that noise has been recognized naturally as an occupational health hazard.[28]

It is estimated that the average background noise level throughout the United States has been doubling each 10 years. At this rate of increase, living conditions will be intolerable within a few years. Such a crescendo of sound results from the steady increase of population and the concomitant growth of the use of power on every hand—from the disposal in the kitchen and the motorcycle in the street to power tools in the factory. Buses, jet airliners, television sets, stereos, dishwashers, tractors, mixers, waste disposers, air conditioners, automobiles, jackhammers, power lawn mowers, vacuum cleaners, typewriters, and printers are but a few examples of noise producers that are deemed desirable to today's high standard of living, but which may very well also prevent man from fully enjoying the fruits of his labors, unless the sound levels at which they operate are altered significantly.

OSHA has established permissible levels of noise exposure, as shown in Table 2-1. When employees are subjected to sound exceeding the exposures listed, feasible administrative or engineering controls must be utilized.

Except in the case of minimizing aircraft noise, the United States lags far behind many countries in noise prevention and control. Virtually all manmade noise can be suppressed, and the same engineer who formulates the idea for a new type of kitchen aid or designs an improved family vehicle must also be capable of solving the acoustical problems that are associated with its manufacture and use. In this regard, as an engineer, you are responsible to generations yet unborn for the consequences of your actions.

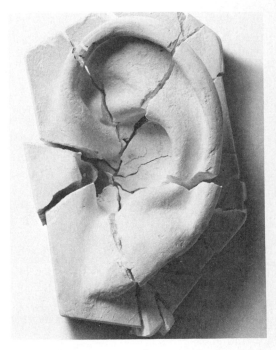

[2-36] If the ear were to shatter or bleed profusely when subjected to abuse from intense or prolonged noise, we might be more careful of its treatment.

---

[28]"Effect of Noise on Hearing of Industrial Workers," State of New York Department of Labor, *Special Bulletin, 166* (New York: Bureau of Women in Industry, 1930).

[2-37] Man stands at the end of a long cycle of energy exchanges in which there is a calculable and irreversible loss of energy at each exchange. A grown adult irradiates heat equivalent to that of a 75-watt bulb. His total energy output, in 12 hours of hard physical work, is equivalent to only 1 kilowatt-hour. He requires daily 2200 calories of food intake, 4½ pounds of water, and 30 pounds of air, and he discards 5 pounds of waste. Considered as an energy converter, man is the least efficient link in his particular "food chain," and for this reason the most vulnerable to catastrophic ecologic change. Such a change can be caused by overloading the energy circuit. There are two new humans added to the globe's population *every second*.

$E = mc^2$

Albert Einstein
Annalen der Physik, 1905
Statement of the mass–energy equivalence relationship

F. D. Roosevelt
President of the United States
White House
Washington, D.C.

Sir:

Some recent work by E. Fermi and L. Szilard, which has been communited to me in manuscript, leads me to expect that the element of uranium may be turned into a new and important source of energy in the immediate future . . . that it may be possible to set up a nuclear chain reaction in a large mass of uranium by which vast amounts of power and large quantities of new radium-like elements would be generated. . . .
Albert Einstein
Old Grove Road, Nassau Point
Peconic, Long Island
August 2, 1939

MAN'S INSATIABLE THIRST FOR ENERGY

In man's earliest habitation of the earth he competed for energy with other members of the earth's ecological environment. Initially his energy requirements were primarily satisfied by food—probably in the range of 2000 calories per person per day.[29] However, as he has been able to make and control the use of fire, domesticate the plant and animal kingdoms, and initiate technologies of his own choosing, his per capita consumption of energy has increased appreciably.

In 1940, it was estimated that the total energy generated in the United States would be equivalent in "muscle-power energy" of 153 slaves working for every man, woman, and child in this country. Today a similar calculation would show that over 500 "slaves" are available to serve each person. In this respect every person is a monarch. This demand for increasing forms of energy has followed an exponential pattern of growth similar to the growth of the world population, *except that the annual rate of increase for nonnutrient energy utilization is growing at a rate* (approximately 4 percent per year) *considerably in excess of the world's growth in population* (slightly less than 2 percent per year). This is brought about by man's appetite for more gadgets, faster cars and airplanes, heavier machinery, and so on.

The principal sources of the world's energy prior to about A.D. 1200 were solar energy, wood, wind, and water. At about this time in England it was discovered that certain "black rocks" found along the seashore would burn. From this there followed in succession the mining of coal (the black rocks) and the exploration of oil and natural-gas reservoirs.[30] More recently, nuclear energy has emerged as a promising source of power. The safe management and disposal of radioactive wastes, however, continue to present problems for the engineer. Renewable energy sources appear to have considerable promise in providing the additional energy needs of the country that will develop over the next two decades. These include wind power, geothermal energy, solar energy, fuel cells, hydropower, wood fuel, ocean thermal energy differences, and energy crops [2-38 to 2-40].

The graph [2-41] provides a record of the history of energy consumption in the United States since 1850. Of course, the future is unknown, and a prediction of our energy sources for the year 2000 and beyond is mere conjecture. It depends to a large extent on the background and experience of the predictor. External factors may also intervene. It may well be, for example, that although fossil fuels seem to be sufficient in quantity, they might be undesirable for expanded use because of their combustive pollutant effects. The solving of such problems represents a number of challenges for the engineer.

---

[29]The calorie is an energy unit. However, the food industry conventionally uses a "calorie" that is 1000 times the size of the "calorie" used by the general scientific community. The 2000 food calories referred to here would be equivalent to 2000 kilocalories expressed in the more common scientific units.

[30]There are evidences that coal was used in China, Syria, Greece, and Wales as early as 1000–2000 B.C.

In the United States, the energy consumption is distributed approximately as follows:

Residential	21 percent
Commercial	14 percent
Automobiles	16 percent
Transportation	8 percent
Industrial and other	41 percent

Considering the fact that currently the five most common air pollutants (carbon monoxide, sulfur oxides, hydrocarbons, nitrogen oxides, and solid particles) are primarily by-products of the combustion of fossil fuels, it behooves the engineer to design and utilize energy sources that are as free from pollution as possible.

In the next two decades the United States will consume more energy than it has in the past 200 years combined. The artful manipulation of energy is an essential component of man's ability to survive in a competitive world environment.

The most accepted index for measuring a nation's aggregate economic output, the total market value of the goods and services produced by a nation's economy during a specific period of time (usually a year) is the gross national product (GNP). Although the GNP is not necessarily a measure of the quality of life, it does represent in the broadest sense a measure of a nation's standard of living. If one plots the GNP of the nations of the world versus their total energy consumption per year per capita, a linear relationship will result [2-42]. A compelling case can be made, therefore, for the hypothesis that the productive utilization of energy has played a primary role in shaping the science and culture of the world.[31]

It would appear that in the long run the earth can tolerate a significant increase in man's continuous release of energy (perhaps as much as 1000 times the current U.S. daily consumption—or more) without deleterious effect.[32] Such increases would, of course, be necessary to accommodate a constantly increasing population. However, extrapolations and statements of this type concerning the future are meaningless unless the short-range problems—the problems of today—are solved. Our society has invested the engineer with a responsibility for leadership that must not fail.

GO-GO-GO

At the present time American motorists are traveling over 1.5 million million miles per year on the nation's highways—an equivalent distance of almost 3 million round trips to the moon. More than one-half of this travel is in urban areas [2-43], where for the most part the physical layouts—the planning, the street design, and basic service systems—were created over 100 years ago. As the population of the nation continues to shift to the urban and suburban areas,

[31] Chauncey Starr, "Energy and Power," *Scientific American*, September 1971.
[32] A. M. Weinberg and R. P. Hammond, "Limits to the Use of Energy," *American Scientist*, August 1970, p. 413.

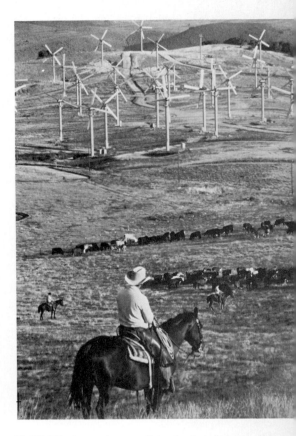

[2-38] Wind power's potential value is in its ability to substitute for the use of oil, coal, or nuclear energy. The most promising form of wind machine is the large, horizontal-axis wind turbine with propeller-type rotor blades.

[2-39] Geothermal energy includes the harnessing of both natural steam and hot water. Some 2400 quadrillion (2.4×10^{18}) Btu of geothermal resources have been identified in this country. This is over 30 times the nation's present total energy consumption.

[2-40] Solar One, a 10-megawatt installation located near Barstow, California, is the nation's first solar-thermal central receiver electric generating station. A field of 1818 heliostats, controlled by computer, tracks the sun throughout the day and reflects its energy onto the receiver, where steam is produced.

"If we can build an atomic bomb, if we can put a man on the moon, why can't we solve our energy problems?" The answer, in a word, boils down to "complexity." Although building the first bomb and putting the first man on the moon were admittedly complicated efforts involving the examination of many alternative approaches, and the coordination of many institutions, there is a certain essential unity about both these efforts. There was, in each case, one mission, one developer, one user. The objective could be clearly, simply, and unambiguously defined. There were no degrees of success short of the goal. Success could be clearly—and dramatically—measured.

By contrast, the energy problem is a multi-headed one. Virtually every person and institution in the country is directly affected. There are multiple possible goals, degrees of success, and pathways to the goal. It might not even be clear just when the goal is reached. There are endless interactions between energy decisions and the environment, the economy, and public welfare. Different regions have different needs, different resources, different limitations, different rules and regulations, and competing interests. There are many questions for public debate. There is, in short, no easy way to identify a goal and reach it.

Nuclear power development seems to have stagnated. Synthetic fuel plants are not yet proven commercially. Renewable resources remain too highly priced to be competitive in most applications. Declining domestic oil and gas discoveries portend decreasing supplies in coming years. . . .

Gail H. Marcus
The Changing Role for Federal Energy R&D, 1983
The Library of Congress

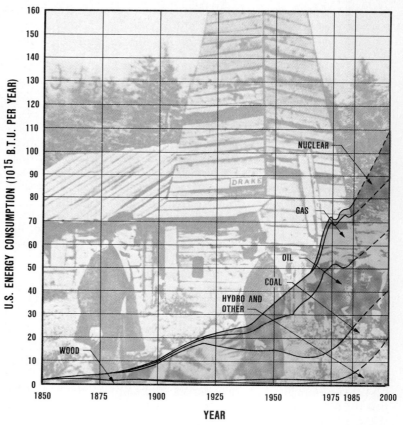

[2-41] Energy consumption in the United States.

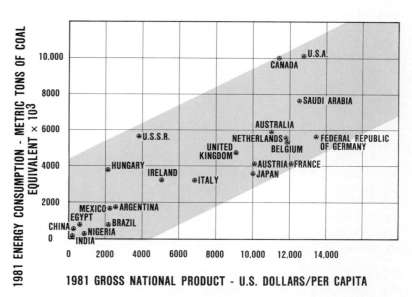

[2-42] Relationship of energy consumption to gross national product.

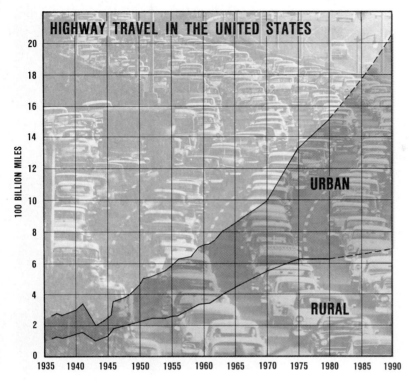

HIGHWAY TRAVEL IN THE UNITED STATES

URBAN

RURAL

100 BILLION MILES

1935 1940 1945 1950 1955 1960 1965 1970 1975 1980 1985 1990

[2-43] Highway travel in the United States.

It is better to more effectively insulate our homes than to compete for oil. It is better to design around a need for a scarce metal than to go to war over it.
Myron Tribus
Roy V. Wright Lecture, ASME, *1971*

It is our task in our time and in our generation to hand down undiminished to those who come after us, as was handed down to us by those who went before, the natural wealth and beauty which is ours.
John Fitzgerald Kennedy
March 3, 1961

many of the *frustrating* problems of today will become *unbearable* in the future. Since 1896, when Henry Ford built his first car, the mores of the nation have changed gradually from an attitude of "pioneer independence" to a state of "apprehensive dependence"—to the point where one's possession of a means of private transportation is now considered to be a *necessity*.

The family which takes its mauve and cerise, air-conditioned, power-steered, and power-braked automobile out for a tour passes through cities that are badly paved, made hideous by litter, blighted buildings, and posts for wires that should long since have been put underground. They pass on into a countryside that has been rendered largely invisible by commercial art . . . . They picnic on exquisitely packaged food from a portable ice box by a polluted stream and go on to spend the night at a park which is a menace to public health and morals. Just before dozing off on an air mattress beneath a nylon tent, amid the stench of decaying refuse, they may reflect vaguely on the curious unevenness of their blessings. Is this, indeed, the American genius?
John K. Galbraith, 1908–
The Affluent Society

[2-44] Automobile travel and parking problems of yesterday!

[2-45] Automobile travel today.

Motor trucks average some six miles an hour in New York traffic today, as against eleven for horse-drawn trucks in 1910—and the cost to the economy of traffic jams, according to a *New York Times* business survey, is five billion dollars yearly!

The Poverty of Abundance

In the next 40 years, we must completely renew our cities. The alternative is disaster. Gaping needs must be met in health, in education, in job opportunities, in housing. And not a single one of these needs can be fully met until we rebuild our mass transportation systems.

Lyndon Baines Johnson
1968

Over 80 percent of the families in the United States own at least one automobile, and over one-fourth can boast of owning two or more. However, because of inadequate planning, this affluence has brought its share of problems for all concerned. Beginning about 3500 B.C. and until recent times, roads and highways were used primarily as trade routes for the transport of commerce between villages, towns, and cities. The Old Silk Trade Route that connected ancient Rome and Europe with the Orient, a distance of over 6000 miles, was used extensively for the transport of silk, jade, and other valuable commodities. The first really expert road builders, however, were the Romans, who built networks of roads throughout their empire to enable their soldiers to move more quickly from place to place. In this country early settlers first used the rivers, lakes, and oceans for transportation, and the first communities were located at points easily accessible by water. Then came the railroads. A few crude roads were constructed, but until 1900 the railroad was generally considered to be the most satisfactory means of travel, particularly when long distances were involved. With the advent of the automobile, individual desires could be accommodated more readily, and many road systems were improvised to connect the railroad stations with frontier settlements. At first these roads existed mainly so that farmers could market their produce, but subsequent extensions were the direct result of public demands for an improved highway system. People in the cities wanted to visit the countryside and people in the outlying areas were eager to "get a look at the big city." Within a few years we became a *mobile* society, but the road and highway system in use today was designed primarily to accommodate the transfer of goods rather than large volumes of people. Because of this, many of these "traffic arteries" are not in the best locations nor of the most appropriate designs to satisfy *today's* demands. Thus attempts to *drive* to work, *drive* downtown to shop, or take a leisurely *drive* through the countryside on a Sunday afternoon are apt to be "experiences in frustration." Vehicle parking is also becoming a critical problem. It takes an acre of parking area, for example, to accommodate 200 subcompact vehicles.

Most cities have made only half-hearted attempts to care for the transportation needs of their most populous areas. Although those owning automobiles do experience annoying inconveniences, those without automobiles suffer the most—especially the poor, the handicapped, the elderly, and the young. Too often the public transit services that do exist are characterized by excessive walking distances to and from stations, poor connections and transfers, infrequent service, unreliability, slow speed and delays, crowding, noise, lack of comfort, and a lack of information for the rider's use. Moreover, passengers are often exposed to dangers to their personal safety while awaiting service. Not to be minimized are the more than 18 million vehicle accidents and the 54,000 fatalities that result annually from motor vehicle accidents. (For perspective, in the past seven years highway fatalities have exceeded the total loss of American lives in the Vietnam War.)

Traditionally, people have moved into a locality, built homes, businesses, and schools and then demanded that adequate transportation facilities be brought to them. We may now live in an era when this independence is no longer feasible; rather, people eventually may be required to settle around previously designed transportation systems. (This is much the same as it was 100 years ago. However, during this period of time the transportation systems have changed greatly.) Engineers can provide good solutions for all these problems *if they are allowed to do so by the public.* However, there will be a cost for each improvement—whether it be a better vehicle design, computerized control of traffic flow, redesigned urban bus systems, rapid-transit systems, highway-guideway systems for vehicles, or some other entirely new concept. In some instances, city, state, or federal taxes must be levied; in others, the costs must be borne by each person who owns private transportation. The quality of our life-style depends on a unified commitment to this end.

THE CHALLENGE OF CRIME

Crime, one form of social pollution, is increasing rapidly in the United States in particular and throughout the world in general. The rate of increase in this country can be attributed variously to the population explosion, the increasing trend to urbanization, the changing composition of the population (particularly with respect to such factors as age, sex, and race), the increasing affluence of the populace, the diminishing influence of the home, and the deterioration of previously accepted value systems, mores, and standards of morality. A survey made by the National Opinion Research Center

[2-46] Parking problems today.

[2-47] Rapid-transit systems have proved their usefulness in many major metropolitan areas.

[2-48] Highway-guideway systems would relieve the driver of the tedious and tiring task of maneuvering his vehicle from one destination to another.

[2-49] Growth of reported crime in the United States.

The state of Arizona uses space-age engineering innovation to catch "water rustlers." (Zane Grey and the *Hashknife Outfit* would be jealous with envy!) By state law, water usage in some areas is regulated. Only land that was irrigated from 1975 to 1980 is allowed irrigation rights. No new land can be irrigated for agriculture. Anyone who does so is prosecuted for water rustling. Instead of using a deputy sheriff to seek out those who break the water law, today's law enforcement officers use Landsat, a satellite that orbits the earth 14 times each day. Although it is 560 miles above the ground, its television sensors transmit pictures of the land area. Agricultural lands that have been irrigated show up as a bright, deep-red color. Then it is a simple matter to overlay the satellite map with a computer map that shows only the authorized water areas. The water thieves are revealed.

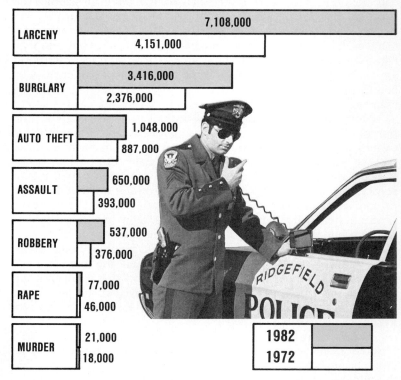

CRIMES REPORTED IN THE U.S.

LARCENY	7,108,000
	4,151,000
BURGLARY	3,416,000
	2,376,000
AUTO THEFT	1,048,000
	887,000
ASSAULT	650,000
	393,000
ROBBERY	537,000
	376,000
RAPE	77,000
	46,000
MURDER	21,000
	18,000

1982
1972

[2-50] Many of us are unconcerned about the mounting epidemic of crime—until it gets personal.

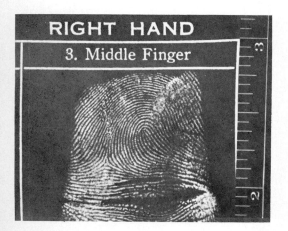

RIGHT HAND

3. Middle Finger

[2-51] Two decades ago fingerprints were symbols associated only with criminals. Today, parents record their children's fingerprints to protect them from criminals.

of the University of Chicago indicates that the actual amount of crime in the United States is known to be several times that reported. A comparison of recent increases in the seven forms of crime considered to be the most serious in this country is shown in Figure [2-49]. A brief examination of these data indicates that the rate of increase of crime is now several times greater than the rate of increase of the population. In fact, crime is becoming such a serious social issue as to challenge the very fabric of our American way of life.

Not all people react in the same way to the threat of crime. Some are inclined to relocate their residences or places of business; some become fearful, withdrawn, and antisocial; some are resentful and revengeful; and a large percentage become suspicious of particular ethnic groups whom they believe to be responsible. A number, of course, seize the opportunity to "join in," and they adopt crime as an "easy way" to get ahead in life. The majority, however, merely display moods of frustration and bewilderment. In all cases, consequential results are detrimental to everyone concerned, because a free society cannot long endure such strains on public and private confidences, nor tolerate the continual presence of fear within the populace.

Traditionally, the detection, conviction, punishment, and even the prevention of crime have been functions of local, state, or federal agencies. Only in rare instances has private enterprise been called on to assist in any significant way, and there has been no concentrated effort to bring to bear on these situations the almost revolutionary advances that have been made in recent years in engineering, science, and technology. Rather, a few of the more spectacular developments have been modified or adapted for police operations or surveillance.

What is needed, and needed now, is a delineation of the vast array of problems that relate to the prevention, detection, and punishment of crime, with particular attention being directed toward achieving *general* rather than *specific* solutions. In this way technological efforts can be concentrated in those areas where they are most likely to be productive. For example, petty thefts may occur more frequently in one area of the city, murders more frequently in some other area, burglaries in another, and so on. What may be needed is a systems analysis of the city to delineate the contributing factors—rather than, for example, equipping all homes with burglary alarms. The engineer can make a significant contribution in such an endeavor.

OTHER OPPORTUNITIES AND CHALLENGES _____

So far we have discussed primarily the societal environment as an area of challenge for the engineer today. Of necessity, many very important challenges have not been discussed, such as the mounting congestion caused by the products of communication media and the threatening inundation of existing information-processing systems, ocean exploration with all of its varied technical problems and yet almost unlimited potential as a source of material, the expertise that the engineer can contribute to the entire field of health care and biological and medical advance, and the attendant social problems that are closely related to urbanization and population growth—such as mass migration, metropolitan planning, improved housing, and unemployment caused by outmoded work assignment.

It is axiomatic that technological advance always causes sociocultural change. In this sense the engineers and technologists who create new and useful designs are also "social revolutionaries." After all, it was they who brought about the obsolescence of slave labor, the emergence of transportation machines that allowed redistribution of the population, the radio and television sets that provide "instant communication," and every convenience of liberation for the housewife—from mixers, waste disposers, dishwashers, ironers, and dryers to frozen foods. They have most recently brought robotics into the forefront of modern manufacturing methods, and no family has been untouched by the new information society, brought about by dramatic advances in solid-state electronics and the computer revolution. The wonderful thing about our modern sophisticated technology is that it gives us options. Unlike our ancestors, whose limited technical means gave them no choice but to do and make things in the same way as their ancestors, we have many different choices we can make regarding where and how and when we should apply our informational expertise. And we can make the decisions on the basis of the perceived social consequences as well as on the technical elements involved.[33] Frequently, society is not pre-

_____
[33]Melvin Kranzberg, "Technological Revolutions," *National Forum,* Summer 1984, p. 10.

[2-52] Using a computer, the location of accidents can be quickly identified, license-plate checks made in a matter of minutes, and verification of criminal records greatly simplified.

We have met the enemy . . . and he is us.
Pogo

The world is very different now. For man holds in his mortal hands the power to abolish all forms of human poverty . . . .
*John F. Kennedy
January 1961*

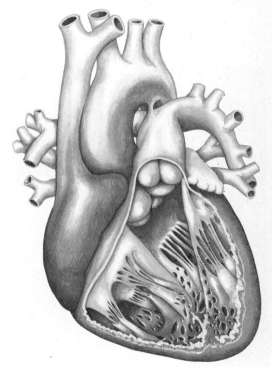
[2-53] If one could substitute for the heart a kind of injection [of blood], one would succeed easily in maintaining alive indefinitely any part of the body. Julian-Jean César La Gallois, 1812

[2-54] An artificial human heart must be many things. It has to be the appropriate size, about a kilogram or less, to fit into a small area of the chest cavity. It must work without failing, beating about 60 times a minute, 40 million times a year. The human heart beats about 3 billion times in an average life span, pumping some 250 million liters of blood.

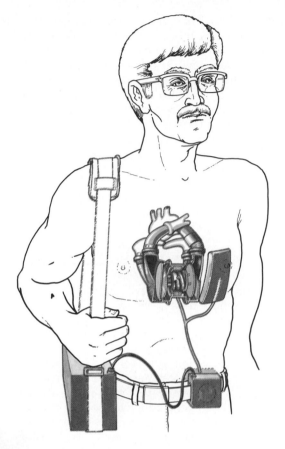

[2-55] Lest we alter our course, we may soon become captives to the media of communication we have created.

pared to accept such abrupt changes—even though it is generally agreed that they are for the overall betterment of mankind. Because of this, as a future engineer, you have a dual responsibility to society. You must not only continue to bring about improvements for the benefit of society, but you must exert every possible means to acquaint society with its responsibility for continual change. Without such an active voice in community and governmental affairs, irrational forces and misinformation can prevail.

AN ENVIRONMENT OF CHANGE

Man has always lived in an environment of change. As he has been able to add to his store of technical knowledge, he has also been able to change his economic structure and his sociological patterns. For centuries the changes that took place during a lifetime were hardly discernible. Beginning about 1600, the changes became more noticeable; and today technological change is literally exploding at an exponential rate. It is interesting to contemplate one's future if a growth-curve relationship such as $k = a(i)^t$ is followed [2-57] (both a and i are constants).

In Figure [2-57] engineering and scientific knowledge is assumed to be doubling every 15 to 20 years. Experience with other growth curves of this nature indicates that at some point a limit will be reached and the rate will begin to level off and then decline. However, in considering the expansion of technology, no one can say with certainty that there is a limit or when this slowing down might occur.

Similar factors are working to provoke changes in educational goals and patterns. In 1900, for example, the engineering student studied for four years to earn a baccalaureate degree, and relatively little change took place in the technological environment during this period. Today, however, due to the accelerated growth pattern of engineering and scientific knowledge, many significant changes will have taken place between the freshman and senior years in college. In fact, complete new industries will be bidding for the services of the young graduate that were not even in existence at the time of his or her enrollment as an engineering student. This is particularly true of the engineering student who continues graduate studies for a master's or a doctorate. It is also interesting to contemplate that at the present rate of growth, engineering and scientific knowledge will have doubled within 20 years after graduation. This places a special importance on continuing lifetime studies for all levels of engineering graduates.

These growth patterns, which are promoting change in all phases of society, are also causing educators and leaders in industry to reappraise educational practices with a view to increasing their scope and effectiveness. From time to time these changes, although not revolutionary, often provoke a sense of progress that shocks those who received their formal education a scant generation before.

THE EDUCATION OF THE ENGINEER

Engineering students who will be best prepared for a career of change should have better-than-average abilities:[34]

[34]Joseph Kestin, Brown University *Engineer*, No. 7 (May 1965), p. 11.

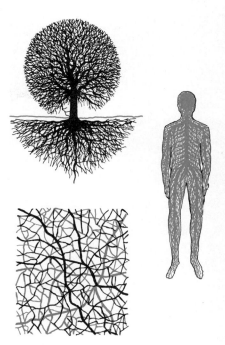

[2-56] In human communities, networks are the interfaces along which the interaction takes place between organic systems—nature, man, society, and, of course, other networks. Rural villages have fairly simple systems of networks, but urban communities interweave many systems of networks at various levels—water supply, sewage and waste disposal, electrical and natural-gas systems, movement of people and goods, telephone, radio, television, and mass printed media. What is important about the vitality of an urban community is not its size or complexity but the degree to which the networks function efficiently. Shown here are analogous network systems in nature, man, and an urban community.

There is no excuse for western man not to know that the scientific revolution is the one practical solution to the three menaces which stand in our way—H-bomb war, overpopulations, and the gap between the rich and the poor nations.

Lord C. P. Snow, 1905

The man who graduates today and stops learning tomorrow is uneducated the day after.

Newton D. Baker

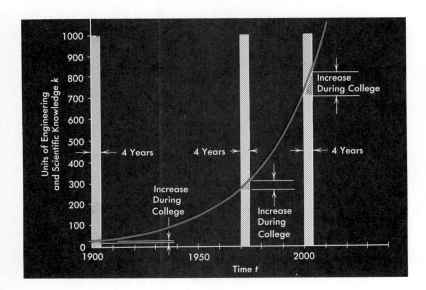

[2-57] Growth of engineering and scientific knowledge.

[2-58] History is the accumulation of the ideas of man that have burned for varying lengths of time.

The average person puts only 25 percent of his energy and ability into his work. The world takes off its hat to those who put in more than 50 percent of their capacity, and stands on its head for those few and far between souls who devote 100 percent.

Andrew Carnegie,
1835–1919

Education is what you have left over after you have forgotten everything you have learned.

Education is the instruction of the intellect in the laws of nature.

Thomas Huxley, 1825–1895

1. To think with imagination and insight.
2. To understand scientific principles and apply analytical methods to the study of natural phenomena.
3. To conceive, organize, and carry to completion appropriate experimental investigations.
4. To synthesize and to design.

In general, engineering programs in colleges and universities have concentrated on providing a broad-based education that is not closely aligned to a specific state of the art. This has been necessary because for one to acquire even a small part of all the factual knowledge now available, a continuous memorization process would be required. It is, therefore, more appropriate to learn the basic laws of nature and certain essential facts that contribute to an understanding of problem solving. Emphasis must be placed on developing mature minds and in educating engineers who can both *think* and get things done. A means of condensing and concentrating the material to be learned is also of paramount importance. A powerful way of doing this is to employ mathematical techniques that can describe technical situations. For this reason mathematics is a most effective tool of the engineer and its mastery early in one's college career will allow for more rapid progress in such subjects as engineering, mechanics, physics, and electrical circuit analysis. In a similar way, if a student learns the principles of physics, this knowledge will bind together such diverse engineering developments as magnetic materials, gas

[2-59] The challenge of engineering lies in providing light for man in the search for truth and happiness—without also bringing about some undesirable side effects.

discharges, semiconductors, and dielectric and optical properties of materials. Similarly, there is no substitute for a mastery of the fundamental principles of other sciences, such as chemistry and biology.

Naturally, the education of an engineer must not end upon graduation from college. The pace of discovery is too great to consider any other course of action than to study and keep abreast of the expanding realm of science and technology. Therefore, in addition to learning fundamental principles of science and engineering in college, the student must develop an intellectual and technical curiosity that will serve as a stimulus for continuing study after graduation.

For the foreseeable future, opportunity for the engineer will continue to expand. Barring a national catastrophe or world war, available knowledge, productivity, and the living standard will probably continue to increase. It is essential that man's appreciation for moral and esthetic values will continue to deepen and keep pace with this technological explosion.

PROBLEMS

2-1. Describe one instance in which the ecological balance of nature has been altered unintentionally by man.

2-2. Plot the rate of population growth for your state since 1900. What is your prediction of its population for the year 2000? Explain the reasons underlying your predictions.

2-3. What can the engineer do that would make possible the improvement of the general "standards of living" in your home town?

2-4. Investigate world conditions and estimate the number of people who need some supplement to their diet. How can the engineer help to bring such improvements about?

2-5. From a technological and economic point of view, what are the fundamental causes of noise in buildings?

2-6. Borrow a sound-level meter and investigate the average sound level in decibels of (a) a busy freeway, (b) a television "soap opera," (c) a college classroom lecture, (d) a library reading room, (e) a personal computer, (f) a riverbank at night, (g) a "rock" combo, (h) a jackhammer, (i) a chain saw, and (j) a kitchen mixer.

2-7. Which of the air pollutants appears to be most damaging to man's longevity? Why?

2-8. Explain the greenhouse effect.

2-9. Investigate how the *smog intensity level* has changed over the past 10 years for the nearest city of over 100,000 population. With current trends, what level would you expect for 1995?

2-10. Describe some effects that might result from a continually increasing percentage of carbon dioxide in the atmosphere.

2-11. Investigate the methods used in purifying the water supply from which you receive your drinking water. Describe improvements that you believe might be made to improve the quality of the water.

2-12. What are the apparent sources of pollution for the water supply serving your home?

2-13. Seek out three current newspaper accounts in which man has caused pollution of the environment. What is your suggestion for remedy of each of these situations?

2-14. Where do some of the highest chemical exposures occur? Give some examples.

2-15. Why should society be concerned with uncontrolled emissions (solids, liquids, or gases) of toxic chemicals into the workplace and environment?

2-16. Estimate the amount of electrical energy consumed by the members of your class in one year.

2-17. Considering the expanding demand for energy throughout the world, list 10 challenges that require better engineering solutions.

2-18. What are the five most pressing problems that exist in your state with regard to transportation? Suggest at least one engineering solution for each.

2-19. List five new engineering designs that are needed to help suppress crime.

2-20. In the United States, what are the most pressing communications problems that need solving?

2-21. List five general problems that need engineering solutions.

2-22. Which three of the renewable energy sources show the greatest promise of development? Why?

2-23. Two decades ago it was widely predicted that solar energy would be used to supply most of the world's energy needs. Why has this not happened?

2-24. Discuss the impact of robotics on the manufacture of automobiles.

2-25. Discuss the obligation of the engineer to check carefully the environmental impact of his designs.

Chapter 3

Study Habits to Maximize Success

In college the pace and standards of education are geared to a superior group of students, not the average you knew in high school. Many students who go to college do not realize how much will be expected of them. Often in high school time could be wasted without serious consequences. At a college or university, there simply is not that much time: courses are more difficult; standards are higher; assignments are more demanding; some classes require extensive library work; and the study materials in both lectures and textbooks are more condensed. The pace is much faster. If you got behind in high school, a friendly, indulgent teacher would probably allow you to "catch up." In college, once you get behind, there is often no chance to catch up.

Many students are so bright that in high school they never needed to employ a systematic approach to study and are at a loss for coping with the sudden academic pressures of higher education. Some have not learned personal discipline and responsibility. To develop skill in being responsible and disciplined requires sustained practice on your part.

Admission to a college or university implies that those authorities recognize your sense of responsibility and that they feel you are capable of independent thought and action, emotional stability, and self-control—in other words, that you are an adult. College life has many rewards and satisfactions, but it also demands obligations. Engineering and technology educators have observed that a high school graduate who has the ability to *read* and *add* possesses the capability to succeed in a college technical program. Most often the secret to such success comes with the development of appropriate and effective *study skills*.

In some cases it is possible to get good grades by being a bookworm or you might choose to be totally involved in social and campus activities and rarely study. However, if you learn to study efficiently you can achieve your educational goals while at the same time working, playing, and enjoying life. The purpose of this chapter is to suggest methods that will help you improve your study skills.

GOALS

My interest is in the future, because I'm going to spend that rest of my life there.

Charles Kettering

There is an axiom that always seems to hold true: *if you fail to plan, you have planned to fail.* By enrolling in college you have indicated to yourself and others that you do have a purpose or goal. It may be either vague and tentative, or sharply defined, but it must be yours. Self-chosen goals are motivating forces and give you direction. But if you are enrolled in college for reasons that are not your own, it is quite possible that you should not be there.

It has been proven that a successful student must have personal goals. If you are having difficulty making a decision about your career, start asking yourself some basic questions, such as "What am I trying to do?" or "How can I do it?" Thinking through these questions and arriving at clear logical answers will give purpose to your decisions and provide an opportunity to review the decisions that you have made. A periodic review of such questions will reaffirm your personal goals and help establish the value of your pursuits. If, after a sincere effort, you find that your tentative career goal is not satisfying, be flexible enough to make adjustments until you are comfortable with your choice.

A career goal is a *long-term* goal. However, to reach that goal, it will be necessary to set a number of *short-term* goals. You should take advantage of the advising services that are available. There are usually a variety of student services on campus to help you plan a program of study. Some of these are academic services, advisement coordinators, career services, vocational counselors, departmental advisors, and peer advisors (fellow students) who have been successful in your chosen field. Do not hesitate to consult any of them. As a member of the college community, you owe it to yourself to take advantage of the available services.

Many departments have a curriculum check sheet listing all the required courses for a degree and a recommended course sequence for completion. It is important to read the college catalog section that pertains to your major. Also, read those pertaining to the required academic standards. Find out what is required of you and prepare to apply yourself.

TIME MANAGEMENT

Time management is simply planning your work and then working your plan. A daily time schedule will be an invaluable tool which will be useful both in college and in future employment. We will learn how to do this later in this chapter. Remember, time is money. The student who continually complains that there isn't enough time in a day is often the type of person who wastes time. If you must work while in college, determine if there is a rule of thumb which indicates a comfortable class load that allows academic success. A general formula often used for school and work load is:

$$\text{Credit hours of enrollment} = \frac{48 - (\text{number of hours employed each week})}{3}$$

Depending on the rigor of the classes, the amount of homework assigned, and the type of work activity, this formula can be relaxed or adjusted. However, it should be a starting point for planning your course schedule. You are now ready to fill out a master schedule of those hours which you have allocated for: attending class, studying, sleeping, eating, working, commuting, household chores, and other important activities.

Plan Your Work

Have you ever looked carefully at how you are spending your time? Now would be a good time to check up on yourself. Keep an accurate record for at least three days of all your activities. How much time do you spend eating? How many hours do you watch TV? Visiting with friends? Working? Now, inventory the time you spend in an average day.

Time Inventory.   Itemize your average activity time per week. Tabulate only the hours you spend on the listed activities.

	Hours/day	*Hours/week*
Sleep	_____	_____
Eating	_____	_____
Classroom	_____	_____
Hygiene	_____	_____
Study	_____	_____
Recreation	_____	_____
Work	_____	_____

Total hours used/week = _____

Total hours available/week   = 168/week

Total hours used/week   = _____

  Hours/week remaining/week = _____

Are you using your time effectively? After making the tabulation above, most students discover that they have a surprising number of "leftover" hours. If this is the case for you, it means that a plan for time allocation will permit time for work, recreation, and social activities. Everyone has 168 hours each week, no more, no less. If this does not seem to be enough, take a serious look at where you can cut back.

Plan to get up before 7 A.M. or you will lose some of the valuable daylight time, which is more conducive to studying. Where possible, leave time to study before recitation or laboratory classes. Also leave some time after lecture classes for immediate review. This will increase your retention of the subject content. In general, you should study in blocks of time—50 minutes followed by a 10-minute break. Also, do not plan to eat just before an afternoon lecture class. If you do, you will be sleepy. Most people find that one hour of afternoon study is equivalent to about $1\frac{1}{2}$ hours of study at night. Therefore, plan to study during the available daylight hours in order to lessen the number of study hours at night. Planning well for daytime study will also help free you for evening distractions, such as watching TV, social engagements, and campus activities. Also, if you study during the day, your overall anxiety level will be reduced and you can relax without being concerned with uncompleted homework.

We sometimes attach too much importance to the quantity of

study and ignore the more important aspect of quality. Discipline yourself to do a good job when you study. This should include taking a break on a periodic basis. Remember, one hour of concentrated, high-quality study is more valuable than three hours of semi-intense effort. Follow your schedule, which includes breaks away from your study area. Too much study for too long a period results in exhaustion, frustration, and anxiety. It is possible to study too much!

In allocating blocks of time on your schedule for study, be sure to indicate the subject you intend to study at that time. This avoids postponement of the more difficult tasks. The course you perceive as most difficult should be given preference.

It is also important that your study schedule be flexible enough to allow for unforeseeable interruptions, but not so flexible that such interruptions can be used as excuses. Figure [3-1] is an example of a typical detailed study schedule.

Work Your Plan: The GPA

The primary purpose of attending college should be to learn as much as possible. The sincere student soon realizes that studying is "hard work" and that it calls for prolonged attention. But learning for learning's sake does not rule out the need to earn good grades. We live in a competitive environment. The ever-present grade-point average (GPA) is important. However, the letters GPA can also stand for *g*oal, *p*lan, and *a*ction. The *goal* is to make a responsible decision concerning a career choice. The *plan* is a carefully considered allocation of time—your schedule. *Action* includes attitude and begins with your constant attendance in class and your unwavering attention to studying the subject. Action also means being prepared with the necessary tools: the assigned textbook, a three-ring loose-leaf notebook, pens and pencils, and whatever special aids (such as a hand calculator) are appropriate.

Attending class calls for certain preparations. Most professors distribute a course outline or syllabus informing you of their expectations. Class attendance is most often required. If the instructor's attendance policy is not announced on the initial handout, ask. Do not be penalized for a lack of common information. Sometimes, there are no makeup exams allowed in case you are absent. It is likely that you will lose points for late assignments. A wise student will find out the rules for the class and for each professor. They are not standardized, and you are responsible for all of them. Keep this information in your notebook along with your notes. Consult it often.

For lecture classes, reading the text assignment *and beyond* beforehand will enable you to have a framework into which you can fit notes and additional information. By reading the text before the lecture, you will be able to identify those parts that are unclear to you, and the lecture material may then clarify or answer your questions. In preparation for a recitation section, it is also important that you read the assigned materials beforehand in order to actively participate and to gain new ideas and to benefit from the experience. Reading the lab assignment will put you steps ahead because you

FALL SEMESTER 1988

Time	Monday	Tuesday	Wednesday	Thursday	Friday	Saturday	Sunday
6:00	Get Up Exercise Get Dressed					→	
7:00	Breakfast Commute					→	Get Up Exercise Get Dressed
8:00	Math	Math	Study Math	Math	Math	Work At Job	Breakfast
9:00	Study Math	Study Math	↓	Study Math	Study Math		Worship
10:00	Chem	↓	Chem	↓	Chem		Worship
11:00	Intro to Eng'r & Technology	P.E./Band	Intro to Eng'r & Technology	P.E./Band	Intro to Eng'r & Technology	↓	Worship
12:00	Lunch				→	Lunch	Lunch
1:00	Study Chem	English	Prepare for Chem Lab	English	Study Chem		Course Review
2:00	↓	↓	Chem Lab	↓	↓		As Needed
3:00	Study Eng'r & Tech.	Study Chem	↓	Study Eng'r & Tech.	Study Eng'r & Tech.		↓
4:00	↓	↓	↓	↓	↓	Commute	
5:00	Dinner						→
6:00							
7:00	Study Math	→					
8:00	Study English	Study English	Write Lab Report	Study English			
9:00	↓	↓	↓	↓			

[3-1] Detailed study schedule.

(continued)

FALL SEMESTER 1988 (continued)

Time	Monday	Tuesday	Wednesday	Thursday	Friday	Saturday	Sunday
10:00		Study Eng'r & Tech.	Study Eng'r & Tech.				
11:00	Go to Bed	⟶					Go to Bed
12:00					Go to Bed	⟶	

will know what to expect and be prepared with some intelligent questions.

What is an intelligent question? Most important, it is one that you want answered. Second, it should be one that is asked in such a way that the answer will give you the desired information. Third, it should serve to confirm the accuracy of your understanding.

Arrive at class a few minutes early, and sit as close to the front and center of the room as is physically comfortable. If you sit in the back of the room you may have difficulty seeing or hearing. If you are prepared, you will have the advantage of being able to focus your attention on the subject, and you will be ready to listen to the speaker. Listening for prolonged periods is a very complex skill requiring attention and concentration and must be practiced. In the classroom, practice listening as though you were going to take a quiz immediately following the lecture, with no opportunity for review. Through concentrated listening you can learn to disregard potential distractions and interferences. Do not expect to have absolute recall or even 50 percent recall unless you have taken notes.

NOTE TAKING

Note taking is essential if you are to remember and retain important knowledge. Making notes will further assure that it will be so. You cannot record every word, so listen for ideas. Record the main ideas, the important points. As you review your notes, record an example or two to refresh your memory. Reproduce diagrams, tables, and illustrations from the board. Many students find that outlining is an efficient method for making notes. Most lecturers have prepared their presentations in an organized fashion, making it easier to outline in a like manner. But if you find that you have more than six or seven major headings, it is likely that you are not differentiating between the professor's major points and the supporting ideas. Re-

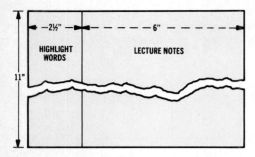

[3-2] Layout of a typical page of notes.

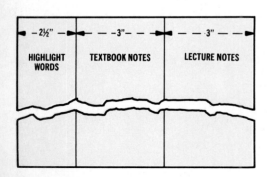

[3-3] Combining textbook and lecture notes.

member, there is always more than one approach to any task. If outlining the lecture does not work for you, try writing telegraphic paragraphs. Telegraphic writing uses abbreviations and symbols and leaves out all the little words. You will capture only the essence of an idea. Remember that it is not easy to listen and write simultaneously. A good note-taking method is to record notes selectively. Practice this technique. If you miss some point in the lecture, check with a classmate or later with the instructor. Write as little as possible while capturing as many facts and principles as you can. This system has been used by successful students for years.

It is suggested that you record your lecture notes on the right three-quarters of your loose-leaf notebook page [3-2]. The notes should be accurate and legible so that they will not have to be recopied. As soon as possible after class (preferably immediately), review your notes. Review each section, and only one major idea at a time. Then, in your own words, write key words or key phrases or perhaps questions in the lefthand column to help you recall the information on the right side. Now cover up the right column and review the lecture content using only reminders: the key words, phrases, and questions in the left column. In this way, you should be able to recall the main points of the lecture and also consider what parts might be used on future exams. Writing a summary of each note page, or perhaps at the end of the last page of notes for the lecture, will be a major step forward in studying for exams.

Is this the only method that works? No, there are others. However, it *is* the general method that works most often. A modification can be used when you are ready to include your study of the textbook [3-3].

Textbooks are valuable pieces of property, new or used. Don't plan to share your textbook with a friend or classmate. This is not being selfish. It just rarely works out to your satisfaction. Plan to purchase the required text well in advance of the first assignment. Become acquainted with the author's purpose and the book in general. First, read the table of contents and briefly survey the chapters. Check the glossary and appendices, if they are included. If you must purchase a used book with highlighting and underlining, do your best to ignore the previous owner's estimation of what was important. *It may not have been!* As you approach your first assignment, consider the following steps:

1. What is my goal for this assignment?
2. Survey the chapter—look at the heads—read the chapter summary or conclusion.
3. Identify the questions to be answered.
4. Some assignments can be read through once using a few marginal notes where necessary. Speed reading a textbook is nearly impossible unless you have one of those wondrous "photographic" memories, which the majority of us do not possess. In general, your reading speed will depend on the purpose and difficulty of the assignment.
5. Most assignments of a technical nature require careful

paragraph-by-paragraph reading that is summarized by notes in your notebook.

6. If highlighting or underlining seems to help—do it; however, make notes of the author's salient points. Follow one of the systems suggested earlier for recording lecture notes ([3-2] or [3-3]).

7. Upon finishing your first reading of the chapter or assignment, consider whether you have answered all the questions you had before you began.

8. Read the text assignment again, if necessary. But remember that it is not the number of times that you read it, it is *how* you read it.

9. Using your notes, summarize and recite the material to yourself— *preferably aloud*. Reciting aloud is a long-established positive reinforcement for retention, and the greater percentage of study time devoted to this practice will result in increased learning.

10. Review. This step is most effective the day after you first read new text material. Review it again the day before an exam.

11. Reflect. This is the summary of your study effort.

EFFECTIVE AND EFFICIENT STUDY _____

A man may have a great mass of knowledge, but if he has not worked it up by thinking it over for himself, it has much less value than a far smaller amount which he has thoroughly pondered.

Schopenhauer

Your study surroundings are a basic ingredient in your plan to study successfully. Choose a place with the fewest visual distractions and the least noise. The place could be your room at home or in the dormitory. There should be space enough for a desk, a sturdy but comfortable chair, adequate lighting, and no TV or radio. Background sounds such as TV or radio are only convenient distractions. Although a quiet spot is not essential for studying, most people can concentrate better without noise. Take advantage of studying in the library. Here you will have access to the reference materials for additional information, explanations, and exercises. Most libraries have been designed to help you succeed in studying by offering special facilities for your use. Sometimes there are special study spaces and rooms available in the student union or activity center. If you study in the same place every day, you will soon become accustomed to your surroundings and can dedicate yourself more easily to the task at hand.

Study to Remember

Memory skills are not a natural endowment. They must be learned and practiced. A number of studies of how students learn and then forget textbook material have shown that up to 50 percent of forgetting takes place immediately after learning. Forgetting continues at a great rate during the following two weeks. After that, forgetting slows down considerably, but by then, there is not much left to forget. Let us redirect our efforts. You should ask yourself, "How can I study to *remember*?" This is important because the student with the highest grades is the one who has remembered and mastered the material. Do the reading assignment *before* the lecture. Jot down those questions you would like clarified by the lecturer. During the lecture use your listening skills and make accurate, concise, legible notes. Copy the illustrations placed on the chalkboard. It is easier to remember what you have read than what you have heard. So, both attentive listening and precise note taking of material that will be presented only once is imperative.

Different courses will make different demands on your study time. For those courses that require you to memorize and/or those which require the acquisition of physical skills, regular or distributed practice for short periods of time is most effective. As an example, to memorize lists of terms, symbols, or parts of the body, distributed practice results in better retention. Distributed practice can be as short as the 10-minute breaks between classes but should not be much longer than 50 minutes. In some situations, lengthy *massed practice* (or continuing until a task is complete) is superior. In creative work, such as writing a term paper, preparing a lab report, or organizing data for a computer class assignment, a longer period of time may be required in order not to interrupt your thinking.

Faced with retaining great masses of information, you must consider organizational options. Such information may best be remembered by organizing it into categories—lists with headings. A system that has proven valuable for long-term storage and recall is oral recitation. "When you say it, then you know it." Recitation reflects what you know.

Studying Language Courses

Many techniques have been developed to aid in learning foreign languages. The following procedures have been found to be helpful.

Learn a vocabulary first. Study new foreign words and form a mental image of them with a conscious effort to think in the new language. As you study, practice putting words together, and if the course includes conversation, say the words aloud. Space your vocabulary study and review constantly, always trying to picture objects and actions in the language rather than in English.

Rules of grammar are to be learned as any rule or principle: first as statements and then by application. Reading and writing seem to be the best ways of aiding retention of grammar rules. Read a passage repeatedly until it seems natural to see or hear the idea in that form. Write a summary in the language, preferably in a form that will employ the rules of grammar which you are studying. Unfortunately, there is no way to learn a new language without considerable effort on your part.

Studying Social Science Courses

These courses can be interesting and satisfying or dull and dry, depending on the student's attitude and interest. Most texts use a narrative style in presenting the material and, as a consequence, the assignment should be surveyed quickly for content and then in more detail for particular ideas. Here the use of notes and underlining is invaluable, and summaries are very helpful in remembering the various facts.

If the course is history, government, sociology, psychology, or a related subject, it contains information that is necessary to help you as a citizen. These subjects will aid you in dealing with other people and give you background information in the evaluation of material that has been specifically designed to influence and control people's thinking. Study each course for basic ideas and information and, unless the instructor indicates otherwise, do exaggerate the importance of detail and descriptive information.

These principles apply also to courses in economics, statistics, and related courses except that they frequently are treated on a more mathematical basis. Here a combination of techniques described above, together with problem-solving procedures, can be helpful. Again, since the volume of words usually is quite large, it is necessary to use notes and summaries.

Studying Technical Courses

In this type of course your study plan should be to direct your study toward understanding the meanings of words and toward grasping the laws and principles involved. Remember also that a word does not always have the same meaning in different courses. For example, the word "work" as used in economics has a meaning quite different from the word "work" as used in physics.

When the definitions of words are obtained, study for complete understanding. Texts in technical courses tend to be concise and extremely factual. A technique of reading must be adopted here for reading each word and fitting it into its place in the basic idea. Except for the initial survey reading of the lesson, do not skim rapidly through the explanations, identifying the particular ideas in each paragraph. If example problems are given, try working them yourself without reference to the author's solution.

After definitions and basic ideas are studied, apply the principles to the solution of problems. You can predict the principles that will be used in solving the problems. For example, in chemistry, a vast number of compounds can be used in equation-balancing problems. However, a very few basic principles are involved. If the principle of balancing is learned, all problems, regardless of the chemical material used, are solved the same way. The objective, then, is to determine the few principles involved and the few problem patterns that can be used. After this, all problems, regardless of their number arrangement and descriptive material, can be classified into one of the problem patterns for which a general method of solution is available. For example, a problem in physics may involve an electrical circuit in which both current and voltage are known and an unknown resistance is to be determined. Another problem may suggest a circuit containing a certain resistance and with a given current in it. In this case a voltage is to be found. The problems are worded differently, but a general principle involving Ohm's law applies to each situation. The same problem structure is used in each case. The only difference appears in where the unknown quantity lies in the problem pattern.

Do not become discouraged if you have difficulty in classifying problems. Practice in problem solving is essential. One of the best ways to aid in learning to classify problems is to work an abundance of problems. It is then more likely that several of the examination problems will be similar to those you have solved before.

Learn to analyze each problem in steps. Examine the problem first for any operations that may simplify it. Sometimes a change in units of measure will aid in pointing toward a solution. Try rewriting the problem in a different form. In mathematical problems, this is often a useful approach. Write down each step as the solution proceeds. This approach is particularly helpful if the solution will involve a number of different principles. If a certain approach is not productive, go back and reexamine the application of the principles to the data. For problems that have definite answers, these techniques usually will provide a means for obtaining a solution.

PROBLEM SOLVING IN ENGINEERING AND TECHNOLOGY

One characteristic of engineers that sets them apart from other technical professionals is their ability to solve applied problems. Within each engineer this characteristic is exemplified by an inquisitive mind that has been forged by the rigors of an educational environment that requires repeated problem solving and analysis. Finally, it is fine-tuned by the practice of the profession.

At some time within the educational process, each student usually realizes that there is a general methodology that is useful in solving a large class of engineering problems, whether these problems are mechanical, electrical, or chemical in nature. Some students recognize this early in their educational careers, whereas others discover it much later. Once you have become aware of this general methodology and can put it to use, your capability in problem solving will greatly increase.

The overall scheme of this methodology embodies the following steps:

1. Define the system (that portion of the universe) to which you are going to devote your attention.
2. Conceptualize the problem in your mind (i.e., try to visualize it in your mind; use simple sketches and think about it).
3. Identify the knowns and unknowns in the problem.
4. Model the problem in some appropriate manner.
5. Make the necessary analyses or conduct the required experiments that will provide needed information.
6. Evaluate the effect that assumptions, made in step 4, have on the final solution. Remodel the problem is necessary.

Each of these steps is discussed and explained in more detail in the sections that follow.

Define the System

This first step is important because it allows you to concentrate your attention on the subject at hand. You do not want to concern yourself with all the problems of the universe; you can solve many of those later. It is important to realize that the physical laws of nature all deal with *something*. That is, they apply to some physical object: a box sliding across the floor, a resistor in an electrical circuit, the temperature of a piece of metal, and so on.

Therefore, in solving technical problems that are amenable to solution through an application of the fundamental physical laws, you will need a clear definition of your system. If you have difficulty defining your system, you will have trouble applying the fundamental laws. This is an extremely important point and, experience tells us, one that can never be overemphasized.

Conceptualize and Sketch the System

During your college studies you will find that various disciplines assign different names to these conceptualized system sketches. In mechanics they are called *free-body diagrams;* in the study of fluid mechanics and thermodynamics (energy) they are often called *control volumes;* in electrical sciences they may be termed *circuit diagrams.* However, the purpose of the idealized representation is always the same: to focus your attention on the key elements of the problem as you start your solution.

Most engineering instructors will require a system sketch as a part of each problem solution. Whether or not it is required, you should make it an automatic part of your solution.

Identify the Knowns and Unknowns

This may seem to be an obvious step in any problem-solving methodology, but it is listed here for additional emphasis. Unless you make this a conscientious part of your overall "problem-solving toolkit," it is likely to be overlooked. The best way to approach this step is to make two lists: one list for the knowns and one for the unknowns.

This step helps you to "size up" the problem. By coming to grips with the specific information that is known and determining what questions are to be answered, the choice of problem solution techniques is usually made much clearer.

Model the Problem

Analytical and/or experimental modeling, whichever may be appropriate for the problem at hand, plays an important function in engineering problem solving. The role of modeling in engineering is discussed in more general terms in Chapter 14.

In general, junior- and senior-level engineering courses will teach engineering modeling. The degree of sophistication of these models will grow as you delve deeper into your studies.

Students often wonder why engineers must study calculus and even higher-level mathematics. The answer is that much engineering modeling is accomplished by using mathematics. If some new process is to be developed or a new device built, it is usually much more economical to do mathematical modeling to refine the design than to set up the process in the laboratory or build the actual device in the model shop. That is, fewer physical models and prototypes have to be built and tested when mathematics is used first to narrow the final design choices. Hardware models are very expensive to build because of their labor, space, and equipment requirements. Perhaps more important, mathematical modeling can cut the time necessary to bring a product or process to market. This places a company in a more visible and competitive position.

Computer-aided engineering (CAE), which is just another form of mathematical modeling, is developing rapidly because it has the potential to simplify modeling, thereby reducing development costs and bringing products to market quicker.

Conduct Analyses and/or Experiments

The execution of this step depends on the level of sophistication of the modeling required to solve the problem. If, for example, only algebraic equations are involved in the mathematical modeling, you must have the same number of independent algebraic equations as there are unknowns in the problem.

Remember Peers' Law
The solution to a problem changes the problem.
John Peers, President, Logical Machine Corp.

Evaluate the Accuracy of the Final Result

There are many factors that control the accuracy of the final solution to an engineering problem. For example, assumptions must always be made in the modeling step and these can greatly influence the result. The correctness of the assumptions you have made must always be confirmed.

Also needing confirmation here are the units of each term. To talk about a temperature of 3 or an area of 406 is nonsense. You must always ask, "3 *what?*" or "406 *what?*" As will be pointed out in Chapter 12, a thorough units analysis should be a standard part of your problem-solving technique.

EXAMPLES OF ENGINEERING ANALYSES

The computation paper used for most calculations is $8\frac{1}{2}$ by 11 inches in size, with lines ruled both vertically and horizontally on the sheet. Usually, these lines divide the paper into five squares per inch, and the paper is commonly known as cross-section or engineering calculation paper. Many bookstores stock paper that has the lines ruled on the reverse side of the paper so that erasures will not remove them. A fundamental principle to be followed in the use of the paper is that the problem work shown should not be crowded and that all steps of the solution should be included.

Several styles of model problem sheets are shown in Figures [3-5] and [3-6]. Notice in each sample that an orderly sequence is followed in which the known data are given first. The data are followed by a brief statement of the requirements and then the engineer's solution.

VOCABULARY

Part of learning to study and to read faster with better comprehension is to become master of the words that you read. No one knows all of the approximately 600,000 words in the English language. But success in college has much to do with the number of words that you do know and how you use them. A good vocabulary usually accom-

[3-4] *"I found the answer but I forgot the problem."*

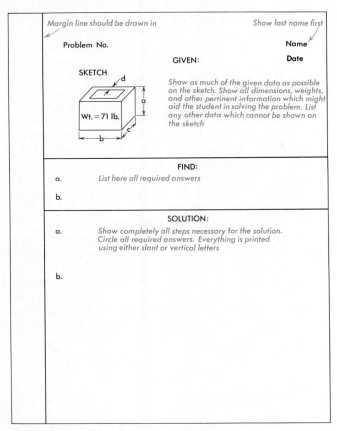

[3-5] Model problem sheet, style A. This style shows a general form that is useful in presenting the solution of problems in engineering and technology.

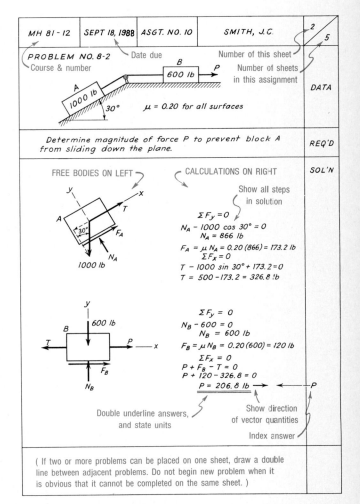

[3-6] Model problem sheet, style B. This style shows a method of presenting stated problems. Notice that all calculations are shown on the sheet and that no scratch calculations on other sheets are used.

You may possess a tool and know its name, but if you don't know how to use it, it is worthless.
Kelly Stadmiller

panies a high GPA. It is a certainty that clear, concise vocabulary is indispensable to the engineer and technologist. Much of the vocabulary you will hear in lectures and study in texts is technical in nature and may not be found in a standard college edition dictionary. You should buy a good standard English dictionary to keep on your desk to use when you study. In the library you will find technical dictionaries for nearly every discipline. You should check your textbook for definitions within the chapter or in the glossary. If you use a dictionary only to find definitions, you have missed most of the worth of the book. In the front of most dictionaries is a section entitled "Guide to the use of the dictionary." Take a few minutes to read it and look up two or three words just for practice. Knowing the denotation (explicit definition) of a word is valuable only if you are aware of the connotation (appropriate usage).

Obtain some 3- by 5-inch lined index cards. Write the new words that you have encountered (check the spelling) and the pronunciation on one side, and the definitions on the other. It is helpful to use the word in a sentence to become familiar with its use. Not only can you study them at your desk, but they are portable, will fit in your pocket or purse, and can be pulled out anywhere for a minute or two of study. Also, you can organize a card file with the different definitions and review them from time to time to refresh your memory.

TESTS AND EXAMINATIONS

We live in a world of exams, and we prepare for each of them in different ways. Whereas some of them take little preparation, others take years of education and training. The college student needs to learn how to prepare for and consistently score high marks on exams. The student who has planned the use of available time well and has studied according to plan will find at exam time that his or her preparation is nearly 80 percent complete. The remaining time should be spent in reviewing the material that has already been mastered.

Begin by studying as if an essay test were scheduled. If you begin by thinking of whole ideas and supporting examples, you will find that you are ready for either an objective test (true–false; matching; multiple choice) or an essay test. This applies to all courses including math and theoretical material. Review those notes that you have carefully organized and mastered throughout the semester, both lecture and textbook. Group items that are similar. Label categories. *Predict exam questions* from those questions that you have asked yourself as you have studied, together with those points that the professor has stressed or items that he or she has suggested may be on the exam. If you can contribute and gain from such an exchange, join a discussion or study group. Review old exams if they are available. These can sometimes be found in the library and often the professor will make old exams available. Preparing for and taking exams is the time when your constant attention to studying will "pay off." Your attention to *g*oals, *p*lans, and *a*ctions will be rewarded.

Cramming is advantageous for the well-prepared student who has been studying and attending class, because it's a form of review. For those who have not taken the time to prepare, cramming is a poor last resort. Immediately before an exam, don't attempt to read the text for the first time or even to reread it. This would be a waste of time. Instead, survey the chapters concerned, read the heads and subheads set in heavy print, and carefully digest the summary or conclusion of each chapter. Because of the anxiety and apprehension of not being prepared, it is nearly impossible for your mind to comprehend and assimilate large amounts of information. This type of stressful situation is threatening and punishing and has little to do with your overall goal of becoming educated. Besides, the rate of forgetting what you have crammed is generally equal to the speed with which you crammed it. Your mind works best in recall if your body is rested. So, to complete your preparation for an exam, be sure to get a full night's sleep the night before the test. Material that has been mastered will not be forgotten after eight hours of sleep.

Test Taking Techniques

The professor will usually tell you the kind of exam that will be given. Skill in the mechanics of taking an exam or knowledge of test-taking techniques has a great deal to do with the grades you will earn, but no amount of test-taking skills can compensate for knowledge. However, the combination of knowledge and test-taking skill will be reflected in your final grade.

Multiple-choice, matching, and true–false questions are means of testing a student's knowledge within a short period of time, and they are quickly and easily graded. True–false questions are those that you must mark as *completely* true or false. Often, the addition of a qualifier word will change a completely true statement into a false one. Be aware of such qualifiers as

all, most, some, none

always, usually, sometimes, never

great, much, little, no

more, equal, less

good, bad

is, is not

Be aware of negatives:

no, not, none, cannot

Watch for negative prefixes:

un-, im-, dis-, non-, in-, ir-, il-

Remember that if any part of the statement is false, the whole statement is false. Matching questions most often appear on tests of history and social sciences when you are expected to associate events and dates or famous people and their contributions. Start by reading the top item in the left-hand column. Then, scan the entire right-hand column for a match. Be certain that the match is correct. Proceed with the remainder of the left column. In this case, postpone guessing until the very last, because one initial wrong answer can result in a string of wrong answers.

Multiple-choice questions consist of a partial statement with four or five options to complete the thought [3-7]. An important reminder: *Carefully read the directions*. If you do not understand them, ask the professor for clarification. Read the partial statement and all of the options before you mark any of them. If you aren't sure of an answer, start by eliminating the false ones. If you can reduce the options to two, you have a 50–50 chance of making a correct selection. Watch for items that are alike except for one word; "all of the above" items, which are frequently correct; and longer statements, which are also usually correct. Sentence completion and fill-in-the-blank questions are no more difficult than true–false, multiple-choice, and matching questions. It is only harder to guess. There's no trick to answering sentence-completion or fill-in-the-blank questions. If you do not have the facts, think of an appropriate

[3-7] Birth of the multiple choice answer.

answer. An empty blank or an incomplete sentence will not earn you any points.

An essay exam usually does not have many questions, but be sure to read the directions. Surveys have shown that graders of essay exams look at four qualities: reasoning ability, factual accuracy, relevance to the question, and organization (in that order). Beginning with organization, step one is to answer the question directly in the first sentence. An introduction is not necessary. Write deductive paragraphs. Make a clear statement of your major idea and follow it with facts and logical support and examples. End with a summary of two or three sentences. Be neat and pay attention to spelling and grammar. These last suggestions do not gain points, but they may help not to lose points, and they do make a good impression on the grader.

Recommended techniques for every exam:

1. Arrive and be in your seat before the period begins.

2. Have whatever paper, bluebook, sharpened pencils, or special instruments that are required or necessary. If you don't know, ask the day *before* the test.

3. Listen carefully to the instructions from the test proctor. There may be changes in the directions or wording of the exam.

4. Survey the entire exam. How many sections of the test are there? How many questions are in each? Is one section worth more points than another? Is any section more difficult for you to answer than another?

5. Determine which parts will take more time, and budget your time accordingly.

6. Read the directions before you begin any part. If you do not understand some point or direction, ask before you begin. Many whole sections of exams have been failed for lack of reading and understanding the directions.

7. *Answer the easy questions first.* Mark the more difficult ones and return to them later.

8. Watch for qualifiers and negatives and underline or circle them as a reminder.

9. If there is no penalty, make an educated guess at the answers to the questions that you have left blank.

10. Take all of the time allotted for the exam. If you finish early, review your answers. Watch for arithmetic mistakes and incorrect algebraic signs.

Cheating, like cramming, has no place in the learning process. Cheating is generally defined as intentionally using or attempting to use unauthorized materials, information, or study aids in any academic exercise. Most colleges and universities have a *code of conduct* that describes prohibited conduct and all forms of student academic dishonesty, including cheating, fabrication, facilitating academic dishonesty, and plagiarism. The worst consequence of "getting caught" is not the disciplinary action which assuredly would follow but that the student who cheats is the victim of his or her own crime.

Students preparing to become responsible professionals in their chosen field need to emphasize development and application of a professional attitude concurrently with completing the course work required to earn their degree. This means developing and applying the same set of ethical values that they will some day apply as they work in industry as professionals. The same degree of honesty should be used in completing course assignments, and in supplying answers to test questions. It is difficult to picture someone who cheats on examinations being totally trustworthy when required to make difficult on-the-job ethical decisions.

TRANSMITTING INFORMATION

In this age of information and communication, the engineer and engineering technologist is expected to be able to express ideas both orally and in writing. Technical writing and public speaking are on the top of the list of subjects most needed for careers in industry (Table 3-1).[1] No matter how good your ideas may be, if you cannot communicate them to others, they are of little value. Since part of the work of an engineer or technologist is writing reports and papers, the opportunity to learn and to practice this skill in school should be exploited.

In general, good writing involves good grammar, correct spelling, and an orderly organization of ideas. The basic rules of grammar should be followed and a logical system of punctuation used. If you are in doubt about correctness, there are a number of handbooks of English to supply you answers. Spell correctly. There is so little room for choice in spelling that there is no excuse for a technical student to misspell words. If you do not know how to spell a word, look it up in the dictionary or reference book and remember how to spell it correctly thereafter. Also, a most important characteristic of good writing is a clear, orderly organization of ideas.

Engineering papers and reports generally are concerned with technical subjects. Frequently, the first paragraph summarizes the thoughts in the whole report in order to give the reader a quick survey without having first to skim through the manuscript. The following paragraphs will outline the contents in more detail. Technical papers frequently include graphs, drawings, charts, and diagrams to support the conclusions reached.

In preparation of written work, some research usually is needed. The library is full of source materials pertaining to your topic. If you cannot locate the information you need, consult with a librarian. To aid in keeping notes in usable form, it is helpful to record abbreviated notes from research works on cards in order that the arrangement of the writing of the paper can be made in a logical order. Usually, in compiling information you do not know how much will be used, so placing notes on cards provides a flexibility of choice that is helpful in the final organization of the paper. The cards can be

---

[1] Adapted from Thomas N. Huckin and Leslie A. Olsen, *English for Science and Technology: A Handbook for Nonnative Speakers* (New York: McGraw-Hill, 1983), p. 4.

TABLE 3-1   SUBJECTS MOST NEEDED FOR ENGINEERING (AND TECHNOLOGY) CAREERS IN INDUSTRY

Rank	Subject	Rank	Subject
1	Management practices	20	Applications programming
2	Technical writing	21	Psychology
3	Probability and statistics	22	Reliability
4	Public speaking	23	Vector analysis
5	Creative and innovative thinking	24	Electronic systems engineering (circuit design)
6	Working with people		
7	Visualization	25	Laplace transforms
8	Reading ability (speed and discernment)	26	Solid-state physics
		27	Electromechanical energy transformation
9	Talking with people		
10	Business practices (marketing, accounting, finance)	28	Matrix algebra
		29	Computer systems engineering
11	Use of computers	30	Operations research
12	Heat transfer	31	Law principles (patents, contracts)
13	Instrumentation and measurements		
		32	Information and control systems
14	Data processing	33	Numerical analysis
15	Systems programming	34	Physics of fluids
16	Economics	35	Thermodynamics
17	Ordinary differential equations	36	Electromagnetics
		37	Human engineering
18	Logic	38	Materials engineering
19	Engineering economic analysis		

either 3 by 5 inches or 4 by 6 inches, with the latter being the better choice because of more available space.

Preparing research papers requires special attention to honesty. You need to be certain that your paper represents *your* personal research effort and that the work supplied by others is properly documented. It is proper to use the words of others in direct quotation or by paraphrasing, as long as you acknowledge the source. However, the preponderance of the content of the paper should be your own thoughts.

Even for brief reports you should prepare an outline of the material you intend to cover. An outline ensures a more logical arrangement of ideas and helps to make the writing follow more smoothly from concept to concept. Write a draft copy first to get your ideas down on paper, and then go over the copy to improve the rough places. Adopting the policy of putting the paper aside for a day or two will give you a more objective approach for the second draft. Even professional writers do not produce a "masterpiece" with the first, or even the second draft. When you are satisfied that you have produced your best effort, the paper is ready to be typed.

An oral presentation is another opportunity to practice your communication skills. Having chosen (or having been assigned) a topic and having completed your research, you are now ready to outline your speech. Avoid writing out the entire presentation because the tendency will be to memorize it. If you do this, you will lose the enthusiasm and spark of a good speaker. Another temptation is to read it aloud to the audience. There is no greater insult to an audience than to have a speaker read to them material that they could very well read for themselves. So, from your detailed outline, make as many notes as you believe are necessary to steer you through the presentation. Now, practice aloud while timing yourself, since most speeches have a time limit. If you do not have a sympathetic roommate to listen to you, deliver your speech to the mirror or into a tape recorder. The tape recorder is an excellent tool to record your speech several times. In this way you can refine it until you are satisfied. Also, by listening to the tape several times, its content will be lodged in your long-term memory and you will be able to present it without notes. Then listen to yourself critically. Everyone experiences the trauma of speaking for the first time in front of a group; however, thorough acquaintance with your topic and adequate practice will alleviate much of your nervousness.

LEARNING AND THINKING

Much of college learning deals with theorems, laws, ideas, principles, and so on. This kind of intellectual work requires mastery of the information and reflection or thinking to make the information meaningful. Thinking is a process which is composed of a number of skills. Learning or mastering facts and principles may be assigned, but finding the implications of those facts and principles requires thinking. A fact does not stand alone. Do you perceive a relationship that allows you to place a particular fact into a larger scheme or organization? Compare this fact with others in the same category. How are they alike? How do they differ? When you are introduced to a new concept, look for practical examples. Could you apply this concept in a situation that is not described in the text or mentioned in the lecture? Can you tie in this new information with your life? Where does it fit? Expand your understanding of abstractions. Can you make them more concrete and meaningful? Learn to draw your own conclusions by asking questions and thinking through possible answers. New material becomes meaningful only when it relates to your existing knowledge.

In summary, your education is one of your most valuable assets. Once you have obtained knowledge, you have a possession that cannot be taken from you. And the more you use it, the more valuable it becomes to you.

PROBLEMS

3-1.
A. What does your college catalog list as requirements for your major field of study?
B. Looking at your academic career, which electives would serve you best?

3-2.
A. Which advising services are available at your institution?
B. Visit them and ask questions important to you.

3-3. Make a master schedule and follow it for a week. If it seems that you are studying all the time (including Sunday night), what changes are necessary to be more efficient?

3-4. Do you know any of your classmates? If not, for mutual support, make a point of becoming acquainted with at least two of them in each class.

3-5. Are you mastering new vocabulary? Assign yourself five new words each week and try the card system for remembering them.

3-6. Have you taken time to reflect on your class assignments? Choose a recent assignment and list meaningful applications that you thought of yourself.

3-7. Analyze how you could improve your score on an upcoming exam.

3-8.
A. Few humans succeed at everything. How do you feel about yourself?
B. Ask others how they are doing. Such exchanges often help one's attitude.

Chapter 4

Career Fields in Engineering and Technology

When a person alibis that he could have amounted to something if it had not been for his race, creed, or religion, one should call attention to Epictetus, the slave who lived in the first century in Greece, and became one of the world's most profound scholars and philosophers. He should be reminded that Disraeli, the despised Jew, became Prime Minister of Great Britain; that Booker T. Washington, who was born in slavery in this country, became one of the nation's greatest educators; and that another slave, George Washington Carver, became one of the greatest scientists of his generation. Lincoln, born of illiterate parents in a log cabin in Kentucky, lived to be acclaimed one of the greatest statesmen of all times.
Phil Conley

Much of the change in our civilization in the past 100 years has been due to the work of the engineer. We hardly appreciate the changes that have occurred in our environment unless we attempt to picture the world of a few generations ago, without electric lights, automobiles, telephones, radios, computers, transportation systems, supersonic aircraft, robotic-operated machine tools, television, and all the modern appliances in our homes. In the growth of all these things the role of the engineer is obvious.

Development in the field of science and engineering is progressing so rapidly at present that within the last 10 years we have acquired materials and devices that are now considered commonplace but which were unknown to our parents. Through research, development, and mass production, directed by engineers, ideas are made into realities in an amazingly short time.

The engineer is concerned with more than research, development, design, construction, and the operation of technical industries, however, since many are engaged in businesses that are not concerned primarily with production. Formerly, executive positions were held almost exclusively by persons whose primary training was in the field of law or business, but the tendency now is to utilize engineers more and more as administrators and executives.

No matter what kind of work the engineer may wish to do, there will be opportunities for employment not only in purely technical fields but also in other functions, such as general business, budgeting, rate analysis, purchasing, marketing, personnel, labor relations, and industrial management. Other opportunities also exist in such specialized fields of work as teaching, writing, patent practice, and work with the national defense.

Although college engineering curricula contain many basic courses, there will be some specialized courses available that are either peculiar to a certain curriculum or are electives. These specializations permit each student to acquire a particular proficiency in certain technical subjects so that, for example, professional identification can be acquired in such fields as chemical, civil, electrical, mechanical, or industrial engineering.

Education in the application of certain subject matter to solve technological problems in a particular engineering field constitutes engineering specialization. Such training is not for manual skills as in trade schools, but rather is planned to provide preparation for research, design, operation, management, testing, maintenance of projects, and other engineering functions in any given specialty.

The principal engineering fields of specialization that are listed in college curricula and that are recognized in the engineering profession are described in the following sections.

AEROSPACE AND ASTRONAUTICAL ENGINEERING

Powered flight began in 1903 at Kitty Hawk, North Carolina. Perhaps no other single technological achievement has had such an influence on our way of life. Through faster transportation, space exploration, and improved communications, almost every aspect of our daily life has been affected. However, not all challenges are associated with spaceflight. Problems associated with conventional aircraft, and the development of special vehicles such as hydrofoil ships, ground-effect machines, and deep-diving vessels for oceanographic research are all concerns of the industry.

Within the past few years many changes have taken place which have altered the work of the aeronautical engineer—not the least of which is man's successful conquest of space. Principal types of work vary from the design of guided missiles and spacecraft to analyses of aerodynamic studies dealing with the performance, stability, control, and design of various types of planes and other devices that fly.

Although aerospace engineering is one of the newer fields, it offers a great variety of opportunities for employment. Continued exploration and research in previously uncharted areas is needed in the fields of propulsion, materials, thermodynamics, cryogenics, navigation, cosmic radiation, and magnetohydrodynamics. It is predicted that within the near future the chemically fueled rocket engine, which has enabled man to explore lunar landscapes, will become obsolete as the need increases to cover greater and greater distances over extended periods of time.

[4-1] Aerospace engineering includes a broad range of engineering activities.

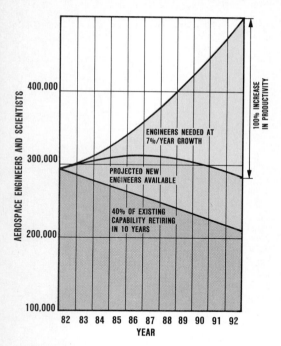

ENGINEERS NEEDED AT
7%/YEAR GROWTH

PROJECTED NEW
ENGINEERS AVAILABLE

40% OF EXISTING
CAPABILITY RETIRING
IN 10 YEARS

[4-2] In the future, one aerospace engineer or scientist must do the work that two do now.[1]

There will be fewer aerospace engineers and scientists available in ten years than we have today. . . . Meanwhile, other major nations press the training of engineers. The Soviet Union, Japan, and even the Peoples Republic of China are graduating more engineers than the United States . . . In view of this trend, we will have to be twice as productive in aerospace ten years from now as we are today.

Robert W. Hager, Vice President, Engineering,
Boeing Aerospace
Astronautics & Aeronautics, May 1983

[4-3] Agricultural engineers at a research center test the safety features of a farm tractor. Agricultural engineers apply fundamental engineering principles of analysis and design to improve our methods of food production and land utilization.

The rapidly expanding network of airlines, both national and international, provides many openings for the engineering graduate. Since the demand for increasing numbers of aircraft of various types exists, there are opportunities for work in manufacturing plants and assembly plants and in the design, testing, and maintenance of aircraft and their component parts. The development of new types of aircraft, both civilian and military, requires the efforts of well-trained aeronautical engineers, and it is in this field that the majority of positions exists. Employment opportunities exist for specialists in the design and development of fuel systems using liquid and solid propellants. Control of the newer fuels involves precision valving and flow sensing at very low and very high temperatures. The design of ground and airborne systems that will permit operation of aircraft under all kinds of weather conditions is also a part of the work of aeronautical engineers.

The aerospace engineer works on designs that are not only challenging and adventuresome but also play a major role in determining the course of present and future world events.

AGRICULTURAL ENGINEERING

Agricultural engineering is that discipline of engineering that spans the area between two fields of applied science—agriculture and engineering. It is directly concerned with supplying the means whereby food and fiber are supplied in sufficient quantity to fill the basic needs of all mankind. By the year 2000 almost 2 billion additional people will populate the earth—a number roughly equal to the world's total population in 1940. This factor, plus the increasing demands of people throughout the world for increased standards of living, provides unparalleled challenges to the agricultural engineer. Not only must the quantity of food and fiber be increased, but the efficiency of production must also be steadily improved in order that personnel may be released for other creative pursuits. Through applications of engineering principles, materials, energy, and machines may be used to multiply the effectiveness of man's effort. This is the agricultural engineer's domain.

In order that the agricultural engineer may understand the problems of agriculture and the application of engineering methods and principles to their solution, instruction is given in agricultural subjects and the biological sciences as well as in basic engineering. Agricultural research laboratories are maintained at schools for research and instruction using various types of farm equipment for study and testing. The young person who has an analytical mind and a willingness to work, together with an interest in the engineering aspects of agriculture, will find the course in agricultural engineering an interesting preparation for his or her life's work.

Many agricultural engineers are employed by companies that serve agriculture and some are employed by firms that serve other industries. Opportunities are particularly apparent in such areas as (1) research, design, development, and sale of mechanized farm

---

[1] *Astronautics & Aeronautics,* May 1983, p. 66.

equipment and machinery; (2) application of irrigation, drainage, erosion control, and land and water management practices; (3) application and use of electrical energy for agricultural production, and feed and crop processing, handling, and grading; (4) research, design, sale, and construction of specialized structures for farm use; and (5) the processing and handling of food products.

He gave it for his opinion, that whoever could make two ears of corn or two blades of grass to grow upon a spot of ground where only one grew before, would deserve better of mankind, and do more essential service to his country than the whole race of politicians put together.

Jonathan Swift, 1667–1745
Gulliver's Travels II, vi

ARCHITECTURAL ENGINEERING

The architectural engineer is interested primarily in the selection, analysis, design, and assembly of modern building materials into structures that are safe, efficient, economical, and attractive. The education received in college is designed to teach one how best to use modern structural materials in the construction of tall buildings, manufacturing plants, and public buildings.

The architectural engineer is trained in the sound principles of engineering and at the same time is given a background supportive of the point of view of the architect. The architect is concerned with the space arrangements, proportions, and appearance of a building, whereas the architectural engineer is more nearly a structural engineer and is concerned with safety, economy, and sound construction methods.

Opportunities for employment will be found in established architectural firms, in consulting engineering offices, in aircraft companies, and in organizations specializing in building design and construction. Excellent opportunities await the graduate who may be able to associate with a contracting firm or who may form a partnership with an architectural designer. In the field of sales an interesting and profitable career is open to the individual who is able to present ideas clearly and convincingly.

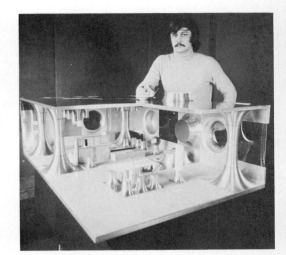

[4-4] Architectural engineers must be equally cognizant of aesthetic and structural design considerations.

BIOMEDICAL ENGINEERING

Biomedical engineering encompasses all aspects of the application of engineering methods to the use and control of biological systems. It bridges the engineering, physical, and life sciences in identifying and solving medical and health-related problems. Biomedical engineers are team players in much the same way as many athletes. For example, engineers, physicists, chemists, and mathematicians routinely join with the biologist and physician in developing techniques, equipment, and materials.

One of the earliest applications of bioengineering in the use of prosthetic devices was made by maimed ancient warriors who made their own wooden limbs. In the Middle Ages "experts" got into the field when armorers fashioned artificial limbs for knights injured in battle. It was a natural by-product of their work since they had to create suits of armor that fit the human form and joints that moved with ease. Goethe wrote of one German knight, Goetz von Berlichengin, who had an artificial hand.[2]

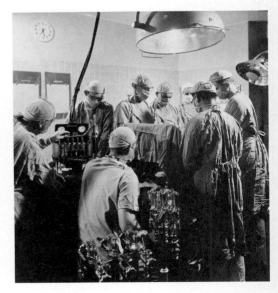

[4-5] The work of bioengineers has made possible the development of many life-lengthening and life-enhancement systems.

---

[2]"How MEs Aid MDs," *Mechanical Engineering*, June 1982.

ARTIFICIAL KIDNEY

[4-6] Schematic drawing of the artificial kidney.

[4-7] Ceramic engineers hold in their hands a very important answer to the impending crisis in the shortage of metallic materials.

Today, the range of the bioengineers' interests is very broad. It would involve, for example, the development of highly specialized medical instruments and devices—including artificial hearts and kidneys and the development of lasers for surgery and cardiac pacemakers that regulate the heartbeat. Other biomedical engineers may specialize more particularly in the adaptation of computers to medical science, such as monitoring patients or processing electrocardiograph data. Some will design and build systems to modernize laboratory, hospital, and clinical procedures.

Those selecting a career in biomedical engineering should anticipate earning a graduate degree since advanced study beyond the bachelor's degree is acutely needed to attain a depth of knowledge from at least two diverse disciplines.

At present biomedical engineering is a small field because few engineers have attained the necessary depth of academic training and experience in the life sciences. Therefore, job opportunities for graduates are excellent. Here indeed is a promising new field for those so inclined.

CERAMIC ENGINEERING

Today, our technological world is amazingly dependent on ceramics of all types. Unlike many other products they appear in every part of the spectrum of life, from beautiful but commonplace table settings, to the protective coatings of electrical transducers or the refractories of space exploratory rocket nozzles, to the spark plugs of a farmer's tractor. Exactly what are ceramics? When did man first find a use for them?

Ceramics are nonmetallic, inorganic materials that require the use of high temperatures in their processing. In the earliest form, clay pottery of 10,000 B.C. has been found to be excellently preserved. The most common of ceramics, glass—an ancient discovery of the Phoenicians (about 4000 B.C.), is a miracle material in every sense. It may be made transparent, translucent, or opaque, weak and brittle or flexible and stronger than steel, hard or soft, water soluble or chemically inert. Truly, it is one of the most versatile of engineering materials. Ceramic engineers are employed by a variety of industries, from the specialized raw material and ceramic product manufacturers to the chemical, electrical and electronic, automotive, nuclear, and aerospace industries.

CHEMICAL ENGINEERING

Chemical engineering is responsible for new and improved products and processes that affect every person. This includes materials that will resist extremities of heat and cold, processes for life-support systems in other environments, new fuels for reactors, rockets, and booster propulsion, medicines, vaccines, serum, and plasma for mass distribution, and plastics and textiles to serve a multiplicity of human needs. Consequently, chemical engineers must be able to apply scientifically the principles of chemistry, physics, and engineering to

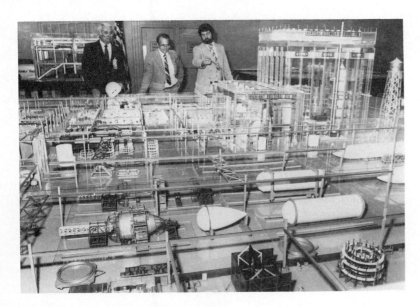

[4-8] Chemical engineers frequently make use of scale models in the design of petrochemical installations.

the design and operation of plants for the production of materials that undergo chemical changes during their processing.

The courses in chemical engineering cover inorganic, analytical, physical, and organic chemistry in addition to the basic engineering subjects; and the work in the various courses is designed to be of a distinctly professional nature and to develop capacity for original thought. The industrial development of our country makes large demands on the chemical engineer. The increasing uses for plastics, synthetics, and building materials require that a chemical engineer be employed in the development and manufacture of these products. Although well trained in chemistry, the chemical engineer is more than a chemist in that he or she applies the results of chemical research and discovery to the use of mankind by adapting laboratory processes to full-scale manufacturing plants.

The chemical engineer is instrumental in the development of the newer fuels for turbine and rocket engines. Test and evaluation of such fuels and means of achieving production of suitable fuels are part of the work of a chemical engineer. This testing must be carefully controlled to evaluate the performance of engines before the fuel is considered suitable to place on the market.

Opportunities for chemical engineers exist in a wide variety of fields. Not only are they in demand in strictly chemical fields but also in nearly all types of manufacturing. The production of synthetic rubber, the uses of petroleum products, the recovery of useful materials from what was formerly considered waste products, and the better utilization of farm products are only a few of the tasks that will provide work for the chemical engineer.

CIVIL AND CONSTRUCTION ENGINEERING

Civil engineers plan, design, and supervise the construction of facilities essential to modern life in both the public and private sectors—

facilities that vary widely in nature, size and scope, space satellites and launching facilities, offshore structures, bridges, buildings, tunnels, highways, transit systems, dams, airports, irrigation projects, treatment and distribution facilities for water, and collection and treatment for wastewater.

Construction engineering is concerned primarily with the design and supervision of construction of buildings, bridges, tunnels, and dams. The construction industry is America's largest industry today. Geotechnic investigations, such as in soil mechanics and foundation investigations, are essential not only in civilized areas but also for successful conquest of new lands such as Antarctica and extraterrestrial surfaces. Transportation systems include the planning, design, and construction of necessary roads, streets, thoroughfares, and superhighways. Engineering studies in water resources are concerned with the improvement of water availability, harbor and river development, flood control, irrigation, and drainage. Pollution is an ever-increasing problem, particularly in urban areas. The environmental engineer is concerned with the design and construction of water supply systems, sewerage systems, and systems for the reclamation and disposal of wastes. City planning and municipal engineers are concerned primarily with the planning of urban centers for the orderly, comfortable, and healthy growth and development of business and residential areas. Surveying and mapping are concerned with the measurements of distances over a surface (such as the earth or the moon) and the location of structures, rights-of-way, and property boundaries.

Civil engineers engage in technical, administrative, or commercial work with manufacturing companies, construction companies, transportation companies, and power companies. Other opportunities for employment exist in consulting engineering offices, in city and state engineering departments, and in the various bureaus of the federal government.

[4-9] This civil engineer is marking a structural member that will be used in the erection of a building that she designed.

Civil engineering is the oldest and one of the most exciting branches of engineering. Civil engineers are the "earth changers."

The Book of Knowledge

[4-10] Man has always suspected that some state of perfection exists where function, beauty, truth, and everlastingness converge. Perhaps the closest he has come to reaching this state is with the bridges he has built.

From "An Essay on Bridges," a *CBS-TV Reports* program, February 1965

ELECTRICAL AND COMPUTER ENGINEERING

Electrical engineering is concerned, in general terms, with the utilization of electric energy. It is divided into broad fields, such as information systems, automatic control, and systems and devices. Electricity used in one form or another reaches nearly all our daily lives and is truly the servant of mankind.

The electrical engineer applies engineering principles, both mechanical and electrical, in the design and construction of computers and auxiliary equipment. The basic requirements of a computer constantly change and new designs must provide for these necessary capabilities. In addition, a computing machine must be built that will furnish solutions of greater and greater problem complexity and at the same time have a means of introducing the problem into the machine in as simple a manner as possible.

There are many companies that build elaborate computing machines, and employment possibilities in the design and construction part of the industry are not limited. Many industrial firms, colleges, and governmental branches have set up computers as part of their capital equipment, and opportunities exist for employment as computer applications engineers, who serve as liaison between computer programmers and engineers who wish their problems evaluated on the machines. Of course, in a field expanding as rapidly as computer design, increasing numbers of employment opportunities become available. More and more dependence will be placed on the use of computers in the future, and an engineer specializing in this work will find ample opportunity for advancement.

The automatic control of machines and devices, such as autopilots for spacecraft and missiles, has become a commonplace requirement in today's technically conscious society. Automatic controlling of machine tools is an important part of modern machine shop operation. Tape systems are used to furnish signals to serve units on automatic lathes, milling machines, boring machines, and other types of machine tools so that they can be programmed to perform repeated operations. Not only can individual machines be controlled but also entire power plants can be operated on a programmed system. The design of these systems is usually performed by an electrical or mechanical engineer.

Energy-conversion systems, where energy is converted from one form to another, are also a necessity in almost every walk of life. Power plants convert heat energy from fuels into electrical energy for transmission to industry and homes. In addition to power systems, communication systems are a responsibility of the electrical engineer. Particularly in communications the application of modern electronics has been most evident. The electrical engineer who specializes in electronics will find that the majority of communication devices employs electronic circuits and components.

Other branches of electrical engineering that may include power or communication activities, or both, are illumination engineering, which deals with lighting using electric power; electronics, ultrasound, which has applications in both power and communications; and such diverse fields as x-ray, acoustics, and seismograph work.

It is well into the 1990s, a time when computers can understand and reply to human voice commands. The place is the Pentagon war room. Suddenly, an alarm goes off, signaling that the country is under attack. One of the generals shouts to a computer, "Is the attack coming from land, air or sea?"
"Yes," the computer responds.
"Yes what?" the general snaps.
"Yes, sir!" the machine replies.

Boston Globe

[4-11] Electrical engineers learn the fundamentals of designing and fabricating microelectronic circuit chips.

[4-12] A human hair is laid across a semiconductor chip that includes transistors and other circuit components which are smaller than 1 micrometer.

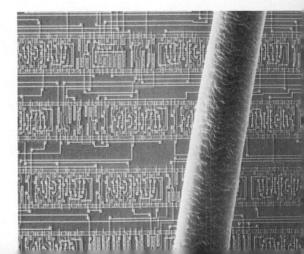

[4-13] Industrial engineers are proficient in discovering ways to economize the manufacturing process.

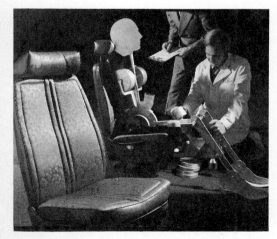

[4-14] The matching of the human being to the machine is a special area of study within industrial engineering.

Employment opportunities in electrical engineering are extremely varied. Electrical manufacturing companies use large numbers of engineers for design, testing, research, and sales. Electrical power companies and public utility companies require a staff of qualified electrical engineers, as do the companies that control the networks of telegraph and telephone lines and the radio systems. Other opportunities for employment exist with oil companies, railroads, food processing plants, lumbering enterprises, biological laboratories, and chemical plants. The aircraft and missile industries use engineers who are familiar with circuit design and employment of flight data computers, servomechanisms, computers, and solid-state devices. There is scarcely any industry of any size that does not employ one or more electrical engineers as members of its engineering staff.

INDUSTRIAL ENGINEERING

Industrial engineers are concerned with the design, improvement, and installation of integrated systems of people, materials, equipment, and energy in a production environment. Whereas other branches of engineering tend to specialize in some particular phase of science, the realm of industrial engineering may include parts of all engineering fields. The industrial engineer then will be more concerned with the larger picture of management of industries and production of goods than with the detailed development of processes.

The work of the industrial engineer is rather wide in scope. Generally, work is with people and machines, and because of this it is important that one be educated in both personnel administration and in the relations of people and machines to production.

The advent of the digital computer and other solid-state support equipment has revolutionized the business world. Many of the resultant changes have been made as a result of industrial engineering designs. Systems analysis, operations research, statistics, queuing theory, information theory, symbolic logic, and linear programming are all mathematics-based disciplines that are used in industrial engineering work.

The industrial engineer must be capable of preparing plans for the arrangement of plants for best operation and then organizing the workers so that their efforts will be coordinated to give a smoothly functioning unit. In such things as production lines, the various processes involved must be timed perfectly to ensure smooth operation and efficient use of the worker's efforts. In addition to coordination and automating of manufacturing activities, the industrial engineer is concerned with the development of data-processing procedures and the use of computers to control production, the development of improved methods of handling materials, the design of plant facilities and statistical procedures to control quality, the use of mathematical models to simulate production lines, and the measurement and improvement of work methods to reduce costs.

Opportunities for employment exist in almost every industrial plant and in many businesses not concerned directly with manufacturing or processing goods. In many cases the industrial engineer may be employed by department stores, insurance companies, consulting companies, and as a city engineer. The industrial engineer is trained in fundamental engineering principles, and as a result may also be employed in positions that would fall in the realm of the civil, electrical, or mechanical engineer.

MARINE ENGINEERING, NAVAL ARCHITECTURE, AND OCEAN ENGINEERING

For many centuries the sea has played a dominant role in the lives of people of all cultures and geographical locations. For this reason in every era the designers of ships have been held in the highest regard for their knowledge and understanding of the sea's physical influences and for their artistry and ability in marine craftsmanship. As our civilization increases in complexity, all peoples of future generations will depend to an even greater extent on vessels of the sea to keep food, materials, and fuel flowing.

Naval architects design marine vehicles as total systems, especially the internal layout. Ship design is a refined art as well as an exacting science since most ships are custom built—one at a time. Many large ships are virtually floating cities containing their own power sources, sanitary facilities, food-preparation center, and recreational and sleeping accommodations. Every service that would be provided to city dwellers must also be provided for the ship's crew. As with aircraft design, the ship's structural members and intricate networks of piping and electrical circuits must fit together harmoniously in the minimum space possible.

Marine engineers must have a broad-based engineering educational background. They will design the mechanical systems that go into the ship, especially the propulsion and auxiliary power machinery.

Ocean engineers are designers of those systems that cannot be called a ship or a boat but still must operate in the marine environment. Some examples are oil rigs that drill offshore, man-made islands, and offshore harbor facilities.

The basic design of seaworthy cargo ships changed very little from 1900 to 1960. However, with the advent of nuclear power and sophisticated electronic computers a new era in ship design has begun to develop. Surface-effect or air-cushion-type vehicles, submarine tankers, and deep-submergence research vehicles have all emerged from the realm of science fiction to enter one of engineering reality. The application of these newer ideas for the shipbuilding industry awaits only a more positive commitment to the task by government and industry. With the shrinking world supply of food and energy, this commitment is certain to come.

MECHANICAL ENGINEERING

Mechanical engineering is concerned with the design of machines and processes used to generate power and apply it to useful purposes. These designs may be simple or complex, inexpensive or expensive, luxuries or essentials. Such items as the kitchen food mixer,

"How thick do you judge the planks of our ship to be?" "Some two good inches and upward," returned the pilot. "It seems, then, we are within two fingers' breadth of damnation."
Rabelais, Works, Book IV, *1548, Chapter 23*

[4-15] The marine engineer's role is significant in helping to solve the world's transportation problems.

[4-16] The mechanical engineer is a specialist in designing and testing a great variety of instruments and machines.

the automobile, air-conditioning systems, nuclear power plants, and interplanetary space vehicles would not be available for human use today were it not for the mechanical engineer. In general, the mechanical engineer works with systems, subsystems, and components that exhibit motion. The range of work that may be classed as mechanical engineering is wider than that in any of the other branches of engineering, but it may be grouped generally under two heads: work that is concerned with power-generating machines, and work that deals with machines that transform or consume this power in accomplishing their particular tasks. Design specialists may work with parts that vary in size from the microscopic part of the most delicate instrument to the massive parts of heavy machinery. Included are mass transit systems that are rapidly becoming a part of our nationwide transportation system. Automotive engineers work constantly to improve the vehicles and engines. Heating, ventilating, air-conditioning, and refrigeration engineers are concerned with the design of suitable systems for making our buildings more comfortable and for providing proper conditions in industry for good working environments and efficient machine operation.

Employment may be secured by mechanical engineering graduates in almost every type of industry. Manufacturing plants, power-generating stations, public utility companies, transportation companies, airlines, and factories, to mention only a few, are examples of organizations that need mechanical engineers. Experienced engineers are needed in the missile and space industries in the design and development of such items as gas turbine compressors and power plants, air-cycle cooling turbines, electrically and hydraulically driven fans, and high-pressure refrigerants. Mechanical engineers are also needed in the development and testing of airborne and missile fuel systems, servovalves, and electro-mechanical control systems.

METALLURGICAL ENGINEERING

In many respects the past 25 years may be said to be an "age of materials"—an age which has seen the maturing of space exploration, nuclear power, digital computer technology, and ocean conquest. None of these engineering triumphs could have been achieved without the contributions of the metallurgical engineer. Metals are found in every part of the earth's crust, but rarely in immediately usable form. It is the metallurgical engineer's job to separate them from their ores and from other materials with which they exist in nature.

Metallurgical engineering may be divided into two branches. One branch deals with the location and evaluation of deposits of ore, the best way of mining and concentrating the ore, and the proper method of refining the ore into the basic metals. The other branch deals with the fabrication of the refined metal or metal alloy into various machines or metal products.

The metallurgist performs pure and applied research on vacuum melting, arc melting, and zone refining to produce metallic materials having unusual properties of strength and endurance. In addition,

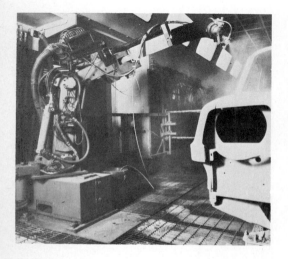

[4-17] Industrial robots, such as this "elephant trunk" painting robot in a Renault assembly plant, are designed by mechanical engineers.

the metallurgist in the aircraft and missile industries is often called upon to give an expert opinion on the results of fatigue tests of metal parts of machines.

MINING AND GEOLOGICAL ENGINEERING

The mining and geological engineer of today who searches the earth for hidden minerals is necessarily a person of quite different stature than the traditional explorer of yesteryear. These engineers must possess a combination of fundamental engineering and scientific education and field experience to enable them to understand the composition of the earth's crust. They must be experts in utilizing very sensitive instruments as they seek to locate new mineral deposits and to anticipate the problems that might arise in getting them out and transporting them to civilization. For this reason it is not unusual to find a mining or geological engineer in a modern office building in New York one week, and the next in Arizona or Zambia— or commuting between an expedition campsite and technical laboratories.

The work of mining and geological engineers lies generally in three areas: finding the ore, extracting it, and preparing the resulting minerals for manufacturing industries to use. These engineers design the mine layout, supervise the construction of mine shafts and tunnels in underground operations, and devise methods for transporting minerals to processing plants. Mining engineers are also responsible for mine safety and the efficient operation of the mine, including ventilation, water supply, power, communications, and

[4-18] Almost every aspect of our life is affected by advances in metallurgy and materials science. For example, teeth can now be straightened because of the development of a special type of steel . . . rust-proof, strong yet ductile, and hard yet smooth . . . unchanged through ice-cold sodas and red-hot pizzas.

[4-19] Geological engineers investigate new sources of essential mineral bodies.

[4-20] The recovery of certain minerals can best be accomplished by open-pit mining.

[4-21] Disposal of spent nuclear fuel requires careful monitoring by the nuclear engineer.

equipment maintenance. Geological engineers are more directly concerned with locating and appraising mineral deposits.

An important part of the mining and geological engineer's work is to keep in mind inherent air- and water-pollution problems that might develop during the mining operation. This involves establishing efficient controls to prevent harmful side effects of mining and designing ways whereby the land will be restored for people to use after the mining operation terminates.

NUCLEAR ENGINEERING

Nuclear engineering is one of the newest and most challenging branches of engineering. Although much work in the field of nucleonics at present falls within the realm of pure research, a growing demand for people educated to utilize recent discoveries for the benefit of humankind has led many colleges and universities to offer courses in nuclear engineering. The nuclear engineer is familiar with the basic principles involved in both fission and fusion reactions; and by applying fundamental engineering concepts, is able to direct the enormous energies involved in a proper manner. Work involved in nuclear engineering includes the design and operation of plants to utilize heat energy from reactions, and the solution of problems arising in connection with safety to persons from radiation, disposal of radioactive wastes, and decontamination of radioactive areas.

The wartime uses of nuclear reactions are well known, but of even more importance are the less spectacular peacetime uses of controlled reactions. These uses include such diverse applications as electrical power generation and medical applications. Other applications are in the use of isotopes in chemical, physical, and biological research, and in the changing of the physical and chemical properties of materials in unusual ways by subjecting them to radiation.

Recent advances in our knowledge of controlled nuclear reactions have enabled engineers to build power plants that use heat

[4-22] Nuclear power offers a great potential to supplying this country's energy needs. Safety of operation and disposal of nuclear wastes are still problems that must be solved in a more reassuring way. The San Onofre Nuclear Power Plant, located along the coast midway between Los Angeles and San Diego, began operation in 1968 and supplies 2650 megawatts of power.

from reactions to drive machines. Nuclear energy plays an important role in our national energy supply. More than 12 percent of today's electrical power is nuclear generated, and this is projected to be 20 percent of the national total by the year 2000. Submarine nuclear power plants, long a dream, are now a reality, and experiments are being conducted on smaller nuclear power plants that can be used for airborne or railway applications.

At present, there are opportunities for employment of nuclear engineers in both privately owned and government-operated plants, for the generation of electricity, and where radioactive waste disposal or processing of nuclear materials is performed. Nuclear engineers are also needed by companies that may use radioactive materials in research or processing involving agricultural, medical, metallurgical, and petroleum products.

PETROLEUM ENGINEERING

In early America, wood was the primary source of energy. Today the major source of energy is petroleum. It is the most widely used of all energy sources because of its mobility and flexibility in utilization. Approximately three-fourths of the total energy needs of the United States are currently supplied by petroleum products, and this condition will likely continue for many years. Petroleum engineering is the practical application of the basic sciences (primarily chemistry, geology, and physics) and the engineering sciences to the development, recovery, and field processing of petroleum.

Petroleum engineering is concerned with all phases of the petroleum industry, from the location of petroleum in the ground to the ultimate delivery to the user. Petroleum products play an important part in many phases of our everyday life in providing our clothes, food, work, and entertainment. Because of the complex chemical structure of petroleum, we are able to make an almost endless number of different articles. Owing to the wide demand for petroleum products, the petroleum engineer strives to satisfy an ever-increasing demand for oil and gas from the ground.

The petroleum engineer is concerned first with finding deposits of oil and gas in quantities suitable for commercial use, in the extraction of these materials from the ground, and in the storage and processing of the petroleum above ground. The petroleum engineer is also concerned with the location of wells in accordance with the findings of geologists, the drilling of wells and the myriad problems associated with the drilling, and the installation of valves and piping when the wells are completed. In addition to the initial tapping of a field of oil, the petroleum engineer is concerned with practices that will provide the greatest recovery of the oil, considering all possible factors that may exist many thousand feet below the surface of the earth.

After the oil or gas has reached the surface, the petroleum engineer will provide the means of transporting it to suitable processing plants or to places where it will be used. Pipelines are providing an ever-increasing means of transporting both oil and gas from field to consumer.

[4-23] Petroleum engineers specialize in discovering and recovering petroleum products.

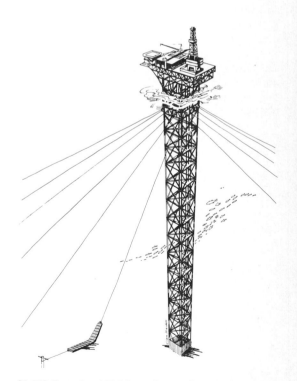

[4-24] Exxon's 1300-foot Lena Guyed Tower is slightly higher than New York City's Empire State Building. It is held in place by 14 piles and 20 guy lines. This permits the tower to move slightly with the wind and waves.

Many challenges face the petroleum engineer. Some require pioneering efforts, such as with the rapidly developing Alaska field. Other opportunities lie closer at hand. For example, it is known that because of excessive costs in recovery, more than one-half of the oil already discovered in the United States *has yet to be brought to the surface of the earth*. It is estimated that even a 10 percent increase in oil recovery would produce 3 billion barrels of additional oil, a worth of over $60 billion.

Owing to the expanding uses for petroleum and its products, the opportunities for employment of petroleum engineers are widespread. Companies concerned with the drilling, producing, and transporting of oil and gas will provide employment for the majority of engineers. Because of the widespread search for oil, employment opportunities for the petroleum engineer exist all over the world; and for the young person wishing a job in a foreign land, oil companies have crews in almost every country over the globe. Other opportunities for employment exist in the field of technical sales, research, and as civil service employees of the national government.

PROBLEMS

4-1. Discuss the changing requirements for aerospace and astronautical engineers.

4-2. Investigate the opportunities for employment in agricultural engineering. Discuss your findings.

4-3. Write a short essay on the differences in the utilization and capability of the architectural engineer and the civil engineer who has specialized in structural analysis.

4-4. Interview a chemical engineer. Discuss the differences in her work and that of a chemist.

4-5. Assume that you are employed as an electrical engineer. Describe your work and comment particularly concerning the things that you most like and dislike about your job.

4-6. Explain why the demand for industrial engineers has increased significantly during the past 10 years.

4-7. Write a 200-word essay describing the challenging job opportunities in engineering that might be particularly attractive for an engineering graduate.

4-8. Explain the importance of mechanical engineers in the electronics industry.

4-9. Describe the changes that might be brought about to benefit humankind by the development of new engineering materials.

4-10. Investigate the need for nuclear and petroleum engineers in your state and report your findings.

Chapter 5

Work Opportunities

During the college years, engineering students will study courses in many subject areas. English courses make it easier to organize and present ideas effectively; mathematics courses improve the analysis or computational skills useful in modeling processes and devices; social science courses assist in finding a place in society as informed citizens; and various technical courses help to gain an understanding of natural laws. In your study of technical courses, you will become familiar with a store of factual information that will form the basis for your engineering decisions. The nature of these technical courses, in general, will influence your choice of a major field of interest. For example, you may decide to concentrate your major interest in a particular field, such as civil, chemical, electrical, industrial, or mechanical engineering.

College courses also provide training in learning facts and in developing powers of reasoning. Since it is impossible to predict what kind of work a practicing engineer will be doing after graduation, the objective of an engineering education is to provide a broad base of facts and skills on which the engineer can rely.

It usually is not sufficient to say that an engineer is working as a *civil engineer*. The work experience may vary over a wide spectrum. As a civil engineer, for example, one may be performing research on materials for surfacing highways, or be employed in government service and be responsible for the budget preparation of a missile launch project. In fact, there are many things that a practicing engineer will be called on to do which are not described by a specialized course of study. The *type* of work that the engineer may do, as differentiated from a major field of specialization, can be called "engineering function." The major functions are research, development, design, production, construction, operations, sales, and management.

It has been found that in some engineering functions, such as in the management of a manufacturing plant, specialization is of lesser importance, whereas in other functions, such as research in microelectronics, specialization may be extremely important. To understand more fully the activities of a practicing engineer, let us examine some of these functions.

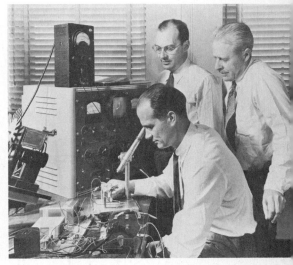

[5-1] Bell Telephone laboratories researchers William Shockley, John Bardeen, and Walter Brattin received the Nobel Prize in 1956 for their invention of the transistor.

RESEARCH

In some respects, research is perceived to be one of the more glamorous functions of engineering. In this type of work the engineer

Research is an organized method of trying to find out what you are going to do after you find that you cannot do what you are doing now.
Charles F. Kettering, 1876–1958

[5-2] Research is an important type of work performed by the engineer. Engineers in research employ basic scientific principles in the discovery and application of new knowledge. This engineer is experimenting with a wind tunnel model of the city of Denver, Colorado, during a simulated atmospheric inversion.

Observation, not old age, brings wisdom.
Plublilius Syrus
Sententiae

The greatest invention of the nineteenth century was the invention of the method of invention.
Alfred North Whitehead, 1861–1947
Science and the Modern World

But in science the credit goes to the man who convinces the world, not to the man to whom the idea first occurs.
Sir Francis Darwin, 1848–1925
First Galton Lecture before the Eugenics Society

delves into the nature of matter, exploring processes to use engineering materials and searching for reasons for the behavior of the things that make up our world. In many instances the work of the scientist and the engineer who are engaged in research will overlap. The work of scientists is usually closely allied with research. The objective of the researcher is to *discover truths*. The objective of the research engineer, on the other hand, usually is directed toward the practical side of the problem: not only to discover but also *to find a use for the discovery*.

The research engineer must be especially perceptive. Patience is also required, since most tasks have never before been accomplished. It is also important to be able to recognize and identify phenomena previously unnoticed. As an aid to educating an engineer to do research work, some colleges offer specific courses in research techniques. However, the life of a research engineer can be quite disheartening. In addition to probing and exploring new areas, much of the work is trial and error, and outstanding results of investigation usually occur only after long hours of laborious and painstaking work. On occasion, the final result is entirely different from the original mission.

Until the last few decades, almost all research was solo work by individuals. However, with the rapid expansion of the fields of knowledge of chemistry, physics, and biology, it became apparent that groups or "research teams" of scientists and engineers could accomplish better the aims of research by pooling their efforts and knowledge. Within the teams, the enthusiasm and competition provide added incentive to push the work forward. Since each person is able to contribute from a particular specialty, discovery is accelerated.

As has been indicated, a thorough training in the basic sciences and mathematics is essential for a research engineer. In addition, an inquiring mind and a great curiosity about the behavior of things is desirable. Most successful research engineers have a fertile and uninhibited imagination and a knack of observing and questioning phenomena that the majority of people overlook. For example, one successful research engineer has worked on such diverse projects as an automatic lawn mower, an electronic biological eye to replace natural eyes, and the use of small animals as electrical power sources.

Most research engineers secure advanced degrees because they need additional background in the basic sciences and mathematics. In addition, this study usually gives them an opportunity to acquire useful skills in research procedures.

DEVELOPMENT

After a basic discovery in natural phenomena is made, the next step in its utilization involves the development of processes or machines that employ the principles involved in the discovery. In the research and development fields, as in many other functions, the areas of activity overlap. In many organizations, the functions of research and development are so interrelated that the department performing this work is designated simply as a research and development (R&D) department.

The engineering features of development are concerned principally with the actual construction, fabrication, assembly, layout, and testing of scale models, pilot models, and experimental models for pilot processes or procedures. Where the research engineer is concerned more with making a discovery that will have commercial or economic value, the development engineer will be interested primarily in producing a process, an assembly, or a system *that will work*.

The development engineer does not deal exclusively with new discoveries. Actually, the major part of work assignments will involve using well-known principles and employing existing processes or machines to perform a new or unusual function. It is in this region that many patents are granted. In times past, the utilization of basic machines, such as a wheel and axle, and fundamental principles, including Ohm's law and Lenz' law, have eventually led to patentable machines, such as the electric dynamo. On the other hand, within a very short time after the announcement of the discovery of the laser in 1960, a number of patents were issued on devices employing this new principle. Thus the lag between the discovery of new knowledge and the use of that knowledge has been steadily decreasing through the years.

A patent is a legal document granting its owner the sole right to exclude others from using, manufacturing, or selling the invention described by the patent claims. In the United States the rights granted by the patent endure for a period of 17 years from its date of issue. Upon expiration of this grant, the information disclosed passes into the public domain and then can be used by anyone. As an exam-

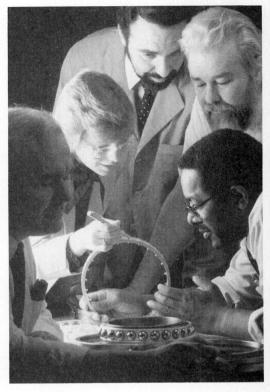

[5-3] Today, significant research is most often accomplished by teams of engineers and technicians.

Nothing is invented and perfected at the same time.
John Ray

The zipper, invented in 1879, was so shrewdly and strongly opposed by button manufacturers that it didn't get on the market for 30 years.
E. V. Durling

Invention, strictly speaking, is little more than a new combination of those images which have been previously gathered and deposited in memory. Nothing can be made of nothing; he who has laid up no materials can produce no combinations.
Sir Joshua Reynolds, 1723–1792

[5-4] Schematic drawing of a conventional refrigeration system.

HEAT ENTERING REFRIGERANT

LOW PRESSURE COOL GASEOUS REFRIGERANT

HIGH PRESSURE COOL LIQUID REFRIGERANT

MECHANICAL COMPRESSOR

EVAPORATOR

WORK DONE ON REFRIGERANT

CONDENSER

EXPANSION VALVE

HIGH PRESSURE HOT GASEOUS REFRIGERANT

HIGH PRESSURE COOL LIQUID REFRIGERANT

HEAT LEAVING REFRIGERANT

In 1899, the director of the U.S. Patent Office urged President McKinley to abolish the Patent Office along with his own job because "everything that can be invented has been invented."

United States Patent [19]
Hosterman et al.

[11]   **4,157,015**
[45]   Jun. 5, 1979

[54] **HYDRAULIC REFRIGERATION SYSTEM AND METHOD**
[75] Inventors: **Craig Hosterman,** Scottsdale; **Warren Rice,** Tempe, both of Ariz.
[73] Assignee: **Natural Energy Systems,** Tempe, Ariz.
[21] Appl. No.: **862,119**
[22] Filed: **Dec. 19, 1977**
[51] Int. Cl.² F25B 1/00; F25D 15/00; F25B 1/06
[52] U.S. Cl. 62/115; 62/119; 62/500
[58] Field of Search .... 62/115, 116, 119, 122, 62/260, 498, 500, 514 R, 467 R; 126/400
[56]   **References Cited**
                U.S. PATENT DOCUMENTS

1,882,256 10/1932 Randel 62/500
2,152,663   4/1939 Randel 62/500
2,191,864   2/1940 Schaefer 62/119
3,789,617   2/1974 Rannow 62/115
3,848,424 11/1974 Rhea 62/115

Primary Examiner—Lloyd L. King
Attorney, Agent, or Firm—Cahill, Sutton & Thomas

[57]   **ABSTRACT**

A refrigerant fluid is entrained within a down pipe of a closed loop water flow circuit to compress the refrigerant fluid from a gaseous state to a liquid state. A separation chamber at the lower extremity of the down pipe separates the liquid refrigerant fluid from the water and the water is drawn off. The water flows upwardly through a return pipe and pump, through a pipe for reintroduction to the down pipe at the lower end thereof. The drawn off liquid refrigerant flows upwardly through a return pipe and through an expansion valve. The refrigerant fluid, converted to a mixture of vapor and liquid, called a "quality mixture of the refrigerant" by the expansion valve, flows through an evaporator to cool a medium, such as air, passing therethrough. The refrigerant fluid, flowing from the evaporator and in a gaseous state at a higher temperature, is introduced to the upper end of the down pipe for reentrainment in the water flowing into the down pipe.

18 Claims, 5 Drawing Figures

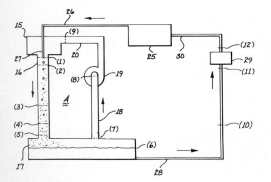

[5-5] Typical patent drawing.

ple of how the process works, let us examine a new entry into the continuing development of mechanical refrigeration systems.

Conventional refrigeration systems require four major components: an *evaporator* (to allow heat to enter the liquid refrigerant and change it to gaseous state), a mechanical *compressor* (to increase the pressure and temperature of the refrigerant), a *condenser* (to cool the refrigerant and change its state from gas to liquid), and an *expansion valve* (to allow the liquid refrigerant to change to a partly gaseous state). Such a system is illustrated in schematic form in Figure [5-4]. Two Arizona inventors have been issued a patent, No. 4,157,015, *Hydraulic Refrigeration System and Method* (HRS), that eliminates the need for a mechanical compressor in the refrigeration cycle.[1] This novel invention uses a pump-driven descending liquid column (such as water) to accept and entrain the vaporized refrigerant (such as Freon) coming from the evaporator. As the water and Freon mixture descends, the gaseous Freon is compressed to its liquid state. This natural compression of the refrigerant eliminates the need for a conventional refrigerant compressor, which is a mechanically complex machine that needs maintenance and is expensive to operate. The patent drawing [5-5] shows schematically how the invention has been designed to function. The numbers on the drawing refer to the specific locations in the system where various changes of state occur. For example, the substantial differences in densities of the condensed Freon and the water at the bottom of the system (17) allows for separation of the two fluids (7 and 28). Heat rejection from the water column at 3 (the condenser function) can be accomplished by heat absorption by the earth, a water body, or by a conventional cooling tower. The expansion valve (29) and the evapora-

---

[1]This patent is owned by Natural Energy Systems, 2042 East Balboa Drive, Tempe, AZ 85282.

tor (25) are standard units. This is a novel refrigeration system because it requires a single moving part for it to operate—the water pump (19). (It is not unusual for submersible water pumps to operate for 20 years or more without maintenance.) The processes that occur at each numbered point on the patent drawing are described in turn in the patent document.

Considerable research has been conducted on this invention, including the construction of a full-scale experimental model. Further development is currently under way. What might be the benefits to humankind when such a new concept in refrigeration can be fully and economically developed? Let us consider some possibilities.

The natural compression of the HRS provides a 15 to 50 percent energy conservation advantage over conventional industrial and residential units that use mechanical compression. Perhaps of even greater significance, however, the HRS greatly simplifies the construction and maintenance of refrigeration systems. It has long been predicted that the introduction of a simple, relatively inexpensive method of producing refrigeration would greatly accelerate both the quality of life and the industrialization of Third World countries, particularly large population areas such as those found in India, China, Africa, Mexico, and South America. The HRS may be the answer these people are seeking.

In most instances the tasks of the development engineer are dictated by immediate requirements. For example, a new type of device may be needed to determine at all times the position in space of an airplane. Let us suppose that the development engineer does not know of any existing device that can perform the task to the desired specifications. Should he or she immediately attempt to invent such a device? The answer, of course, is "usually not." First, the files of available literature should be searched for information pertaining to existing designs. Such information may come from two principal sources. The first source is library material on processes, principles, and methods of accomplishing the task or related tasks. The second source is manufacturers' literature. It has been said humorously that "There is no need to reinvent the wheel." A literature search may discover a device that can accomplish the task with little or no modification. If no device is available that will do the work, a system of existing subassemblies may be set up and joined to accomplish the desired result. Lacking these items, the development engineer must explore further into the literature, and, using results from experiments throughout the world, formulate plans to construct a model for testing. Previous research may point a way to go, or perhaps a mathematical analysis will provide clues as to possible methods.

The development engineer usually works out ideas on a trial or "breadboard" basis, whether it be a machine or a computer process [5-8]. Having the parts or systems somewhat separated facilitates changes, modifications, and testing. In this process, improved methods may become apparent and can be incorporated. When the system or machine is in a workable state, the development engineer must then refine it and package it for use by others. Here again, ingenuity and a knowledge of human needs are important. A device that works satisfactorily in a laboratory when manipulated by skilled technicians may be hopelessly complex and unsuited for field use.

[5-6] Literature searches can be frustrating and nonproductive . . .

If you want to be happy for an hour, get drunk.
If you want to be happy for three days, get married.
If you want to be happy for eight days, kill your pig
    and eat it.
If you want to be happy forever, invent a machine
    useful to your fellowmen.
                                    Old Chinese Proverb (Revised)

Never get to the point where you will be ashamed to ask anybody for information. The ignorant man will always be ignorant if he fears that by asking another for information he will display ignorance. Better once display your ignorance of a certain subject than always know nothing of it.
                                    Booker T. Washington, 1856–1915

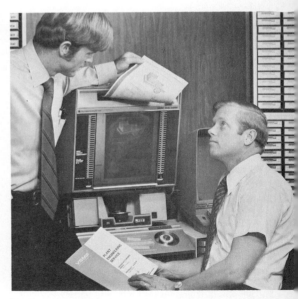

[5-7] . . . or satisfying and successful.

[5-8] The development engineer frequently uses a "breadboard" model to test the design's operational characteristics.

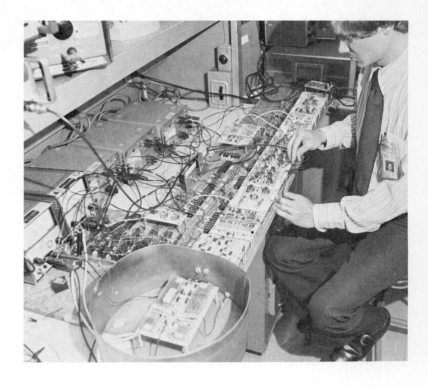

Were it not for imagination, Sir, a man would be as happy in the arms of a chambermaid as of a Duchess.
Samuel Johnson
Boswell's Life, May 9, 1778

The development engineer is the important person behind every pushbutton.

The education of an engineer for development work is similar to the education that the research engineer will expect to receive. However, creativity and innovation are perhaps of more importance, since the development engineer is standing between the scientist or the research engineer and the members of management who provide money for the research effort. The economic value of certain processes over others to achieve a desired result must be recognizable. It is also important to be able to convince others of the soundness of any conclusions reached. A comprehensive knowledge of basic principles of science and an inherent cleverness in making things work are also essential skills for the development engineer.

DESIGN

In our modern way of life, mass production has given us less expensive products and has made more articles available than ever before in history. In the process of producing these articles, the design engineer enters the scene just before the actual manufacturing process begins. After the development engineer has assembled and tested a device or a process and it has proved to be one that it is desirable to produce for a mass market, the final details of making it adaptable for production will be handled by a design engineer.

The design engineer must anticipate all manner of problems that the user may create in the application of the machine, or use of the

Thousands of engineers can design bridges, calculate strains and stresses, and draw up specifications for machines, but the great engineer is the one who can tell whether the bridge or the machine should be built at all, where it should be built, and when.
Eugene G. Grace

structure. The design must prevent user errors, accidents, and dissatisfaction. This is especially true for products that will be integrated into larger systems, or be used by customers under widely varied circumstances. For example, a car designed in Detroit must survive and operate in arctic Alaska, desert Arizona, humid Florida, and atop Pike's Peak.

In bridging the gap between the laboratory and the production line, the design engineer must be a versatile individual. This requires a mastery of basic engineering principles and mathematics, and an understanding of the capabilities of machines. It is also important to understand the temperament of the people who operate them. The design engineer must also be conscious of the relative costs of producing items, for it will be the design that will determine how long the product will survive in the open market. Not only must the device or process work; in many cases, it must also be made in a style and at a price that will attract customers.

As an example, let us take a clock, a simple device widely used to indicate time [5-10]. It includes a power source, a drive train, hands, and a face. Using these basic parts, engineers have designed spring-driven clocks, weight-driven clocks, and electrically driven clocks with all variations of drive trains. The basic hands and face have been modified in some models to give a digital display. The case has been made in many shapes and, perhaps in keeping with the slogan "time flies," it has even been streamlined! In the design of each modification the design engineer has determined the physical structure of the assembly, its aesthetic features, and the economics of producing it.

Of course, the work of the design engineer is not limited solely to performing engineering on mass-produced items. Design engineers may work on items such as bridges or buildings where only one item is made. However, in such work they are still fulfilling the design process of adapting basic ideas to provide for making a completed product for the use of others. In this type of design, engineers must be able to use their training, in some cases almost intuitively, to arrive at a design solution that will provide for adequate safety without excessive redundancy. The more we learn about the behavior of structural materials, the better we can design without having to add additional materials to cover the "ignorance factor" area. Particularly in the aircraft industry, design engineers have attempted to use structural materials with minimum excess being allowable as a safety factor. Each part must perform without failure, and every ounce of weight must be saved.

Of course, to do this, fabricated parts of the design must be tested and retested for resistance to failure due either to static loads or to vibratory fatiguing loads. Also, since surface roughness has an important bearing on the fatigue life of parts which are subjected to high stress or repeated loads, much attention must be given to specifying that surface finishes meet certain requirements.

Since design work involves a production phase, the design engineer is always considering costs as a factor in our competitive economy. One way in which costs can be minimized in manufacture or construction is to use standard parts, and standard sizes and dimensions for raw material. For example, if a machine were designed using nonstandard bolt threads or a bridge designed using nonstand-

(a)

(b)

[5-9] The design engineers pictured in part (a) are evaluating a cost-reduction proposal for the connector base plate assembly shown in part (b).

[5-10] The availability of solid-state circuitry has revolutionized the design of clocks and watches.

The fact is, that civilization requires slaves. The Greeks were quite right there. Unless there are slaves to do the ugly, horrible, uninteresting work, culture and contemplation become almost impossible. Human slavery is wrong, insecure, and demoralizing. On mechanical slavery, on the slavery of the machine, the future of the world depends.

Oscar Fingle O'Flahertie Wills Wilde, 1854–1900
The Soul of Man Under Socialism

Laws of Thermodynamics

1. You cannot win.
2. You cannot break even.
3. You cannot get out of the game.

Anonymous

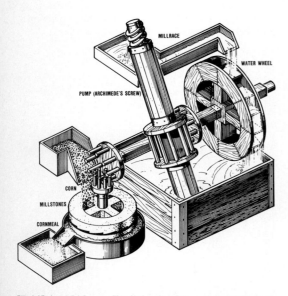

[5-11] In 1618, an English physician named Robert Fludd proposed a perpetual-motion machine to grind grain. The design violates the principle of energy conservation, more formally stated in the first law of thermodynamics.

ard steel I-beams, the design probably would be more expensive than needed to fulfill its function. Thus the design engineer must be able to coordinate the parts of a design so that it functions acceptably and is produced at minimum cost.

The design engineer soon comes to realize also that there usually are many acceptable ways to solve a design problem. Unlike an arithmetic problem with fixed numbers which gives one answer, real design problems can have many answers and many ways of obtaining a solution, *and all may be acceptable*. In such a case the engineer's decision becomes a matter of experience and judgment. At other times it may become just a matter of making a decision one way or the other. Regardless of the method used, the final solution to a problem should be a conscious effort to provide the *best* method, considering fabrication, costs, and sales.

What are the qualifications of a design engineer? Here creativity and innovation are key elements. Every design will embody a departure from what has been done before. However, all designs must be produced within the constraints of the reality of the physical properties of available materials and by economic factors. Therefore, design engineers must be thoroughly knowledgeable in fundamental engineering in a rather wide range of subjects. They must be able to apply the natural laws of nature appropriately to ascertain whether proposed ideas are feasible. In addition, they must be familiar with basic principles of economics, both from the standpoint of employing people and using machines. As they progress upward into supervisory and management roles, the employment of principles of psychology and economics becomes of even more importance. For this reason they usually will have more use for management courses than will research or development engineers.

PRODUCTION AND CONSTRUCTION

In the fields of production and construction, the engineer is more directly associated with the technician, mechanic, and laborer. The production or construction engineer must take the design engineer's drawings and supervise the assembly of the object as it was conceived and illustrated by drawings or models.

Usually, production or construction engineers are associated closely with the process of estimating and bidding for competitive jobs. In this work they employ their knowledge of structural materials, fabricating processes, and general physical principles to estimate both time and cost to accomplish tasks. In construction work the method of competitive bidding is usually used to award contracts, and the ability to reduce an appropriate amount from an estimate by skilled engineering practices may mean the difference between a successful bid and one that is either too high or too low.

Once a contract has been awarded, it is usual practice to assign a "project engineer" as the person who assumes overall responsibility and supervision of the work from the standpoint of materials, labor, and money. The engineer will supervise other production or construction engineers, who will be concerned with more specialized features of the work, such as civil, mechanical, electrical, or chemical engineering. Here the project engineer must complete the details of the designers' plans. Provision must be made to provide the specialized construction tools needed for the work. Schedules of production and/or construction must also be set up and questions that technicians or workers may raise concerning features of the design must be answered. Design engineers will need to be advised concerning desirable modifications that will aid in the construction or fabrication processes. In addition, the project engineer must be able to work effectively with construction or production crafts and labor unions.

Preparation of a schedule for production or construction is an important task of the engineer. In the case of an industrial plant, all planning for the procurement of raw materials and parts will be based on this production schedule. An assembly line in a modern electronics plant is one example that illustrates the necessity for scheduling the arrival of parts and subassemblies at a predetermined time. As another example, consider the construction of a multistory office building. The necessity for parts and materials to arrive at the right time is very important. If they arrive too soon, they probably will be in the way, and if they arrive too late, the building is delayed, which will cause an increase in costs to the builder.

Qualifications for production or construction engineers includes a thorough knowledge of engineering principles. In addition, they must have the ability to visualize the parts of an operation, whether it be the fabrication of a microprocessor or the building of a concrete bridge. From an understanding of the operations involved, they must be able to arrive at a realistic schedule of time, materials, and manpower. Therefore, emphasis should be placed on courses in engineering design, economics, business law, and psychology.

[5-12] The construction industry is the largest industry in the United States.

Observe due measure, for right timing is in all things the most important factor.

Hesiod, ca. 700 B.C.
Works and Days

OPERATIONS

In modern industrial plants, the number and complexity of machines, the equipment and buildings to be cared for, and the planning needed for expansion have brought forward the need for specialized engineers to perform services in these areas. If a new manufacturing facility is to be constructed, or an addition made to an existing facility, it will be the duty of a plant engineer to perform the basic design, prepare the proposed layout of space and location of equipment, and to specify the fixed equipment such as illumination, communication, and air conditioning. In some cases, the work of construction will be contracted to outside firms, but it will be the general responsibility of the plant engineer to see that the construction is carried on as it has been planned.

After a building or facility has been built, the plant engineer and an appropriate staff are responsible for maintenance of the building, equipment, grounds, and utilities. This work varies from performing routine tasks to setting up and regulating the most complex and automated machinery in the plant.

To perform these functions, the plant engineer must have a wide knowledge of several branches of engineering. For land acquisition and building construction, civil engineering courses will be needed.

[5-13] The plant engineer is a critical element in the safe and efficient operation of the modern industrial plant.

For work with power generation equipment, mechanical and electrical backgrounds are essential. For work in specialized parts of the plant, knowledge may be needed in such fields as chemical, metallurgical, nuclear, petroleum, or textile engineering. Work activities will be in one or more of the following areas: plant layout and design, construction and installation, maintenance–repairs–replacement, operation of utilities, or plant protection and safety.

Plant engineers should be able to compare costs of operating under various conditions and set schedules for machines so that the best use will be made of them. In the case of chemical plants, they will also attempt to regulate the flows and temperatures at levels that will produce the greatest amount of desired product at the end of the line.

In the dual role as a plant and operations engineer, it is important to evaluate new equipment as it becomes available to see whether additional operating economies can be secured by retiring old equipment and installing new types. In this function the engineer must frequently assume a salesperson's role in order to convince management that it should discard equipment that, apparently, is operating perfectly and spend money for newer models. Here the ability to combine facts of engineering and economics is invaluable.

Plant and operations engineers must be able to work with people and machines and to know what results to expect from them. In this part of their work, a knowledge of industrial engineering principles is valuable. In addition, it is desirable to have a basic understanding and knowledge of economics and business law.

SALES

An important and sometimes unrecognized function in engineering is the realm of applications and sales. As is well known, the best designed and fabricated product is of little use unless a demand either exists or will be created for it. Since many new processes and products have been developed within the past few years, a field of work has opened up for engineers in presenting the use of new products to prospective customers.

Discoveries and their subsequent application have occurred so rapidly that a product may be available about which even a recent graduate may not know. In this case, it is the responsibility of the engineer in sales who has intimate knowledge of the principles involved to educate possible users so that a demand can be created. In this work the engineer must assume the role of a teacher. In many instances the product must be presented primarily from an engineering standpoint. If the audience is composed of engineers, the sales engineer must "talk their language" and answer their technical questions. But if the audience includes nonengineers, the sales engineer must present the features of the product in terms that can be easily understood.

In addition to acquiring a knowledge of the engineering features of a particular product, sales and application engineers must also be familiar with the operations of the customer's plant. This is important from two standpoints. First, they should be able to show how

[5-14] Engineers must be articulate and able to present their ideas with clear, concise presentations.

their product will fit into the plant, and also they must show the economics involved to convince customers that they should buy it. At the same time, they must point out the limitations of their product and the possible changes necessary to incorporate it into a new situation. For example, a new bonding material may be available, but in order for a customer to use it in an assembly of parts, a special refrigerator for storage may be necessary. Also, the customer would need to be informed of the necessity for proper cleaning and surface preparation of the parts to be bonded.

A second reason that the sales and application engineer must be familiar with a customer's plant operations is that it is here that many times new requirements are generated. By finding an application area in which no apparatus is available to do the work, the sales engineer is able to report back to the company that a need exists and that a development operation should be undertaken to produce a device or process to meet the need.

Almost all equipment of any complexity will need to be accompanied by introductory instructions when it is placed in a customer's plant. Here the application engineer can create goodwill by conducting an instruction program outlining the capabilities and limitations of the equipment. Also, after the equipment is in service, maintenance and repair capabilities by competent technical personnel will serve to maintain the confidence of customers.

Sales and applications engineers should have a basic knowledge of engineering principles and should, of course, have detailed knowledge in the area of their own products. Here the ability to perform detailed work on abstract principles is of less importance than the ability to present one's ideas clearly. A genuine appreciation of people and a friendly personality are desirable personal attributes. In addition to basic technical subjects, courses in psychology, sociology, public speaking, and human relations will prove valuable to the sales and applications engineer.

Usually, an engineer will spend several years in a plant learning the processes and the details of the plant's operation and management policies before starting out to be a member of the sales staff. As a sales engineer you represent your company in the mind of the customer. Therefore, you must present a pleasing appearance and give the customer a feeling of confidence in your engineering ability.

MANAGEMENT

Results of recent surveys show that the trend today is for corporate leaders in the United States to have backgrounds in engineering and science. It has been predicted that within five years, the *majority* of corporation executives will be persons who are trained in engineering and science as well as in business and the humanities, and who can bridge the gap between these disciplines.

Since the trend is toward more engineering graduates moving into management positions, let us examine the functions of an engineer in management.

The basic principles used in the management of a company are generally similar whether the company objective is dredging for oys-

[5-15] Sales should not be made unless the product will serve the needs of the client.

[5-16] Sales engineers must have a basic knowledge of engineering principles and detailed technical knowledge in the area of their own products.

The first principle of management is that the driving force for the development of new products is not technology, not money, but the imagination of people.

David Packard
Quoted in Industrial Engineering, August 1984, p. 61

Man's greatest discovery is not fire, nor the wheel, nor the combustion engine, nor atomic energy, nor anything in the material world. It is the world of ideas. Man's greatest discovery is teamwork by agreement.

J. Brewster Jennings

The question "Who ought to be boss?" is like asking "Who ought to be the tenor in the quartet?" Obviously, the man who can sing tenor.

Henry Ford

[5-17] It is important for the engineer in management to listen to the advice and counsel of subordinate engineers.

ters, building diesel locomotives, or producing microcomputers. These basic functions involve using the capabilities of the company to the best advantage to produce a desirable product in a competitive economy. The use of the capabilities, of course, will vary widely depending on the enterprise involved.

The executive of a company, large or small, has the equipment in the plant, the labor force, and the financial assets of the organization to use in conducting the plant's operations. In management, the engineer must make decisions involving all three of these items.

A generation ago it was assumed that only persons trained and educated in business administration should aspire to management positions. However, now it has been recognized that the education and other abilities which make a good engineer also provide the background to make a good management executive. The training for correlating facts and evaluating courses of action in making engineering decisions can be carried over to management decisions on machinery, personnel, and money. In some cases, the engineer is technically strong but may be less experienced in the realm of business practicability. Therefore, it is in the business side of an operation that the engineer usually must work harder to develop the necessary skills.

The engineer in management is concerned more intimately with the long-range effects of policy decisions. Where the design engineer considers first the technical phases of a project, the engineer in management must consider how a particular decision will affect the employees who work to produce a product and how the decision will affect the people who provide the financing of the operation. It is for this reason that the management engineer is concerned less with the technical aspects of the profession and relatively more with the financial, legal, and labor aspects.

This does not imply that engineering fundamentals should be minimized or deleted. Rather the growing need for engineers in management shows that the type and complexity of the machines and processes used in today's plants require a blending of technical and business training in order to carry forward effectively. This trend is particularly strong in certain industries, such as aerospace and electronics, where the vast majority of executive managerial positions are occupied by engineers and scientists. As other industries become computer intensive and automated, a similar trend in those fields also will become apparent.

The education that an engineer in management receives should be identical to the basic engineering education received in other engineering functions. However, young engineers usually can recognize early in their careers whether or not they have an aptitude for working with people and directing their activities. If you have the ability to "sell your ideas" and to get others to work with you, probably you can channel your activities into managerial functions. You may start out as a research engineer, a design engineer, or a sales engineer, but the ability to influence others to your way of thinking, a genuine liking for people, and a consideration for their responses will indicate that you probably have capabilities as a manager.

Of course, management positions are not always executive positions, but the ability to apply engineering principles in supervisory work involving large numbers of people and large amounts of money is a prerequisite in management engineering.

OTHER ENGINEERING FUNCTIONS

A number of other engineering functions can be considered that do not fall into the categories previously described. Some of these functions are testing, teaching, and consulting.

As in the other functions, there are no specific curricula leading directly toward these types of work. Rather, a broad background of engineering fundamentals is the best guide to follow in preparing for work in these fields.

In testing, the work resembles design and development functions most closely. Most plants maintain a laboratory section that is responsible for conducting engineering tests of proposed products or for quality control on existing products. The test engineer must be qualified to follow the intricacies of a design and to build suitable test machinery to give an accelerated test of the product. For example, in the automotive industry, not only are the completed cars tested, but also components, such as engines, brakes, and tires, are tested individually to provide data to be used in improving their performance. The test engineer must be able also to set up quality control procedures for production lines to ensure that production meets certain standards. In this work, mathematics training in statistical theory is helpful.

A career in teaching is rewarding for many persons. A desire to help others in their learning processes, a concern for some of their personal problems, and a thorough grounding in engineering and mathematics are desirable for those considering teaching engineering subjects. In the teaching profession, the trend today is toward the more theoretical aspects of engineering, and a person will usually find that teaching is more closely allied with research and development functions than with others. Almost all colleges now require the faculty to obtain advanced degrees, and a person desiring to be an engineering teacher should consider seriously the desirability of obtaining a doctorate in his or her chosen field.

More and more engineers are going into consulting work. Work as an engineering consultant can be either part time or full time. Usually, a consulting engineer is a person who possesses specific skills in addition to several years of experience, and may offer services to advise and work on engineering projects either part time or full time.

Frequently, two or more engineers will form an engineering consulting firm that employs other engineers, technicians, and computer-aided design specialists, and will contract for full engineering services on a project. The firm may restrict engineering work to rather narrow categories, such as the design of irrigation projects, power plants, or aerospace facilities, or a staff may be available that is capable of working on a complete spectrum of engineering problems.

On the other hand, as a consulting engineer you may prefer to operate alone. Your firm may consist of only your own skills such

[5-18] Teaching is a rewarding activity of engineering. Frequently, the engineering professor is the first person to introduce the student to the ethics and responsibilities of the profession.

He who knows not, and knows not that he knows not, is a fool. Shun him.
He who knows not, and knows that he knows not, is simple. Teach him.
He who knows, and knows not that he knows, is asleep. Waken him.
He who knows, and knows that he knows, is wise. Follow him.

Arabic Apothegm

FINANCIAL PRINCIPLES & MANPOWER UTILIZATION

Management
Industrial
Sales
Operation & Maintenance
Construction & Production
Design
Development
Research

ABSTRACT SCIENTIFIC PRINCIPLES

[5-19] Application of principles in various engineering functions.

that, in a minimum time, you may be able to advise and direct an operation to overcome a given problem. For instance, you may be employed by an industrial plant. In this way the plant may be able to solve a given problem more economically, particularly if the required specialization is only occasionally needed by the plant.

As may be inferred, a consulting engineer must have specific expertise to offer, and must be able to use his or her creative ability to apply individual skills to unfamiliar situations. Usually, these skills and abilities are acquired only after several years of practice and postgraduate study.

Consulting work is an inviting part of the engineering profession for a person who desires self-employment and is willing to accept its business risks to gain an opportunity for financial reward.

ENGINEERING FUNCTIONS IN GENERAL

As described in previous paragraphs, training and skills in all functions are basically the same, that is, fundamental scientific knowledge of physical principles and mathematics. However, it can be seen that research on one hand and management on the other require different educational preparations.

For work in research, emphasis is on theoretical principles and creativity, with little emphasis on economic and personnel considerations. On the other hand, in management, primary attention is given to financial and labor problems and relatively little to abstract scientific principles. Between these two extremes, we find the other functions with varying degrees of emphasis on research- or managerial-oriented concepts.

Figure [5-19] shows an idealized image of this distribution. Bear in mind that this diagram merely depicts a trend and does not necessarily apply to specific instances.

To summarize the functions of the engineer, we can say that in all cases the engineer is a problem identifier and solver. Whether it be a mathematical abstraction that may have an application to the design of a space station or a meeting with a bargaining group at a conference table, it is a problem that must lie identified and reduced to its essentials and the alternatives explored to reach a solution. The engineer then must apply specialized knowledge and inventiveness to select a reasonable method to achieve a result, even in the face of vague and sometimes contradictory data. The engineer has been able, in general, to accomplish this as evidenced by a long record of successful industrial management and productivity.

PROBLEMS

5-1. Discuss an important scientific breakthrough of the past year that was brought about by an engineering research effort.

5-2. Discuss the differences between engineering research and engineering development.

5-3. Interview an engineer and estimate the percentage of his

or her work that is devoted to research, development, and design.

5-4. Discuss the importance of the engineer's design capability in modern industry.

5-5. Investigate the work functions of the engineer and write a brief essay describing the function that most appeals to you.

5-6. Discuss the importance of the sales engineer in the total engineering effort.

5-7. Interview an engineer in management. Discuss the reasons that many engineers rise to positions of leadership as managers.

5-8. Compare the engineering opportunities in teaching with those in industry.

5-9. Investigate the opportunities for employment in a consulting engineering firm. Discuss your findings.

5-10. Discuss the special capabilities required of the engineer in construction.

5-11. Perpetual motion powered by electricity was often favored as an idea by nineteenth-century inventors. Figure [5-P11] illustrates one of these designs. Discuss how it was supposed to work. Why did it fail to work?

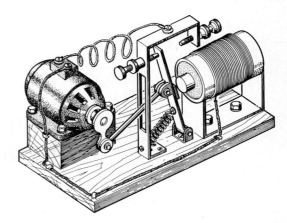

[5-P11] Proposed perpetual-motion machine.

Chapter 6

Professional Responsibilities of Engineers and Technologists

[6-1] By virtue of education and experience, the engineer is better equipped than most people to foresee and appreciate problems as well as to identify and assess alternatives.

Profession—The pursuit of a learned art in the spirit of public service.

American Society of Civil Engineers

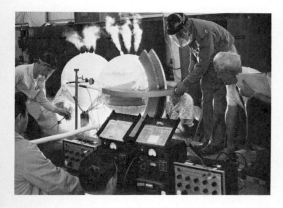

[6-2] Professionalism includes the establishment of performance standards and safety criteria.

ENGINEERS AND TECHNOLOGISTS AS PROFESSIONAL PERSONS

Who is a professional? As generally used in the sense of the learned professions, a professional person is one who applies certain knowledge and skill, usually obtained by college education, for the service of people. In addition, a professional person observes an acceptable code of conduct, uses discretion and judgment in dealing with people, and respects their confidences. Also, professional persons usually have legal status, use professional titles, and participate in a professional organization.

Knowledge and skill above that of the average person is a characteristic of the professional. Where a worker will have specific skills in operating a particular machine, a professional person is considered able to apply fundamental principles that are usually beyond the range of the average worker. The knowledge of these principles as well as the skills necessary to apply them distinguishes the professional. The engineer, because of an education in the basic sciences, mathematics, and engineering sciences, is capable of applying basic principles for such diverse things as improving the construction features of buildings, developing processes that will provide new chemical compounds, or designing canals to bring water to arid areas.

An important concept in the minds of most persons is that a professional person will perform a service for people. This means that service must be considered ahead of any monetary reward that a professional person may receive. In this respect the professional, individually, should recognize that a need for personal services exists and seek ways to provide a solution to these needs.

Engineering may be considered to be a profession insofar as it meets these characteristics of a learned professional group:

- Knowledge and skill in specialized fields above that of the general public.
- A desire for public service and a willingness to share discoveries for the benefit of others.
- Exercise of discretion and judgment.
- Establishment of a relation of confidence between the engineer and client or the engineer and employer.

100

■ Self-imposed standards for qualifications (such as accredited schools, registration laws, and the formulation and conduction of licensing examinations.

■ Acceptance of overall and specific codes of conduct.

■ Formation of professional groups and participation in advancing professional ideals and knowledge.

■ Recognition by law as an identifiable body of knowledge.

With these as objectives, students should pursue their college studies and training in employment so as to meet these characteristics within their full meaning and take their places as professional engineers in our society.

SOCIAL RESPONSIBILITY: THE PIVOTAL ROLE OF ENGINEERS AND TECHNOLOGISTS

Society has a special challenge for the engineering profession: use acquired knowledge for the benefit of humankind without endangering the surrounding environment or adding risks to the lives and safety of individuals. The response of engineers to this challenge will be on continuous public display because they design, produce, and operate many physical systems and products which have a direct impact on both individuals and society as a whole. When flawed, large-scale projects such as dams, bridges, and jet aircraft are apt to lead to serious accidents. Flawed designs of industrial process equipment may cause illness or injury to thousands of workers who must use the machines and ancillary chemicals. Apparent beneficial engineering operations such as power generating plants can also lead to environmental damage, depletion of resources, and other unique problems. Computers and robots used in factories to increase production and improve quality may also require a redistribution of specific worker manual skills. Another major concern of society today is the development of nuclear weapons and space warfare techniques which could place our entire civilization in danger. All of these resulting conditions must be dealt with by society.

The engineering profession plays a pivotal role in this important function of controlling technology for the benefit of society. Because of their education and experience, engineers are especially equipped to anticipate problems and to evaluate alternatives. As technology becomes more complex, society makes engineering increasingly responsible for this control.

Development of the Role of Engineering

For centuries society seemed assured that technologists could (and would) regulate themselves to assure a high level of competence and responsible behavior. However, as technology became more complex, society called for more safeguards and regulation. Design catastrophes such as boiler explosions, mine cave-ins, and building collapses began to change the image of engineering from one where its members are always considered to be a part of a respected guild, to one where its participants needed more than self-regulation.

Members of a profession—A group of people who have dedicated themselves to unselfish services to humanity through the application of knowledge and skills possessed by the group.

E. C. Easton
"An American Engineering Profession,"
Journal of Engineering Education, *April 1962*

Technology can have no legitimacy unless it inflicts no harm.

Admiral H. G. Rickover
"Humanistic Technology," Mechanical Engineering,
November 1982

Leonardo da Vinci writes in his autobiographical notes that he has discovered how to build a submarine which has a special application for naval warfare. This underwater boat would allow the user to sneak into a busy harbor without being seen and drill holes in the bottom of ships. History records, however, that during his lifetime he withheld disclosure of this invention because he believed that it would be an abomination to mankind.

Victor Paschkis
Conference on Engineering Ethics, *American Society of Civil Engineers, New York, 1975.*

[6-3] Yeah . . . I know it leans a little to the starboard. Actually, that's intentional . . . to compensate for the spongy land I bought on sale. According to a new method of calculating that I've worked out, it will right itself in about two months.

[6-4] The Tacoma Narrows Bridge was designed for static loading . . . but not for dynamic loading.

There is a difference, however, between "state of the art" design (which ultimately may prove to be unsafe) and risk design. There is much about the universe and the laws of nature that we do not yet understand. As engineers, we are expected to execute designs using principles that are based on the most current understanding of natural physical laws. We would expect no less of our family physician, whom we would expect to treat a body ailment using the most modern understanding of medicine and its effect on the human body. However, even the best of designs *under these circumstances* may fail. The collapse of the Tacoma Narrows Bridge at Puget Sound, Washington, in 1940, only four months after its completion, is an example. The bridge had two spans of 2800 feet with a width of only 39 feet, and the deck was stiffened throughout its length by means of 8-foot-deep plate girders. The design seemed to be adequate because similar designs had not failed in over 50 years of use. Such was not the case, however. The location of the bridge and its particular configuration made it susceptible to severe torsional vibration and aerodynamic instability. Ultimately, in winds of only 42 mph, the vibrations became so violent that the deck was torn away and crashed into the water [6-4]. Since that time a number of design methods have been developed to circumvent this problem, among them to shape the deck like a shallow airfoil. But, in 1940, who would have thought that a bridge should be designed like an airplane?

Risk design involves allowing sufficient tolerance above the point of failure, even under the most unusual circumstances, so that failure will not occur. This involves knowledge and judgment as well as conjecture as to what unexpected and unforeseen events might occur. The tolerance above the expected failure level is called the *factor of safety*. However, if the design is to work at all, it must also be compatible with the natural laws of nature. For example, a highway bridge might have a factor of safety of 2, meaning that the bridge is designed to hold up to two times the load that is anticipated to cause failure.

This reasoning would not work in the case of airplane design. If the component parts of the airplane were all oversized by two times, the airplane would be so heavy that it would not fly. It is more customary, then, for airplane designers to use factors of safety of 1.02, or even less for many of the structural components. The bridge is overdesigned by 200 percent, and some parts of the airplane by only 2 percent. Now, knowing this, will you be more nervous when you fly to Hawaii for a vacation? Actually, there is little cause for concern. More and more, emphasis is being directed today to overestimating expected design loads rather than using a true factor of safety.

How can society be assured that appropriate engineering decisions are being made? Because of people's concerns about the consequences of these decisions, almost a century ago engineers in the United States began to impose design regulations through technical societies. These activities, which first became significant near the end of the nineteenth century, were aimed at standardizing engineering methods, and at creating a controlled professional environment for engineers. These early efforts led to more formalized

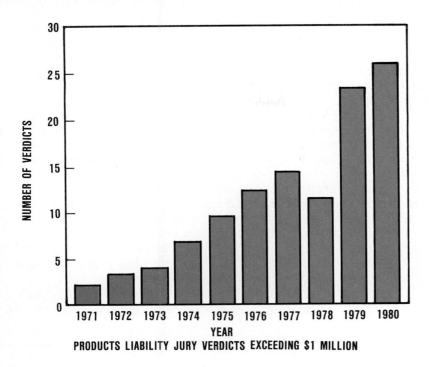

NUMBER OF VERDICTS

YEAR

PRODUCTS LIABILITY JURY VERDICTS EXCEEDING $1 MILLION

[6-5] In the early 1960s, product liability lawsuits comprised only a small portion of the litigation in this country. Then came the deluge, exemplified by a nearly sixfold increase between 1971 and 1976. Between 1974 and 1981, the number of suits grew at an average annual rate of 28 percent, more than three times faster than other civil suits.[1]

regulation through licensing requirements and the formation of ethical codes.

This formalized regulation within the profession was the major method for control of technology until after World War II (1950), when society, becoming increasingly concerned about the impact of technology, began to demand improved performance from those in the engineering profession. This demand took the form of new public laws in the areas of environmental protection, occupational safety and health, and consumer protection. For example, the Occupational Safety and Health Act (OSHA) of 1970 was a significant piece of legislation aimed at providing every working person in the nation safe and healthful working conditions and preserving our human resources. Among the functions included was the authority to prescribe regulations that would protect workers from potentially toxic substances or harmful physical agents that might endanger their safety and health.

By imposing legal controls, society is sending a message that engineers must be held accountable for the consequences of their designs. The degree of this accountability depends on the particular duties of the engineer and the nature of the technical problems being solved. However, there is little doubt that the intent is to hold engineers legally responsible for their actions—or inactions.

One important outcome of this imposition of legal control is the increased emphasis on technical competency and on having a high

Responsibility n.—state of being responsible; that for which any one is responsible; a duty; a charge; an obligation.

Webster's Dictionary

[1] Data from *Personal Injury Valuation Handbooks: Injury Valuation Reports, Current Award Trends,* Vol. 1, No. 258 (Solon, Ohio: Jury Verdict Research, Inc., 1982).

The study of ethics is the study of human action and its moral adequacy.

Kenneth E. Goodpaster, Ethics in Management,
Boston: Harvard Business School, 1984

It is not enough that you should understand about applied science in order that your work may increase man's blessings. Concern for man himself and his fate must always form the chief interest of all technical endeavors, concern for the great unsolved problems of the organization of labor and the distribution of goods—in order that the creations of our mind shall be a blessing and not a curse to mankind. Never forget this in the midst of your diagrams and equations.

Albert Einstein, 1879–1955
Address, California Institute of Technology, 1931.

How can a person talk of pursuing the good before knowing what the good is?

Socrates,[2] 469–399 B.C.

level of current knowledge related to specific assignments. For example, recent court decisions indicate that the engineer is expected to seek out information on all known hazards associated with a particular project *before beginning any design.* The level of responsibility is determined by what a prudent and knowledgeable engineer would have done.

Working in complex organizations, engineers may believe that their decisions have no impact, but the necessary control of technology starts with knowledgeable individuals taking responsible actions. At all levels, judgments are made which have widespread effects.

In many cases the engineer is the first to realize that there may be ecological considerations or perhaps a problem with user safety, and that a design change is needed. Such an early discovery would be very important to reveal because, as the project develops, corrective measures become more difficult to make and costs for changes increase rapidly. Engineers are also frequently asked to give their expert opinions to those who make policy decisions.

ETHICS: A DECISION-MAKING PROCESS

The satisfactory completion of a design project requires that the project engineers work together as a team, each making decisions related to their own design responsibilities but which may also affect the efforts of other team members. Effective team operation also requires good communications, clear assignments, and a recognition of the feelings and needs of others.

Making decisions that have interpersonal, moral, and legal implications can be more complex and difficult for engineers than might be the case for other professionals (such as physicians), because most engineers are not self employed. In fact, over 95 percent of graduate engineers are employees of consulting firms, industrial corporations, or governmental agencies. The result is that in some cases the engineer is faced with a dilemma—individual ethics versus job security. As an engineer, you will have an obligation to protect the well-being of society, act in accordance with a strong moral code, and at the same time be a reliable employee.

To operate effectively in such a system of interrelated decisions, engineers need to base their actions on a code or pattern of behavior that is built around a moral point of view which can be used to decide between good and bad, right or wrong, virtue or vice. Professional engineering societies have developed and are continuing to develop codes of ethics as they make decisions that affect their employers, society in general, and their fellow engineer members. These developed codes provide valuable guidelines, but ethical conduct must always be an individual attribute and a matter of personal conscience [6-6]. It requires a lifelong process of learning and examination to set values that determine responsible and ethical action.

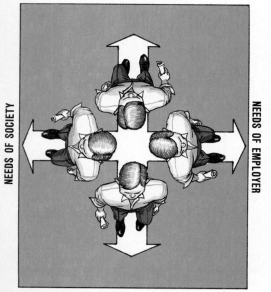

PERSONAL CONSCIENCE

NEEDS OF SOCIETY

NEEDS OF EMPLOYER

PROFESSIONAL ETHICS

[6-6] The engineer is pulled in different directions.

[2]Quoted by L. L. Nash, "Ethics Without the Sermon," *Harvard Business Review,* November–December 1981, p. 79.

Because ethical decisions usually involve relations with other people and affect peoples' lives, they cannot be reduced to an analytical process such as solving a physics problem. However, there is a place for a reasoning pattern in ethics which is built around a moral point of view, and it starts with an ideal as basic as the Golden Rule. Using this moral guide, an engineer can make judgments and decisions with a well-defined set of ethical values. This set of ethical values might include such personal actions as:[3]

- Obey the law.
- Keep promises, contracts, and employment agreements.
- Respect the rights of others.
- Be fair, do not lie or cheat.
- Avoid harming others.
- Prevent harm to others.
- Help others in need.
- Help others in the application of these values.

For many decision situations, a personalized set of values such as the above will be sufficient. Such a value system functions as a pattern of behavior which demands much from the engineer, but it can also work as a base to provide inner satisfaction and be an effective guide as the engineer works with others.

Sometimes, however, there may be conflicts between certain items within the individual set of values, or interpersonal conflicts may arise with people who do not have the same set of values or may attach different weights to the various values.

For example, suppose that an engineer on an automobile design team is responsible for the catalytic converter. The present design frequently fails to meet governmental standards. The engineer has been developing and testing a new technique which shows promise of making the emission control system even more effective than is needed to meet the minimum government emissions standards. The engineer strongly recommends to the project engineer that the improved design be selected. However, the engineer estimates that an additional two weeks would be required to adequately test the new technique for verification of the anticipated results. The supervisor in charge of this phase of the design is anxious to meet the overall design schedule and does not want to delay the project. The supervisor therefore approves the old system design with an emission control device which has proven to be only marginally satisfactory in past applications. The action taken results from a disagreement of values between the two individuals and can create a conflict for the engineer as a course of action is being considered.

In this case, the catalytic converter design engineer may be forced into a second level of ethical decision making where additional judgmental thinking is involved [6-7]. Here the search will be for additional criteria that can be used to select the applicable ethical values, clarify their applicability in this particular circumstance, and

---

Golden Rule

Do unto others as you would have others do unto you.

All people have a special dignity or worth.

It takes a wise person to give the answer to a technical question that involves the conflicting rights and desires of a number of people. Yet the engineer is often required to give such an answer and on very short notice.
Philip L. Alger et al., Ethical Problems in Engineering *(New York: Wiley, 1984)*

---

[3]Kenneth E. Goodpaster, *Ethics in Management* (Boston: Harvard Business School, 1984).

[6-7] Block diagram representing the steps involved in the process of making ethical decisions.

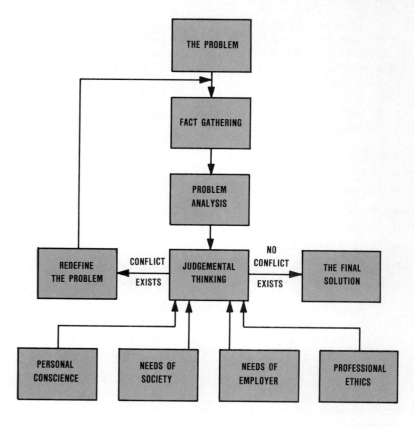

[6-8] Three key areas . . .

Studies indicate that most people will accept voluntary risks in the range of one fatality risk in 1000 (equivalent to driving a car 300,000 miles or smoking 1400 cigarettes in a lifetime.) On the other hand, many people believe and several government agencies require that risk imposed involuntarily must not exceed one fatality in 1,000,000.
Dennis J. Paustenbach, "Risk Assessment and Engineering in the 80's," Mechanical Engineering, *November 1984, p. 55*

resolve any conflicts between the values. Such judgmental thinking requires looking at *three key areas:*

1. What action represents the strongest duty or responsibility?
2. What action most fairly respects the rights of all people involved?
3. What action maximizes all benefits versus all costs?

In struggling to find the proper ethical position in this situation, the engineer should start with the detailed problem analysis and include the extensive fact gathering. To establish neutrality and reduce defensive emotions, an attempt should be made to look at the other side of the problem. (In this example, how does the supervisor see the problem?) Before making a final decision the engineer needs to determine how everybody will be affected by his or her decision. What would you have done?

USING RISK ANALYSIS TO AID IN MAKING ETHICAL DECISIONS

All people in their daily lives are regularly involved in activities that require weighing benefits against possible risks. They analyze the consequences and then proceed with risk-taking actions such as driving a car, sky diving, surfing, smoking, skiing, or even overeating.

Similarly, engineers in their professional capacity make risk/benefit decisions. Although these two types of decisions require similar thinking processes to arrive at logically selected actions, there is a major difference in the decision outcomes. In one case, people voluntarily accept actions involving a self-determined degree of risk. In the other case, engineers make judgments about the degree of risk to which someone else will be involuntarily or unknowingly exposed. Engineering decisions such as determining safety factors in a bridge design, or deciding on the location of a waste disposal facility, directly expose people to risk.

The far-reaching effects of engineering decisions involving involuntary risks and the complexity of today's risk-related problems emphasize the need for carefully making reasonable and ethical judgments. This requires a systematic risk analysis process which includes anticipating hazards, determining the degree of acceptable risk, and then selecting the best risk/benefit alternative. The various steps in the risk analysis process can be grouped into two major areas: *risk assessment*, which includes the steps required to determine the numeric probability of an adverse effect, and *risk management*, which includes the steps required to analyze risk consequences and implement selected action.[4]

Risk Assessment

The end result of risk assessment is an expression of risk that defines the probability of harm which could result from a given action or situation. This could range from an expression that states the risk associated with placing wastes in a selected disposal site, to a number that defines the reliability of equipment design. The expression should combine all elements included in the risk assessment, such as extent or concentration of the hazard, effect or consequence of exposure to the hazard, and possibility of exposure. All of these factors are combined to describe the nature and magnitude of the risk along with all uncertainties associated with the entire process. Uncertainties can result from the manner in which the process is implemented, or basing results on unreliable or insufficient data.

When involved in assessing risks, engineers must carefully implement the process and allow for possible inadequacy of information and test data. They must recognize that assessment cannot be based on blind dependence on what is assumed to be good information or objective data. Judgments must be made concerning evaluation and application of the data. There may be an inner tension created by the gap between what is known and what must be assumed.

One risk element that often falls in the gap between what is known and what must be assumed is the possibility of human failure. When assessing risks, engineers need to consider carefully how possible human errors could cause failures or compound the consequences of design failures. An example of how an improper human reaction to an equipment failure can create serious consequences is

[6-9] Surfing, like manned exploration of space, involves risk.

[6-10] Risk analysis.

[4]A. Alan Moghissi, "Risk Management—Practices and Prospects," *Mechanical Engineering*, November 1984.

Two Key Questions for Engineers in Risk Analysis

■ Does a significant risk exist?
■ Are there feasible alternatives for avoiding, preventing, or decreasing risk?

Engineers must be trained in risk assessment and must be equipped to confront ethical dilemmas as they appear.

Lay Wilkinson, President of Union Carbide Corp.
Engineering Times, December 1984

Typical Expressions of Risk

■ Certain radiation emission from a nuclear power plant equals specified health risk.
■ Exposure to specified concentration of benzene over a 70-year lifetime equals specified added fatality risk.

As far as we know, our computer has never had an undetected error.

Conrad H. Weisert, Union Carbide Corporation
Quoted by OMNI, September 1979

All substances are poisons. The right dose differentiates a poison and a remedy.

Philippus Paracelsus, 1493–1541

illustrated in the poison-gas leak that occurred at a Union Carbide Corp. pesticide plant in Bhopal, India, during the night of December 4, 1984, causing the world's worst industrial accident.[5] Deadly methyl isocyanate gas diffused undetected through this city of over 600,000 people for over 40 minutes. Over 1700 people were killed and 300,000 others were left with long-term serious injuries. What went wrong, the equipment or the people?

Subsequent investigations of this accident revealed that the gas leak was caused by a large increase in pressure inside one of three storage tanks that contained the liquefied gas which was used to manufacture insecticides. When the pressure became too intense, a valve opened, allowing gas to escape into a "scrubber" mechanism. This was a safety feature designed to neutralize any leaking gas and vent it as harmless gas into the atmosphere. Apparently, one of the scrubbers was not operable, due to a maintenance problem. The pressure of the operating scrubber was so high and the flow of gas so fast that the "scrubber" could not neutralize the gas *before* it escaped into the atmosphere. Backup safety procedures called for two workers to be assigned to monitor the tank gas pressure and to cool the tank by spraying it with water whenever the escape valve opened. However, the workers panicked, ran away, and failed to cool the tank. This allowed the deadly fumes to escape for over 45 minutes before supervisors with protective gas masks could arrive and cool the tank. The government of India has charged that the plant's designers are at fault because "the plant's design and manufacturing controls were not foolproof to ensure absolute avoidance of violent chemical reactions and leakage of toxic gases." The company contends that "the workers at the Bhopal plant were poorly trained and failed to prevent the accident." Who do you think was responsible for this disaster? Do you believe there were elements of risk which were not properly considered in the risk assessment process?

Risk Management

Using the risk value obtained from the risk assessment procedure to decide what action to take is called *risk management*. It involves weighing the calculated risk against benefits, desired results, or requirements and then selecting a cost-effective alternative to achieve the best risk/benefit ratio. Safety, physical, and chemical risks are all present to some degree in the work environment. It is the designer's responsibility to see that these risks are reduced to the lowest level possible.

In some cases, risk management may involve weighing a calculated reliability or risk value against a criterion of acceptability or standards established by industry, government (OSHA), or a certifying agency. The risk management process is based on the premise that no action is totally risk free; there are no rewards without risks, no operating systems without potential failures.

---

[5]This unfortunate catastrophe is described by the authors using the most current Associated Press accounts available at the time of this text's publication. It is likely that the matter will be in litigation for several years. Also, the long-term effects of the atmospheric pollutants could significantly increase the reported casualties.

The improper perception of risk by the people affected by the risk is a serious obstacle to effective risk management. Society tends to perceive risks in two broad categories: high probability/low consequences and low probability/high consequences. A typical high probability/low consequence action is cigarette smoking, which has been proven to result in a high percentage of lung cancer. However, because cancer appears many years after the smoker starts to smoke, and the smoker is the prime victim, society has been slow to impose even minimum restrictions. A typical low probability/high consequence action is the generation of energy with nuclear reactors. Society reacts to this action with extensive and costly regulations. NASA's loss of the space shuttle Challenger has been viewed as a national disaster because the previous 24 successful flights had given the public a false impression that little risk was involved.

Another major factor in the risk management process is the impact of financial costs. This can involve determining costs related to taking required actions, costs involved in various trade-offs, and costs involved if no actions are taken to reduce the risks. Lost along with the $1.2 billion space craft Challenger were a $100 million communications satellite and a smaller $10 million payload that was to have studied Halley's comet.

In disasters such as that which occurred in Bhopal, India, management may be considered to be personally liable and be required to pay the consequences. Immediately after the Bhopal disaster, five plant officials were arrested on negligence charges (they were eventually released) and everyone involved will have to live with the uneasy concern that a more realistic assessment of risk could have averted the disaster. Corporate management makes key economic cost/benefit decisions and then expects the engineer to use his or her technical expertise to make certain that these decisions, along with other risk/benefit decisions, are properly implemented.

MAJOR TECHNOLOGICAL AREAS REQUIRING ETHICAL DECISIONS

Atmospheric Contamination

A major challenge for technology today is to provide for society's increasing demand for energy without polluting the air that we must have to live. Engineers functioning in their jobs are faced with ethical decisions related to this technological challenge.

Air pollution poses a special challenge for engineers because it is a complex problem resulting from an interaction of individual pollutants that originate from a combination of sources. Coal-burning power plants, smelters, and motor vehicles all produce pollutants which can individually cause severe damage or can interact with other pollutants or substances in the atmosphere to form still another damaging agent. For example, hydrocarbons and nitrogen oxides produced mainly by gasoline-powered vehicles combine in the presence of sunlight to produce ozone, a known cause of damage to crops, forests, and human health. As discussed in Chapter 2, carbon dioxide, a direct by-product of all fossil-fuel combustion, is possibly

[6]*Emergency Medicine,* March 1970, p. 61.

LOCAL INJURY
erythema   depilation   vesiculation
necrosis   gangrene

ACUTE RADIATION SYNDROME
nausea   vomiting   diarrhea
anemia   leukopenia

CHRONIC INJURY
anemia   leukemia   cataracts
neoplasia
shortening of life span (?)
genetic mutation (?)

HOW TO HANDLE A RADIATION VICTIM

BEGIN EMERGENCY PROCEDURES IMMEDIATELY
resuscitate and stabilize patient
get detailed history
give symptomatic-treatment of systemic and skin reactions

DETERMINE INTENSITY OF IRRADIATION AS SOON AS POSSIBLE
map the involved anatomic area

DECONTAMINATE PATIENT
remove clothing and store for analysis
remove penetrating missiles
clean wounds surgically and seal with plastic
wash—do not shower—patient
shampoo and cut—do not shave—hair

PREPARE PATIENT FOR EVACUATION TO RADIATION MEDICAL CENTER
dress   in hospital gown
wrap   in blankets
shield   with plastic sheet

[6-11] What exposure to radiation can do.[6]

Recent studies warn that within 60–80 years the carbon dioxide concentration will be twice existing levels. This higher concentration is expected to cause a global temperature rise sufficient to raise sea levels, diminish water supplies, and alter rainfall patterns.
Sandra Pastel
Air Pollution, Acid Rain and the Future of Forests
Washington, D.C.: Worldwatch Institute, 1984

TABLE 6-1   SOURCES OF SULFUR AND NITROGEN OXIDE EMISSIONS IN THE UNITED STATES

Source	Percent of Total	
	Sulfur Dioxide	Nitrogen Oxides
Electric utilities	66	29
General industries	22	22
Smelters	6	1
Motor vehicles	3	44
Homes, businesses	3	4

Source: Adapted from Sandra Pastel, *Air Pollution, Acid Rain and the Future of Forests* (Washington, D.C.: Worldwatch Institute, 1984), pp. 15–16.

Because acid rain, ozone and the build-up of carbon dioxide have common origins, they can have common solutions.
Sandra Pastel
Air Pollution, Acid Rain and the Future of Forests
Washington, D.C.: Worldwatch Institute, 1984

a long-term serious pollutant because of its heavy concentration and potential for causing extensive worldwide climate change.

Sulfur and nitrogen oxides emitted during the burning of fossil fuels and the smelting of metallic ores, unlike carbon dioxide, do not stay in the atmosphere, but eventually return to the earth in some form to produce their own damaging effects (Table 6-1). Some stay as gases in the region of the pollutant source, where they are visible in the haze that hangs persistently over many cities. Some are deposited in dry form on surfaces such as leaves or needles, where they react with moisture to form damaging acids. Some of the sulfur and nitric oxides combine with water vapor existing in the air to form sulfuric and nitric acids, which are considered to be the major constituents of acid rain, a newly recognized pollutant now being intensely studied and evaluated because of its potential for worldwide environmental damage.

Environmental specialists, industrialists, and legislators continue to debate about the causes of acid rain and the need for corrective action, while the pollution effects continue to accumulate. It is estimated that at least 180 lakes in the Adirondack Mountains in New York have become so acidic in content that fish and almost all other forms of aquatic life can no longer survive. Over 100 miles of streams in the Monongahela National Forest of West Virginia are in a similar condition. Killing fish and other aquatic life may not be the only effect of acid rain; cattle and game animals that eat waterside vegetation and birds that feed on fish and aquatic insects may also be affected by acidic food or even experience a total loss of food sources. Damage to trees in the United States is still confined to a few species at high altitudes in localized areas, but the Black Forest in Germany has sustained damage in many species, in various types of soils, and in a variety of altitudes. Scientists in both Germany and the United States are continuing to study both the extent of direct damage and tree rings to determine the short-term and long-term effects of acid rain on forests.[7]

Although we do not fully understand all of the reasons why forest destruction is occurring, the effects of acid rain and other

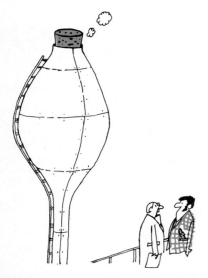

[6-12] ". . .quite frankly Johnston when I asked you to solve the emission problem from that stack, I expected a somewhat more sophisticated solution. . ."

[7]Bob Thomas, "Acid Rain," *Arizona Republic,* June 1983.

pollutants are apparently stressing sensitive forests beyond their ability to survive. Weakened in this way, many trees lose their resistance to natural calamities such as drought, insect attacks, and frost. In some cases the pollutants alone cause direct injury or decline in growth. The mechanisms are complex and may take decades of additional research to provide us with a complete understanding.[8]

In 1963, the U.S. government passed the Clean Air Act, which recognized that since the largest portion of air pollution is caused by urban centers that cross state borders, federal financial assistance and leadership is essential.

At various times since 1967 the law has been amended to emphasize research efforts, establish national emission standards for motor vehicles, establish economic approaches for controlling pollution, and allow citizen action suits against the Environmental Protection Agency (EPA, the agency designated to implement and enforce the law).[9]

Although this law provides a basis for national action, its effectiveness depends on responsible action by engineers and other technical professions. This responsible action must start with research to discover or verify the complex mechanisms producing the damage. This is especially important in the case of acid rain, where so many possible sources and pollutants are involved. A related area of research is the investigation of technologies that could be used to reduce pollution produced by combustion.

Another important responsibility of engineers related to solving problems of this nature is to clarify and translate technical information into a form that is understandable to the decision makers in order that it can be used as a basis for judgment. Decisions related to the control of air pollution are particularly difficult to make because they require a close cooperation between states, regions, and even nations. For example, erecting a high smokestack to reduce localized pollution in one area may result in gases being carried downwind causing acid rain pollution many hundreds of miles away. Also, some regions or nations may be reluctant to clean up their own area if pollution continues to come in from neighbors who have taken no action. For example, pollutants originating from electric utilities and smelters in the coal-using states of Ohio, Indiana and Illinois are thought to contribute to the extensive damage throughout Ontario, Canada, and into Nova Scotia.[10] The smelters at Copper Cliffs, Ontario, Canada, also contribute greatly to this problem.

In addition to the complexity of causation and wide-ranging effects of air pollution, economic factors also bring about their own problems for the decision makers. On one side they might hear, "Implementing this change to reduce pollution will increase the cost of the end product." On the other side they might also hear, "Environmental damage is a cost which should be included in the price

---

[8]Sandra Pastel, *Air Pollution, Acid Rain and the Future of Forests* (Washington, D.C.: Worldwide Institute, 1984), p. 7.

[9]*Environment Reporter (Federal Laws)* (Washington, D.C.: The Bureau of National Affairs, Inc., 1984).

[10]Sandra Pastel, "Protecting Forests from Air Pollution and Acid Rain," in *State of the World, 1985* (New York: W. W. Norton, 1985).

[6-13] One consequence of the drive to purify urban air over the last two decades has been the construction of tall smokestacks, to better disperse pollutants into the atmosphere. These smokestacks, along with high levels of emissions, sent pollutants traveling hundreds of kilometers before returning to the earth's land and waters.

[6-14] How much is clean air worth?

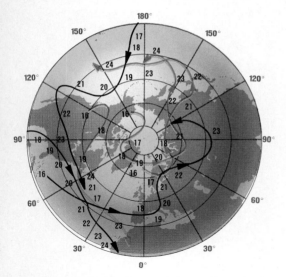

[6-15] Movement of air around the world can be determined by identifying specific air parcels and tracking them. The tracks shown above were identified by Edwin Danielsen, National Center for Atmospheric Research. The parcels in the troposphere are shown in black and those in the stratosphere are shown in color. The numbers represent successive days in April, 1964.

that consumers pay. Those creating the pollution must pay for it." Which advice is correct? An amendment to the Clean Air Act passed in 1984 provides for economic approaches to controlling air pollution, such as incentives for positive actions.

However, the engineer designing systems involving any aspect of pollution control or who may be placed in a position of evaluating various solutions, will be faced with using economic factors in the decision-making process.

In addition to making technical decisions related to their job responsibilities, engineers have an opportunity to respond to pollution problems as interested, responsible, and knowledgeable citizens.

The Clean Air Act states that any person may commence a civil action against the EPA where there is alleged to be a failure to perform any act or duty required under the act. The "What would you do?" case below is taken from an actual EPA case. In the actual situation, EPA allowed the manufacturer to proceed with their proposed alternative plan. However, several individuals have sued to force the EPA to order a recall of the inferior equipment.

Chemical Contamination of Underground Water Supplies

Pollution of groundwater caused by toxic chemicals leaking from waste storage sites is also rapidly becoming a major environmental concern in this country [6-16 and 6-17]. It affects every state in the

An Example of a Decision in Engineering Ethics: What Would You Do?

When engines exceed pollution control standards, *recall* is the only remedy outlined in the Clean Air Act.
U.S. Appeals Court Ruling, October 26, 1984

Court mandates recall of vehicles for emission-standards violations.

Headline
Arizona Republic, *October 27, 1984*

Assume that you are a chemical engineer working for the Environmental Protection Agency responsible for investigating motor vehicle engine pollution control standards. You have found that one group of engines produced by a major automobile manufacturer for the current model of automobiles now being sold exceeds pollution control standards. You report your findings to your superior, who notifies the manufacturer to determine the direction of corrective recall action that will be implemented.

The manufacturer submits a plan to you proposing to leave the current engines alone, but to make sure that next year's models would be especially designed for lower emissions to make up for the excessive levels emitted by the current models. It is argued that their plan would save them $11.8 million and in addition would save the owners $25.8 million in fuel costs because the repairs necessary to improve emission performance would increase fuel consumption. They also argue that their proposal would be better for the environment because emissions from all of next year's models would be below the standard, while not all of the current recalled models would be brought in for repair.[11]

The proposal from the manufacturer comes to you for your recommendation. What do you recommend: an acceptance of the manufacturer's proposal, or initiation of immediate recall action? Include your reasons in your recommendation.

[11]Adapted from "Court Mandates Recall of Vehicles for Emission-Standard Violations," *Arizona Republic,* October 27, 1984.

nation and threatens the water supply of half of the nation's population. It is especially dangerous because its effects are gradual and less apparent than the effects of other pollutants, such as smog or the sight of dead trees caused by acid rain.

Pollution of underground water supplies can confront an engineer with ethical considerations similar to those related to the acid rain problem discussed above. Like acid rain, water pollution is a complex problem resulting from many types of pollutants emanating from a wide range of sources. Water-contaminating chemicals, which can cause cancer and other less traumatic illnesses, vary from pesticides originating in agricultural areas to volatile organic compounds that often are used in high-tech industries. Also, like acid rain, these contaminants can affect wide areas many miles distant from the pollutant source. The resources often affected are huge underground water reservoirs, called *aquifers,* that underlie much of the country. These aquifers now provide water for half of the nation, including more than 80 percent of the rural population which live in areas previously considered free from possible pollution sources.

Similar to the problems associated with acid rain, laws and regulations alone have proven to be inadequate for solving the problem. There is a need for action by responsible and knowledgeable people at key points in the decision making process. For example, a federal law[12] passed in 1976 aimed at controlling toxic wastes requires operators of the thousands of waste facilities around the country to monitor leakage from their installations. The law allows for issuance of an interim permit that will be valid until a monitoring system can be established. However, the final permit that would be issued after monitoring is established would require more stringent and usually more expensive anticontamination efforts. Therefore, many operators postpone establishing monitoring equipment until forced to do so by a court order. The U.S. Government Environmental Protection Agency has delegated responsibility for administering and enforcing this program to the state agencies, which in many cases have been unable or unwilling to force the operators to obey the law.

In addition to making decisions concerning the disposal of wastes that cause water pollution, engineers can play a key role in making decisions at the point where the wastes are generated. Organizations generating the waste know more than anyone else about their wastes and the processes that generate it. Therefore, they can be most effective in investigating possibilities for reducing the toxicity and quantity of their wastes. One way to begin is by investigating possibilities for in-plant process changes. An example of a possible process change is in the steel-finishing industry, where before steel products are painted, they must be treated to remove rust. The most commonly used method is to dip the steel in an acid bath. This generates large amounts of waste acid. Most of the rust, however, could be removed mechanically, thereby reducing the volume of acid required.[13] Another area for investigation at the source could be the possibility of recycling or reuse of materials to reduce wastes.

---

[12]Toxic Substances Control Act (15 USC 2601-2629), October 11, 1976.

[13]David Anderson and Beth Fentrup, "How Should We Dispose of Hazardous Wastes?" *Civil Engineering,* April 1984, p. 42.

The solution to a coming water crisis may be more elusive and expensive than the energy crisis.
Representative Mike Synar, Oklahoma, Chairman of House Government Operations Environment Subcommittee Quoted in Arizona Republic, *October 26, 1984*

[6-16] In the past, municipal water supplies were subject to contamination because of improper disposition of animal wastes . . .

[6-17] . . . today the culprit is more likely to be improperly discarded man-made chemicals.

Contamination of underground water supplies is bad and getting worse because state and federal laws and programs do not provide sufficient protection.
*Congressional Office of Technology Assessment
Quoted in* Arizona Republic, *October 26, 1984*

**WATER-TAINT MONITORING 'HAS FAILED'
TOXIC-DUMP CHECKS FLAWED, EPA ADMITS**

By PHILIP SHABECOFF

New York Times

WASHINGTON—The key government program to monitor toxic waste contamination of underground water supplies is not working, according to a report drafted by Environmental Protection Agency officials.

The report says that a large majority of site operators are not doing the job as required, that many of the states have proved unable or unwilling to make the operators obey the law and that the agency itself had been deficient in overseeing the states and in ensuring that standards for protecting the water supplies are met.

Arizona Republic, *October 16, 1984*

The cheapest way of disposing of waste is to dig a pit and throw the waste in it. The rationale for such facilities is that dilution is the solution to pollution. You mix a little waste with a lot of clean air, rainwater or groundwater and your waste problems are solved. But dilution only gives the illusion of being a solution.
David Anderson and Beth Fentrup, "How Should We Dispose of Hazardous Wastes?" Civil Engineering, *April 1984, p. 43*

Another decision area for engineers is in the selection of the most effective disposal method for the remaining waste. Such a decision may often involve research and special technical investigation, and in general keeping current on new techniques and procedures. It may also involve resisting pressure to take the path of least resistance or of lowest cost instead of the ethical path.

CONTROLLING OCCUPATIONAL HEALTH AND SAFETY HAZARDS

A demanding challenge for engineers today is helping society control the effects of toxic agents on people working in various industrial environments. The effects of industrial toxic agents maybe more apparent and intense than toxic agents in the general environment because workers may be exposed to higher concentrations over long and continuous periods. Unless controlled, this continuous exposure could result in occupational illnesses ranging from skin disorders to possibly cancer, depending on the extent of exposure to the toxic agent. The number of agents that could harm people working in various industrial environments is large and growing. The National Institute for Occupational Safety and Health has developed a registry that lists over 60,000 toxic substances which are manufactured or used in some workplaces. Within this group, they have identified approximately 1500 suspected cancer-causing agents.[14]

Increased awareness of the number of toxic agents and their effects has resulted in an increasing demand for protection. Since 1965, the U.S. government has enacted over 30 pieces of legislation aimed at controlling the use of toxic substances [6-18]. Similar legislation has been introduced in other industrialized countries.[15]

[14]Mary Jane Bolle, *Effectiveness of the Occupational Safety and Health Act: Data and Measurement Problems* (Washington, D.C.: Congressional Research Service, Library of Congress, 1984).

[15]Dennis J. Paustenback, "Occupational Safety—Are You Professional?" *Mechanical Engineering*, March 1984, p. 80.

What Would You Do?

Assume that you are an engineer working for a high-tech company that makes extensive use of a trichloroethylene (TCE), an industrial solvent that causes cancer and affects the liver, kidneys, and nervous system. Your company has established a storage site and you are responsible for designing a monitoring system for detecting possible leakage. You submitted your design to your supervisor for approval two months ago, but to date have not received any indication that the corrective design solution has been reviewed. Two weeks ago when you questioned your supervisor about the status, he said it was still in the "approval cycle." What will you do now? (If you decide to wait another two weeks and again find that no decision has been made, then what will you do?)

By enacting this legislation, society is saying that the health and safety aspects of any project should concern the engineer as much as its efficiency, timeliness, overall quality, or cost. To implement the legislation, industrial management looks to engineering to recommend the most cost-effective alternative for meeting the requirements for minimizing health risks.

Typical recommendations could involve the substitution of a less hazardous substance, changing a processing technique, selection of a control technique, or selection of various types of personal protection equipment. Engineers would also be involved in the creation of safety and monitoring procedures.

A major difficulty in controlling occupational health hazards is the uncertainty of information concerning both the hazards and the control techniques. Most occupational illnesses are clinically indistinguishable from general chronic-type diseases, and diseases such as cancer may develop long after the worker leaves the hazardous work environment. Also, because most of the data concerning effects of toxic substances is based primarily on animal tests, it may not be applicable for controlling human diseases. This uncertainty concerning hazards results in an uncertainty concerning the relative effects of various control techniques and procedures.

Today's engineers, however, cannot hide behind this possible lack of reliable information. Design engineers must know occupational and environmental exposure limits and how to design their equipment to meet or exceed these limits. This requires also that the engineer work closely with health and other engineering personnel and to be familiar with current technical literature on the subject.

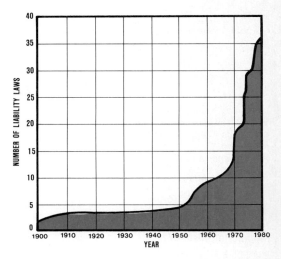

[6-18] The increase in the number of federal laws regulating hazardous materials in the United States over the past 20 years has been dramatic.

The science of preventive medicine involves a study of its causes and prevention. The knowledge and application of preventive measures takes us into regions that are engineering rather than medical.

E. B. Phelps et al.
Public Health Engineering, *Vol. 1 (New York: Wiley, 1984)*

What Would You Do?

Assume that you are an engineer working for the Environmental Protection Agency and are responsible for cleaning out a site that has been cited as a major pollution source in the immediate area. Cleaning the site requires moving contaminated materials to a new site in an adjoining state. People living in the area around the new site are registering major objections to this move and local authorities there claim they have test data which show that the new site is already a pollution problem. Authorities in your agency claim that the new site is now safe and that the added contaminants will not create a problem. What will you do?

- Assume the information from your agency is correct and proceed with the cleanup as planned?
- Obtain information available from the new site to determine if moving new material into the site will create new problems or add to an already existing problem?

If you do investigate and determine that moving contaminants have the potential of creating additional pollution problems at the new site, but you also calculate that the overall potential dangers would be much less at the new site than those existing at the site being cleaned, what action would you take?

Caveat venditor (let the seller beware) is replacing the old adage *caveat emptor* (let the buyer beware).
Kenneth E. Goodpaster,
Ethics in Management
(Boston: Harvard Business School, 1984)

Today, it is reasonable to expect that if a manufacturer sells his product, he will be strictly liable in almost every circumstance should a user be injured and elect to sue to recover damages.
Verne L. Roberts
1984 Institute for Product Safety, Durham, N.C.

Intentional Misuse???

In 1979, a 14-year-old boy died after intentionally inhaling the freon propellant from a can of Pam (an aerosol sprayed on cooking pots to prevent sticking) in order to experience a tingling sensation in the lungs. The can carried a warning: "Avoid direct inhalation of concentrated vapors. Keep out of reach of children."

The boy's mother in suing the manufacturer of Pam, maintained that this warning was inadequate, since the company knew that 45 other teenagers had previously died from inhaling the fumes.

The jury, ignoring the fact that the product had been deliberately misused, awarded her $585,000.
"At issue: Product Liability," Shell News,
Vol. 52, No. 4 (1984)

Two 19th century reports from the Caribbean describe the ready availability of an axe in every sugar mill to amputate a slave's arm should it be caught in the in-running nip point of the rollers used to crush the sugar cane.
C. G. A. Oldendorp, Der Geschicht der Mission der Evangelische Bruder von Inseln S. Thomas, S. Croix, S. Jan, Barby Johann Jacob Bossart, 2 vols., 1777

At the heart of this problem is a changing focus on the liability statutes. At one time, the issue was one of moral responsibility—had the maker or seller of a product been negligent in its construction or sale? Now the changing focus has turned the spotlight on the product itself—is there a condition associated with the product that creates an unnecessary hazard or danger? This switch in emphasis led to the concept of "strict liability," which allows courts and juries to assign liability even though there has been no *fault* or lapse of moral responsibility. Liability can now result if the court deems that a product's design, construction, operating instructions or safety warnings make its use unreasonably hazardous.
Shell News, Vol. 52, No. 4 (1984), p. 18

Society's Demand for Safety

Engineers today work within a society which expects technology to provide all that is needed for an improved life-style and at the same time demands a high degree of protection to accompany this luxury. This demand for safety in everything from consumer products to airplane travel takes the form of regulations, codes, and laws aimed at controlling all of the design, production, construction, and service functions required to provide the good life.

The emphasis on product safety has increased rapidly during the past 30 years. Consumerism, a social movement emphasizing the rights of buyers and users, has resulted in the passage of a Consumer Product Safety Act which holds the manufacturer liable if the product contains a flaw (arising from production or design) that makes it unreasonably dangerous. Working with this legislation, courts have moved toward strict liability, holding manufacturers responsible for any product defect that results in an injury.

Concurrently with increasing emphasis on safety for consumer products, there is also increasing emphasis with regard to safety for products and machinery used in an industrial setting. Pressure for improved machinery design and safety procedures has come from various sources, including the worker, safety and health groups, and employers who recognize that a safe working environment has many benefits, including improved production. This recognition of the value of safety is a major change from the approach used in early industrial plants, where workers were considered expendable and the major objective was to make certain that injuries did not interfere with production.

Machine safety design requirements are based on codes and regulations that emphasize identification and elimination of hazards rather than specifying safety requirements for the use of each specific machine. The codes emphasize designing to make certain that machine operators cannot proceed with their work unless the required safety conditions prevail [6-19].

Another area of safety emphasis is public transportation, especially automobiles and airplanes. Consumer advocate organizations, such as the privately funded Center For Auto Safety, have been very active in pressuring for required safety legislation and monitoring performance. The major government agency involved in regulating automobile safety is the National Highway Traffic Safety Administration (NHTSA) of the Department of Transportation. The major emphasis in airplane safety today is for all accidents to be "survivable." Airplane accidents are termed survivable if the fuselage is not severely damaged on impact or if there is at least one survivor or if exit from the plane is possible. Because most of the fatalities with this type of accident are fire related, safety research is being aimed at techniques or materials which could prevent fires or reduce their effects. This involves fuel additives, protective face hoods, interior materials, and seat construction. It is a challenging area with many performance, cost, and safety trade-offs.

Meeting the requirements for safety in areas such as consumer products, industrial machinery, transportation, and construction continues to be a challenge, even with the existing increased empha-

TABLE 6-2 INJURY RATES BY INDUSTRY FOR 1982 (RATE PER 100 FULL-TIME WORKERS)

Construction	14.5
Agriculture	11.3
Manufacturing	9.9
Transportation	8.4
Trade (wholesale and retail)	7.1
Services	4.8

Source: Mary Jane Bolle, *Effectiveness of the Occupational Safety and Health Act: Data and Measurement Problems* (Washington, D.C.: Library of Congress, 1984).

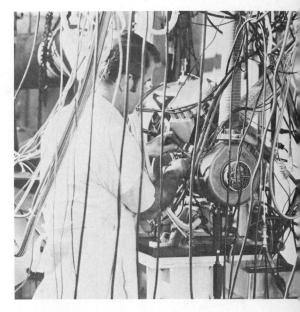

[6-19] The safety of the workplace must be a concern for members of the technical team.

sis and regulations. During 1983, according to the National Transportation Safety Board, 43,000 people were killed on U.S. highways. Government statistics indicate that each year approximately 30,000 Americans are killed as a result of accidents related to faulty consumer products,[16] and injuries continue at unsatisfactory rates in most areas of industry (Table 6-2).

Safety problems will continue and could even become more severe as technology becomes more complex and the increase in population results in added demands for products and services.

Society expects engineers to play a major role in meeting the requirements for safety. Along with an ethical responsibility, meeting safety requirements is also a serious legal responsibility. As suggested above, safety legislation and governmental regulations are tending to hold engineers legally liable for the consequences of their designs. Any design flaw that contributes to an injury could lead to legal claims against the engineers involved with designing or developing the product [6-20].

DEVELOPING AN ENGINEERING RESPONSE TO THE DEMAND FOR SAFETY

Use Risk Analysis to Select the Best Alternative

To provide a safe product or project, engineering teams must use a systematic risk analysis procedure which starts by identifying all potential hazards [6-10]. This requires determining where the product will be used, how it will be used, and who will use it. The identified hazards can then be used along with applicable standards, regulations, and codes to determine the basic safety requirements. These are weighed against other factors, such as cost, schedule, and technical feasibility, to determine the most feasible alternative. How this selection was made is a critical consideration in product liability cases. If there is a safety problem, the outcome of any lawsuit could depend on whether another safer alternative was available to the designers.

[6-20] Along with the ethical responsibility, meeting safety requirements is also a serious legal responsibility.

[16]Kenneth E. Goodpaster, *Ethics in Management* (Boston: Harvard Business School, 1984), p. 107.

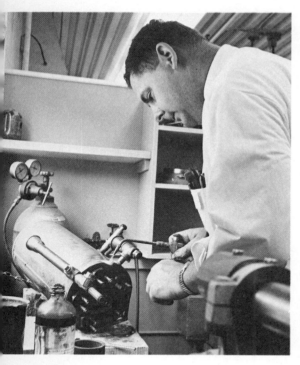

[6-21] The worker's eyes and hands should be protected.

[6-22] Protective eye shields and protective clothing can do much to reduce injury.

Example 6-1

An example of how the selection of a design alternative affects the safety record of a product is the Pinto, a subcompact automobile manufactured by Ford for six years starting in 1970. Pintos were sold in spite of recognized safety problems which ultimately resulted in a costly recall program.

An early design decision concerned the selection of the location for the gas tank to achieve maximum safety. The main hazard to be considered was the possibility of fires caused by rear-end collisions of the vehicle. The design team selected a location under the rear floor and behind the rear axle, which was the normal location in most other automobiles at that time. A location above the rear axle was considered but not selected, because it would increase the threat of fire in the passenger compartment. Also, it required installation of a filler pipe which was more apt to be damaged during a collision. In addition, an over-the-axle location would not be suitable for station wagons and hatchback models which were to be considered for production after the Pinto two-door sedan (the first model to be designed) was in production.

A major problem in selecting this location and designing for maximum fuel-system safety was the lack of adequate and consistent safety standards concerning rear-end collisions. Ford originally designed the tank in accordance with an internal company standard, referred to as "a 20-mph fixed barrier regulation." The National Highway Traffic Safety Administration (NHTSA) proposed an alternate rear-end fuel system integrity standard approximately 18 months after the Pinto design program had been started. Although not strictly required to do so, Ford tested against this standard and altered the fuel tank design before production began. The NHTSA standard was not fully implemented until six years after the first Pinto was produced.

The final enactment of the NHTSA standard resulted from pressure by consumer groups and Congress who were becoming increasingly concerned about safety problems with the Pinto. Rear-end collisions were causing fires which resulted in serious injuries and, in several cases, fatalities. In the six years from proposing the new standard to enacting it, NHTSA was hampered by the lack of relevant and meaningful statistical information which could be used to specifically identify hazards and serious consequential results.

After years of defending the safety of the Pinto, Ford finally conceded that NHTSA had identified specific design features which needed to be changed to reduce the risk of gas leakage caused by rear-end collisions. They agreed to recall over 1.5 million automobiles for replacement of the fuel filler pipe and installation of a shield across the front of the fuel tank.

Ford's estimate for the cost of this recall program was over $20 million. Table 6-3 and Figure [6-23] summarize this example.

Remember the User

When designing for safety, engineers should be aware of errors the user can make as well as errors the designers can make. In addition to being an ethical responsibility, preventing user errors or reducing their effects is becoming a very serious legal requirement.

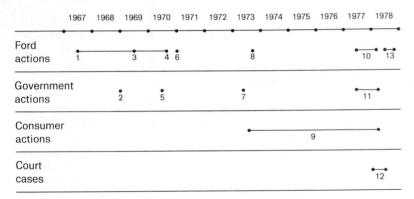

[6-23] Story of the Ford Pinto.

Procedural Problems: Rolls-Royce, the luxury car manufacturer, was forced to recall 2000 Silver Shadows in 1978 because one owner reported the brakes had failed.

The *Almanac of Investments* reports the company claimed its autos never broke down, they merely "failed to proceed."

It is an easy task to formulate a plan of accident preventing devices after the harm is done, but the wise engineer foresees the possible dangers ahead and embodies all the necessary means of safety in the original design.

J. H. Cooper, "*Accident-Preventive Devices Applied to Machines,*" ASME Transactions, Vol. 12 (1891), p. 249

TABLE 6-3   STORY OF THE FORD PINTO

1. *June 1967:* Ford starts design and development of the Pinto.
   - Assembles a special team of engineers dedicated to development of a Pinto model within 40 months.
   - Establishes management design constraints: must not weigh an ounce over 2000 pounds; not cost a cent over $2000.
   - Makes key decision after trade-off studies: locate gas tank under rear floor, behind the rear axle. (No government standards on gas tank design existed during the early design stage.)

2. *January 1969:* A government agency (National Highway Traffic Safety Administration—NHTSA) proposes the first rear-end integrity standard.
   - Requires that a stationary vehicle should leak less than 1 ounce of fuel per minute after being hit by a 4000-pound barrier moving at 20 mph (called the 20-mph moving barrier).

3. *June 1969:* Ford tests and alters the Pinto design to meet the proposed government standard.

4. *August 1970:* First Pinto rolls off the assembly line and sales begin.
   - The vehicle meets the schedule, weight, and cost goals.

5. *August 1970 (after production of the first Pinto):* NHTSA proposes new requirements.
   - Calls for changes from the existing 20-mph moving-barrier standard to a more severe fixed-barrier standard.
   - Indicates a possibility of a long-term requirement for a 30-mph fixed-barrier standard.

6. *August 1970–January 1971:* Ford considers whether to test vehicles against the existing or the proposed governmental standard.
   - Decides to continue designing to meet the original 20-mph moving-barrier standard.
   - Decides against making gas tank modifications to meet a possible future 30-mph fixed-barrier standard.

7. *August 1973:* NHTSA proposes a new 30-mph moving-barrier standard effective 1976 for all 1977 models.
   - Also adopts a fuel system standard applicable to rollover accidents.

8. *Late 1973:* Ford agrees to modify rollover standards after testing and making controversial cost/benefit claims.

9. *Late 1973–May 1978:* Consumer actions place pressure on NHSTA to force Pinto recall actions.

10. *September 1977–May 1978:* Ford publicly refutes consumer claims and defends safety record.

11. *September 1977–May 1978:* NHTSA investigates Pinto fuel tank systems and determines that safety problems require recall actions.

12. *February 1978–March 1978:* Series of court actions focuses attention on Pinto safety problems.
    - *February 1978:* California jury assesses $125 million in punitive damages in a case involving rupture and explosion of the fuel tank on a 1972 Pinto.
    - *March 1978:* Pinto owners in Alabama and California file class action suits demanding that Ford recall all Pintos built from 1971 through 1976 and requiring modification of their fuel system.

13. *March 1978–June 1978:* Ford initiates a recall and modification program.
    - *March 1978:* Recalls 300,000 1976 Pintos.
    - *June 1978:* Recalls 1.5 million Pintos.
    - Agrees to replace fuel filler pipe and install a polyethylene shield across the front of the fuel tank. Total estimated cost for modification program $20 million after taxes.

Source: Adapted from Kenneth E. Goodpaster, *Ethics in Management* (Boston: Harvard Business School, 1984).

In 1974, a Pennsauken, New Jersey, police officer was severely injured when his patrol car—a Dodge Monaco—spun off a rain-slicked highway and slammed sideways into a steel pole 15 inches in diameter. The policeman sued the Chrysler Corporation, contending that the car's design was unsafe because it did not have a rigid steel body.

Chrysler argued that the car's flexible body design provided maximum passenger protection in front- or rear-end collisions, by far the most numerous types of accidents, and that rigid construction would add enough weight to appreciably reduce fuel efficiency and increase operating costs. Therefore, to meet federal regulations for both fuel economy standards and front-end collision survivability, Chrysler maintained, the design was optimal.

The jury awarded the plaintiff $2 million.
"At Issue: Product Liability," Shell News,
Vol. 52, No. 4 (1984)

[6-24] The DC-10 jet aircraft has experienced some design defects, particularly with regard to the operation of the cargo door.

[6-25] Passenger-restraint systems were tested recently in a full-scale crash test of a Boeing 720 aircraft. In the 20-year period since the jet age began in 1959, 933 jetliner accidents occurred worldwide. Less than 30 percent of the 12,668 passengers aboard were killed. For comparison, in 1983 more than 43,000 people were killed on American highways.

User experience with cordless telephones, a new consumer convenience product, illustrates the need to be aware of safety problems that can occur when possible user errors are not adequately considered. For convenience and portability, the "ringer" in most cordless telephones is located in the earpiece, where it must be loud enough to attract attention to incoming calls. In most cases a "standby/talk" switch is provided to allow the user to disable the "ringer" when making a call. There is the possibility, however, of the user holding the telephone to an ear without switching to "talk." In this case, the ring of an incoming call can cause damage to nerve endings in the ear.

The Federal Consumer Product Commission became aware of this problem and ran tests which they said showed that sound levels produced by the ringer present the potential for unacceptable levels of injury as well as the probability of occurrence resulting in an unacceptable risk. They issued a consumer warning to remind users to switch from "standby" to "talk" when placing a call.[17]

A possible outcome from this safety problem could be lawsuits on a national level because of the extensive use of cordless telephones. After these safety problems became apparent, manufacturers of cordless telephones placed warnings directly on the phones or provided instruction sheets to accompany them.

There Is No Such Thing as an Insignificant Detail

Engineers working on teams designing large complex products or structures need to remember that overall safety is determined by the safety of each individual component. The selection and application of each component should be the result of a risk analysis made to identify possible hazards and to select the best alternatives for eliminating or minimizing these hazards. In most cases this requires a thorough analysis to determine how the components interact and how user operation of one component can affect overall safety.

Example 6-2

An airplane is an example of a complex mechanism where the design of what might otherwise be considered to be an insignificant item could affect overall safety. The failure of one switch, a break in an oil line, or a ruptured tire could result in a serious accident. On March 3, 1974, a Turkish Airline DC-10 jet aircraft crashed after taking off from the Paris, France, airport, killing 346 people [6-24]. Safety investigations made after the crash determined that the accident was caused by a faulty locking mechanism on a cargo door. At an altitude of approximately 10,000 feet, the cargo door burst open, creating a pressure differential within the aircraft which caused the floor of the passenger compartment to collapse. Hydraulic control lines running under the floor panel to the rear control surfaces were crushed, causing the aircraft to be uncontrollable.

There were several interrelated safety factors involved in this accident. The DC-10 cargo doors open outward and are closed be-

---

[17]Chuck Hawley, "Harmful Earful," *Arizona Republic,* October 15, 1984.

fore takeoff with a latching mechanism that is operated from the outside by the baggage handler. In the design used on the crashed aircraft, proper operation of the latch was very dependent on adjustments made during maintenance. Also, it was possible for the baggage handler and flight crew to be convinced that the door was properly latched when in actuality certain of the latching pins were not fully engaged. Other critical factors included strength of the floor panels and the location and construction of the control lines. All of these conditions were considered in the design changes that have subsequently been implemented by the manufacturer to prevent similar accidents from occurring.

Make Certain That Design Changes Do Not Reduce the Safety of the Original Design

Engineers working on design teams often need to make design changes that are required to meet new design requirements, incorporate new technology, or to reduce costs. After the product has been in use, field experiences related to performance or safety (such as the DC-10 accident) may indicate that changes in the design are required. These changes must be made with the same care and control used to complete the original design.

Example 6-3

An example of how an improperly implemented change can affect safety is the catastrophe that occurred in Kansas City, Missouri, on July 17, 1981. A Hyatt Regency Hotel fourth-floor skywalk collapsed and caused a second-story skywalk to fall with it to the hotel lobby, killing 114 persons and injuring more than 200 others [6-28].

City records, visual examinations by experts, and photographic evidence indicate that sometime during the construction process a design change was made that altered the weight distribution on the skywalk supporting structures and doubled the resulting stress on beams supporting the fourth-floor skywalk[18] [6-29].

This accident points out that when changes are made, all safety factors need to be considered. Also, it is essential to check to make certain that all details of the design are properly implemented.

Maintain Design Integrity

Designing for safety must be more than just good intentions. A project team dedicated to meeting safety requirements must carefully implement design decisions through the assembly or construction processes. Inspections and tests are completed at critical stages and any detected discrepancies are corrected. Changes in construction procedures or in the design are made whenever test results or field experiences show that a basic problem exists. This type of program requires a management/engineering team that is dedicated to product integrity.

---

[18]Rick Alm and Thomas G. Watts, "Critical Design Change Is Linked to Collapse of Hyatt's Skywalk," *Kansas City Star*, July 21, 1981.

[6-26] and [6-27] A Boeing 720 aircraft skids across a dry lake bed (December 1, 1984) after being guided to a remote-controlled crash. The purpose was to test the effectiveness of a new, specially treated kerosene fire-retardant fuel as well as improved aircraft safety features.

You can go a long way towards checking structural viability when you review the plans. Basically the success of determining the structural integrity is in the plan.
William Bullard, Planning Director, Independence, Missouri, Kansas City Star, *July 21, 1981*

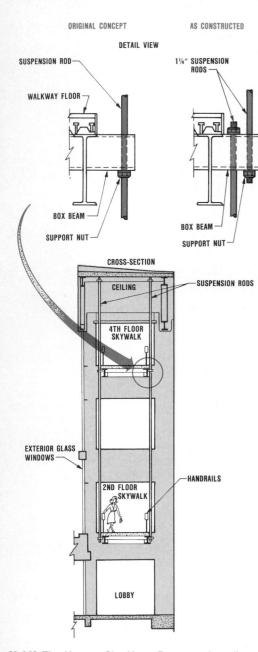

ORIGINAL CONCEPT

AS CONSTRUCTED

DETAIL VIEW

SUSPENSION ROD

1¼" SUSPENSION RODS

WALKWAY FLOOR

BOX BEAM

SUPPORT NUT

BOX BEAM

SUPPORT NUT

CROSS-SECTION

CEILING

SUSPENSION RODS

4TH FLOOR SKYWALK

EXTERIOR GLASS WINDOWS

2ND FLOOR SKYWALK

HANDRAILS

LOBBY

[6-28] The Kansas City Hyatt Regency skywalk.

[6-29] The before-failure and after-failure drawings show the design change that allowed the entire walkway to collapse. The original design called for the walkway to be supported by continuous rods [6-28].

Improper implementation of the steps required for product integrity can lead to serious contract problems, economic sanctions, and lawsuits. National Semiconductor Corporation, a major supplier of microchips to the U.S. government, paid over $1.7 million in civil and criminal penalties in March 1984. The firm had been indicted on charges which claimed that microchips sold to the government between 1978 and 1981 had been inadequately tested. The indictment included criminal charges of mail fraud and making false statements regarding the testing program. The government originally proposed to ban National Semiconductor from doing any more business with the Defense Department. However, this serious economic threat was dropped after the firm took several corrective actions: creation of an independent quality auditing group, reassignment of the several managers who were responsible for quality control during the problem years, institution of a company policy requiring dismissal of any employee who fails to follow government regulations in the future, and the creation of an internal company hotline which allows any employee to report anonymously any wrong doing.[19]

POSSIBLE SOURCES OF SUPPORT FOR ENGINEERS AND TECHNOLOGISTS

Organizational Support

Engineers and technologists need not believe that they stand alone as they make critical ethical decisions related to their job responsibilities. Because society emphasizes the need for correct decisions and recognizes the difficulty of decision making, there are many sources of support for the engineer. The first of these is the company organization for which the engineer works. In addition to ethical consider-

[19]*Electronic Buyers News,* August 13, 1984.

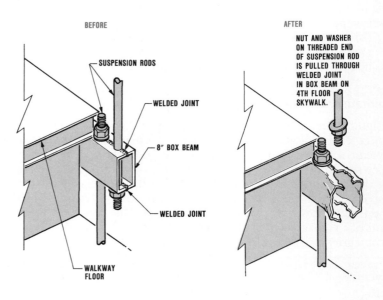

BEFORE

AFTER

NUT AND WASHER ON THREADED END OF SUSPENSION ROD IS PULLED THROUGH WELDED JOINT IN BOX BEAM ON 4TH FLOOR SKYWALK.

SUSPENSION RODS

WELDED JOINT

8" BOX BEAM

WELDED JOINT

WALKWAY FLOOR

ations, the possibility of costly lawsuits and damage to reputation make today's business firms especially interested in making correct decisions.

In most cases disagreements with company management over ethical approaches are settled in an amicable manner since all concerned want to find the best solution to the problem. However, occasionally a case involving several issues, priorities, and conflicting values may result in a dispute between an engineer and project supervision. In this case, the first choice for the engineer is to try to solve the dispute using established organizational channels and procedures for resolving conflicts.

If an impasse develops with organizational management, the engineer is faced with choosing between three alternatives: give up, appeal within the organization, or report the problem to persons outside the organization for possible resolution of the problem. The easiest alternative is the first one—total surrender. Before taking this easy way out, an engineer needs to ask the question, "Can I live with myself if I stand by and let something go on that could result in serious injury to others or perhaps have other serious consequences?" The last alternative, called "whistle blowing," is a last-resort alternative because of the obvious negative effects it will have on present working conditions and on your future career possibilities within the company. You may wish to seek outside counsel. Before resorting to outside help, however, you should carefully review the validity of the position you have taken and the importance of the issue. Writing a position statement can help to focus your thinking on the facts. In any case, take the position that is ethically proper.

Technical Societies

Engineers and scientists have organized a number of technical societies in various fields of specialization (Table 6-4). The first societies were originally formed over 100 years ago to allow engineers to band together to exchange ideas, improve their technical knowledge, initiate design regulation, and to learn new skills and techniques

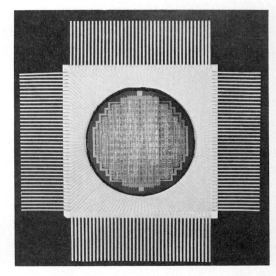

[6-30] Valid testing of semiconductor components has become a concern of the government. If a critical chip fails in a satellite or undersea detection device, the security of the United States might be at risk.

What to Do in Case of an Impasse

1. Prepare a position statement which focuses on the facts and considers all facets of the issue.
2. Use all available appeal procedures within the organization.
3. Seek help and advice from professional friends, colleagues, and technical societies.
4. Go outside if the conflict has significant ethical or safety implications and cannot be resolved inside the organization.

What Would You Do?

Assume that you are an engineer in charge of the design of a microwave oven that has been designed for general consumer use. Quality control tests in your organization show that radiation leakage from the ovens is in excess of U.S. government standards. Although your research shows that the level of leakage from your ovens is substantially below the true hazard level identified by health professionals, you have made a design change to make certain that all units now in production and all future units will meet the government standard. However, 10,000 units which could violate the government standard are now in the stores for Christmas sales. Should you recommend that your firm initiate a recall program, keep quiet, or take some other action?

TABLE 6-4   ENGINEERING AND TECHNICAL SOCIETIES

Code	Name	Address	Year Organized	Total Membership
AcSoc	Acoustical Society of America	335 East 45th St. New York, NY 10017	1929	5,750
APCA	Air Pollution Control Association	P.O. Box 2861 Pittsburgh, PA 15230	1907	7,921
AAAS	American Association for the Advancement of Science	1776 Massachusetts Ave., N.W. Washington, DC 20036	1848	139,000
AAEE	American Academy of Environmental Engineers	P.O. Box 269 Annapolis, MD 21404	1955	2,400
AAES	American Association of Engineering Societies	345 East 47th St. New York, NY 10017	1979	n.a.
AACE	American Association of Cost Engineers	308 Monongahela Bldg. Morgantown, WV 26505	1956	6,100
AAPM	American Association of Physicists in Medicine	335 East 45th St. New York, NY 10017	1958	2,200
ACI	American Concrete Institute	22400 West 7 Mile Road Detroit, MI 48219	1906	14,716
ACM	Association for Computing Machinery	11 West 42nd St., 3rd Floor New York, NY 10036	1947	58,000
ACS	American Ceramic Society, Inc.	65 Ceramic Dr. Columbus, OH 43214	1898	8,152
ACS	American Chemical Society	1155 16th St., N.W. Washington, DC 20036	1876	131,764
AES	Audio Engineering Society, Inc.	60 East 42nd St. New York, NY 10165	1948	9,600
AIAA	American Institute of Aeronautics and Astronautics	1633 Broadway New York, NY 10019	1932	35,448
AIA	American Institute of Architects	1735 New York Ave., N.W. Washington, DC 20006	1857	42,132
AIChE	American Institute of Chemical Engineers	345 East 47th St. New York, NY 10017	1908	62,000
AICE	American Institute of Consulting Engineers	345 East 47th St. New York, NY 10017	1910	420
AIME	American Institute of Mining, Metallurgical, and Petroleum Engineers, Inc.	345 East 47th St. New York, NY 10017	1871	99,734
AIP	American Institute of Physics	335 East 45th St. New York, NY 10017	1931	61,000
AIPE	American Institute of Plant Engineers	3975 Erie Ave. Cincinnati, OH 45208	1954	8,675
AMS	American Mathematical Society	201 Charles St. Providence, RI 02940	1888	20,392
ANS	American Nuclear Society	555 N. Kensington Ave. LaGrange Park, IL 60525	1954	14,650
APS	American Physical Society	335 East 45th St. New York, NY 10017	1899	32,781

continues next page

TABLE 6-4 ENGINEERING AND TECHNICAL SOCIETIES (CONTINUED)

Code	Name	Address	Year Organized	Total Member-ship
APHA	American Public Health Association	1015 15th St., N.W. Washington, DC 20005	1872	28,268
ASAE	American Society of Agricultural Engineers	2950 Niles Ave. St. Joseph, MI 49085	1907	11,300
ASCE	American Society of Civil Engineers	345 East 47th St. New York, NY 10017	1852	92,747
ASEE	American Society For Engineering Education	Eleven Dupont Circle Washington, DC 20036	1893	10,060
ASEM	American Society for Engineering Management	301 Harris Hall, University of Missouri, Rolla, MO 65401	1979	1,011
ASM	American Society for Metals	Metals Park, OH 44073	1913	45,104
ASQC	American Society for Quality Control	230 W. Wells St. Milwaukee, WI 53203	1946	43,345
ASHRAE	American Society of Heating, Refrigerating and Air-Conditioning Engineers, Inc.	345 East 47th St. New York, NY 10017	1894	49,000
ASLE	American Society of Lubrication Engineers	838 Busse Highway Park Ridge, IL 60068	1944	3,850
ASME	American Society of Mechanical Engineers	345 East 47th St. New York, NY 10017	1880	111,645
ASNE	American Society of Naval Engineers, Inc.	1452 Duke St. Alexandria, VA 22314	1888	6,800
ASSE	American Society of Safety Engineers	850 Busse Highway Park Ridge, IL 60068	1911	19,000
ASTM	American Society for Testing and Materials	1916 Race St. Philadelphia, PA 19103	1898	28,692
AWRA	American Water Resources Association	St. Anthony Falls Hydraulic Laboratory Miss. River at 3rd Ave. S.E. Minneapolis, MN 55414	1964	2,485
AWWA	American Water Works Association, Inc.	6666 W. Quincy Ave. Denver, CO 80235	1881	29,475
CEC	Consulting Engineers Council of the United States of America	1155 15th St., N.W. Washington, DC 20005	1959	2,300
IES	Illuminating Engineering Society of North America	345 East 47th St. New York, NY 10017	1906	7,665
IEEE	Institute of Electrical & Electronics Engineers, Inc.	345 East 47th St. New York, NY 10017	1884	240,068
IES	Institute of Environmental Sciences	940 East Northwest Highway Mt. Prospect, IL 60056	1956	2,122
IIE	Institute of Industrial Engineers	25 Technology Park/Atlanta Norcross, GA 30092	1948	39,000
ITE	Institute of Transportation Engineers	525 School Street, S.W. Suite 410 Washington, DC 20024	1930	6,911

continues next page

TABLE 6-4   ENGINEERING AND TECHNICAL SOCIETIES (CONTINUED)

Code	Name	Address	Year Organized	Total Member-ship
ISA	Instrument Society of America	67 Alexander Dr. P. O. Box 12277 Research Triangle Park, NC 27709	1945	36,000
NACE	National Association of Corrosion Engineers	P. O. Box 218340 Houston, TX 77218	1945	14,000
NAPE	National Association of Power Engineers, Inc.	176 West Adams St., Suite 1914 Chicago, IL 60603	1882	8,900
NICE	National Institute of Ceramic Engineers	65 Ceramic Dr. Columbus, OH 43214	1938	1,904
NSPE	National Society of Professional Engineers	2029 K Street, N.W. Washington, DC 20006	1934	79,370
ORSA	Operations Research Society of America	Mount Royal and Guilford Aves. Baltimore, MD 21202	1952	6,700
SESA	Society for Experimental Stress Analysis	14 Fairfield Drive Brookfield Center, CT 06805	1943	2,900
SIAM	Society for Industrial and Applied Mathematics	117 S. 17th St., 14th Floor Philadelphia, PA 19103	1952	6,000
SAE	Society of Automotive Engineers	400 Commonwealth Dr. Warrendale, PA 15096	1905	42,060
SES	Society of Engineering Science	c/o Dept of Engineering Science Virginia Tech, Blacksburg, VA 24061	1963	400
SME	Society of Manufacturing Engineers	P.O. Box 930 Dearborn, MI 48121	1932	71,714
SNAME	Society of Naval Architects and Marine Engineers	One World Trade Center Suite 1369 New York, NY 10048	1893	13,400
SPE-AIME	Society of Petroleum Engineers of AIME	6200 N. Central Expressway Drawer 64706 Dallas, TX 75206	1913	54,413
SPE	Society of Plastics Engineers, Inc.	14 Fairfield Dr., Brookfield Center, CT, 06805	1941	23,500
SWE	Society of Women Engineers	345 East 47th St. New York, NY 10017	1950	14,000

[6-31]. In addition, the National Society of Professional Engineers is concerned primarily with the legal and registration aspects of the entire field of engineering.

These technical societies can provide various degrees of support for engineers as they are faced with making critical ethical decisions in the profession. Advice from sympathetic, objective, and experienced colleagues can be very valuable. Some societies may provide direct support for engineers involved in organizational disputes resulting from ethical issues. This support may include an investigation of the circumstances surrounding the situation and preparation of a report which can be used to help resolve the conflict. In some cases legal support may be provided.[20]

Another major area of support provided by technical societies is the formation of codes of ethics. These provide guidelines for individual members by indicating how general principles of ethical behavior can be applied in specific circumstances, and establishing publicly approved standards of behavior. Present society codes place heavy emphasis on the responsibilities of professional integrity and give less attention to the rights of engineers or to the protection of engineers who may take job risks instead of standing firm for ethical principles.

ENGINEERING REGISTRATION

An important element for any professional is to gain respect and trust of the public. One way to do this is through state or national registration, which provides a license to practice. It is also a commitment to public service and to conducting a trustworthy career. Licensing implies that the professional will be guided by ethical codes and imposes standards of competence and integrity. Registration encourages continued efforts to remain up to date, thus upgrading the profession.

Registration could become even more important if legal liability and consumer protection groups require engineers to go through a licensing procedure which periodically documents their competence and good judgment.

The first step in obtaining professional engineering (PE) registration can be accomplished at or near the end of your senior year in college. This is passing the Engineer-In-Training (EIT) examination on engineering fundamentals. It includes problem solving from a number of basic subjects, beginning with mathematics, physics, and chemistry and extending to mechanics, engineering economics, thermodynamics, and electrical network theory. If you have mastered the principles of engineering economy in Chapter 13 of this text, for example, you should be able to pass that portion of the EIT examination. Other steps in the registration process include four or more years of pertinent professional work experience and passing a professional examination.

PROFESSIONAL RESPONSIBILITY AND THE ENGINEERING STUDENT

Engineers and technologists do not automatically become responsible professionals the day they receive their baccalaureate degrees, start work on their first professional assignment, join a professional society, or meet the minimum requirements for professional registration. Becoming a responsible professional requires continuous ac-

[20]Stephen H. Unger, *Controlling Technology: Ethics and the Responsible Engineer* (New York: Holt, Rinehart and Winston, 1982).

[6-31] Participation in technical society activities are important in the professional development of the engineer.

Three things are to be looked to in a building: that it stand on the right spot; that it be securely founded; that it be successfully executed.
Johann Wolfgang von Goethe, 1749–1832
Elective Affinities

In an examination those who do not wish to know, ask questions of those who cannot tell.
Sir Walter Raleigh, 1552–1618
Some Thoughts on Examinations

[6-32] Fundamental tenets of professionalism are developed prior to graduation from college.

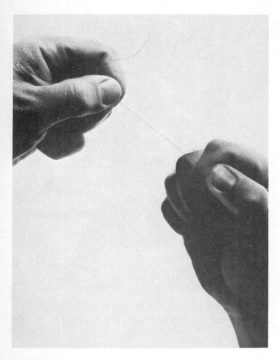

[6-33] Designing to the limits of the system is analogous to the stretching of a thread. On the one hand, the engineer strives for the most efficient and least expensive design . . . while on the other hand society insists on a tolerance to insure safety and minimum risk.

quisition of the technical knowledge and background required to help solve specific problems, and development of a professional attitude that is needed to work effectively with others in making critical ethical decisions.

Students preparing to become responsible professionals in their chosen field need to emphasize development and application of a professional attitude concurrently with completing the course work required to earn their degree. This means developing and applying the same set of ethical values that they will someday apply as they work in industry as professionals. Engineering students need to complete course requirements with the same integrity, diligence, and persistence that will someday be required to meet on-the-job responsibilities. The same degree of honesty should be used in completing course assignments and in supplying answers to test questions. Begin to develop desirable work habits and personal discipline in completing project assignments. It is difficult to picture someone who cheats on examinations being totally trustworthy when required to make difficult on-the-job ethical decisions.

Preparing student research papers requires special attention to honesty. Students need to be certain that their papers represent their personal research effort and that the work supplied by others is properly documented. Developing a proper work ethic in preparing student papers will carry over after the student graduates.

Developing the ability to work with others is another important area in preparing to be a responsible professional. Working on team projects associated with some courses, engineering students have an opportunity to develop their ability to assume personal responsibility for their share of the team effort and to interact with others. This will involve developing communicating skills, sharing technical knowledge, and making certain that each team member gets his or her earned acknowledgment for work done. Participation in student technical organizations and extracurricular activities can also give you an opportunity to develop teamwork skills.

PROBLEMS

6-1. Discuss the similarities and dissimilarities of engineering and other professions. What do you think is the most significant common factor? Why is this so?

6-2. Discuss the factors that make engineering a unique profession. What are the reasons for this uniqueness?

6-3. How should these common and unique factors (see Problems 6-1 and 6-2) affect the approach you use as you prepare to enter the profession of engineering?

6-4. Interview three graduate engineers or engineering technologists, each practicing in a different field. Obtain their answers to these questions:

- Why do you consider yourself a professional?
- What do you consider to be the major challenges you encounter as you practice your profession?
- What do you consider to be the most important factors in preparing to be a professional?

Prepare a summary of the results of the interviews. In what ways were the answers received from the three professionals similar? What major differences did you see? How do you explain these differences?

6-5. Discuss the reasons why a professional person such as a registered engineer, an attorney, or a physician will not bid competitively on the performance of a service.

6-6. In interviewing for permanent employment, a senior student in a California engineering school agreed to visit on two successive days a company in Chicago and a company in Detroit. Upon her return home both companies sent her checks to cover her expenses, including round-trip airfare. Discuss the appropriate actions that should have been taken by the student.

6-7. The majority of all engineering designs require some extension of the engineer's repertoire of scientific knowledge and analytical skills. How can the engineer determine whether or not this extension lies beyond the "areas of his or her competence"?

6-8. Assume that you are working in your first professional position for Sillwell Co., a firm that has developed an inexpensive household specialty that they hope will find a huge market among housewives. They want to package this product in 1-gallon and ½-gallon sizes. A number of container materials appear to be practical—glass, aluminum, treated paper, steel, and various types of plastics. As an engineer assigned to the manufacturing department, you have completed a container-disposal study which shows that the disposal cost for 1-gallon containers can vary by a factor of 3, depending on the weight of the container, whether it can be recycled, whether it is easy to incinerate, and whether it has good landfill characteristics.

The company's marketing specialist believes that the container material with the highest consumer appeal is the one to use, although it presents the most severe disposal problem. He states that the sales potential would be at least 10 percent less if the easiest-to-dispose-of container were used because it would be less attractive and distinctive.

The results of your study have been forwarded in a report to company management, but you are concerned that your study is not being properly considered and that management will follow the recommendations of the marketing expert. Do you think you have an ethical responsibility to present a stronger case before management and make specific recommendations as to which container should be used? If you think you should make a stronger case, how would you proceed? What action would you take if after considering your recommendation, the company implements the marketing expert's recommendation?

6-9. Jerry Williams is a chemical engineer working for a large diversified company on the east coast. For the past two years he has been a member—the only technically trained member—of a citizen's pollution-control group working in his city.

As a chemical engineer, Williams has been able to advise the group regarding what can reasonably be done about abating various kinds of pollution, and he has even helped some smaller companies to design and buy control equipment. (His own plant has air and water pollution under good control.) As a result of Williams's activity, he has built himself considerable prestige on the pollution-control committee.

Recently, some other committee members started a drive to pressure the city administration into banning the sale of phosphate-containing detergents. They have been impressed by reports in their newspapers and magazines on the harmfulness of phosphates.

Williams believes that banning phosphates would be misdirected effort. He tries to explain that although phosphates have been attacked in regard to the pollution of the Great Lakes, his city's sewage flows from the sewage-treatment plant directly into the ocean. And he feels that nobody has shown any detrimental effect of phosphate on the ocean. Also, he is aware that there are conflicting theories on the effect of phosphates, even on the Great Lakes (e.g., some theories put the blame on nitrogen or carbon rather than phosphates, and suggest that some phosphate substitutes may do more harm than the phosphates).

In addition, he points out that the major quantity of phosphate in the city's sewage comes from human wastes rather than detergent.

Somehow, all this reasoning makes little impression on the backers of the "ban phosphates" measure. During an increasingly emotional meeting, some of the committeemen even accuse Williams of using stalling tactics to protect his employer, who, they point out, has a subsidiary that makes detergent chemicals.

Williams is in a dilemma. He sincerely believes that his viewpoint makes sense and that it has nothing to do with his employer's involvement with detergents (which is relatively small, anyway, and does not involve Williams's plant). Which step should he now take?

A. Go along with the "ban phosphates" clique on the grounds that the ban won't do any harm, even if it doesn't do much good. Besides, by giving the group at least passive support, Williams can preserve his influence for future items that really matter more.

B. Fight the phosphate foes to the end on the grounds that their attitude is unscientific and unfair, and that lending it his support would be unethical. (Possible outcomes: his ouster from the committee, or its breakup as an effective body.)

C. Resign from the committee, giving his side of the story to the local press.

D. Other action (explain).

6-10. Assume that you are a chemical engineer working for a small chemical company. You have been assigned the task of taking periodic samples of the effluent in a river resulting from drainage from an overflow pipe. The sampling location was selected by a representative of the state health department.

The sampling program consistently indicates a pollution rate well within the allowable limits. You notice, however, that the sampling site has been incorrectly chosen and does not detect pollution resulting from a discharge that flows through a deep pipe not visible from the surface.

Revealing the existence of the overlooked pipes could expose your company to major expense required to lower the actual pollution rate to within the limits of the discharge permit.

What do you think is the proper action for you to take? Do you think you would be legally liable if you remained silent? What would you do if you notified your company but the sampling location was not changed?[21]

6-11. Smith and Jones worked together for three years on a major research project and had nearly completed a paper for joint presentation to the national meeting of their engineering society. Smith was fired by their company for poor work habits and insubordination. Since Smith no longer works for the company, the management wants Jones to complete the paper and present it at the national meeting with no credit to Smith. What action should Jones take:

A. Complete the paper and present it as the only author.

---
[21]Adapted from Stephen H. Unger, *Controlling Technology: Ethics and the Responsible Engineer* (New York: Holt, Rinehart and Winston, 1982).

B. Complete the paper and present it with acknowledgment to Smith (risking management displeasure).

C. Stop preparation of the paper.

D. Take some other action.

Discuss your choice.

6-12. Select one of the student engineering societies at your school for investigation. Attend at least one meeting; talk to several officers and members. Prepare a short report answering these questions: What are the major objectives of this society? Why do the members you talked to belong to the society? What do they expect to obtain from the meetings? How are they working to improve their society? Would you like to belong to the society? Why or why not?

6-13. Determine the requirements in your state for obtaining a license as a professional engineer. What engineering fields of specialization are recognized by your state registration board for licensing?

6-14. Review a copy of an Engineering Code of Ethics and discuss the following questions:

- What items are included which help to provide a practical standard of behavior for practicing engineers?

- Are there any irrelevant items which could be eliminated? Why do you think they are irrelevant?

- Are there items which are confusing and that might be misinterpreted? Suggest ways for rewriting which could improve clarity.

- Could you use this code as a standard for yourself as you study to be an engineer? What areas would be the most difficult to comply with?

6-15. Interview some OSHA personnel in your area. What are the most prevalent violations? What are the most serious violations in terms of the number of people that are affected? In your opinion are any of these violations caused by poor engineering design?

Chapter 7

Freehand Drawing and
Visualization

Drawing is the ability to translate a mental image into a visually recognizable form. Simply, it requires the use of an instrument to mark or stain a surface. Today there are three general methods of producing drawings: freehand drawing, projection drawing, and computer-aided drawing (CAD). There are many variations of these three types, such as artistic drawing, design drawing, engineering drawing, architectural drawing, technical illustration, and cartography. Each drawing type has its own inventory of fundamental tenets that must be mastered by the beginner if competency is desired.

Of the general methods of drawing, the freely drawn artform, *freehand drawing,* is the oldest. Prehistoric man, wishing to record the results of a triumphant hunt, drew pictures of his experiences on rocks or on the walls of caves [7-1]. Until after the Middle Ages this was the only type of drawing. Drawing became more formalized when *projection drawing* was developed in France in the eighteenth century by Gaspard Monge (1746–1818). It was first used to simplify the design and construction of military fortifications. Descriptive geometry and engineering drawing were developed from these principles [7-2].

Engineering drawing, together with the principal of *tolerances,* unlocked man's ability to produce interchangeable parts in the manufacturing process. Engineering drawing is a precise discipline based on a thorough understanding of the principles of orthographic projection. Since this form of drawing values the accuracy that results from the application of projection theory, it has been greatly enhanced by the development of *computer-aided drawing.* Along with the graphical description of an object, computer-aided drawing creates an extensive data base detailing the attributes of the object. A computer is not only very adept at producing and manufacturing a graphical image, but also very efficient at data recording and manipulating for use in design and manufacturing. Where product drawings are generated by computer-aided drawing, the production machines that manufacture the products can receive their operating instructions directly from the data base in the computer. This results in fewer errors and a considerable savings in man-hours of work.

Today, using the power and diversification of the computer, it is possible to transfer freehand "idea drawings" into computer drawings, where they can be scrutinized simultaneously by many people. Objects can be easily rotated, sectioned, and viewed from myriad positions. Revisions can be readily made, if necessary, and then the final design can be permanently recorded in the computer's memory

[7-1] Prehistoric drawing of hunters.

[7-2] Engineering drawing is a mechanically precise form of drawing compared to freehand drawing.

[7-3] Computer drawings are used extensively in industry.

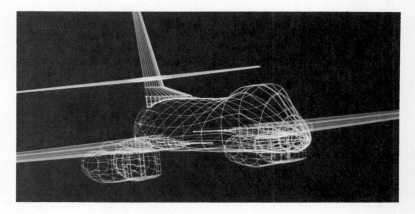

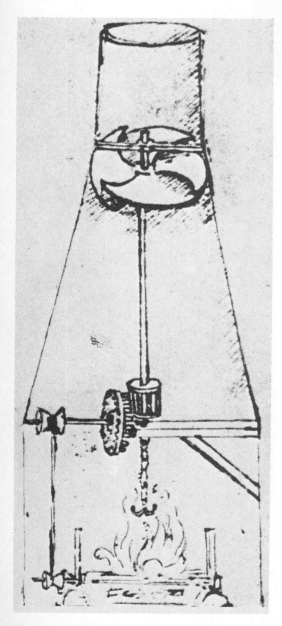

[7-4] Today, 400 years after Leonardo da Vinci drew them, his imaginative design drawings are still easy for the average person to understand.

[7-3]. Consequently, in industry more and more drawings of various types are being generated on computers rather than by engineers and drafters using pencil and paper. Engineering drawings can best be developed if engineers know how to select and use the appropriate computer software. This saves both time and money, and the result is much less susceptible to error.

Engineering drawing is not the best medium to use when a creative design engineer wants to convey an idea to nontechnical people. Even a superior knowledge of engineering drawing is not effective in this environment. A drawing form is needed that can be understood by all members of today's management teams, which frequently include both technical and nontechnical personnel. For most people, freehand pictorial drawing is the most easily and universally understood when it is necessary to realistically represent ideas that are intended to show "how things work."

One of the earliest "engineers" who used this medium to record his design ideas was Leonardo da Vinci (1452–1519). This man was a genius of great versatility. Although he is still acclaimed today as one of the greatest of all artists, his contributions as an architect, engineer, and inventor are perhaps even more impressive. In retrospect, it is interesting to note that the common thread that ran through his work in these seemingly diverse fields was his excellence in freehand drawing and visualization. In many cases, the only keys we have today to his brilliance are the idea drawings that he made [7-4]. His ability to produce realistic sketches makes the visualization of his thoughts possible [7-5]. Although most of us do not possess the mental talents of da Vinci, we can achieve skill in freehand drawing through proper training and experience. We can develop our ability to *draw with our pencil* those things that exist only in our imagination. As with da Vinci's sketches, our sketches can be used by others to see and understand the images that exist only in our minds. The objective of the authors is to instill within each engineering student the desire, confidence, and some of the proficiency demonstrated in the freehand drawing of Leonardo da Vinci.

This chapter has been written for the student who has had little or no formal instruction in freehand drawing. Its purpose is to develop a person's proficiency in freehand drawing and visualization. This will greatly improve your ability to record quickly the essence of your ideas as you follow the sequential steps in the design process (see Chapter 14).

The authors make no claim that the methods used here to teach freehand drawing are unique or that they are more effective than the time-honored techniques that have been developed throughout the ages by artists from every culture.[1] Indeed, we have drawn heavily on the experience of both traditionalists and advocates of the right-brain methodology in developing our teaching methods. This chapter is not intended to air pedagogical arguments or to produce artists. Of one thing we are certain, however; the methods described here *will work for you* if you:

1. Put aside any negative experience that you may have had with drawing. Be willing to learn.
2. Believe that you can learn to draw with skill.
3. Acquire the ability to perceive images accurately with your eyes.
4. Master a few basic skills in using a pencil to create images.
5. Follow each recommended drawing exercise in sequence. Don't leave any out. Do more if you have time or experience difficulty in improving your drawing skill.

Several of the illustrations used in this chapter are examples of student work. In each instance the students have consented for their work to be shown to serve as realistic examples of freehand drawing skills.

If you are an average engineering student, you will have emphasized in your prior schooling those subjects that were said to be good preparation for an engineering career: mathematics, physics, chemistry, biology, foreign language, and perhaps typing or computer programming. These are certainly all appropriate subjects, but primarily they are subjects that require intense stimulation and interaction with the left hemisphere of your brain [7-6]. The development of the right hemisphere of your brain (which is more concerned with music and art awareness, insight, imagination, and three-dimensional forms) has probably lagged behind. Engineering drawing, which is a subject that in former years was a requirement for most engineering students, is also a subject that draws heavily on interaction with the left hemisphere of the brain. On the other hand, freehand drawing, being free of technical symbols, is dominated by the right hemisphere of the brain. Because of the nature of your prior schooling, it is understandable that your proficiency in drawing and visualization may not equal your skills in mathematics and science. Since some of the exercises developed in this chapter are based on the conclusions above, let us briefly examine the scientific basis for the two-hemisphere brain theory.

THE BRAIN

The human brain is a 1500-gram information processing organ of miraculous configuration. For protection, it is housed in the skull. For efficiency, it is located near the eyes, nose, ears, and mouth. It

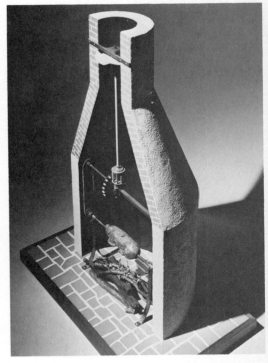

[7-5] It is easy today to build a model of one of Leonardo's conceptual drawings made over 500 years ago.

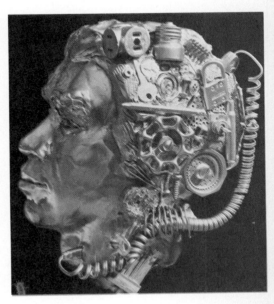

[7-6] The left hemisphere of the brain processes the verbal, mathematical, and scientific information used extensively in engineering. The right hemisphere processes information related to artistic awareness, imagination, and intuitive insight.

[1]Kimon Nicolaides, *The Natural Way to Draw* (Boston: Houghton Mifflin, 1941).

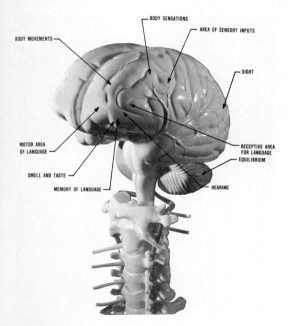

[7-7] The brain is the command center of the body.

The soul never thinks without a mental picture.
Aristotle, 384–322 B.C.

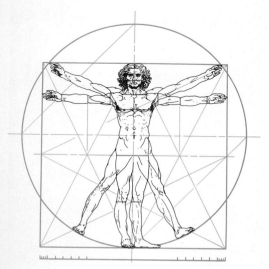

[7-8] The right hemisphere of the brain controls the body actions of the left side. Conversely, the left hemisphere of the brain controls the body's right-side actions.

also sits atop the remaining part of the central nervous system located within the spinal cord, which collects information from the muscles and skin [7-7]. The brain's fundamental working unit is the *neuron*. Each neuron is different from all the others. It is estimated that there are at least 10 billion neurons in each person's brain. Their function is to communicate with each other through electrical and chemical messages in a vast neuronal system network whose total number of connections exceeds 10^{15}.[2]

For reasons that no psychologist or neuroscientist fully understands, our brain is divided into two apparently symmetrical halves or hemispheres. In appearance, each hemisphere seems to be a mirror reflection of the other, but they are not precisely alike. In terms of capability, they are also not equally proficient in processing different types of information. Midway below the top of the brain, the two hemispheres are connected by a web of nerve fibers called the *corpus callosum*. This connection allows the two hemispheres to process information between them and even for one side to take over (temporarily or permanently) the normal duties of the other side.

The human body is a remarkable machine. Although it does not have interchangeable parts, it does have some back up (pinch hitter) capabilities. If one part fails or weakens, another similar part can carry some of the original responsibility. For example, we have two arms, legs, eyes, ears, lungs, kidneys, and cerebral hemispheres. Over time, however, the brain's hemispheres, unlike other organs, develop specializations of their own. As one grows older, each hemisphere is less able to carry out the functions of the other.

The design of the nervous system is such that each cerebral hemisphere receives and transmits information primarily from the opposite half of the body [7-8]. Thus the left hemisphere of the brain controls the actions of the right side of the body—the right hand, the right leg, and so on. The right hemisphere specializes in control of the left-side body components. However, there are other differences between the two hemispheres of the brain. In general, the right hemisphere is more proficient with visualizing and remembering events and faces and with other spatial and emotional functions such as visual construction tasks, artistic awareness, musical appreciation, and intuitive insight. The left hemisphere is more dominant with regard to written and spoken language and motor skills, analytical reasoning, and sequenced goal-oriented activities [7-9].

These conclusions about the specialized information-processing capabilities of the two cerebral hemispheres were made possible by extensive research by a number of neurosurgeons and experimental psychologists over the past 40 years. In the late 1940s a neurosurgeon, William Van Wagenen, performed the world's first split-brain surgery, severing an epileptic patient's corpus callosum. As soon as the patient's seizure charges were unable to cross the corpus callosum, the formerly untreatable epileptic spasms stopped. More important, there was apparently no detectable impairment in the patient's mental functions. In 1981, Roger Sperry was named a recipient of the Nobel Prize in Psychology and Medicine for his

[2]Richard M. Restak, *The Brain* (New York: Bantam Books, 1984).

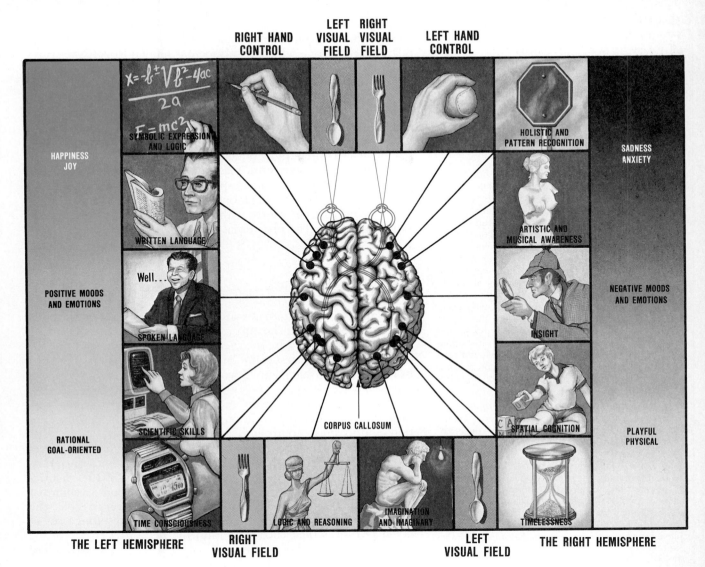

RIGHT HAND CONTROL   **LEFT VISUAL FIELD**   **RIGHT VISUAL FIELD**   **LEFT HAND CONTROL**

HAPPINESS JOY

SYMBOLIC EXPRESSION AND LOGIC

WRITTEN LANGUAGE

POSITIVE MOODS AND EMOTIONS

Well...

SPOKEN LANGUAGE

SCIENTIFIC SKILLS

RATIONAL GOAL-ORIENTED

TIME CONSCIOUSNESS

LOGIC AND REASONING

CORPUS CALLOSUM

IMAGINATION AND IMAGINARY

HOLISTIC AND PATTERN RECOGNITION

SADNESS ANXIETY

ARTISTIC AND MUSICAL AWARENESS

INSIGHT

NEGATIVE MOODS AND EMOTIONS

SPATIAL COGNITION

PLAYFUL PHYSICAL

TIMELESSNESS

THE LEFT HEMISPHERE   **RIGHT VISUAL FIELD**   **LEFT VISUAL FIELD**   **THE RIGHT HEMISPHERE**

[7-9] The right and left hemispheres of the brain are assigned different primary functions.

split-brain research on epileptic patients.[3] This research further clarified the two modes of thinking in a split-brain person. In contrast, however, another English psychologist and Nobel Prize recipient, Sir John Eccles, disputes Sperry's belief that split-brain patients have two separate minds and separate spheres of consciousness.[4]

The purpose of this little discussion is to show that even the most learned of men still disagree on specific details concerning how the brain works. However, there is now general agreement that for

[3]R. W. Sperry, "Hemisphere Disconnection and Unity in Conscious Awareness," *American Psychologist,* Vol. 23 (1968).

[4]J. Eccles, *The Brain and Unity of Conscious Experience: The 19th Arthur Stanley Eddington Memorial Lecture* (Cambridge: Cambridge University Press, 1965).

The main theme to emerge . . . is that there appear to be two modes of thinking, verbal and nonverbal, represented rather separately in left and right hemispheres, respectively, and that our educational system, as well as science in general, tends to neglect the nonverbal form of intellect. What it comes down to is that modern society discriminates against the right hemisphere.

Roger W. Sperry
"Lateral Specialization of Cerebral Function in the Surgically Separated Hemispheres," 1973

Right-hemisphere vs. Left-hemisphere Mental Processes

Over twenty years ago an anthropologist, Thomas Gladwin, contrasted the nonverbal right-hemisphere check and balance mode of thinking of the South Pacific Trukese navigators with the more rigidly structured left-hemisphere mode of thinking of navigators in the European tradition.

As an example, a native navigator of a multi-man sailing canoe will begin a 100-mile voyage across open ocean with a clear mental image of the relative positions of his home island, other islands in the area, and his destination island. This is often a tiny dot of land less than a mile across, and visible from any distance only because of the height of the coconut trees growing on its sandy soil. He has no navigational instruments of any type. Rather, he relies on dead reckoning and he continually adjusts his direction to compensate for changing wind and wave conditions. He sets his course by the rising and setting of stars, having memorized for this purpose the knowledge gleaned from generations of observations of the directions in which stars rise and fall through the seasons. A good navigator can tell by observing wave patterns, for example, when the wind is shifting its direction or speed, and by how much. Although the entire process is mental and involves an incredible number of complex decisions, the Truk navigator cannot describe in words how it all was accomplished. For the most part he has used right-hemisphere of the brain mental processes.

On the other hand, Western navigators carefully plan their voyage in advance. A course is plotted on a chart and this, in turn, provides the criteria for decision. Unless the navigator is sailing a direct point-to-point course, he does not carry in his mind a physical sense of where he is going. Once the plan of the trip is conceived and put into operation, the navigator need only carry out each step consecutively to arrive at the pre-planned designation. If asked, he can describe precisely how each navigation instrument (map, compass, sextant, etc.) was used to assist in helping achieve his goal. The Western navigator uses primarily left-hemisphere of the brain mental processes.

Thomas Gladwin, "Culture and Logical Process," in Explorations in Cultural Anthropology, *(New York: McGraw-Hill, 1964), pp. 167–77.*

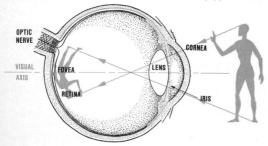

[7-10] Sight is achieved when the eye conveys a sense of vision to the brain. Coded information in the form of electrical impulses is arranged in a way that the objects being seen are represented to the brain.

healthy individuals the verbal, analytic left-hemisphere part of the brain is dominant for most individuals in the Western world. Similarly, the nonverbal, global hemisphere is known to perceive reality in its own way and to experience and process information independently. It is this underdeveloped hemisphere that we must work with if we want to develop our visualization and freehand drawing skills. It is the authors' belief that the ultimate objective is to arrive at a state where both hemispheres work together with equal intensity and proficiency. We must be whole-brain engineers—not entirely left brain or entirely right brain.

So much for the scientific reasons why *most of us* need some help in strengthening right-brain responses in learning to visualize, to understand unusual combinations and arrangements, and to draw with our pencil what we see with our eyes, or what we perceive in our imagination. The first step is to learn to *see* in this way. Let us begin.

SEEING AND ILLUSION

We learn about the external world through interaction of our brain and our senses—sight, hearing, smell, taste, and feeling. Of these, sight is the most used by the average person, and it is also the one that gives us the greatest confidence. At one time or another most of us have made the statement, "I won't believe it until I see it." Sight is achieved when the eye conveys a sense of vision to the brain. We use sight to detect light intensity, recognize images and patterns, estimate distances (depth perception), distinguish colors, perceive motion, and to aid in controlling our bodily actions.

The eye is a remarkable organ of the body. Each of the two eyes accepts light rays through its lens and projects them onto the retina. The two visual pathways cross behind the retinas [7-10], each response passing to the opposite hemisphere of the brain—a shift that makes unified, three-dimensional (stereoscopic) vision possible. The pictorial image has been inverted by the lens. The resulting upside-down picture is then encoded within the neuron lacework of the brain for storage and deciphering. What we see, then, is the result of both the visual stimulus that reaches the brain and the brain's interpretation of the stimulus.

We perceive scenes differently depending on our sociological and psychological conditioning. Several people observing the same street corner accident will frequently swear under oath that conflicting events occurred. Perceiving what we see improves with practice. How perceptive are you? One measure of a person's ability in this area is how successfully and quickly you can identify embedded figures in a more complex figure. Figure [7-11] is an example of such a problem. See how long it takes you to identify the location of the small hexagonal area within the larger pyramidal figure. It is interesting that when this type of test is taken by members of different cultures, some of the fastest solutions were provided by Eskimo hunters. They had an easy time with this type of problem because they had been solving similar and more complex tasks by distinguishing polar bear shapes on distant ice packs.

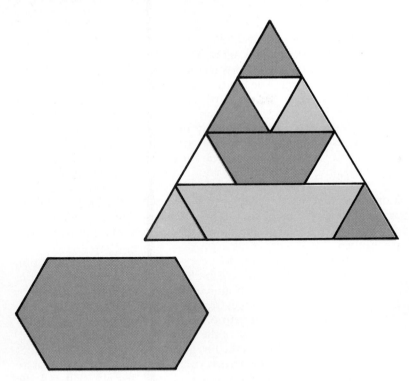

[7-12] M. C. Escher's "Waterfall."

[7-11] Find the hidden hexagon.

There are occasions where the eye's visual signals and the brain's interpretative messages are in conflict. Because of this, it is not unusual for us to see what our brain tells us we are seeing—not what our eyes actually envisage. Such visual conflicts are called *illusions*. The drawing "Waterfall" by M. C. Escher [7-12] is an example of such a conflict. The water falling onto the millwheel seems to be the source of perpetual motion. On examination we see that all of the water that flows onto the waterwheel eventually runs uphill until it can drop again and provide the energy necessary to grind the grain. Our brain warns us that such a condition is a violation of the natural laws of nature. Even knowing this, our eyes continue to deceive us.

In 1970, M. Gardner[5] created an illusion by incorporating an impossible figure into a picture [7-13]. (A more modern version of Gardner's illusion appeared recently as a three-legged blue-jeans advertisement.) In 1960, C. F. Cochran[6] used his "freemish" crate to demonstrate effectively that *seeing is not necessarily believing* [7-14 and 7-15].

If we are to become proficient in drawing shapes that represent real or imagined images, we must understand something of illusionary representation. The drawing of the duck and rabbit [7-16], first used in 1900 by the psychologist Joseph Jastrow, and the 1915 car-

[5]M. Gardner, "Of Optical Illusions, from Figures That Are Undecidable to Hot Dogs That Float," *Scientific American*, Vol. 272 (May, 1970), p. 124.
[6]C. F. Cochran, Letter to *Scientific American*, Vol. 214 (June 1966), p. 8.

[7-13] An impossible situation. What did the artist do to make this illusion?

[7-14] How was this crate constructed?

[7-15]   . . . by illusion.

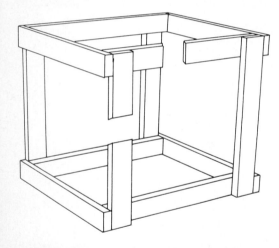

toon drawing of the "Wife and Mother-in-law" by W. E. Hill [7-17] are other examples of ambiguous illusions that are difficult for the eye and brain to assimilate agreeably.

In viewing, the eye reports to the brain both familiar and unfamiliar states of being. The brain measures this information against a data bank (memory) of past experience. It then communicates the comparison to you so that you can act accordingly. For example, if you are driving a car on the highway and you suddenly see a blinking red light in your rear view mirror, your first split-second reaction would likely be to put your foot on the car's brake pedal. If you have been exceeding the speed limit, your overall feeling will probably be one of despair. On the contrary, if you are in the process of being kidnapped by a bank robber, your feeling will more likely be one of apprehension, if not ecstatic relief. Similarly, what appears to be attractive or beautiful to one person may be seen as repulsive or ugly to another.

The brain is responsible for representing a person's perception of the outside world at any specific point in time. For this reason, even though the eyes of two individuals may see the same scene or action, the resulting messages from the two brains may be quite different [7-18]. The competition between foreground and background relationships was one of the earliest conflicts to be identified. Generally, the eye will see in the foreground an outline or object that we will call the *figure*. It has structure and is viewed as being separate from its boundless, shapeless surroundings that we will call the *ground*. Knowing the orientation of the figure with respect to the ground is also important. Figure [7-19] is an example of this rela-

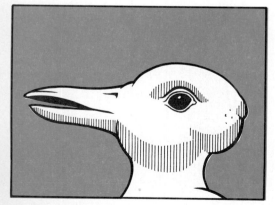

[7-16] Duck or hare?

[7-17] Wife or mother-in-law?

tionship. The white area representing the vase will be recognized by most people as being the figure and the surrounding black area as being the ground. Now turn the figure upside down. The two facing blue faces, which were considered to be the ground before, now peer at each other in front of a lighted doorway. Your brain will allow you to alternately see the vase or the two faces but not both at the same time. If you hide one-half of the drawing by covering it with a sheet of paper, the remaining face is still very easy to identify, but the visible white area has no meaning. Your brain is not accustomed to identifying a one-sided vase.

There are other optical illusions that should be introduced to those who want to learn how to draw. In the late nineteenth century, psychologist Wilhelm Wundt described the simplest of visual illusions—that a vertical image looks longer than a horizontal image of equal length. Figure [7-20] illustrates this. Note that the vertical height of the top hat appears to be greater than the width of the brim, yet they are equal in length. The Gateway Arch in St. Louis is another example of this illusionary distortion. It is 630 feet tall and 630 feet wide [7-21]. Artists of ages past have known that a tree looks much shorter when it has been felled than when it is standing. Figure [7-22] is actually square but would look more like one if the top and bottom lines were covered.

In 1889, Franz Müller-Lyer presented a configuration of two lines of equal length that produced a false impression [7-23]. The two horizontal lines are of equal length, although they do not appear to be so. See how many variations of this phenomenon you can contrive. As a variation, the minute hand and hour hand in [7-24] are the same length.

The mind is susceptible to error by drawing conclusions too soon and using too few data. This is particularly true when we see a few members of an apparent series and conclude that all the other

[7-19] Vase or twins?

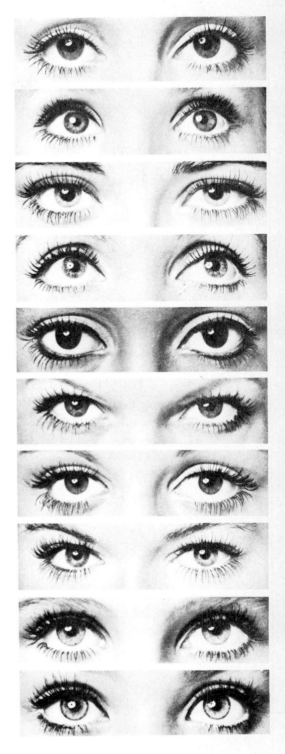

[7-18] Ten pairs of eyes viewing the same scene or action will convey 10 different messages to their respective brains.

[7-20] Square top hat.

[7-21] St. Louis Gateway Arch.

[7-22] How many pancakes should you remove to have a square stack?

members will follow in turn. J. Fraser's spiral [7-25] presents us with a dramatic example of this problem.[7] Only a tracing pencil will convince us that we are not looking at a spiral at all. We are seeing a series of concentric circles.

This is not a chapter on illusions, but it is important to understand that we must learn to draw what our eyes actually see—not what our brain tells our eyes they are seeing. Later in the chapter where appropriate, other illusions—and how they affect our perception of what we are seeing—will be discussed. We are now ready, however, to try our hand at drawing.

FREEHAND DRAWING

Anyone can learn to draw! The reason some people can pick up a pencil and draw anything with realism is because their hand is able to duplicate what their eye is seeing. They do not have to think about telling the hand what the eye is seeing. It just seems to be a natural process. Fortunately for most of us, however, this is an acquired ability that can be developed and enhanced through a series of exercises that will strengthen the coordination of the eye–brain–hand team. Experience has shown that it is much easier to draw well than it is to play expertly on a musical instrument. Neither skill is acquired by inheritance. Like learning to master music, learning to draw requires both instruction in the fundamentals and directed practice. Learning to draw is basically learning to see. It is necessary to practice until the eye can accurately perceive images that the brain will correspondingly translate into hand movement.

[7]J. Fraser, "A New Visual Illusion of Direction," *British Journal of Psychology*, Vol. 2 (1980), p. 307.

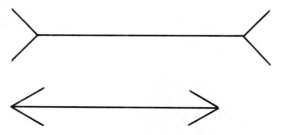

[7-23] Which horizontal line is longer?

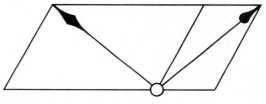

[7-24] Is the hour hand as long as the minute hand?

[7-25] A spiral—or concentric circles?

The drawing exercises in this chapter will help you to represent things as they *appear to the eye*—not how they actually exist in reality. You must learn to make the transition from a condition where your brain controls your eye and tells it what it *should be seeing* to a preferable one where your eye reports to your brain what it is *actually seeing*. Remember, there is a vast difference in what we know about our surroundings and that which we actually see.

[7-26] Frogs are practically blind, and their eyes are stationary in their sockets. However, they have other remarkable abilities. They receive, process, and relay information *important to the frog* and filter out everything else. For example, a juicy bug flying toward him is important; one flying away is not. The frog's eye signals his brain if:
(a) An object is flying toward him,
(b) The object is "bug size,"
(c) The object is flying at "bug speed,"
(d) The object is within range.
Everything else is ignored. The frog's eye also makes other life and death decisions for him without bothering his feeble brain. A sudden shadow, for example, will trigger a danger signal causing the reflexive jumping mechanism to function. His eye–brain dominance would make him a good candidate to learn to draw—if he could only hold a pencil.

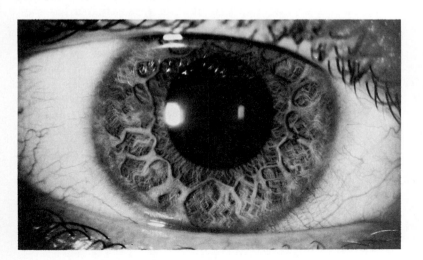

[7-27] The human eye, unlike that of the frog, moves. It also negotiates with the brain to interpret the meaning of transmitted light. Your proficiency in drawing depends on how well your hand can replicate the image that the eye is seeing, not the image that the brain thinks the eye should be seeing.

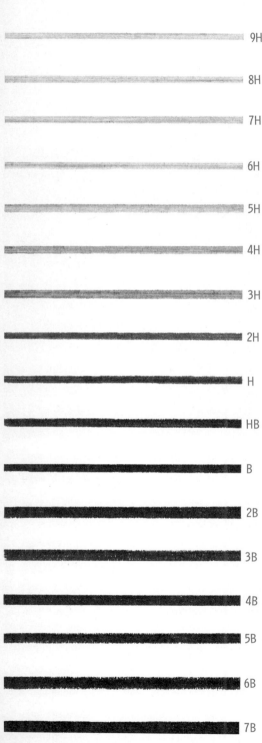

9H

8H

7H

6H

5H

4H

3H

2H

H

HB

B

2B

3B

4B

5B

6B

7B

[7-28] Pencil lead weights.

Once we can master the skill of seeing, we can develop proficiency in translating this into a two-dimensional image. Since we see in three dimensions, the task of accurately transferring an image to a two-dimensional form on paper may seem overwhelming. Do not despair. There are several techniques that we can use to help make the transition easier. Here, illusion becomes our friend—not our enemy. Before beginning the first drawing exercise, we must obtain appropriate drawing materials.

MATERIALS FOR DRAWING

It takes very little to begin drawing. Basically you need only two items: a pencil and a piece of paper. This sounds simple enough, but it is useful to understand something about your materials before you begin. First, the *pencil*. Graphite pencil leads are divided into two categories, soft (the B series) and hard (the H series). An HB pencil falls midway between the two types [7-28]. The most commonly used weight for freehand drawing is 2B. We will use this pencil to begin our drawing exercises. Rather than changing pencils to obtain gradations in line intensity, we will learn to vary the pressure on the pencil. Applying light pressure will produce fine, lightly textured lines. Heavy pressure will produce thicker, denser strokes in the areas you desire them. However, as you progress you will find that a variety of pencil weights can add interest to your drawings. A soft pencil will make a darker, thicker line, and a harder pencil will make a faint but sharper line. You will find that the softer pencils are much easier to draw with. This is why it is recommended that you begin with a 2B pencil. Do not use a lead holder or an ultrathin mechanical pencil. If you use a wooden pencil, sharpen it so that it has a conical-shaped point. Always have three or four sharpened pencils available before you begin to draw.

You will quickly discover that there are a number of different types of drawing paper to choose from. Drawing papers may be purchased in pads (spiral or glue bound) or as loose sheets. A good size is 11 by 14 inches. The beginner should start with a drawing paper that has a little "tooth" to it. This means that it has some surface texture that will let the graphite from the pencil adhere to it easily. The "slicker" the paper, the more difficult it will be to draw on it with a pencil.

You may want to purchase some standard, inexpensive newsprint paper to practice drawing lines and to do some of the quick gesture drawing exercises. However, the majority of your drawings should be done on good-quality, white, bond drawing paper. Remember, all of these materials are relatively inexpensive, and unwanted drawings can be discarded without emotion. Always assume, however, that you will keep every drawing you begin. Every drawing should be a potential "keeper."

For the beginner, *erasers* offer a temptation for indecisive and sloppy work. They should not be used! Do not even buy one. If a line is drawn in the wrong place, redraw a better one. Do not concern yourself with the original line. It will fade into insignificance as you proceed. If not, start over. After you have mastered the fundamentals in this chapter and have gained considerable proficiency, you will find that a *kneaded eraser* can serve as an invaluable drawing tool. By pressing the paper with a kneaded eraser, unwanted lines and tones can be removed without destroying the crispness of the other linework. At that point you may also want to acquire an *erasing shield*. Now, let's get back to the pencil.

HOLDING THE PENCIL

The manner in which you hold a pencil is a personal matter. By now you probably have acquired a consistent and comfortable way to hold a pencil. The most effective method is illustrated in Figure [7-29]. You may wish to try this position if you do not now use it. Use the bottom of the little finger as a steadying point on which to slide the hand.

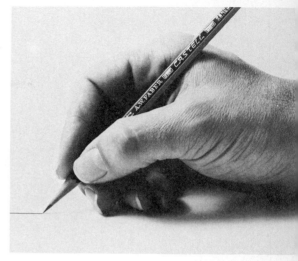

[7-29] Hold the pencil comfortably.

KEEPING A SKETCHBOOK

The artist's sketchbook is an important tool used in learning to see. It helps to make drawing more than just a classroom activity and brings it into your everyday life. It is a method for recording your ideas, thoughts, and experiences, as well as providing a convenient method to practice your drawing skills. You will be interested to note your improvement from the onset of your sketchbook to its final pages.

For maximum effectiveness your sketchbook should be used daily. This may seem difficult at first, but you will notice that in time you will place a gradual dependency on it for practicing new methods and freely expressing your ideas. It can be considered to be a journal, and you may want to make comments on new techniques or elaborate on ideas for later development. In selecting a sketchbook, keep in mind that it should be an easy size to carry, but not so small that it is difficult to make full-sized drawings. Make sure that its construction is sturdy and able to withstand the punishment of being carried with your other books. A variety of styles are available at your bookstore, ranging from a hardback, bound volume to a spiral bound, or cardboard cover style. A good size to purchase would be 9 by 12 inches or 11 by 14 inches. Be sure to date each page of your sketchbook, not only to note the progress of your improvement, but for future reference to keep in mind when and where your ideas and drawings were made.

Now we are ready for our first exercise.

MAKING A LINE

A line drawing is the most direct means of expression, and lines are the basis for all pencil drawings. They define the shapes of objects and give meaning to form. Lines alone can be interesting and expressive, and variation in line character can add relevance and feeling to any drawing. As you draw lines, a variation in line quality can be achieved by varying the pressure used on your pencil. In Figure

[7-30] notice the expressive quality that is achieved by the variation of line weights. After practice, this will become second nature to you. You will find that your hand will automatically vary the pressure on the pencil as you are drawing. This will not be a deliberate attempt on your part to change the line widths, but rather it will happen instinctively as you move the pencil across the paper. When this happens it will be encouraging, because it is a spontaneous response of the hand to what the eye is seeing.

Exercise 7-1

(a) Holding your pencil in a normal drawing position, make a series of horizontal lines running all the way across the page. The pencil's movement should not be the result of the movement of your fingers. Rather, the pencil will be moved by the simultaneous motion of the hand and forearm. Your wrist will be locked and your fingers will move very little. Vary the pressure on your pencil as it glides across the page. This should cause the line to have varying width and *value* (relative lightness or darkness). Make several pages of these horizontal lines. Experiment with making the lines vary in value in the same places on the page so that an optical illusion of motion is created. The lines may even begin to look like waves even though they are actually straight lines, with the only variation being in line width and value [7-31].

(b) Now make a series of vertical lines using the same technique. Try to keep the spacing between lines equal to develop control over your pencil and the paper. Once again, vary the line pressure to create an optical illusion of lines that are in motion running across the page. The vertical lines may be more difficult to draw because you are used to making and seeing lines running left to right across a page. This unfamiliar arm movement may also require more practice to achieve a page filled with straight vertical lines.

(c) The last series of lines are made in the shape of circles and ellipses. First, make a series of circles in horizontal rows running across the page. Try varying the pressure on the pencil in different positions on each circle. Some will have a dark area at the top of the circle, some at the sides and some at the bottom. Now, continue by making rows of ellipses, varying the angle of each. Make several rows of ellipses that slant toward the left side of the paper as well as the more familiar form, to the right side.

Do not get discouraged if your pages of lines are not neat. These are exercises to help you develop a *mastery* over the pencil and paper. Beauty is not our main concern. Let the pencil and paper know that you are in charge! They will soon become eager servants waiting to perform whatever you ask.

This exercise is designed to help train your eyes to see rather than to rely on what your mind believes to be true. Many familiar objects do not look the same when they are taken out of a familiar context. Salvador Dalí was a master at being able to take familiar objects and place them in an unfamiliar setting to cause a disturbed feeling [7-116]. Notice how much longer you spend looking at one of his paintings as you try to recognize all the objects. In many cases our eyes are telling us one thing while our mind is telling us another. It is important to train your mind to rely on your eyes and to transfer what the eyes see to the hand, and in turn to the pencil.

[7-30] Line can add spatial dimension. The accent and variation of line used here help to give it a plastic quality. Juan Gris, Portrait of Max Jacob, 1919

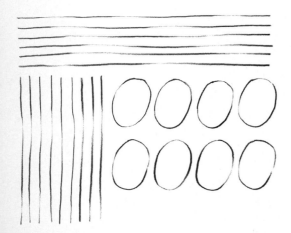

[7-31] Learning to vary the pressure on your pencil is an essential skill that is learned through repetition of line exercises.

Exercise 7-2

(a) Find a line drawing of a familiar person, such as the one of Albert Einstein [7-32]. If you have seen his picture before, he is easily recognized and named because your mind identifies the familiar features of his face and tells you it is Albert Einstein. Place the line drawing in front of you *in the upside-down position* and begin to copy it on your paper. Your drawing will also be upside down since it will match the drawing in front of you. Start at the top of the drawing and progress downward. Try to copy each line on your drawing as it appears on the original drawing. Keep in mind the length and shape of each line and its relationship to the other lines and to the edges of the paper. It may help you to define your drawing space. Sometimes it is difficult for beginners to visualize that the edge of the paper can also be the edge of the drawing. Once you get started with your drawing you will find that you are more interested in copying the lines and shapes rather than in identifying the familiar features of Albert Einstein. Do not turn your drawing upright until you are completely finished.

(b) Figure [7-33] is an upside-down photograph of a well-known American actor. In its unnatural orientation you may have difficulty recognizing the person as John Wayne. However, do not concern yourself with identification. Again your task is to *copy the upside-down image* on the photograph onto your drawing paper. Proceed just as you did in copying the drawing of Albert Einstein. Copy each line, reproducing line connections, angles, and contour shapes. You will find that it will be more difficult to copy the wrinkles and contours on the photograph than it was to copy those on the Albert Einstein drawing. You will have to make your eyes work harder in transferring information to the brain, and the brain must work harder to sort out the characteristics for duplication. Of course, the photograph has subtleties of shading that were not present in the line drawing. Finish the drawing in one sitting. When you have completed the drawing, turn it right-side up. You will be amazed at the quality of your effort.

(c) Once you have completed parts (a) and (b) you may want to try another by first tracing a photograph to get it into line form. Then turn it upside down and try to redraw it as in part (a). You will be surprised at how proficient you can become at recreating a drawing, because you are taking more time to see what is actually there.

(d) This exercise should be done in groups of four or more. Acquire an 8- by 10-inch photograph of a familiar person. Cut the photo up into as many equal-size pieces as there are people in your group. Mix the pieces up and let each person in the group pick a piece. Each person will now make a line drawing of that section of the photo. Make sure that each person is making the line drawing on the same-size piece of paper so that the proportions of each drawing will be compatible. Your piece of the photograph may show only one eye and a piece of the nose, but try to be as accurate as possible in recreating your piece. After each person is finished, place all the line drawings together to recreate the photograph. See how accurately your familiar person is recreated in the drawings. It will surprise you to see what details some people emphasize and others ignore. This points out that we all see differently and recognize subtleties in expression to different degrees, but we all recognize the importance of seeing.

[7-32] Turn me upside down.

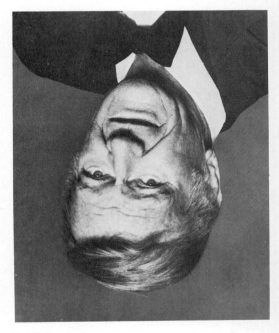

[7-33] Draw what you see.

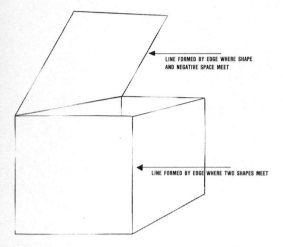

[7-34] Draw the visible edges.

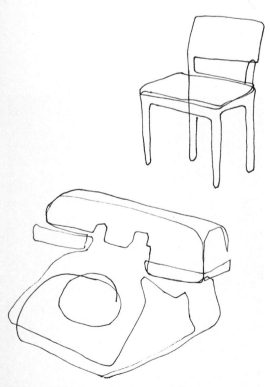

[7-35] Contour drawings project the essence of character of the objects seen.

CONTOUR DRAWING

The simplest line has length, direction, width, and value. Often the line is used to describe the bounds of contours, separating each area or volume from adjacent ones. In this role, it is its most expressive self. The preceding exercises have helped us to develop a sensitivity for drawing the line as an isolated entity. Now we will use the line to create contours.

A contour line is merely a line that describes an *edge*. When two shapes meet, they form an edge, which we also call a *contour*. For instance, the front face and side face of a box meet and form an edge. When we draw all the edges that we can see, we have a contour of the box [7-34]. A contour drawing then consists of all the edges where shapes meet, including the edge created where air (space) meets an object. If you open the box, you will notice that there is an edge made by the top of the lid as it meets the open air as well as other edges that are made by the sides of the box meeting each other. Contour drawings never show value, texture, or shading. They are pure line drawings and consist only of the edges of an object.

Contour drawing will do more to improve your drawing ability than any other exercise. This is because it emphasizes seeing more than any other drawing technique. Done properly, contour drawing will give you the confidence and understanding to draw anything. In this chapter we will progress through three stages of contour drawing: pure contour drawing, modified contour drawing, and cross contour drawing.

First, we want to develop a distinct kinship between the object you are drawing, your eyes, and your hand. Your eyes must learn to focus on and follow along the features of the object as if they were touching it. In this sense you may consider your eyes as a substitute for your hands. This technique will take time and practice, but it will open a whole new world for you once you have become accomplished at contour drawing [7-35].

Pure Contour Drawing

Pure contour drawing develops the eye–hand coordination that is invaluable in being able to draw. Some of the following techniques may seem frustrating at first, but they are designed to enhance your coordination and make it a natural response in later drawings. All contour drawings are to be done very slowly. At first you may feel like it is taking forever just to complete one small drawing. As you become more accomplished and really begin to learn to see, the elapsed time of drawing will not seem so significant. There are a few important points or "rules" to remember when working on contour drawings. These are:

1. *Do not look at your paper.* Since you will not be making any lines that your eye does not see, and you are keeping your eye on the object at all times, there is no need to look at your paper until you have completed the entire drawing. This will become your greatest temptation, *but resist it!*

2. *Keep drawing at the same rate of speed.* Your eye should be moving consistently along the edges of the object and your hand (with pencil) should be following at the same rate. Your hand should not move faster or slower than your eye.

3. *Place the object you are drawing very close to you.* You need to be able to see details easily, and it is also important that your eye is not distracted by other objects that may be between you and the object being drawn.

4. *Make sure you will not be interrupted for at least 30 minutes.* This will allow you to create an environment for total concentration. People talking, music, or television can be a major distraction to the concentration necessary for pure contour drawing.

Keeping the guidelines above in mind, we are ready to move on to your first pure contour drawing.

There is only one right way to draw and that is a perfectly natural way. It has nothing to do with artifice or technique. It has nothing to do with aesthetics or conception. It has only to do with the act of correct observation, and by that I mean a physical contact with all sorts of objects through all the senses.
    Kimon Nicolaides, The Natural Way to Draw, *1941*

Exercise 7-3

Your first contour drawing will be of your hand. You may think this will be easy, since your hand is a very familiar object to you. We are still going to make your mind work extra hard in seeing every detail in that hand. *We will do this by having you draw with the opposite hand to the one that you use to write.* If you are normally right-handed, in this exercise you will use your left hand to hold the pencil, and vice versa [7-36]. Because it will take tremendous concentration for you to instruct your "uncoordinated" hand to make each line the way your eye tells it to, you will find that you will be forced to draw at a much slower pace, but with surprising accuracy. Keep in mind all the previously discussed guidelines—do not look at your paper, do not let your eyes get ahead of your hand, and keep the pencil moving at the same rate. If you can not seem to resist looking at your paper during your first contour drawings, turn your body around so that you are forced to look away from your paper if you look at the hand you are drawing. Since this drawing position is very awkward and the paper may have a tendency to move, you may wish to secure the edges of your paper to the drawing table with tape. Now let's begin the first drawing.

Hold your nondrawing hand slightly below eye level. Spread your fingers slightly. Pick a central point on that hand for your eye to rest on. Pretend that your eye is actually touching the hand. Position your pencil somewhere in the center of your paper. Do not look back down at your pencil until the drawing is completed. Slowly begin to trace the edge of your hand with your eye. At the same time your drawing hand will be moving the pencil on the paper as if it were your eye. Try to observe very closely all the little details of wrinkles and bends along the edge of your hand. Perceive in your mind that your pencil is recording everything exactly as you are seeing it. Do not pause or pick up your pencil during the entire drawing. Keep concentrating on making the pencil follow what the eye is actually seeing. This will be especially difficult since you are drawing with the opposite hand that you are used to writing with and coordination skills will be lacking. It is possible, however, to concentrate hard enough to make a teamwork relationship develop between the eye, mind, and hand. Once you have completed your drawing,

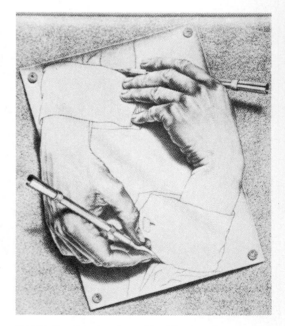

[7-36] Practice drawing with either hand.

[7-37] Drawings made with your nondrawing hand will be crude and should not be judged as to accuracy.

[7-38] The silhouette of a hand (a) may represent either a right hand (b) or a left hand (c).

(a)   (b)

(c)

you can look at your paper. What you see will probably not resemble your hand very closely [7-37]. This is to be expected, especially since you were drawing with the opposite hand to the one you normally use. Few people realize that the mere outline drawing of an outspread hand is ambiguous [7-38]. It is impossible to tell whether it is a right hand seen from the back or a left hand seen from the front. You will also probably notice small details that you never knew existed on your hand, or at least details you never knew could be so graphic.

Exercise 7-4

Now we will repeat Exercise 7-3. This time switch and use the hand for drawing that you are normally used to using. If you are right-handed, you can go back to using your right hand, and if you are left-handed, you may once again use your left hand. Now, complete another contour drawing of your opposite hand. (This time it will be the opposite hand that you will be drawing.) Hold your hand slightly below eye level [7-39]. You may wish to pose your hand differently for this drawing, such as a clenched fist or in a pointing position [7-40]. Once again, pick a point on the outside of your hand for your eye to rest on. Slowly follow along that edge with your eye, matching each eye movement with a corresponding pencil movement on the paper. Remember, you are still not allowed to look at your paper, and you should not let the eye move any faster than the hand can move or the hand faster than the eye. Keep the pace slow and consistent, and do not pause at any point during the drawing. This drawing should take anywhere from 15 to 30 minutes if done correctly. You may find that this process becomes tedious as you try to move slowly around the hand, including each little bump and wrinkle. But, as time passes, you will find that you have become enamored with seeing,

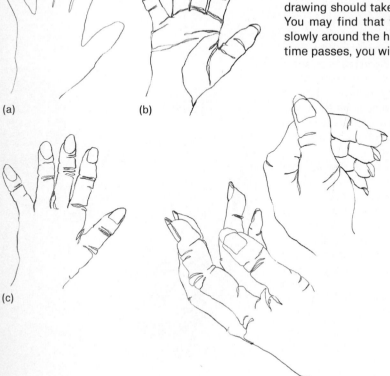

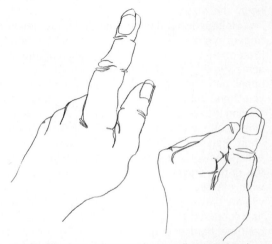

[7-39] Practice drawing your hand in various positions.

[7-40] Clinch your fist or point your finger and make a contour drawing using your normal hand to draw.

perhaps for the first time. Most students have found that this form of drawing takes more energy and concentration than any of the other types of drawing. It is for this reason that it is vital that you emphasize pure contour drawing in your daily practice drawings. Many professional artists begin each session with pure contour drawings just to increase their level of awareness and concentration.

Modified Contour Drawing

Modified contour drawing is done exactly the same way as pure contour drawing with the exception of one guideline. Now you may look at your paper occasionally, but *only* for the purpose of establishing proper lengths and proportions of lines and making sure that angles are correct. You will also need to glance at your paper from time to time if you need to reposition your pencil to begin a new line on the object. You will still make your pencil follow what your eye is following and, although you are allowed to look at your paper, *you should never be looking at your paper while your pencil is moving.* Since your pencil should only be following what your eye is seeing on the object, it is impossible to make a mark on the paper while looking at the paper and not the object. Be sure you do not use the freedom of being able to look occasionally at your paper as a crutch. It is only intended to be a guide in establishing more realistic relationships.

Exercise 7-5

Take off one of your shoes, or even better, trade shoes with another person in your class. Place the shoe on the table or desk immediately in front of your paper. Pick a point on the shoe to place your eye. Slowly start along the edge of the shoe with your eye, and just as in pure contour drawing, follow that same edge with your pencil. When you get to the end of the edge, you may want to check the length of the line you just made and then reposition your pencil for the next contour. To do this you may glance down at your paper briefly. Do not make any marks on your paper while you are looking at it. Once the pencil is repositioned, begin again. Repeat the same procedure until the entire shoe has been drawn [7-41]. This drawing may take as long as 1 hour if done properly. Remember that you are still only drawing contours—no shading or texture should be indicated on this drawing. Concentrate on what you are actually seeing and its relationship to the whole object. When you have completed your drawing, you should note a dramatic difference in the accuracy of this drawing when compared to your previous pure contour drawings. There will still be inaccuracies in proportion and angles, but these will be improved through additional techniques and exercises that we will do later. Notice that your drawing is actually beginning to look like the object you viewed [7-42]. Now you are learning to see, and in turn you are learning to draw.

[7-41] Contour drawings are made up of lines described by an edge.

Exercise 7-6

Take a sheet of paper, preferably torn from a spiral notebook. Crumple this paper in a wad and then let it slowly open out. Now place it in front of your drawing paper. This will be the subject for our second

[7-42] Contour drawing is a good method to simplify and at the same time add a certain emotional quality.

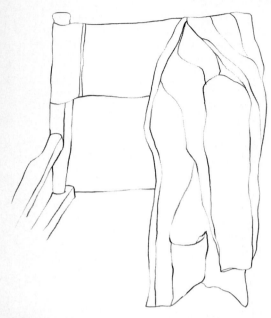

[7-44] Crumpled paper provides a variety of edges for contour drawing.

[7-43] Notice that varying the line pressure on the pencil will add character to the continuous-line drawings.

modified contour drawing. You may find the piece of paper more challenging than the shoe, since it is unique unto itself. No two papers crumple exactly the same, and it will not be as familiar as the shoe might have been. Also, if you have torn the paper from a spiral notebook, you may find all the detail around the torn edge to be quite time consuming to draw if done accurately. Remember, only look at your drawing to reposition your pencil or to check on proportion and angles. Be sure to include all the contours of the different surfaces that have been created when you crumpled the paper.

Exercise 7-7

Now we need to really challenge our ability to see. Take an object that has a lot of detail in it. Some good examples are: an ear of shucked corn, a woven basket, a hand egg beater, a pine cone, or any complex plant or flower. If you browse through the vegetable section of a grocery store, you can find numerous suitable items (artichokes, ginger root, pineapple, etc.) that would be perfect subjects for a challenging modified contour drawing. Spend at least 1 hour making your drawing. Try to see every detail in the object. Remember, you are not to include any shading or shadows, but you should define all the edges (contours) of the details. You will be amazed at the accuracy of your drawing. Notice also that your drawings are becoming more and more expressive. Try to keep in mind your very first exercise of varying the hand pressure placed on the pencil. You should now be automatically creating interesting lines without even thinking about it. Make several other contour drawings using complex objects. When you feel proficient at contour drawing, continue to challenge yourself by combining several objects into entire still-life settings. The more contour drawings you can complete, the more accomplished you will become.

[7-45] Plants make good subjects for contour drawing.

[7-46] Even the simplest of objects make interesting contour drawings.

Cross-Contour Drawing

Cross contours are drawings that help to explain the volume or third dimension of an object. Many students mistake a contour drawing for an "outline" drawing. The two are not synonymous. An outline drawing is very "flat" and shows only the outside lines of an object.

It lacks character. On the other hand, a contour drawing defines the edges of *all* the surfaces, inside the object as well as outside [7-47]. The cross-contour drawing helps to make this more obvious. Cross-contour drawing can define an object using *no* outside edge contour lines. Rather, the object will be defined by showing inside contours only.

Exercise 7-8

Once again you will use your hand for a subject. Hold it out in front of you so that you can still follow all the contours with your eye. You may need to study it by placing your elbow on the desk top. Fix your eye on any outside point, just as you did in the first contour drawing. This time, instead of moving your eye along the edge of the hand, move it across the hand at a right-angle path to the outside edge. At the same time your pencil will just be recording on the paper what your eye is actually seeing. Move your eye in this manner until it reaches the other side of the hand. At this point, move your eye back across the hand to the side where you first began. There are no visible contour lines on your hand to guide you, but there is a contour path along which your eye followed. A cross-contour line will show all the individual dips into the hollow places and rise over all of the little bumps and raised veins [7-48]. Students sometimes benefit from actually making drawn black ink lines across their hands to show the path of the cross contour and to get the feel for the shape of the hand. Continue crossing back and forth over your hand until the shape of the hand appears. You will have made no *outside* contour lines to define the hand. Instead, all the cross-contour lines will have defined the extent of the hand by their stopping and starting points. This is also a good time to practice line quality since you should feel a definite variation on the pressure placed on your pencil. More pressure should be applied as you push into hollow places and less pressure as you climb over bumps.

Exercise 7-9

Now use an inanimate object as a subject for a cross-contour drawing [7-49]. You probably have several that you used for the modified contour drawings. Make a cross-contour drawing of one of the same objects. Now compare the two drawings. Your cross-contour drawing should define the outside shape of the object just as accurately as the original contour drawing. If you are having trouble with cross-contour

[7-47] Contours add definition and depth to shape.

[7-48] A cross-contour line defines on the drawing the image of depth as seen by the eye.

[7-49] A drapery is a good subject for cross-contour study. The sea-like roll made of curving surfaces is defined in its up-and-down movement.

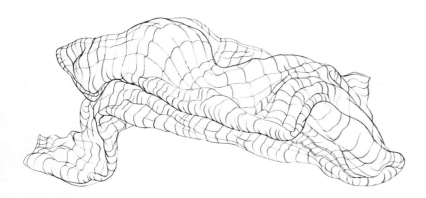

[7-50] Do you sense motion or a third-dimensional quality by looking intently at this drawing?

[7-51] Your drawing can produce a sense of action and drama.

drawing, select an object on which you can easily draw across or around (a crumpled box, a detergent bottle, an old shoe) and can actually draw black lines across or around the object. Now make your eyes slowly follow those lines. As your eyes follow those lines, your pencil should be transferring the corresponding image onto paper. A quick way to get an impression of cross-contour drawing is to crumple a piece of lined notebook paper. See what happens to the lines once they are made to go around all the folds and bends in the paper. Practice cross-contour drawing until you are confident that you are able to perceive the volume and mass of an object as well as the outside shape.

LEARNING TO SEE MOVEMENT: GESTURE DRAWING

Gesture drawing is another approach to *learning to see*. The hand continues to duplicate what the eye is actually seeing. A gesture drawing records not what an object looks like or what the object is, but the movement or action of the object. Figure [7-50] is an optical illusion of motion that was produced by A. Michael Noll of Bell Telephone Laboratories. He used computer-generated line patterns to produce wavelike effects that appear to have a convincing three-dimensional reality.[8] In gesture drawing you are going to be relating to what the object is *doing*, its motion and action [7-51]. The practice of gesture drawing will not only help to enhance your ability to see, but just as important, will add feeling and energy to your drawings.

Gesture drawings, just like contour drawings, still include only what is actually seen by the eye. The details of the subject will be ignored and only the action or movement of the object will be included in your drawing. Do not get caught up with making the object actually look like what it is but try to make the movement recognizable. In Figure [7-52] the action of hitting the tennis ball is more important than the identification of the athlete.

If a gesture drawing appears to be a meaningless scribble after your first effort, it does not matter. The main concern is that it should capture the essence of what was happening at the moment that you were making the drawing. Courtroom drawings are an excellent example of how gesture drawing is put to use. They seem to be alive and take on strong personality just by virtue of their gestural quality. Courtroom artists have become very skilled at being able to see quickly and record the action taking place at the moment.

Exercise 7-10

Just as contour drawing makes you believe that you are actually touching the object with your eyes, gesture drawing makes you believe you are actually doing whatever the object is doing. Your pencil will follow the movement exactly, as if it were making the movement happen. You will place your pencil on the paper and not pick it up until the gesture drawing is completed. All linework should be one continuous

[8]John McCarthy, "Information," in *Scientific Technology and Social Change* (San Francisco: W. H. Freeman, 1974), p. 229.

line. There is no time to pick up your pencil and place it back down on the paper. In most cases the action would be completed by the time you could look at your paper and back up at the model.

Before beginning a gesture drawing, place your pencil at a starting point on the paper that will allow enough room to show all the action taking place. A good way to begin gesture drawing is to draw from a model. This can be a classmate or a roommate making quick poses of some action. Start with simple movements such as walking, jogging in place, raising arms, bending down, or clapping hands. These are very simple and not strenuous to the person doing the modeling. Time your drawings and gradually work into longer periods of time. Start with 15-second drawings [7-53]. At first this may seem impossible, and on your first few drawings you will probably hardly have your pencil on the paper by the time the 15 seconds goes by.

It does not matter where you begin your drawing as long as you have picked one point for your eye to start and then to move it along the lines of movement. Remember that you are recording movement only, not detail. Your first drawings will be unrecognizable scribbles [7-54].

[7-52] Capture the motion . . . .

[7-53] Gesture drawing . . . capture the mood or action in 15 seconds.

[7-54] Your first gesture drawings may appear to be meaningless stick figure scribbles.

[7-55] A gesture drawing that was completed in 3 to 5 minutes.

However, be sure that your "scribbles" resemble the movement that actually took place. As you increase your time on each drawing (progressively add time in 10-second intervals) you will be surprised at how your drawings will begin to take on the characteristics of a person. Continue making pages of gesture drawings until you have completed a series of drawings varying in time from 15 seconds to 2 minutes. It will be necessary to have someone time each drawing, since you cannot be looking at a clock and the model simultaneously.

When starting a gesture drawing, it may be helpful to look quickly for the longest line in the object. In the case of a human subject, you would look for the spine. Then decide what kind of line is being created—vertical, horizontal, diagonal, or a curve. Where is the center of weight being transferred? Just remember to keep your eye and hand moving together, never letting your pencil leave the paper or letting your eye leave the subject.

Exercise 7-11

Now that you have warmed up with the quick "scribble" gestures, you can move to gesture drawings that are done over a slightly longer period of time. Increase the time allowed for each action to 3 to 5 minutes. Your model may have difficulty holding one pose for 5 minutes, so slow down the model's action and perhaps repeat it several times. Also, pick actions that involve more than one type of motion (walking several steps and then kicking the foot, or perhaps clapping the hands). This will cause your final drawing to show a series of gestures drawn one on top of another to show all the action that has taken place. Be sure to start your drawing at an appropriate point on the paper so that the complete range of the action can be recorded. You now have a drawing that

[7-56] The sequence of actions is the most important thing to capture with your eye.

includes a series of actions. Make sure that you do not use the extra time to worry about details of the subject, only record the essence of the movement. We only want to see what the subject is *doing*, not if the subject has hair or is wearing shoes. After completing several drawings, go back and recall each of the actions that took place in each particular drawing. You will be amazed at how your drawings are beginning to become recognizable as records of particular action. Try letting your classmates or friends guess at the action that took place in each of your gesture drawings.

Exercise 7-12

In this exercise we will be combining contour drawing and gesture drawing. Gesture drawing includes more than drawing objects that are moving. It also includes describing the *character* of an action [7-57]. In this sense a plant growing up the side of a wall has gesture, and a crumpled rug on the floor has gesture, even though you do not see the action as it happens. It is important to capture this *implied action* in your drawings. By combining contour drawing for accuracy of detail and gesture drawing to capture the "gesture" or character of the object, we can actually make the subject "come alive." This type of drawing may be identified as a "quick contour" drawing.

Pick a point of focus on a stationary human model just as you did in contour drawing. You will still be looking only at the model and not at your paper. You will be moving your eye much faster and your pencil will be moving much faster than in contour drawing. As you move faster over the contours, your eyes will eliminate much of the detail and you will retain only the "gesture" or pose of the model, not the detail [7-58]. Make sure that you are still making a conscious effort to draw only what you see. Try to move at the same pace throughout the drawing. Draw the entire figure, but do not worry about personal characteristics (hair, eyes, long or short nose, and so on) unless they have a direct relationship to the "feel" of the pose (such as long hair blowing back and forth). Now look back at your original scribble gesture drawings and compare them to your recently drawn quick contour drawings. You should be able to

[7-57] Gesture drawings capture the character of the action.

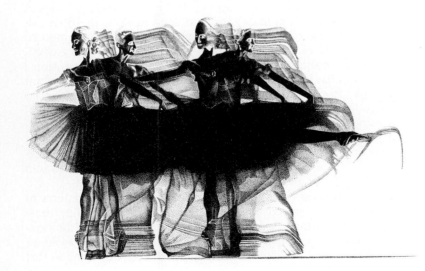

[7-58] Gesture drawings may be made in a variety of styles.

detect the same feeling of action in your quick contour drawing as you have recorded in your moving gesture drawings.

For additional practice in gesture drawing, take a pad of paper and pencil to a playground or park. Watch the children at play or people enjoying a picnic. Make some quick gesture drawings as well as several quick contour drawings. More and more your drawings should begin to clearly define the characteristics of each of the different subjects used in the drawings.

LEARNING TO SEE IN SPACE

One of the more difficult transitions in drawing is to take a three-dimensional subject and then transfer it to a two-dimensional medium and have it still appear to be a three-dimensional subject. What we want to do is create a three-dimensional illusion. We are still going to draw exactly what we see, but we are going to use various illusionary techniques to represent what we see. This will help us to overcome what our mind is telling us that we perceive, rather than what our eyes are actually seeing.

Several methods have been developed to represent three-dimensional space accurately on a two-dimensional surface. A commonly used method is called *linear perspective*. European artists developed this technique during the Renaissance period, and various other forms of perspective have been developed from this method. It is only an approximation of the view seen by the eye. Although perspective can be studied in depth and broken down into a precise technique, you, as a beginning drawing student, will only need to become aware of the various types of perspective and how to apply them to your drawings.

Linear Perspective

In attempting to depict a three-dimensional image on a two-dimensional piece of paper, illusions again play a role in guiding our brain in interpreting what our eye is seeing. In Figure [7-60] the ends of the two parallel vertical lines appear to bend inward. In 1861, Edward Hering discovered that when parallel lines are superimposed on a radial field of lines, they will appear to lose their parallelism. This is because the radial lines are seen both as a two-dimensional pattern on a flat background, and as vanishing lines to a one-point perspective (three-dimensional) drawing. The brain knows that no actual object can be simultaneously two-dimensional and three-dimensional. If you tip the book away from you so that the three-dimensional aspect disappears, the two lines will again appear straight and parallel. This is rather convincing proof that the convergence of lines within our field of vision are important factors in influencing the brain's interpretation of what is seen.

Most artists do not use formal projection system procedures to establish linear perspective in their drawings. Because they draw what their eyes actually see, artists can create a drawing by just looking. However, it is important for us to become generally aware of some of the formal principles involved in perspective drawing so that

[7-59] Inanimate objects can have a gestural quality which gives them an implied action.

[7-60] Tilt away from you to restore parallelism.

we can develop the ability to create proper perspective in our beginning drawings.

If you were to stand in the middle of a set of railroad tracks and look down the tracks into the distance, you would notice that the two rails appear to meet in the distance [7-61]. You know for a fact that this condition could not possibly be true since the train continues to run on parallel tracks to its final destination. If this scene were to be drawn relying on what our mind tells us rather than what our eyes actually see, we would draw railroad tracks as two parallel lines running the length of the paper. Such a representation, however, would destroy the illusion of space and take away the "perspective" feeling of the drawing. Instead, let's consider what our eyes are actually seeing.

In perspective drawing it is again possible to use illusionary tactics to deceive the brain [7-62]. Equal-sized figures do not appear to be equal [7-63]. In Figure [7-63] the naturally appearing perspective lines of the background are drawn to suggest a three-dimensional scene. The people are actually the same size individually, but the person on the extreme right appears to be considerably larger. The converging lines of the background confuse the brain as it attempts to make a definitive judgment. This is possible because the brain already has a large data base that offers convincing evidence that equal-sized figures *must* appear to diminish as their images recede into the distance.

If you were to look directly down a road such as shown in Figure [7-64] at the Dallas–Fort Worth Regional Airport, you would see that the light poles seem to be getting progressively shorter as they recede into the distance. At the farthest extreme, the last pole would appear as a point. You will also notice that this point rests on a line that separates the sky and the earth. This is what the eye actually sees, although reality does not exist in this form. In linear perspective, the line where the sky and earth meet is called the

[7-61] Parallel tracks appear to meet on the horizon.

[7-62] The brain has stored the information that from experience the railroad rails are parallel. It will not allow the eye to conclude that the two railroad ties are identical in size . . . which they are.

[7-63] Which guard is tallest?

[7-64] The last pole vanishes as it rests on the horizon.

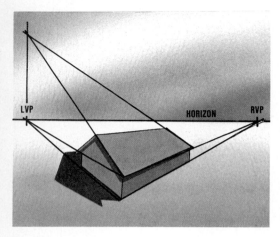

[7-65] Different planes vanish to different vanishing points—on, above, or below the horizon.

[7-66] Bird's-eye view.

[7-67] Worm's-eye view.

horizon line. The point at which the light poles appear to disappear is called the *vanishing point*. Because the illusion of objects disappearing as they recede into the distance appears with absolute regularity, the sizes and spaces between objects can be mathematically determined, if necessary. However, for simple drawing purposes in this chapter, we will be using "sight" perspective and not relying on any mathematical or projective systems. It will be helpful, however, to understand something about how the system works.

Parallel edge lines of horizontal and vertical surface planes vanish to right and left vanishing points lying on the horizon line [7-65]. The parallel edge lines of angled surfaces (such as the pitched roof of a house) also vanish to points on the horizon or to points directly above or below these points. In Figure [7-65] we can see the horizontal edge lines vanishing to points on the right and left ends of the horizon line. The angled lines of the roof plane vanish to a point directly above the left vanishing point. So much for theory, now let's get back to drawing.

The position of the viewpoint in a perspective drawing is determined by where the viewer is located in relationship to the horizon line. If the viewer is positioned above the horizon line, as if looking out of an airplane, the view is considered to be a "bird's-eye" view [7-66]. If the viewer is positioned below the horizon line, as if standing in a deep depression, the view is considered to be a "worm's-eye" view [7-67]. If the viewer is positioned at eye level with the horizon line, the view is said to be an "eye-level" view. All parallel lines will disappear to the same vanishing point.

Perspective drawings also vary according to whether the objects have one, two, or three vanishing points. The most common case is two vanishing points. For example, if you look at the near corner edge of a rectangular box, you will be able to see two of its vertical sides [7-68]. The top and bottom horizontal edge lines of each side will seem to be disappearing in different directions, thus creating two vanishing points. A drawing of this condition would be considered to be a *two-point perspective* [7-69]. *One-point perspectives* have only one vanishing point [7-70], and *three-point perspectives* have three vanishing points [7-71].

Although the vanishing points and the horizon line may not be located within the boundaries of your drawing paper, you must remember that they are always present. To avoid distortion it may be necessary to locate the vanishing points of your drawing some distance off your paper. This results in a more subtle use of linear perspective and therefore makes the drawing more believable.

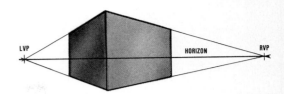

[7-68] Draw a box, using a left and right vanishing point.

[7-69] Two-point perspective.

[7-70] One-point perspective.

[7-71] Three-point perspective.

[7-72] Dürer's wire grid.

Albrecht Dürer (1471–1528) was one of the first artists to develop a device for helping to cope with perspective. In drawing the human figure, perspective is usually referred to as *foreshortening*. Dürer developed a windowlike wire grid that enabled him to foreshorten his visual image as he peered through it at the model [7-72]. The model's head, which is much farther away from the artist, appears to be smaller than it is in reality, and in turn, the feet and legs appear to be much larger than they really are because they are closer to the artist. Dürer then drew his drawing on a piece of paper that matched the wire grid exactly. He copied what he saw through the wire grid, line for line, angle for angle. When Dürer completed his drawing, he had created a foreshortened drawing of the model. Again, this was a tool used to help the artist see since he could not rely on what his mind told him. This is why the Renaissance artists developed the system of linear perspective, to help them overcome what their minds were telling them about what their eyes were actually seeing [7-73]. Through various exercises you can develop a sense

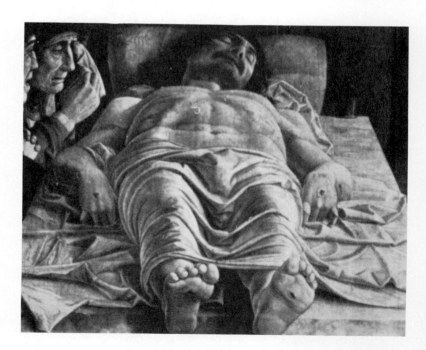

[7-73] The Crucified Christ . . . a famous painting showing the use of foreshortening.

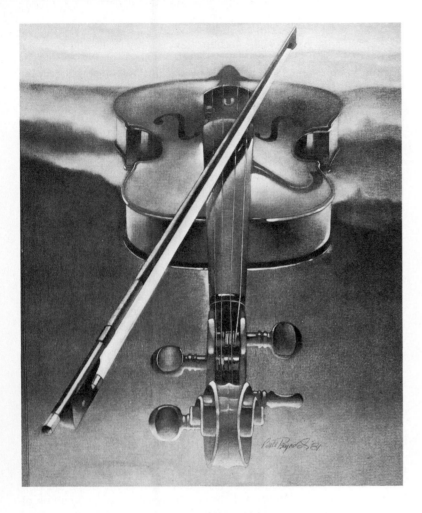

[7-74] Foreshortening adds a sense of realism.

of perspective as a type of second nature. Just as you learned to see detail, you will begin to see perspective in the objects that you draw [7-74 and 7-75].

Exercise 7-13

The horizon line is the foundation reference used for perspective drawing. Remember, it is assumed to be straight and horizontal. Therefore, at a point located approximately in the middle of your paper, draw a horizon line. This is where we will assume that the earth and sky meet. To make a simple one-point perspective drawing, you will only need one varnishing point. Locate it near the center of the horizon line by making a small dot. Now we are going to draw a simple cube. Draw the front face of the cube (this will appear to you as a square) as if you were looking at it directly. Place it either above, directly on, or below the horizon line, but place it to one side of the vanishing point. For this drawing we will want to see at least one additional side of the cube, so make sure that the vanishing point is not covered by any part of the cube. Since we know that all horizontal parallel lines disappear together to the vanishing point, we can now draw lines from each of the corners

[7-75] By use of foreshortening, even common objects can be given a dynamic and interesting quality.

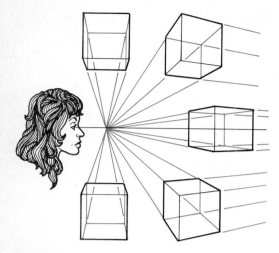

[7-76] View several different box locations.

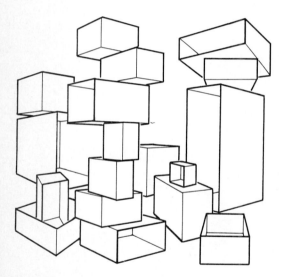

[7-77] Group of boxes.

of the cube's front face to the vanishing point. This will allow us to create the sides of the cube. In our drawing we will assume that the vertical lines still remain vertical. We can now complete the cube by drawing the remaining edges. We have created the illusion in drawing form of the image that our eye is describing. The cube is three-dimensional. Our mind still tells us that all the sides of a cube when measured should be equal, but in reality, if we drew the cube with its receding sides equal in length to the front edges, we would not have created the illusion of a cube at all.

Now repeat this same exercise using two-point perspective. You should place an additional vanishing point on your horizon line. This time, rather than looking at the front face of the cube, you will be looking at the vertical edge created by the two sides of the cube. Therefore, you will begin your drawing using a line representing the point of intersection of the two sides, rather than a square. Connect the top and bottom points of the line with the two vanishing points to create the two sides to your cube. Once again, add the vertical lines for the sides and the parallel lines to create all the remaining edges of your cube.

Try making several cube drawings using a worm's-eye view, a bird's-eye view, and an eye-level view [7-76]. Remember that an eye-level view will be obtained when the cube overlaps the horizon, a worm's-eye view when the cube is above the horizon, and a bird's-eye view when the cube is below the horizon. Next challenge yourself by drawing perspective drawings of groupings of boxes of several different sizes that overlap one another. You may also want to try several groupings of boxes positioned at many different angles, necessitating many different vanishing points [7-77].

Exercise 7-14

In these next drawings you will be able to tap your creative imagination. Using a visible horizon line and two vanishing points, you will draw an imaginary city street scene. In this case we will place the vanishing points off the edges of the drawing paper to avoid distortion of the drawing. The viewpoint can be either worm's-eye, bird's-eye, or eye-level. The more dramatic scenes are created by using either a worm's-eye or bird's-eye view. Try to imagine that you are on a street corner looking down two different streets. You will begin by drawing the vertical corner edge of a building. Remember, each side of the building will then disappear to the vanishing points. Once the first building is drawn, you can continue to draw other buildings along each street. Let your imagination run wild. The city scene may be one of the future, or perhaps it may be a scene from the old Wild West. Add all the little special details that would help describe your city, such as store-front signs, windows, doors, parking meters, hitching rails, cars, sidewalks, and bricks or other surface textures. Keep in mind that all these objects will also shrink to the same vanishing points as your buildings. If you have even one item that you forget to draw in perspective, your drawing will suddenly become less believable. After you have completed your drawing, you may want to lightly erase (did you really throw your eraser away?) your horizon line and vanishing points, since these were merely used originally as guidelines to help you draw in perspective. You will be amazed at how dramatic your drawing looks.

[7-78] This drawing of an ancient city street uses one-point perspective at eye level.

Exercise 7-15

This exercise emphasizes the idea of foreshortening in the use of perspective when drawing the human figure. You will be practicing learning to use perspective as a "second nature" rather than as a formal method of determining distance. Adjust your paper and pencil as if you were going to make a contour drawing. Now, stretch your foot out in front of you so that it is easily visible. If you are seated, you will not have to place your foot out far to be able to see the top of your shoe. You will be applying the principle of foreshortening as you make this drawing. Start by drawing the top of your leg, which is much closer to your eye than your shoe. Now complete the drawing of your shoe. If you are careful about drawing only what you see, your shoe should appear to be much smaller than the top of your leg and knee. Another easy variation of this drawing would be to use your hand and arm instead of your foot. In this case stretch your arm out as far in front of you as possible. Begin to draw what you see, from the top of your arm all the way out to your hand and fingers. Remember, draw only that which you are actually seeing. Your upper arm will appear to be much wider and larger than your hand and fingers, which are farther away. Both of these drawings will give you a rather dramatic demonstration of foreshortening.

Exercise 7-16

Another excellent method for discovering how to capture a feeling of perspective is to draw several "corner" drawings. All we need for a model is the corner of a room. If you are in a classroom, you may want to choose a corner that has either windows or perhaps a chalkboard attached to the wall close to the corner. Place a piece of furniture in the corner (an extra desk and two or three chairs). Give yourself several interesting subjects to draw by adding small details such as a stack of books, coats, pencils, trash can, and so on. Now, sitting back away from the corner, begin to draw all these items as you see them. You will begin by first drawing the corner, including the edges of the walls. Keep in

[7-79] The Egyptians were one of the first to use the principle of overlapping shapes to represent depth.

[7-80] Overlapping shapes create the illusion of depth . . . but how far apart are they?

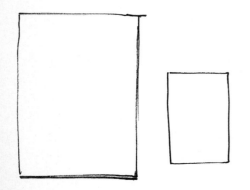

[7-81] If the sizes are identical . . . the larger one must be closer.

mind that you will need a horizon line and vanishing points. (Remember, these may not be visible on the paper.) Draw *only* what you are actually seeing. The items closer to you will appear larger than the items in the back that are directly connected to the wall. Later your may want to produce some drawings of corners in rooms at home, or you may prefer to stand out on a street corner and draw a perspective drawing looking down the street. The world about you is full of exciting possibilities that through the use of dramatic perspective can be represented with dynamic and impressive drawings.

Perspective Illusions

In addition to linear perspective, there are several other visual methods that are used to create the illusion of three-dimensional space. These methods are not as formal as linear perspective and are more directly related to drawing "what you see" rather than what actually exists. The first method involves the use of *overlapping shapes*. When an object or part of an object is hidden behind another object, it is interpreted by your mind as being located farther away than the shielding object [7-79 and 7-80].

However, it is not always evident just how far away the farthest object is located. You might not know, for example, whether a few millimeters or several meters separate the shapes. Sometimes this arrangement can cause a deceptive illusion. There are instances when artists use this type of illusion as a part of their art.

Another method of creating perspective illusions is that of *relative size*. If you see two objects that appear to be identical and one appears to be larger than the other, it is natural to assume that the objects are actually the same size and that the one appearing larger is closer [7-81]. This illustrates the power of the brain to convince the eye that the larger object is closer. You are not always aware of your brain's work since it functions automatically and frequently beyond the level of consciousness. Your mind stores a large amount of information concerning the relative size of objects, and it is particularly adept in making instantaneous judgments about the size of the objects that your eye sees. For example, if you see a traffic sign and an airplane at the same time, and the sign appears to be larger, you ordinarily will assume that it is closer to you [7-82]. The cumulative effect of overlapping shapes and relative size increases the illusion of depth [7-83 and 7-84].

Another condition affecting the illusion of depth is *aerial positioning*. This method is based on the fact that the human eye is located about 1.6 meters above the ground. This gives the human an advantage over most animals by providing an increased ability to see over objects and into the distance. Closer small objects therefore appear lower in your field of vision and objects of the same size that are farther away appear higher [7-85]. Look across the room. The chairs that are close appear lower in your field of vision and those farther away appear higher [7-86]. Elevation in the field of view is one of the most firmly established of the illusionary depth cues used by artists working in the European tradition.

[7-82] Where would the sign be located if it were the same actual size as the airplane?

[7-83] Combining overlapping shapes and relative sizes . . . increases the illusion of depth.

[7-84] The brain interprets the combination of overlapping shapes and relative sizes as a function of distance from the observer.

[7-86] The nearest chair will appear lowest in the visual field.

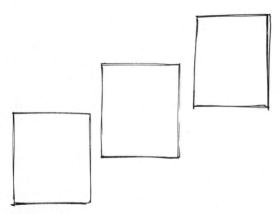

[7-85] Aerial positioning: closer objects appear lower in the field of vision.

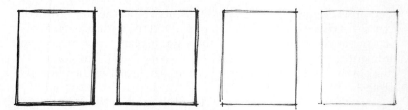

[7-88] Distant shapes are less distinct. This increases the illusion of depth.

[7-87] More distant objects appear less distinct.

Another method of illusion involves the clarity of the atmosphere. This includes the effects that air, smoke, fog, and haze have on the appearance of objects. As an object recedes into the distance, detail becomes less defined and more obscure [7-87]. If the light level remains constant, more distant surfaces tend to appear as middle-tone grays. Careful visualization of grass on a golf course will help us to understand this principle. In the near foreground, you can see individual blades of grass with contrasting dark shadows and brightly lit surfaces. Some 10 meters away, the grass surface appears as a uniform fine texture with suggestions of darks and lights. As the fairway recedes, continues on to the next hill, and approaches the horizon, the individual grass blades are no longer identifiable and only shades of green-gray appear. The saturation, or intensity, of the green hue also recedes with distance. Taking these atmospheric effects into account will help to add increased illusion of depth in a drawing [7-88]. It is most effective, however, to combine all the methods discussed of creating illusions—linear perspective, overlapping shapes, relative size, aerial positioning, and atmospheric effects [7-89].

Exercise 7-17

Place several simple objects at varying distances from you. Put the smallest item closest to you and the largest items much farther away. An example would be to place a pencil very close to your drawing paper. Place your notebook some distance away. Now draw these two objects. You know that in reality the pencil is a much smaller object than the notebook, but because of its proximity to your eyes, it will appear to be much larger. This is the way these objects should appear on your drawing. Also note that in your vision field the notebook may appear to be above the pencil. Therefore, you will place it above the pencil in your drawing. Try several drawings like this using various objects. In each instance you should still be drawing using the continuous-line drawing method. Look only at the item while you are drawing, and leave out any shading or shadows.

Exercise 7-18

To gain experience with atmospheric perspective, we will draw a landscape or at least an outside drawing [7-90]. (The principle of atmospheric perspective is also applicable for indoor scenes.) Find a landscape in which you can see for several miles. Practice drawing objects very close to you in the scene that you are drawing. Include as much

[7-89] Which depth illusion techniques are used here?

[7-90] Atmospheric perspective gives the illusion of distance.

detail as possible. Then add additional items in the background. These will not include very much detail, and you may wish to use aerial positioning as well. Of course, it is still important to remember that you should use linear perspective in all these drawings. The other methods of creating depth illusion are also important to enhance the feeling that you are viewing a real three-dimensional scene.

Exercise 7-19

Overlapping shapes is a very easy illusion to use. A simple still-life arrangement will provide you an excellent example of overlapping shapes. Select "everyday" items that you are familiar with, such as a bowl of fruit, a stack of books, a closet full of clothes, or dishes stacked on the kitchen counter. All of these scenes should be drawn to give the illusion of three-dimensional space by using the method of overlapping shapes. Try to find as many items as you can that naturally shows overlapping, and make several drawings.

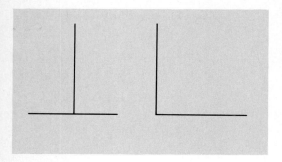

[7-91] Which horizontal line is the longer?

SEEING IN PROPORTION

You may notice in your contour drawings that you have difficulty determining sizes of objects or parts of objects as they appear in relationship to other objects. This is caused by your inability to see "in proportion." Proportion is the size of the parts when measured in relationship to the size of the whole. When you first made a drawing of your hand, you may have drawn a thumbnail that ended up being half as large as the drawn size of your thumb. You did this because it was difficult to determine how large to make the thumbnail in relationship to the rest of the thumb and in turn in relationship to the rest of the hand. Once again your mind has dictated a different arrangement than your eye is actually seeing.

In looking at a doorway, you know the height is supposed to be much longer than the width. *Because your mind knows this,* you may have difficulty in drawing a doorway any other way. Until you learn to draw what your eye is seeing, you may draw doorways that are "out of proportion." There are several illusions that point this out clearly. In Figure [7-91] which horizontal line is the longer? They appear to be different lengths, but if you make a quick measurement, you will see that they are in reality the same length. At this point it is important to convince your eyes to see what you know to be true. In the case of the Müller-Lyer illusion [7-23], you established a sense

[7-92] King Tut's School of Freehand Drawing.

of proportion by measuring the lines to assure yourself that they were actually the same length. This is the same method that an artist uses to establish proportion in drawings.

It would be much too time consuming and impractical to use a ruler and protractor to measure the accuracy of every line and angle drawn. Therefore, a simple method of measurement estimation has been developed. This method has commonly been referred to as *sighting* [7-92]. It is not a precise form of measurement and would never be used in making a scaled drawing. However, it is a quick form of measurement that improves the accuracy of the drawing. Sighting will help you determine relationships between height and width as well as to establish comparative relationships between angles. Once again, let's use the example of the door.

Take your pencil and hold it out directly in front of your face, using a straight, locked arm. Align it with the vertical edge of the door. Sight past the pencil with one eye closed and position the pencil in line with the side of the door that you wish to measure. Slide your thumb along the pencil to adjust its length, matching the apparent distance on the door [7-93]. Now that you have established this vertical measurement, turn your pencil horizontally and align it in a similar way with the width of the door. (Do not move your thumb from its original measured vertical distance on the pencil.) Now continue to "sight across" your pencil and establish the relationship between the height and width of the door [7-94]. Does the door appear to be twice as tall as it is wide, or perhaps does the width measurement appear to be slightly more than one-third the door height? Using this method of "sighting" over your pencil, you should be able to establish comparative relationships between any types of objects or even parts of objects. For instance, hold your thumb directly out in front of you. Now, using the method of sighting over a pencil, establish the size relationship between your thumbnail and your thumb.

It is also possible to establish comparative relationships between angles using the sighting method. Hold your pencil straight out in front of your eyes. Find an angle on the object you are drawing and adjust the pencil so that it lays parallel to or on top of one of the edges [7-95]. Now look at your pencil. At what angle from the horizontal is it tilted? As an alternative, you can hold your pencil horizontally and use it as a base reference [7-96]. Continue to check angles in this manner to establish the relationship of all the angles in the drawing. Remember that a false angle will raise a controversy between your eye and brain, causing an illusion.

Be sure to check the angle and proportions of all of your lines using the sighting method. Once you have developed skill in using this method, many drawings that you previously may have perceived to be "too difficult" are now suddenly easily drawn. You now have a method to make all the objects "fit" together harmoniously in a drawing.

Exercise 7-20

This exercise will help you use sighting to establish proportions and angles. Tape several rectangular pieces of paper of different sizes at various heights on the wall in front of the classroom. You may want to

[7-93] Sight over a marker to estimate distances and proportions . . . hold your arm straight with elbow locked.

[7-94] Compare the width of the door with its previously measured height.

[7-95] A marker held parallel to an inclined edge helps to estimate the angle of decline or incline.

[7-96] A marker held horizontally helps estimate angles.

tape them reasonably close together. Also, tape them in such a way that the top and bottom of each sheet is parallel to the floor and ceiling. Now using the sighting method, make a composite drawing of all the sheets of paper as they hang on the wall. Establish the various sizes in relationship to each other and to their locations on the wall. Include any windows, chalkboards, or bulletin boards that may be on the wall. Once you have completed this drawing, reposition all the papers at various angles to one another. Again use the sighting method to establish the edge angles as well as the paper sizes. Repeat this exercise until you feel confident using the sighting method in establishing sizes as well as angles.

Exercise 7-21

Align several books of various sizes in front of you. Adjust them so that several are standing and slightly opened and several are lying flat on the table. Now draw the set of books as you see them. Use sighting to establish the varying sizes and angles of the books in relationship to one another. This exercise should be repeated several times until you are confident that your drawings are accurate enough to be believable.

SEEING POSITIVE AND NEGATIVE SPACES _____

Everything we look at has both positive and negative space. Sometimes, as seen in several illusion examples, we have difficulty establishing which is the positive and which is the negative space [7-97]. Generally, the eye will search out the positive space, and the brain will take charge and interpret the result. The illusion of vases and faces dramatically points this out [7-19]. When viewed one way, with the white area as a solid background, the drawing appears to be two faces looking at each other. When your eye concentrates on the black area as being the center of attention, the image of a vase appears. However, either way, there is still both positive and negative space. It is this combination of positive and negative space that makes up the composition of every drawing.

[7-97] Which way are the ducks flying?

The composition of a drawing may be viewed as the arrangement that you have chosen for all the elements of your drawing (both positive and negative). Therefore, you are placing all the positive and negative spaces on the drawing surface. Although we are conditioned through what our mind has learned (to search first for the positive spaces), readjusting our eye's priorities in the simple illusion of the vases and faces shows us that proficiency in drawing can change this ingrained habit. This illusion also illustrates the importance of viewing both positive and negative space images with equal importance.

[7-99] Negative space can be used to create the existence of an "implied" image. Here you *assume* the lady is in bed because of the white space that suggests the existence of bed covers.

[7-100] A few lines suggest the existence of a chair. The absence of detailing of the chair actually adds interest to the drawing and concentrates your attention on the old man . . . which is the intent of the person who made the drawing.

Many times the artist uses negative space to create the existence of an "implied" positive image. Because of this we can see that it is just as important to consider the spaces *that you are not drawing* as it is to give attention to the solid object (spaces) *that you are drawing* [7-98 to 7-100]. It is important that you carefully fit all the pieces of each drawing together, including both the positive and negative spaces. Because we are readjusting your learned behavior, the exercises used here to help you learn to view negative space may seem a bit frustrating at first. Practice will help you to start to perceive many different types of negative spaces, rather than just the traditional view of seeing only solid objects. Suddenly you may find yourself looking at the spaces created by the stadium bleachers rather than the bleachers themselves. Photographers have a finely developed sense for perceiving negative space since they spend a lot of time viewing both shadows and streams of light. The area created by a small ray of light or even a reflection can help to make a very dramatic photograph.

Exercise 7-22

In this exercise we are going to use the example of the faces–vase illusion. You are going to recreate such an illusion *from your imagination.* If you are right-handed, draw a subject's profile on the left side of the paper. If you are left-handed, draw a profile on the right side of your paper. Make up your own version of a profile. Let your imagination run free. You can make a strange looking face or even put warts on the end of the person's nose. Next make horizontal lines at the top and bottom of the profile. Once this is done, repeat the profile in reverse on the opposite side of the paper. Your second profile should match exactly your first one in order to complete the drawing. Now take your pencil and blacken in all the area made inside the vase. What is the result? Have you drawn a vase or two faces? What you originally drew as two profiles has become a vase.

Exercise 7-23

On a sheet of paper draw a square, a circle, a triangle, and a rectangle. Do not let them touch. Now make the negative spaces created by these shapes stand out by drawing around each shape. This will create a shape for the negative space. Using your pencil, darken in the negative space. This should help you begin to visualize that a negative space forms one or more shapes. It should also be clear to you now that positive and negative shapes actually share the same edges. When you draw the edge for one, you have in effect created the same edge for the other. Make several more of these shape drawings to help you become aware of the negative shapes that you are creating.

Exercise 7-24

You are now going to *create something by drawing nothing.* Find yourself a chair or desk and place it at some distance in front of you but positioned so that you have an unblocked view of it. Now in your imagination place the image of the chair on your paper in such a way that several edges of the chair touch the outside edges of the paper. The result will be a rather large drawing since this one chair will fill the whole page. Begin to determine *only the negative spaces* created by the chair. You will have to look at the chair for quite some time before your mind begins to relate *only* to the negative spaces. It may help you to cut a small rectangular window in the center of a piece of cardboard and hold it in such a way that the chair is framed within that rectangle. This will help you to limit the outside edges that correspond with the outside edges of your drawing paper. The window also acts as a "frame" for the negative shapes. Now draw all the negative shapes that you see. Once you have drawn all the negative spaces, you should be able to see an accurate outline drawing of the chair. Remember, all edges made by the negative spaces are also edges of the positive spaces. Once you have completed the drawing, take your pencil and darken all the negative shapes. The result should be a solid white chair on a dark background.

Exercise 7-25

Repeat several negative space drawings using arrangements of selected items. An example might be a grouping of books, pencils, crumpled paper, and a decorative plant. Household items may also be used, such as a coffee pot, silverware, soda bottles, and so on. Just make sure that you are arranging them to create a number of negative space images between the items. Remember, if you have trouble relating the arrangement of the items to the outside edges of the paper, use the cut rectangular window to help establish the outside edge relationships. Continue to darken the negative shapes to help you visualize them as being just as important as the positive shapes.

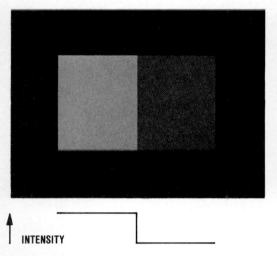

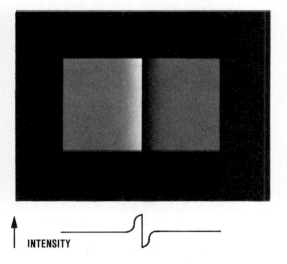

INTENSITY

INTENSITY

[7-101] These two patterns look much the same. But the distributions of light, plotted below each one, show that the right one is an illusory edge created by a sudden local change in intensity in the middle of an area of equal intensity.

LEARNING TO SEE VALUE

Value is a term used to describe how light or how dark each surface of an object appears to a viewer. The relationship between the values of two areas is called *value contrast*. For most everyone, our world is made up of colors—some light, some dark. It is not often that we get the opportunity to see objects in strictly black and white or even only in shades of gray. Although the world is very colorful, each color takes on a specific value depending on the intensity of that particular color and the light being reflected from it. A yellow color has a much lighter value than a deep burgundy or even a black, but it has a darker value than pure white. The value of this specific yellow would change however, depending on the amount of light being reflected from it. A yellow lemon sitting in a dark corner would appear to be a much darker value than would a deep red apple positioned in front of it that had a bright light shining directly on it. This gives us a good example of value contrast.

Again the illusionary capacity of our eye–brain team must be taken into account. Some 100 years ago, an Austrian physicist, philosopher, and psychologist, Ernst Mach, discovered that if two juxtaposed uniform surfaces are seen reflecting different amounts of light, the border between them appears to be lighter than the light area on the side of the latter, and vice versa. The left and right areas in Figure [7-101] appear to be much the same, except in brightness.[9] The distribution of light plotted below the figure shows that the outer edge is illusory, created by a sudden change in intensity in the center of an area of equal intensity. Cover this apparent edge with your pencil and see how the two areas suddenly reflect the same brightness.

Value can add dramatic impact to a drawing. We use differences in value to help us interpret and recognize objects. A strong contrast of value on a person's face can give it a very dramatic, strong impact,

[7-102] This drawing by Alfred Leslie illustrates the dramatic effect that can be achieved by use of strong contrast.

[9]R. L. Gregory and E. H. Gombrich, eds., *Illusion in Nature and Art* (New York: Scribner, 1973), p. 22.

[7-103] Chiaroscuro by Rembrandt.

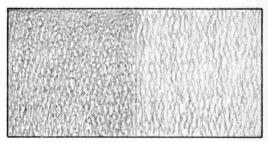

Low value contrast . . . light values adjacent.

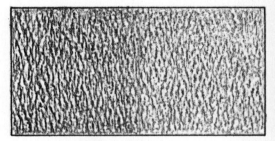

Low value contrast . . . heavy values adjacent.

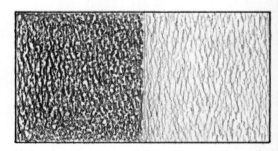

High value contrast . . . light and heavy values adjacent.

[7-104]

perhaps emphasizing rough features [7-102]. Softer values with minimum contrast can make a face look sweet or delicate. One of the first artists to make use of strong value contrast and dramatic lighting was Rembrandt (1606–1669). A strong use of contrast between the light and dark values is called *chiaroscuro* [7-103]. This method of painting became very popular during the Renaissance period.

Value can also help us to imply lines. If you place a very dark value next to a very light value, creating a strong value contrast, there is no need to draw a line to indicate an edge. The eye will automatically see a line where the two values meet [7-104 and 7-105]. Value contrasts can also define textures and patterns. For example, a burlap bag does not have the look and feel of burlap until its texture is defined through the use of value contrast. Values will actually define texture by indicating how lights and darks are created by the surface of an object. This creates a simulated tactile quality [7-106]. Several French painters and early Americans of the nineteenth century such as Harnett perfected a method of using values in what are called *trompe-l'-oeil*, or "fool-the-eye" paintings. Their ability to "see values" is so acute that it is hard to distinguish between real and simulated textures [7-107]. Although you may never be able to develop such a keen sense of value perception, it is important to be able to see values in drawing.

Exercise 7-26

The first step in learning to see value is to make a value scale. This will also help you to learn the subtleties that may be achieved by using only one density of pencil lead. Begin by drawing nine vertically connected boxes on your paper. They do not need to be perfect, but they should be aligned adjacent to each other. Leave the top box blank. This represents the lightest value possible, the blank paper. Next, go to the

[7-105] Value contrast decreasing with distance provides the illusion of deep space.

[7-106] Values are used to simulate a texture, such as a wooden surface.

[7-107] William M. Harnett (1848–1892) was a master at producing drawings that had a photographic quality of realism.

bottom box and using your pencil, darken it as much as you possibly can. Make sure that you get maximum coverage from your pencil. This represents the darkest possible value that you can achieve in your drawing. Now progressively darken each box in value, moving down the value scale (from light to dark) from the first box. Progressively darken the value in each box from the one immediately above it. Try to be consistent in the amount that each box is darkened to give you a smooth, uniform scale [7-108].

You can now move on to experimenting with various pencils ranging from a soft pencil (6B), which will give you a very black final value, to a very hard pencil (6H), which will give you a very light, subtle final value. If you want to play with values, there are many ways to experiment, such as making a dark, black value with soft lead and then erase in gradations to obtain the various shades. You could also take a dark pencil and smear the lines with your fingers.

Sometimes artists will use white pencils along with their graphite pencil to increase the values in a drawing. In such a case it might even be possible to obtain a value that would be viewed as being lighter than the plain white paper. Colored paper, such as cream or buff, may also be tried to allow the white pencil to be noticeably lighter in value than the paper. You may wish to refer to the value scale when drawing. This will help you determine appropriate values as you proceed with your drawing.

There are several methods of creating value with a pencil. Smudging a drawn line with the finger has already been mentioned. Rubbing a line with a kneaded eraser is another. Two other common methods are crosshatching and stippling. *Crosshatching* uses a series of short parallel lines to create a pattern and then going back on top of them with another series of parallel lines drawn at approximately a right angle to the first set. Multiple layers of parallel lines can be used to intensify the value [7-109]. You will develop your own style of

[7-108] Value scale . . . the value contrast between each of the pairs of shaded blocks is equal.

[7-109] Crosshatching uses a series of short parallel lines to create a pattern.

crosshatching as you progress with your practice drawings. When a light value is desired, you will draw the lines farther apart and make fewer layers of them. For a darker value, draw the lines closer together and layer them heavily one on top of the other.

Stippling uses dots to create value [7-110 and 7-111]. With the point of your pencil make a series of dots very close to one another until the area becomes gray or even black. By varying the number and distribution of dots, you will determine the value that you desire. Both crosshatching and stippling are valuable tools when you are trying to show textures by using values.

Exercise 7-27

Gather as many white items as possible for this drawing. Examples might include an egg, a baseball, a pillowcase, a cleansing tissue, a piece of paper, a plate, a sock, and a bar of soap. Use your imagination: you will be surprised at how many white items you will be able to find. Place all these items together. Now draw the assembled collection using only subtle values to indicate the various shapes. Your drawing will end up as a very soft drawing, and it will express subtle changes in value [7-112]. This type of drawing is sometimes called "high key." Take your time and really study the subtleties of each value. A variation of this exercise would be to find as many black items as possible and arrange them together for a similar value drawing. Look around and you can probably find other items that have very similar values. These can be used to make another subtle value drawing, this time in "low key."

[7-110] Stippling uses dots to create value. The value desired can be achieved by varying the number and distribution of dots.

Exercise 7-28

Take a piece of fruit such as an apple, lemon, or orange, and place it in front of you [7-113]. Now using values only, draw the object. Do not use any lines to define the existing edges seen by your eye, only values. You might want to continue making value drawings using other objects. A glass jar, for example, may reflect a lot of light and therefore it may be easier to define the edges using values. If you have a lamp or strong light, you can also achieve some dramatic effects through the use of strong value contrast.

Exercise 7-29

For centuries artists have used draperies to study value. The shadows and changes in values that are created by the folds and curves in a soft piece of cloth provide interesting subject matter. Take a plain sheet or white towel and drape it over a chair in front of you. If you have a lamp or strong light, this will help to create additional value or at least stronger values. If you do not have extra lighting, don't worry, the values still exist. Just look for them. Now draw the piece of cloth using the values that have been created by the folds and bends. One piece of cloth should provide subject matter for several hours of drawing. Just keep changing the way the cloth is draped over the chair.

DRAWING FROM YOUR IMAGINATION _____

The first sentence of this chapter states that "drawing is the ability to translate a mental image into a visually recognizable form." The exer-

[7-111] A range of values can be achieved by stippling.

cises that followed were designed to help train your mind to block out preconceived images and to relate to your hand what your eye was actually *seeing*. By now you will have discovered that you can draw, if you learn to *see*. We are now going to go a step further and involve your imagination. Everyone has an imagination and possesses the ability to create mental images. This creative imagination is put to use in a variety of ways. Some people sing creatively, tell jokes creatively, cook creatively, dance creatively, or even solve engineering problems creatively.

Drawing from your imagination can involve several different aspects of visualization. You may draw exactly what you are seeing in your mind (daydreaming or fantasizing), you may add emotion or character to what your eye is actually seeing, or you may make a drawing entirely from your imagination that serves as your statement (belief) or opinion about something [7-114]. Whatever the purpose of this type of drawing, it will involve the use of your imagination.

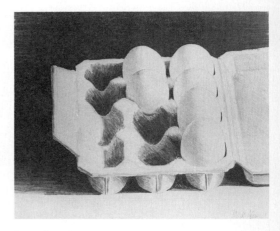

[7-112] Eggs make good models.

DRAWING FROM A DREAM

One of the most common examples of drawing from a dream, or daydream, is the simple "doodle." Most everyone at one time or another has practiced the art of doodling. In many homes a notepad is kept next to the telephone for message taking. If you examine these notes you will find various forms of doodling: around the telephone numbers; a person's name will be embellished with flowers; geometric designs will run up or across the page; or even simple lines and circles will be drawn around the message for emphasis. These are all forms of imagination drawing.

Exercise 7-30

Make a geometric shape in the middle of your drawing paper (square, triangle, circle, etc.). Now make the shape become mechanical. Add gears, wheels, chains, cranks, or whatever it would take to make

[7-113] Vegetables make good subjects for drawing.

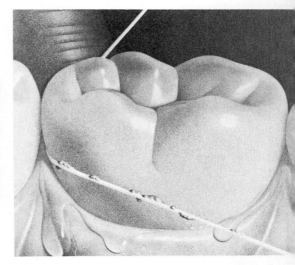

[7-114] If you were a piece of dental floss, imagine how you would spend Thanksgiving afternoon.

[7-115] Ordinary household objects can be altered in their normal functions to accomplish strange or unusual tasks.

this shape become a mechanical object. Let your imagination run wild. Do not let yourself be inhibited by thoughts of what function your little mechanical object will provide. Just make it mechanical. Now make it interact with another shape. Is it pushing a square, or pulling a circle? Now cause your second shape to interact with another of the shapes. Continue drawing in this manner until you have portrayed a chain reaction "happening" around your paper. Draw all the little shapes so that they are different, and make them perform different actions with the other shapes. You may want to give some of your shapes human characteristics, such as eyes, mouths, or teeth.

Exercise 7-31

Draw a familiar, ordinary household object such as a radio, telephone, water glass, or chair. Make a simple line drawing that does not portray any emotional qualities. Now place your object in an unfamiliar environment that it would not normally be found in [7-115]. You can also make the object's size deviate from that which would seem realistic. For example, your telephone may be sitting in the middle of the Gobi Desert, or the radio could be rolling down the freeway. Salvador Dalí was an expert at placing objects in unfamiliar places or distorting them in unusual ways, such as the melting clock [7-116]. Many times this type of drawing becomes a catalyst for new inventions or improved ways to use an old familiar object.

Exercise 7-32

Draw the essence of a dream. Close your eyes and let your mind wander. You may want to sit back in your chair and spend a few minutes letting your mind fantasize [7-117]. Now start to draw. This form of fantasy should eliminate any preconceived ideas concerning size, proportion, function, and convention. Dreams let us depart entirely from the world around us and can represent a spectrum of emotions from playful to horrifying. Let your pencil translate onto the paper what your mind is fantasizing.

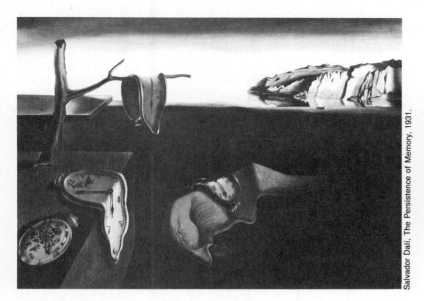

Salvador Dalí, The Persistence of Memory, 1931.

Salvador Dalí, The Persistence of Memory, 1931.

[7-116] Salvador Dalí was a master at giving authenticity to the improbable.

[7-117] Dreams can provide subjects for imaginary drawings.

DRAWING FROM EMOTION

Transferring emotion into a drawing adds character and life to the drawing. Emotions expressed in the drawing can give the added emphasis needed to create a responsive emotion from the person viewing the drawing. It is important that your drawings reflect your emotions; otherwise, they will become static and boring [7-118].

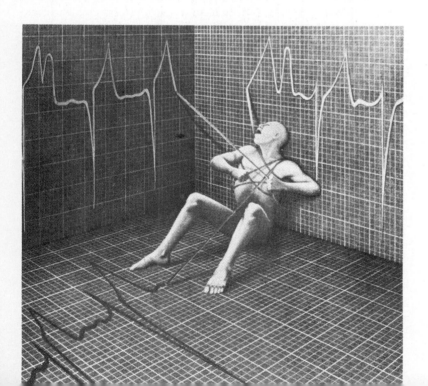

[7-118] Drawings can be a good outlet for expressing your emotions.

[7-119] Walk softly . . . carry a big stick.

Exercise 7-33

Make three drawings of a tree that is growing in three different weather conditions. Draw what you would perceive the tree to look like on a sunny afternoon, on a stormy night, and during an early morning sunrise in winter. Put into the drawing what you would feel like in each of these three situations. Try to evoke those same feelings into that of the tree. Ask someone to decide which tree drawing belongs to each one of the three conditions. Use values in your drawing to emphasize the "mood" of the weather.

Exercise 7-34

After all the contour drawings, your hand should now be a very familiar object. Let's draw your hand again, but try to draw it to represent three states of emotion. How would your hand look when you are angry (clenched fist)? How would you hold your hand in a very relaxed or even a sleeping position? Now, how would you hold your hand(s) if you are very nervous? There are many other emotions that can be expressed by your hand. Draw your hand in other unusual emotional situations.

DRAWING TO MAKE A STATEMENT

Political cartoonists are asked daily to make statements or to give an opinion through their drawing. It is usually very easy for the reader to see at a glance the point the cartoonist is trying to convey [7-119]. Muralists for years have used their art form to make statements and express opinions concerning governments, poverty, and famine. Mexico City is a prime example, where artists have expressed their opinions to the government and to the people of Mexico.

Exaggeration is a method often used by artists of all types to emphasize their opinion. To use exaggeration effectively, however, you must rely heavily on your imagination.

Exercise 7-35

Take an ordinary object with which you may be exasperated or frustrated (a ringing telephone, an inoperative electric can opener, a car that always breaks down, a door that always gets stuck). Now draw this object trying to convey the opinion that you have about it.

In contrast, draw something that gives you a lot of pleasure (a soft pillow, your favorite blanket, your favorite baseball glove). Just by looking at the drawing an observer should be able to tell the difference in your opinion of each of these objects.

THREE TYPES OF ENGINEERING DRAWING

Engineers and engineering technologists have long made use of drawings of various types. The type of drawing we have been exploring in this chapter could best be described as *communication drawing*, for its main purpose is to communicate ideas to others [7-120]. Learning to express yourself through drawing is an essential communication tool. Drawing can also be used as a tool to develop ideas. We call this second type of drawing *ideation drawing*. A third type of

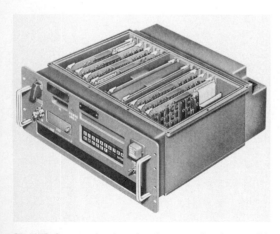

[7-120] Communication drawings can be drawn with such skill that often they are confused with photographs.

drawing, *documentation drawing,* also known as technical drawing, engineering graphics, or drafting, will be discussed in Chapter 9. To understand the function of each of these types of drawing, we need to examine the roles of the engineer and engineering technologist in industry today.

Team effort is the most visible characteristic in contemporary engineering, manufacture, and construction. For this reason, general communication skills are more important than ever before. Good verbal communication skills are very important and are being stressed increasingly in college programs. These necessary skills can be effectively supplemented, however, with skill in communication drawing. The old adage that *a picture is worth a thousand words* is still relevant. *Communication drawing* is that type of drawing whose primary objective is to convey or communicate general information to others.

Documentation drawing does this also, of course, but it uses a highly specialized vocabulary and communicates to only a trained group of technical specialists. True communication drawing should therefore be as universal as possible in order to reach as wide an audience as possible. The type of drawing employed by Leonardo da Vinci in his many notebooks, and which you have learned to emulate in this chapter, comes closest to embodying the true spirit of good communication drawing. We have demonstrated that acquiring the skill to draw from observation can be learned by anyone using the step-by-step exercises in this chapter.

Applications of communication drawing in modern industry are most often found when information transfer occurs across disciplines or when communication is directed to the less technically trained. Communication drawings can be simple or complex. They can vary from a quickly executed line drawing to a formal presentation drawing employing the sophisticated use of artistic media and techniques. Specialists such as graphic artists and delineators are often called upon to execute the more formal communication drawings but technical personnel in the manufacturing environment have many occasions to produce less formal communication drawings in day-by-day interactions.

A second type of drawing, ideation drawing, used by engineers and technologists is often neglected, but is the most important of all in terms of increasing one's productive efficiency [7-121]. Ideation drawing serves as a window to the mind. Ideation drawing includes any drawing that is produced primarily to aid in thinking, visualizing, or conceiving an idea, product, or device. Often, ideation drawing is used to make visual notes to oneself. It differs from communication drawing in that it is not meant to show to others, and may take on a shorthand, cryptic character.

The techniques used to produce ideation drawings are similar to those used in communication drawing. However, ideation drawings are less formal and faster to produce. They are used to formulate, test, and visualize ideas during the critical early design phases of a project. Quick execution is important, and a large number of ideation drawings may be produced in a short period of time.

Most ideation drawings will be discarded or kept for the private notes of the designer. As an idea starts to take form, an ideation

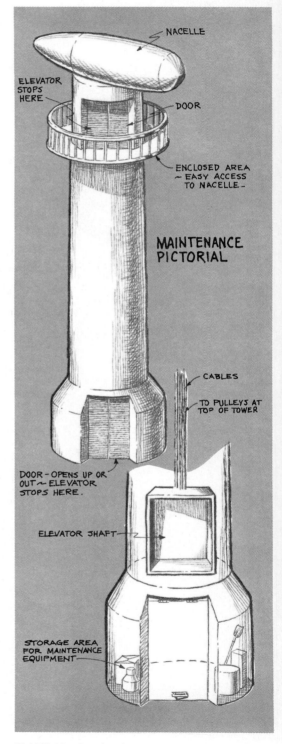

[7-121] Ideation drawing.

THE CREATIVE PERSONALITY

RISK TAKER
COMPLEX
CURIOUS
IMAGINATIVE
SENSE OF HUMOR
TOLERATES AMBIGUITY
SYNERGISTIC

[7-122] Characteristics of creative people.

drawing may gradually be transformed into a more finished communication drawing. The ability to work in this way is a valuable asset to all creative people. Many prominent scientists, inventors, and engineers, including Albert Einstein and Thomas Edison, depended heavily on their ability to convey the essence of mental images by use of a drawing while pursuing their theoretical and experimental work.

CREATIVITY AND IDEATION DRAWING

There is a strong correlation between one's creativity and ideation drawing skills. Research into creativity has identified a number of common characteristics of creative people, regardless of their field of endeavor. Such people are often curious, willing to defer evaluation of ideas until they are more fully developed, have a sense of humor about themselves and others, and are able to develop a number of ideas or variations on a theme before moving on [7-122].

Investigators have identified two unique qualities of creative people. They are called fluency and flexibility. *Fluency* is the ability to conceive many variations within a particular category (before running out of ideas). *Flexibility* is the ability to change categories often. For example, a frequently used test of creativity is to ask a person to name all possible uses of a common building brick. Most people will select a restricted category of uses such as "building material," and then list all uses of a brick when used as building material. Sooner or later, the person's idea bank "runs dry" and it is not possible to think of new uses within that category. A category switch is then made, perhaps to uses of a brick as a weight. Then all such uses in that category are listed. In analyzing the results of such tests, researchers consider the number of variations within a category as a measure of ideational fluency, while the number of category shifts that are made are a measure of ideational flexibility. Creative people are both fluent and flexible in their thinking. Consequently, the designer who is both fluent and flexible will be able to sift through a large number of potential solutions to a given problem, which is a great advantage in ultimately reaching a useful, innovative solution.

As you work through the following exercises, you should have a goal of increasing your drawing speed. Ideation drawings should be executed naturally and easily, almost without conscious effort. Producing a large number of such drawings helps you to explore a given problem more fully. Also, *avoid using an eraser!* Use of an eraser is evidence that you are critically evaluating your work as you proceed. This practice is counterproductive! Researchers of the creative process agree that such evaluation inhibits the free flow of innovative thought. You are not your best when you create and evaluate at the same time. Instead, evaluation should be delayed until later.

Study the drawings you produce and allow the forms with which you are working to shape your thinking. The value of ideation drawing is to be able to watch a form develop before your eyes and then continue to change as your thought processes unfold. Use your drawing ability to explore three-dimensional space and help you understand the design variations available to you.

A valuable technique is to use sequential tracing paper overlays as you refine and develop your drawing. When you wish to make a change or try a different idea, place a fresh piece of paper over your original drawing and trace only the parts you wish to retain. Omit undesired details and strengthen and refine forms as you proceed. Repeat this process as many times as necessary to investigate alternative ideas or try different combinations [7-123].

Ideation drawing is of special value in arriving at innovative solutions to problems when they involve one or more configurations or geometric forms. A number of variations may be explored during the design process and recorded with quick drawings. This skill is an extension of the drawing from observation techniques developed earlier in this chapter. The main difference is, of course, that when you are drawing to represent a mental image, the object is not physically present for study and correction of errors. This disadvantage can be countered by learning to quickly and skillfully draw a few standard geometric shapes such as cones, cubes, spheres, and prisms. You should practice drawing such shapes over and over again in many varied positions so that their appearance in any position can be remembered vividly. These standard shapes, called *primitives*, may then be used as building-block components for the development of imagined objects.

Exercise 7-36

Obtain several models of primitives (geometric shapes) to use in practice drawing from observation. Good sources for materials to be used to make these models are stores that sell model and hobby supplies. Precut, inexpensive Styrofoam shapes are usually available in such stores for use in flower arrangements. You should be able to obtain at least a cube, sphere (or hemisphere), rectangular prism, cone, and torus (doughnut shape). Models that you can easily hold in your hand (about 4 to 6 inches across) will work best. Now, working with one model at a time, fill several pages with drawings of each shape, using the modified contour technique described on page 149. After drawing one view, turn the model to a different position, and draw another view of it. Continue this process until you feel that you are thoroughly familiar with the shape and contours of the model. Now, choose another model and repeat the process [7-124].

You may wish to modify certain of the shapes you have purchased to make them more interesting to draw. You can easily cut Styrofoam with a fine saw or sharp knife. Be careful to hold the material so that you avoid injury! Be selective. For example, a sphere makes a dull contour drawing subject: Every view looks the same. Cut your sphere in half to create two hemispheres, and you will find that making a contour drawing of a hemisphere in different positions is a challenge [7-125].

As you gain more confidence in your abilities to draw the models, imagine that they are transparent. Draw the surface contours of the model on the sides that are hidden from you (hidden edge) as well as those on the sides that you can actually see. Your result will be similar in appearance to the "wireframe" models that can be generated by computer graphics programs. Add enough cross-contours to help define the form of each shape [7-126]. As you become more confident, increase the rapidity of your pencil strokes.

Next, add some shading to clarify the form. In communication draw-

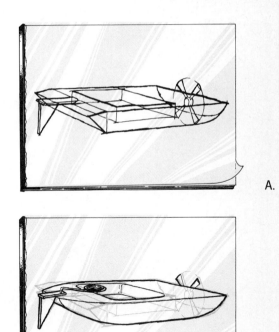

A.

B.

C.

D.

[7-123] Steps in design development using sequential overlays.

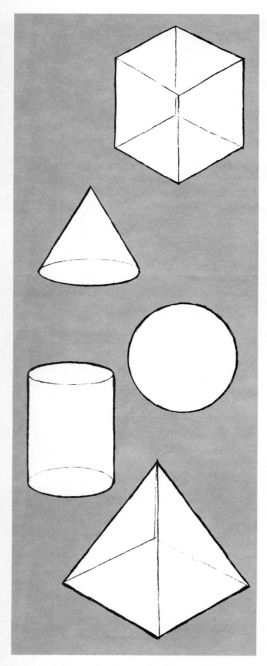

[7-124] Contour drawings of solid primitives.

[7-125] Contour drawings of a hemisphere.

ing, you were able to take a great deal of time to produce a shaded drawing (pages 174 to 178). In this case, however, you should define the form with a minimum of shading that has been rather quickly applied. Remember, to be most useful, ideation drawing should be executed rapidly [7-127].

Until now, you have been drawing from observation, just as you did in the earlier parts of the chapter. When you feel you have gained some confidence in drawing primitives, put the models out of sight. Now, in turn, draw one view of each of the shapes from memory. When you are finished, compare your drawings from memory with the drawings you made as you looked at the models, and finally with the actual models themselves. How did you do?

Exercise 7-37

When you become skilled at drawing primitives singly in a variety of positions, you are ready to combine them to serve as a basis for drawing more complex forms. In your imagination, envisage a new model that is made by combining three different primitives into a single

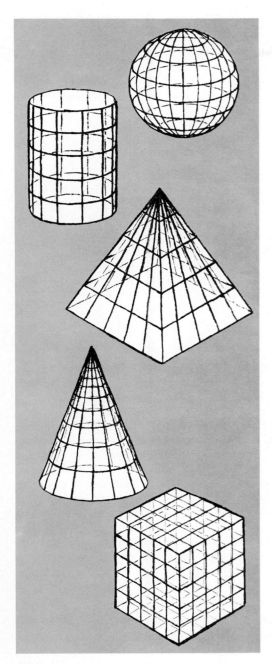

[7-126] Wireframe contour drawings of solid primitives.

[7-127] Quick shading techniques.

shape (e.g., use a cube, a cone, and a torus). In your pencil representations, use both single line and shading. You may change the relative sizes of the primitives, but avoid distorting their forms. However, you may merge or intersect the forms. Now make a wireframe drawing of the imagined model, including hidden edge contours and cross-contours

[7-128] Cluster of primitives.

where appropriate. When using shading to describe form, the presence of hidden edge lines may be suppressed.

This exercise involves drawing from memory, but depends heavily on Exercise 7-36 to establish "good memory" images of the primitives. As you work, gradually increase the speed of your drawing technique without losing form description. Now repeat this exercise using different combinations of primitives.

Exercise 7-38

Select two different primitive shapes (e.g., a cone and a cylinder). Think of at least four iterations of each shape as being joined to form a cluster of eight shapes. For example, select four cones and four cylinders, all touching or intersecting to form a cluster. Mentally rotate each iteration of the primitive into a new position that is different from the other iterations. Draw in contour line (wireframe) and add shading. Work rapidly. If you have difficulty, refer to the actual models of the primitives and compare them with your drawing. Repeat the exercise with other combinations of primitives until you are comfortable drawing the primitives in any position and combination [7-128].

Exercise 7-39

Mentally create a fantasy object by combining at least four different primitives into a single cluster. Each primitive may be used more than once and may be drawn at different scales and orientations to create a composite drawing of 15 to 20 different shapes. Every primitive should touch or intersect at least one other. Let your imagination run free in visualizing the final form of the composite. For example, you may think of a space station, a molecule, a robot, or some other imaginary concept that appeals to you. First, develop the drawing in wireframe. Draw it large scale so that it fills a large sheet of paper. Now add shading as needed to help describe the form. Work visually. That is, let the appearance of the drawing lead you to make changes and additions as you proceed. You may use the primitives as either positive or negative shapes. That is, a cylinder may be either a solid rod or a hole. Remember, work rapidly and make necessary changes as you proceed.

Exercise 7-40

This exercise should help you to extend your abilities to imagine primitive shapes in many different positions. Imagine that a primitive model (e.g., a cylinder or a cube) is setting on the edge of a table when it is knocked off and falls to the floor, bouncing several times. Draw the shape as it falls and bounces, using a stop-motion technique similar to multiple-exposure photography [7-129]. This means that you will need to draw the shape in several different successive positions as it turns and tumbles in its fall through space. Work rapidly and allow the successive shapes of the primitive to overlap. Convey a feeling of motion similar to that which you obtained in gesture drawing (page 152). As you draw, consider the three-dimensional form of the shape in each new position.

If you have difficulty in visualizing a particular view of a primitive, refer again to the solid model of the primitive and determine where you went astray. Repeat the exercise with different primitives until you are confident that you can describe motion in this way. For a more difficult challenge, draw two or more shapes, for example a cone and a cylinder, falling at the same time and bouncing independently.

Exercise 7-41

So far, you have combined primitives into clusters without distorting their original form. Although many shapes in nature or industry that you will want to draw will be similar to the shapes of the primitives, you will want to be able to create even greater diversity of form as you envisage new ideas. One way to enhance this ability is to start with a primitive shape and then to modify it by cutting away portions. A second approach is to start with a primitive and then stretch it (as if it were made of elastic material) into new shapes. Try the "cutting away" approach first.

Draw a rectangular prism using lightweight wireframe edge contour lines. Modify the drawing by making several cuts in succession. Darken the new cut lines. Leave the original edge contours in place to show how the new form relates to the original primitive. Now, draw the new shape in several different positions. Lightly draw the original edge contours of the enclosing primitive to assist you in maintaining proportions and in understanding how the new shape will appear in different positions [7-130]. Repeat the exercise using different primitives as starting points. As your drawing skill improves, you will begin to view familiar objects around you in a new way, identifying the enclosing primitives from which they might have been carved.

Exercise 7-42

This exercise will give you practice in plastically modifying primitive shapes. Start by drawing a primitive (for example, a cylinder) in the upper left-hand corner of a large sheet of paper. Now draw a different primitive, for example, a sphere, in the lower right-hand corner of the paper. Make several drawings in succession, showing the gradual transformation of the first primitive into the second. Each drawing in the series should extend the transformation a small amount [7-131]. Your goal should be to create the impression that there is a constant rate of change from the beginning form to the ending form. Repeat this exercise with different sets of primitives.

Exercise 7-43

This exercise will focus your attention on fluency and flexibility in visual thinking. Select a sphere (or a portion of a sphere), a cylinder, and a rectangular prism. The scale and relative proportions of the primitives may be varied. Also, you may duplicate any or all of the primitives as long as the total number of shapes does not exceed six. For example, a prism, a hemisphere, and four cylinders would be permissible. Solids may be partially embedded in other solids [7-132].

Combine these primitives so that they represent a drawing of a vehicle (such as a truck). Repeat the process but now use the primitives to represent a variation of the same type of vehicle (e.g., a military tank). Keep repeating the process until you have drawn at least six (more if possible!) variations of a vehicle.

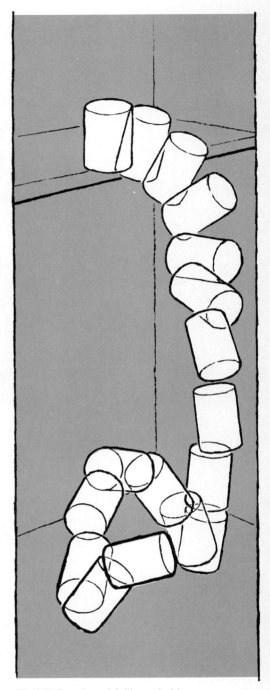

[7-129] Drawing of falling primitive.

A.

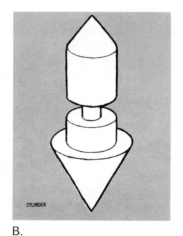

B.

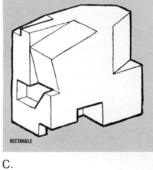

C.

D.

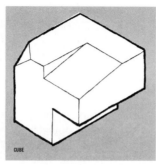

E.

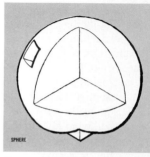

F.

[7-130] Drawing of cutaway primitives.

Now, select a different type of vehicle (such as a velocipede) and repeat the process. Try to be as fluent (variations) and as flexible (different categories) as possible. Work visually. That is, do not list types first and then draw them. Rather, let your drawings suggest new variations and categories as you work.

Finally, choose an appealing design from among those you have created. Use the sequential overlay drawing technique to refine your design. Add further detail and modify forms. In this sequence you have started by producing ideation drawings and then converted one of them into a communication drawing using the overlay technique [7-133].

Exercise 7-44

Elaborative thinking, a common behavior of creative people, is the ability to take existing ideas or forms and then transform them into new combinations. This exercise will allow you to explore elaborative visual thinking. Start with a series of composite forms [7-134]. These forms are more complex than the solid primitives with which you have been working. Use these composite forms as a visual "vocabulary" that you can use to design a vehicle. The forms may be used in any combination, and may be scaled, duplicated, and placed in any position you desire.

You are not restricted in the type of vehicle that you design. Let the composite forms suggest a particular vehicle type to you as you work. It may operate in any kind of environment: air, water, land, outer space, nonterrestrial planets, and so on. The vehicle may utilize any kind of propulsive power: wheels, tracks, propellers, thrusters, walkers, or anti-gravity devices. Allow your imagination full reign. Work out your ideas starting with quick ideation sketches.

When you have developed a number of ideas, stop and choose the one that seems to offer the most promise. Develop this idea further, elaborating on the form and adding additional detail. Use the sequential overlay technique to develop the design.

Prepare a communication drawing of the design, showing the vehicle operating in its environment. Include enough of the environment as possible to explain the method of locomotion and scale of the vehicle. The inclusion of human figures is an effective way to provide the viewer with a sense of scale. Remember, your vehicle need not look like any you have actually seen, but to communicate successfully, it should be recognizable by others as a vehicle when they view the finished drawing. Show it to friends and ask them what they see! Figures [7-135 and 7-136] are examples of student drawings produced in a similar exercise.

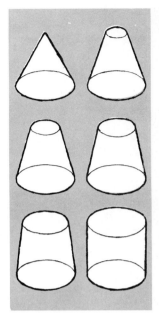

[7-131] Plastic transformations of a shape.

[7-133] Communication drawing of a vehicle.

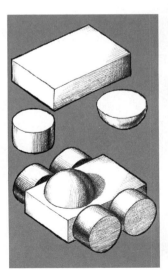

[7-132] Example of vehicle composite.

A.   B.   C.

D.   E.   F.

G.   H.   I.

J.   K.   L.

[7-134] Vocabulary of composite forms.

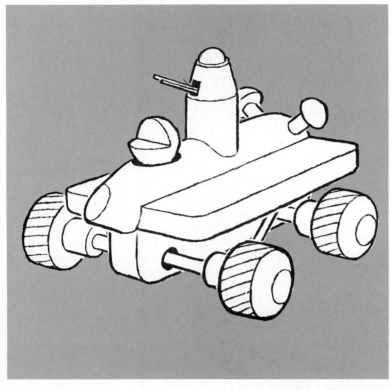

[7-135] Imaginary vehicle design, line.

[7-136] Imaginary vehicle design, shaded.

Chapter 8

Communicating with the Computer

Mathematicians and inventors have been working for centuries on perfecting our ability to make various mathematic computations in quick and accurate ways. Spurred on by societies that become ever more technical, these originators perused their inventions because of an attitude perhaps best stated by Gottfried Wilhelm Leibniz (1646–1716): "It is unworthy of excellent men to lose hours like slaves in the labor of calculation which could safely be relegated to anyone else if machines were used."

Perhaps the longest lived among the gadgets that were invented to aid in computations was the abacus [8-1], which is still in use today by shopkeepers and merchants in East Asia. However, the slide rule [8-2], perhaps the computational aid with the second-longest lifetime and the tool most often used by engineers for decades, has now been relegated to technology museums. During the tenure of these two devices, a large number of mechanical calculators have appeared only to be replaced by better inventions. One of these mechanical devices is shown in Figure [8-3].

However, the time line of computer development [8-4], which we do not purport to be complete in any sense, shows that the technological advances made in just the last 40 years far outnumber those in the last 400 years. The progression has been from mechanical devices, to electromechanical devices, to fully electronic devices. The progression in size has been as interesting as the progression in speed and capability, going from relatively small devices to large devices that required a whole room or more, and returning back to small devices again.

During this evolutionary time, the definition of what might be called a computer has matured to the point where the word now means more than a device with just the ability to do computations; that is, a computer is more than just a calculator. Inherent in today's definition of a computer is also the ability to store an instruction set or program as well as the ability to carry out the instructions and manipulate data. The earliest example of a machine with a stored instruction set was Falcon's loom [8-5].

In this chapter we give a short description of both the "machinery" of modern computers and the necessary instruction sets that make them work for us. It is not entirely necessary for everyone to know how the computer works; probably far more important is that everyone knows how the computer can be useful for them. Therefore, you could skip this chapter and proceed with the balance of the book, which is devoted to exploring the usefulness of the microcomputer to engineers and technologists. However, we recommend that

[8-1] The abacus was one of man's first calculating machines. It has been in use for at least 5000 years and is still in use in East Asia.

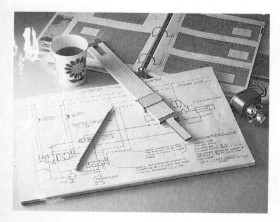

[8-2] The slide rule, now replaced by the hand-held calculator, was the primary computational tool of the engineer for decades. It evolved from John Napier's invention of logarithms in 1600.

engineers and technologist do spend some time here so that they can better understand both the capabilities and limitations of computers and so that they can converse with fellow technical specialists.

COMMON ELEMENTS OF ALL COMPUTING MACHINERY: HARDWARE, SOFTWARE, AND FIRMWARE

Can you imagine a computing system that would accept your information, perform the required calculations, but never provide you the answer? Or, how many computers might be sold that could perform intricate and difficult calculations but not allow you to input unique information for your own problem? These questions and others lead us to the conclusion that there must be several important components for any useful computing system, whether it be the human brain, a mechanical invention, or an electronic device. These are:

- A device for inputting information (numbers, text, sight, taste, hearing, etc.).
- A device for storing information (including the input data, intermediate numbers generated during calculations, and the final results).
- A device to process or execute all the necessary calculations.
- A device to communicate the results to the user.
- A set of instructions that provides the directions to the devices for performing all the tasks.

Without any one of the items listed above, a computing system is severely limited in its usefulness.

The *physical* devices that are used to perform the foregoing tasks comprise the *hardware* of the system [8-6]. As computer systems have advanced through the years, the term *computer hardware* has come to refer to a diverse and ever-increasing number of devices. Devices for entering information into the computer range from crude mechanical card readers interpreting holes punched in cards, to sophisticated devices using laser technology and holography which "check you out" at the supermarket. Later in this chapter we discuss many of the more common devices that make up most of today's computing systems.

The one remaining element of the computing system, however, is the set of instructions that directs all of this hardware to solve some user specified problem. This *instruction set* is a most important component of the computer system, for without it, the computer would simply be a collection of hardware unable to solve any kind of problem. This set of instructions is generally of two types: *software,* instructions that are loaded into the machine by the operator, and *firmware,* instructions that are an inextricable part of certain elements of the hardware. Both software and firmware enable the user to direct the computer to solve a wide range of problems simply by specifying the appropriate instructions from the computer's instruc-

[8-3] Charles Babbage's difference machine was a forerunner of modern computing systems (circa 1822).

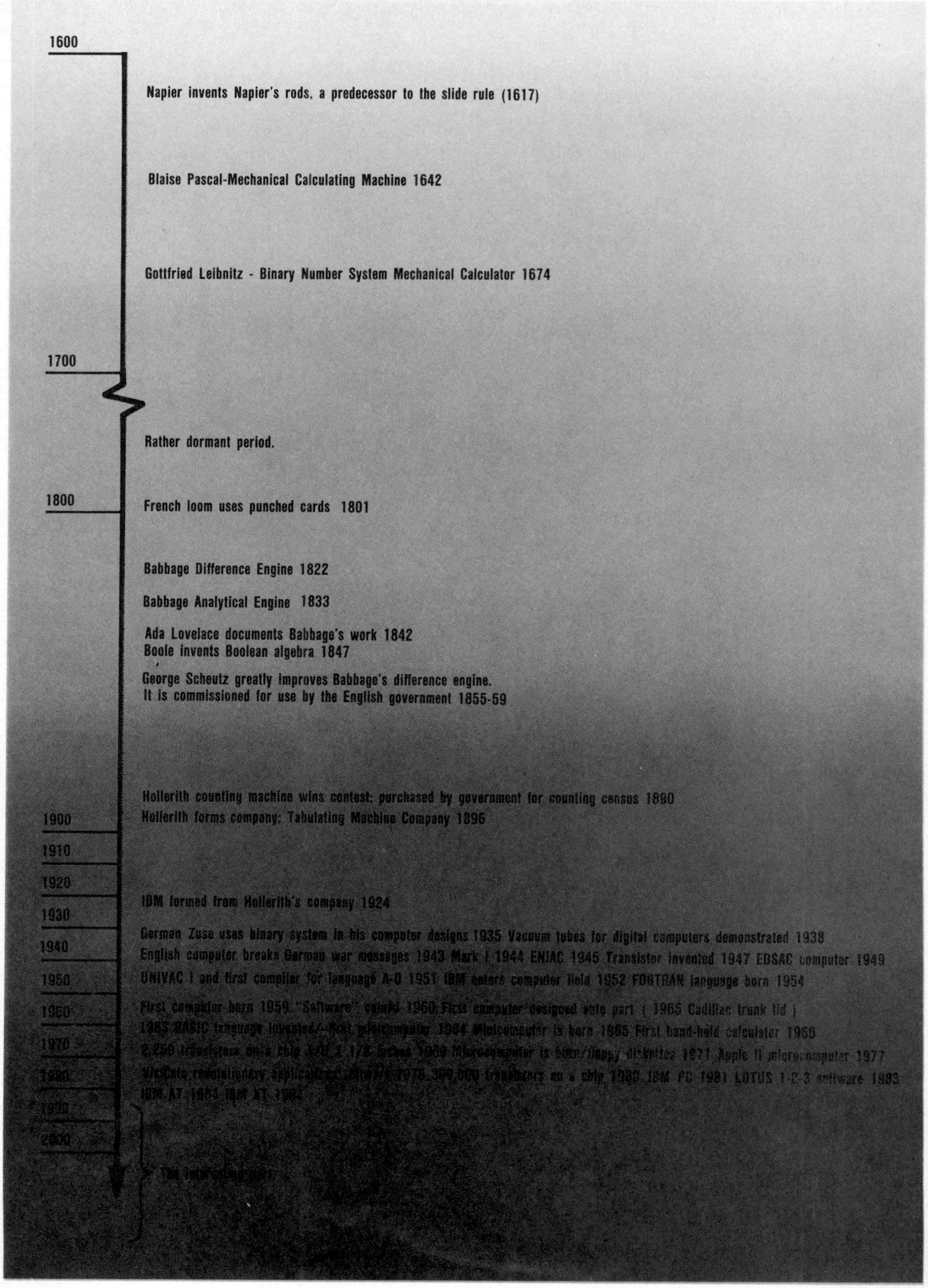

1600

Napier invents Napier's rods, a predecessor to the slide rule (1617)

Blaise Pascal-Mechanical Calculating Machine 1642

Gottfried Leibnitz - Binary Number System Mechanical Calculator 1674

1700

Rather dormant period.

1800   French loom uses punched cards 1801

Babbage Difference Engine 1822

Babbage Analytical Engine 1833

Ada Lovelace documents Babbage's work 1842
Boole invents Boolean algebra 1847

George Scheutz greatly improves Babbage's difference engine.
It is commissioned for use by the English government 1855-59

Hollerith counting machine wins contest; purchased by government for counting census 1890

1900   Hollerith forms company: Tabulating Machine Company 1896

1910

1920

1930   IBM formed from Hollerith's company 1924

1940   German Zuse uses binary system in his computer designs 1935 Vacuum tubes for digital computers demonstrated 1938
English computer breaks German war messages 1943 Mark I 1944 ENIAC 1945 Transistor invented 1947 EDSAC computer 1949

1950   UNIVAC I and first compiler for language A-0 1951 IBM enters computer field 1952 FORTRAN language born 1954

1960   First computer born 1959. "Software" coined 1960. First computer designed auto part (1965 Cadillac trunk lid)
1963 BASIC language invented. First minicomputer 1964 Minicomputer is born 1965 First hand-held calculator 1968

1970   2,250 transistors on a chip with 1 1/2 inches 1969 Microcomputer is born/floppy diskettes 1971 Apple II microcomputer 1977
1980   VisiCalc revolutionary application software 1978. 300,000 transistors on a chip 1980. IBM PC 1981 LOTUS 1-2-3 software 1983
IBM AT 1983 IBM XT 1982

1990

2000

The age of _____?

[8-4] A time-line of computer development.

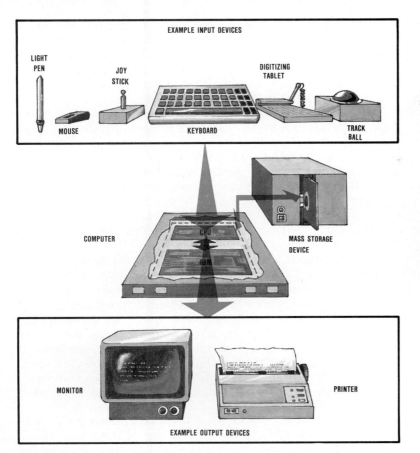

EXAMPLE INPUT DEVICES

LIGHT PEN
JOY STICK
DIGITIZING TABLET
MOUSE
KEYBOARD
TRACK BALL

COMPUTER
CPU
RAM
MASS STORAGE DEVICE

MONITOR
PRINTER

EXAMPLE OUTPUT DEVICES

[8-5] The first machine to be controlled by punched cards was Falcon's loom (1728). The holes in the cards were the instruction set that told the loom what to do.

[8-6] Computer hardware.

tion set in the proper order. The process of establishing the software instruction set is known as *programming*.

We will now discuss electronic digital computing systems since they are currently the most widely used. The word digital represents the fact that these devices use *binary digits*—digits that can represent only one of two values. Binary digits, often referred to simply as *bits* (from *B*inary dig*ITS*), can be considered to be the "atoms" of the information world. They represent the most elemental particle or building block of information processing. All information in digital computers is made up of varying length *strings* of such bits.

The physical representation of binary digits in computers or in their peripheral equipment takes many forms: a hole may or may not be punched in a particular location on a paper tape or card,[1] a particular place on a magnetic record can be magnetized in one of the two polarities—north or south, a transistor can be either turned on (conducting at full current) or turned off (conducting no current), or a switch can be set in either the *off* or *on* position.

---

[1] Punched paper tapes and cards, once an inseparable part of digital computing, have both almost totally succumbed to advancing storage technologies. They will soon take their place, along with the historical computer devices discussed in the first section of this chapter, in science and technology museums.

ARITHMETIC	$+$, $-$
BIOLOGY	♀ ♂
ARISTOTLE	*either ⁓ or*
DESCARTES	x , y
HAMLET	"To be or not to be"
CHESS	P·K4 , P·Q3
PHOTOGRAPHY	◨ ◧
LIGHT BULB	💡 💡
AVIATION	flaps up , flaps down
BASEBALL	**strike , ball**
ACCOUNTING	CREDIT , DEBIT
STOCK MARKET	*Bull* , *Bear*
CRAP TABLE	come , no come
BOWLING	⊠ ▭
SWITCH	ON ⌒ ⌒ OFF
ASTRONAUT	go · no go
TEST ANSWER	true , false
COURSE GRADE	P , F
MARRIAGE	𝔜𝔢𝔰 , NO
COMPUTER LANGUAGE	1 , 0

[8-7] Many things in life are binary.

I never could make out what those damned dots (decimals) meant.
Lord Randolph Spencer Churchill
Quoted in Lord Randolph Churchill II, *by Winston Churchill.*

Regardless of the physical representation of the bits, the mathematical representations of the two binary states are constructed only with the two numeric digits 0 (zero) and 1 (one) [8-7]. Internally, digital computers then perform all of their tasks by reading, interpreting, and outputting multidigit sequences or strings of these binary zeros and ones.

Digital computers typically handle these strings of bits or binary digits in groups called *bytes*. The 8-bit string 01101000 represents one possible 8-bit byte. *Bytes are almost universally defined to consist of 8 bits.* The term *word* is often used interchangeably with the term byte, although it is usually used to represent a collection of several bytes that have some significance. In the section on central processors below, we will discuss the use of other length strings of bits, usually always multiples of 8 such as 16, 24, and 32. These binary strings of bits, or sequences of one or more bytes, are often referred to as *words*.

The counterpart to the digital computer is the *analog computer*. Analog computers do not rely on digits, binary or otherwise, to physically represent information within the computer or its attendant hardware. Such physical representation is accomplished through the use of continuously variable signals—signals that are not limited to only one or two discrete values. Examples are the rotation or partial rotation of shafts in mechanical devices, variable resistors (as opposed to the transistor in digital computers), and voltage levels in electrical devices. For most applications, electrical analog computers, once a strong competitor of digital computers, are now overshadowed by digital devices.

In the next section we discuss in more detail the major parts of the hardware that make up the modern digital computer. This is important since the hardware places overall limitations on what can be accomplished on a particular system. We believe that you need to be familiar with its function in order to better understand what is happening when you try to communicate with any device. You will also be introduced to some of the important computer jargon so that you may talk intelligently with other users.

The "black box" approach presented here ignores the intricate inner workings of the computer, such as the internal logic of the processor. Our goal is not to turn you into a computer scientist but to transform you into an intelligent computer user. If we stimulate you to further study, we will have fulfilled our intent.

GENERAL DIGITAL COMPUTER HARDWARE

The Central Processor

Figure [8-8] is a schematic of the operational arrangement of most digital computers. The heart and soul of the system is the *central processor unit* or *CPU*. It is in this device that the most giant of strides has been made in reducing the size, cost, and electrical power consumption and in increasing the speed of computing.

The CPU is the home of the *arithmetic and logic unit,* or *ALU,* which performs basic arithmetic and logical operations (addition, subtraction, comparison of numbers, etc.). It is where the "number crunching" is done. However, it must be told by some other element in the system where to get the numbers or information it is to use, exactly what it is to do with that information, and where it should put the results.

The ALU performs its operation on strings of binary digits temporarily stored in a very limited number of "registers"[2] and places the results in similar "registers." In modern machines, the actual tasks of the CPU are carried out in the thousands of transistor switches [8-9] and *gates*[3] in the miniature circuits which comprise the CPU. The creation and design of these complicated circuits are the objects of "logic simulation" studies in the fields of electrical and computer engineering. Ironically, such studies are possible only with the aid of the modern computer. "We build on what has gone before."

When a user "runs a program," the *control unit (CU)* in the CPU is the administrator whose task it is to coordinate all the operations that must go on throughout the entire system. This includes coordinating the incoming and outgoing information or data, as well as providing all the necessary instructions to the ALU. The CU obtains both the data to be used and the instructions to be executed from the high-speed memory. It also places results back into the high-speed memory.

High-speed memory is divided into equal parts with each part having an *address* that is specifiable as a string of binary digits. Stored in these addressable parts of memory are binary strings of numbers that represent either instructions to the machine or data (this includes data to be used in some calculation or data that resulted from some operation). The CU addresses each location in memory over the address bus and retrieves instructions and data over the data bus. Of course, it is important that the CU get a valid, executable instruction when it expects one and a valid data string when it expects one.

A most important part of the task of the CU is keeping straight what information in high-speed memory is data and what locations contain instructions to be executed. Since the computer can only store and process groups of bytes consisting of 0's and 1's, the CPU's instruction set must consist of combinations of these values, just as the data must consist of similar combinations.

You can imagine the consequences if the control unit were to lose track of what addresses in high-speed memory contain data and what addresses contain instructions for execution. If the CU begins to fetch instructions from locations containing data, or if it has problems interpreting what is supposed to be instructions, the computer usually quits functioning, and we experience what is often called a "system crash." The only way of recovering is to restart the machine from scratch.

As soon as the computer is turned on or restarted, the CU in the

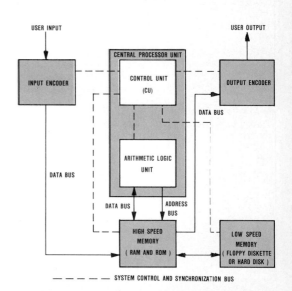

[8-8] Schematic diagram showing the operational arrangement of most digital computers.

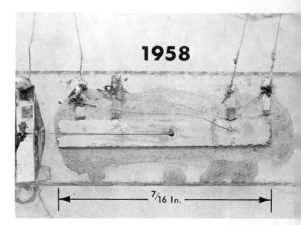

[8-9] The first transistor to be used as a design component was a crude device by today's manufacturing standard, but it worked . . . and it opened a whole new world of solid-state electronics.

---

[2]A register is nothing more than a temporary storage location that the ALU can use during arithmetic operations, or locations that the CPU uses to temporarily store bytes of data when it needs to "fetch" from or store to the high-speed memory.

[3]A gate is an electronic circuit that provides an output based on the input of one or more signals.

[8-10] The Mark I, developed at Harvard University in 1944, was the first fully automatic electromechanical computer. It consisted of over 1 million components, 500 miles of electric wire, and 3000 noisy electromechanical switches or relays. At 51 feet long and 8 feet wide, it was large enough that "you could walk around inside her." It was programmed by changing the coded holes punched into a paper tape read by the machine. It could perform one multiplication or add three eight-digit numbers per second.

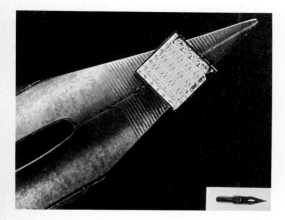

[8-11] In 1969, placing 64 complete electronic memory circuits on a chip of silicon was considered to be a major accomplishment. Such a chip is shown for size comparison on the nib of a pen. In 1980 as many as 300,000 transistors on chips less than 1 mm^2 were available. By the year 2000, chips with more than 1,000,000 transistors should be available.

CPU will attempt to fetch an instruction from a fixed address in high-speed memory and begin execution. This location in memory is usually a location that has a sequence of instructions that is permanently stored there for specifically starting up the computer system. This area of memory is usually called "read-only memory" (ROM) since it can only be read and not erased or written over with new information. Once these instructions begin executing, they direct the CPU to load further instructions into high-speed memory and eventually perform the bulk of the task of starting up the computer. This process is often called "booting" the computer since it "pulls itself up by its bootstraps" by loading the necessary instructions to get itself started. At the end of the booting process, there are enough instructions loaded into memory and initially executed to allow the machine to load, on command, a user's program.

It should be pointed out that all instructions retrieved from memory for execution must consist only of items that the CPU was designed to interpret. Each CPU has its own way of interpreting the sequences of 0's and 1's that represent an executable instruction; thus each CPU has its own unique instruction set. Software written for one CPU will generally not execute on another CPU unless care is taken in CPU design. New CPUs are often assigned so that software that executes on a popular earlier design will also execute on it.

CPUs are classified both by the number of bits that can be used in addressing memory over the address bus and by the number of bits they can shuttle across the data bus. Both of these buses can be thought of as multiwire cables or multilane highways, with each wire or lane carrying a bit [the bits in the form of voltage pulses (bit = 1) or no voltage pulses (bit = 0), are transmitted in parallel over this cable or highway, one bit per wire or lane] [8-12].

Perhaps we should explore briefly how the computer makes sense out of the strings of binary numbers with which it works. Let us assume that we are dealing with a machine that works with single bytes or 8-bit strings (its word length, or length of binary string that it will hold in its registers, is 8 bits). In the decimal (base 10) system there are 256_{10} numbers encompassed by the smallest 8-bit binary number (00000000) and the largest (11111111) (see Table 8-1 for

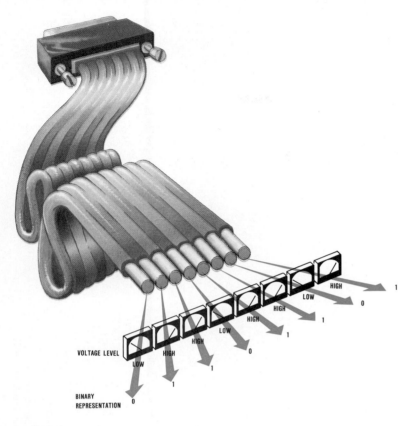

[8-12] Eight parallel wires, each transmitting a bit, can be used to send 1 byte (8 bits) of information.

the decimal (base 10), hexidecimal (base 16), and octal (base 8) equivalents of the binary numbers 00000000 through 01111111).[4]

When the bytes are to be interpreted as instructions, there could be 256_{10} different instructions possible in the instruction set for an 8-bit word. There is no standard for what *instruction* is associated with each binary number; as we have just pointed out, each CPU has its own built-in, unique relationship between binary numbers and instructions.

The representation of data by the eight binary digits available in a one-byte string is considerably more standard, although not universal. Here the problem is one of representing all the useful characters (numbers, lower- and uppercase letters of the alphabet, and other assorted symbols) necessary for carrying out useful tasks. The most widely used standard is that adopted as the American Standard Code for Information Interchange (ASCII). Table 8-1 lists 128_{10} ASCII codes in decimal (base 10), binary (base 2), hexidecimal (base 16), and octal (base 8), together with the symbols they represent. However, the IBM Corporation has adopted its own standard for its mini and mainframe computers called the EBCDIC character set. Table 8-2 compares these two systems for the uppercase A–Z.

[8-13] Wrestling with a computer his own size, this ant attacks an NSC-3200 32-bit microprocessor with computing functions requiring an entire roomful of equipment in the 1960s.

---

[4]Although Table 8-1 shows the code for data and not instructions, it can be used to convert between number bases. For example, $00101101_2 = 45_{10} = 2D_{16} = 55_8$. The subscripts here denote the number base being used. The $_{10}$ is for base 10 or decimal, $_{16}$ is for base 16 or hexidecimal, and $_2$ is the base 2 or binary numbers.

TABLE 8-1 AMERICAN STANDARD CODE FOR INFORMATION INTERCHANGE (ASCII) CODES AND THEIR INTERPRETATIONS FOR DATA

| ASCII Code | | | | Character | Mnemonic | Better |
Dec	Hex	Oct	Binary	or Control	Name	Description
000	00	00	00000000	^@	NUL	Null
001	01	01	00000001	^A	SOH	
002	02	02	00000010	^B	STX	
003	03	03	00000011	^C	ETX	
004	04	04	00000100	^D	EOT	
005	05	05	00000101	^E	ENQ	
006	06	06	00000110	^F	ACK	
007	07	07	00000111	^G	BEL	Bell
008	08	10	00001000	^H	BS	Backspace
009	09	11	00001001	^I	HT	Horizontal tab
010	0A	12	00001010	^J	LF	Line feed
011	0B	13	00001011	^K	VT	Vertical tab
012	0C	14	00001100	^L	FF	Formfeed
013	0D	15	00001101	^M	CR	Carriage return
014	0E	16	00001110	^N	SO	Shift out
015	0F	17	00001111	^O	SI	Shift in
016	10	20	00010000	^P	DLE	
017	11	21	00010001	^Q	DC1	
018	12	22	00010010	^R	DC2	
019	13	23	00010011	^S	DC3	
020	14	24	00010100	^T	DC4	
021	15	25	00010101	^U	NAK	
022	16	26	00010110	^V	SYN	
023	17	27	00010111	^W	ETB	
024	18	30	00011000	^X	CAN	Cancel
025	19	31	00011001	^Y	EM	
026	1A	32	00011010	^Z	SUB	Cancel
027	1B	33	00011011	^[	ESC	Escape
028	1C	34	00011100	^\	FS	
029	1D	35	00011101	^]	GS	
030	1E	36	00011110	^^	RS	
031	1F	37	00011111	^_	US	
032	20	40	00100000			Space
033	21	41	00100001	!		
034	22	42	00100010	"		Quotes
035	23	43	00100011	#		
036	24	44	00100100	$		
037	25	45	00100101	%		
038	26	46	00100110	&		
039	27	47	00100111	'		Apostrophe
040	28	50	00101000	(		
041	29	51	00101001	)		
042	2A	52	00101010	?		

| ASCII Code | | | | Character |
Dec	Hex	Oct	Binary	
043	2B	53	00101011	+
044	2C	54	00101100	, (Comma)
045	2D	55	00101101	- (Hyphen)
046	2E	56	00101110	. (Period)
047	2F	57	00101111	/
048	30	60	00110000	0
049	31	61	00110001	1
050	32	62	00110010	2
051	33	63	00110011	3
052	34	64	00110100	4
053	35	65	00110101	5
054	36	66	00110110	6
055	37	67	00110111	7
056	38	70	00111000	8
057	39	71	00111001	9
058	3A	72	00111010	: (Colon)
059	3B	73	00111011	; (Semicolon)
060	3C	74	00111100	<
061	3D	75	00111101	=
062	3E	76	00111110	>
063	3F	77	00111111	?
064	40	100	01000000	@
065	41	101	01000001	A
066	42	102	01000010	B
067	43	103	01000011	C
068	44	104	01000100	D
069	45	105	01000101	E
070	46	106	01000110	F
071	47	107	01000111	G
072	48	110	01001000	H
073	49	111	01001001	I
074	4A	112	01001010	J
075	4B	113	01001011	K
076	4C	114	01001100	L
077	4D	115	01001101	M
078	4E	116	01001110	N
079	4F	117	01001111	O
080	50	120	01010000	P
081	51	121	01010001	Q
082	52	122	01010010	R
083	53	123	01010011	S
084	54	124	01010100	T
085	55	125	01010101	U

| ASCII Code | | | | Character |
Dec	Hex	Oct	Binary	
086	56	126	01010110	V
087	57	127	01010111	W
088	58	130	01011000	X
089	59	131	01011001	Y
090	5A	132	01011010	Z
091	5B	133	01011011	[
092	5C	134	01011100	\
093	5D	135	01011101	]
094	5E	136	01011110	^
095	5F	137	01011111	_ (UnderScore)
096	60	140	01100000	` (bkwd ')
097	61	141	01100001	a
098	62	142	01100010	b
099	63	143	01100011	c
100	64	144	01100100	d
101	65	145	01100101	e
102	66	146	01100110	f
103	67	147	01100111	g
104	68	150	01101000	h
105	69	151	01101001	i
106	6A	152	01101010	j
107	6B	153	01101011	k
108	6C	154	01101100	l
109	6D	155	01101101	m
110	6E	156	01101110	n
111	6F	157	01101111	o
112	70	160	01110000	p
113	71	161	01110001	q
114	72	162	01110010	r
115	73	163	01110011	s
116	74	164	01110100	t
117	75	165	01110101	u
118	76	166	01110110	v
119	77	167	01110111	w
120	78	170	01111000	x
121	79	171	01111001	y
122	7A	172	01111010	z
123	7B	173	01111011	{
124	7C	174	01111100	|
125	7D	175	01111101	}
126	7E	176	01111110	~ (tilda)
127	7F	177	01111111	‾

In the "ASCII Code" columns, "Dec" denotes decimal number; "Hex" denotes hexidecimal number; "Oct" denotes octal number; "Binary" denotes binary number.
In "Character or Ctrl" column, ^ denotes "Control" (e.g., ^G is Control-G).
"Mnemonic Name" column gives ASCII standard abbreviations for ASCII control codes 0 through 31_{10}.
"Character" columns give the control code or common keyboard character the ASCII code represents.

Of course, more numbers and larger instructions can be represented by larger strings of binary numbers. For example, with a 16-bit word (smallest number 0000000000000000_2 and largest number 1111111111111111_2) there are $65,536_{10}$ different numbers possible.

Many low-cost, high-quality "home computers" make use of 8-bit microprocessors (8-bit words) that can both address memory and transfer data in 8-bit, or one-byte, strings. The emergence of 16-bit microprocessors stimulated the use of microcomputers in the business world because they would address more memory (they can hold much larger numbers in their address registers than 8-bit CPUs) and they could accommodate a much larger instruction set. Thus larger and more complex software could be accommodated. The 16-bit processor was also responsible for awakening the engineering community to the power of the small computer in the first significant way. Intel Corporation's 8088 (used in the IBM PC and the host of PC compatibles) and the Motorola Incorporated's 6800 processors have been the most popular for this type of machine. However, these particular 16-bit processors communicate over an 8-bit data bus. Thus they must move two bytes (8 bits each) to make use of the 16-bit capability in the processor.

The Intel Corporation's 80286 microprocessor, which is popular in many machines, has 24 address channels (three-byte words), giving it much more memory addressing capability than 16-bit CPUs and a much larger potential instruction set. Again, larger, more complex programs can be accommodated. However, the 80286 moves data over the data bus in 16-bit, or two-byte, strings of binary numbers.

CPUs having 32-bit addressing capabilities are also common. Some (like the Motorola Incorporated's 68000 microprocessor) use a 16-bit data bus and some (like the Motorola 68010 and the Intel 80386) use a full 32-bit data bus. The Digital Equipment Company uses its own 32-bit processor in its popular VAX line of medium-size computers. CPUs of even larger addressing capability are used in the super minicomputers and in mainframe computers.

TABLE 8-2 EXTENDED BINARY-CODED-DECIMAL INTERCHANGE CODE (EBCDIC) FOR THE CAPITAL LETTERS (SEE TABLE 8-1 FOR COMPARISON TO ASCII)

Character	EBCDIC Binary
A	11000001
B	11000010
C	11000011
D	11000100
E	11000101
F	11000110
G	11000111
H	11001000
I	11001001
J	11010001
K	11010010
L	11010011
M	11010100
N	11010101
O	11010111
P	11011000
Q	11011001
R	11011010
S	11100010
T	11100011
U	11100100
V	11100101
W	11100110
X	11100111
Y	11101000
Z	11101001

Storage

High-Speed Memory. High-speed memory nearly always contains a section of volatile memory commonly referred to by any of the following names: *random access memory (RAM), read/write memory, main memory,* or *core.* It is from this memory that the CPU gets its instructions for performing its work. That is, the user's program must be loaded into this memory before it can be executed.

The word "volatile" is used to emphasize that information stored in RAM is lost when the computer is turned off or suffers a power failure. This type of memory usually requires electrical power in order to function.

Random access memory (RAM) implies that any part of memory can be accessed as easily and as quickly as any other, as long as addresses are specified [8-14]. Only the address of the memory location is required to read or write to that location. This leads to very

[8-14] RAM is much like "pigeonhole" mail sorting bins. Each pigeonhole has an address and the sorter can proceed directly to the desired address. The CPU is the sorter in the case of the digital computer.

fast access times and increased computation speeds. The name "read/write" highlights the fact that you can both read from and write to this type of memory. The last commonly used name, "core," stems from the era, not too long ago, when high-speed memory was made of ferritic (iron-based) "cores," doughnut-shaped objects that could be magnetized by an electrical current in a wire running through the hole. Today, semiconductor material is used in lieu of ferritic material.

Research in storage techniques is continually increasing storage density. The 256-kilobyte memory chip, capable of storing 256 kilobytes of information, is now the common memory chip. Used in multiplex, these chips can provide very large memories. Soon 512- and 1024-kilobyte chips will be the common-size memory chip.

High-speed memory almost always includes some *read-only memory (ROM)*. ROM, like RAM, can be randomly addressed and accessed by the CPU, but information cannot be written to it. ROM does not normally lose its contents when power is shut off. What is stored in ROM can be reaccessed when power is restored.

ROM contains software that is encoded into a storage element which is a part of the hardware. This encoding is either done when the ROM is manufactured or added by special machinery in the field. ROM usually has stored, as a minimum, the basic instruction sets needed by the CPU to start the boot process and perform very basic operations.

Some brands of hardware have elaborate instruction sets in ROM for executing predetermined program steps. It was pointed out earlier that software embedded in the hardware is called firmware. For programs that are executed often, placing them in firmware offers fast-loading software without the need for external storage.

Low-Speed Memory. We are using the term "low-speed memory" in this text for a wide range of storage devices that are used to save or hold files, data, and programs. A synonymous term often used is *mass storage*. Included in the definition are all storage forms for which the individual storage elements are not directly addressable by the CPU. This means that when material is retrieved from or placed into the low-speed storage, the interaction with the CPU is much slower than when material is temporarily stored in RAM. Mass storage is also distinct from semiconductor RAM because it does not lose its contents when electrical power is removed. It is nonvolatile. Although equipment does fail occasionally, the contents of the low-speed storage can normally be recovered when power is restored.

Punched cards and punched paper tape served for decades as a means of mass storage. Magnetic storage media have improved so much in recent years that they are the predominant type of low-speed memory.

There are essentially two different types of common magnetic mass storage devices. These are:

1. Devices that can be removed from the computer.
2. Devices that are essentially a fixed part of the hardware.

Floppy Disks.   The most common type of removable storage is the *diskette* (sometimes referred to as *floppy disk* or soft disk). These consist of a thin, flexible, circular disk which is coated with a magnetic material and encased in a permanent paper or fiber jacket. When mounted in the "diskette drive" on the computer and rotated within the paper jacket, the diskette's magnetic states are sampled or changed by the drive's read/write head. The head glides along the diskette surface accessing the diskette through a slot in the paper jacket [8-15].

Floppies are convenient to use because of their size, their storage density, and their low cost. However, because of their relatively low spin speeds when mounted and the wide dimensional tolerances they must have, their access time (the time required to read or write to the disk) is relatively slow.

Floppies store their information within "sectors" on concentric "tracks" as demonstrated in Figure [8-15]. Unfortunately, the exact format of stored information has been standardized only on 8-in. (20-cm)-diameter floppies, recording on only a single side, under the CP/M operating system.[5] For example, $5\frac{1}{4}$-in. (approximately 13 cm)-diameter floppies, another common size, are used at various storage densities from 120 kilobytes per diskette (approximately $120,000 \times 8$ bits) to 1.2 megabytes of information per diskette. A formatted 360-kilobyte diskette would hold, for example, the characters (letters, numbers, punctuation, etc.) from about 100 pages of typed, single-spaced, $8\frac{1}{2}$- by 11-in. paper.

New diskettes must be "formatted" for the particular brand of hardware on which they will be used. This "formatting" process, which sets the number of sectors and tracks, is carried out by executing a software program. For all but the 8-in. (20-cm) CP/M-based floppies, this formatting is unique to each type of hardware, making it difficult if not impossible to move a formatted floppy from one brand of hardware to another as a means of transferring data.

Part of one of the tracks on a diskette always contains a directory where information is stored on the file name, size, date of modification, and exact storage location on the diskette of each file. The system updates this information automatically when diskette activities (e.g., when files are saved or copied to the diskette) are necessary. An operating system command (operating systems are discussed later in this chapter) is usually provided to enable a user to query this directory for a list of the information in the directory.

Floppies generally store their information in a manner that allows *random access* to the stored material. That is, when a certain file is required by the user, the system permits that file to be accessed through the information stored in the directory, without reading through all the other files stored on the diskette.

Disk Packs.   Removable disk "packs" are often used on larger computers. These usually consist of several magnetically coated disks, similar to phonograph records, that are attached to a cylindrical hub. The hub is attached to a "drive" and spun about its axis at a high rate of speed, typically 40 revolutions per second. Read/write

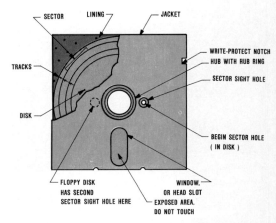

[8-15] Floppy disk.

---

[5]Operating systems are discussed later in this chapter.

[8-16] Magnetic disk drive with disk pack.

heads, passing near the surface of the disks, perform read and write operations with appropriate parts of the magnetic surfaces. It is common for any position of a disk to come under a magnetic read/write head every 25 msec, producing data transfer rates of 500,000 bytes per second.

The disk packs can be demounted, so that others may be mounted and used. Figure [8-16] shows a typical magnetic disk drive unit with a disk pack mounted. Disk packs, like floppy disks, are random-access-type storage.

Magnetic Tape. Magnetic tape, either mounted on reels or in cartridges, is another form of removable magnetic storage. On large computers, this is the form of demountable storage used most often. Magnetic tape libraries of large computer centers often have thousands of tapes on file.

Magnetic tapes have their widths divided into what are called "tracks," most often nine in number, with each of the parallel tracks running the length of the tape. At any one location along the tape, the bits of information can be encoded across the tape with one bit per track. All but one of the bits are used in making up a byte while the remaining one is used for error checking. A tape reading head, monitoring the tracks, simultaneously senses all the bits across the tape.

The layout of the data on a magnetic tape is said to be *sequential* rather than random access. That is, one byte follows another along the tape. If the computer is required to search a magnetic tape for certain information, it must start at the beginning of the tape and progress through the data until the desired information is reached. Thus sequential searches will be slower than finding data in the individually addressed locations in RAM.

The number of frames per inch (fpi)—often called bytes per inch, or simply *bpi*—is called the *tape density*. Typical tape densities are 800, 1600, and 6250 bpi. When the tape density is 6250 bpi and the tape is running at a speed of 75 in., the data transfer rate (read or write) is nearly 500,000 bytes per second. At this rate it would take about 0.2 s to transfer the information contained in this chapter.

A simple calculation would show that a 2400-ft tape, using a density of 6250 bpi, could hold 180,000,000 bytes, or about half of the information in a 30-volume encyclopedia. However, because of blank gaps that must be left periodically to accommodate acceleration and deceleration of the tape, much of the tape is unused. True storage in practice is well below the maximum possible of 180,000,000 bytes.

Hard Disks. Nonremovable disks (sometimes referred to as hard disks) offer improved performance over floppies. Since they are generally made for permanent installation, they have closer tolerances and can spin at higher speeds [8-17]. Thus reads and writes with the disk can be accomplished at much higher speeds. Hard disks operate in much the same fashion as the removable disk packs discussed above.

On small computers, it is not uncommon to find hard disks as small as approximately 13 cm × 15 cm × 4 cm in size (5 in. ×

[8-17] For clarity, the cover has been removed from the head/disk assembly of this hard disk. The disks on which data are stored can be seen.

7 in. × 1.5 in.) and capable of storing 20 megabytes (approximately 20,000,000 × 8 bits) of information (about two volumes of a 30-volume set of encyclopedias). Of course, much larger units are common on larger computers.

Like the removable storage floppies, hard disks have directories stored on the disk, so that current information (file names, exact location on the disk, size, and date of last changes) on the files stored can be maintained.

Input Encoders

We have already discussed the computer world as being a binary one—either on or off, 0 or 1. Our physical world, however, is essentially an analog one. That is, engineers and technologists talk about such things as the stress level in a bridge column, the voltage level in a circuit, or the temperature level in some object. The observable or measurable quantities with which engineers deal are nearly all continuously variable.[6] They are not restricted to discrete values. Therefore, if these measurable quantities are to be represented in computations on a digital computer, some means must exist for converting the inputs of physical measurables into digital signals. This is the purpose of the input encoders. They take nonbinary data and encode it into binary representation.

Keyboards. *Keyboards* are the most often used data encoders, their source of input data being the human brain. They represent digital input devices more than the other input encoders discussed below, since keys are either pushed, or they are not. There is no "in between" [8-18].

When any one of the keys on the keyboard is activated, signals representing the binary string of the appropriate keyboard character are generated within the computer. When the characters are to be shown on the display screen,[7] for example, the binary string generated (e.g., the ASCII codes shown in Table 8-1) on keyboard activation is sent to a character ROM that determines which little dots on the screen will be illuminated.

The *cursor control keys* that allow the user to move the active or input area around the screen are also found on most computer keyboards. The *cursor* is that character, cross-hair, or symbol on the screen that shows you where the input or active area is located.

Many keyboards have *function keys,* sometimes referred to as softkeys, that can often be programmed to do special things. These might be used to carry out several commands at once, save data to disk, execute programs that are located in RAM, or activate certain features of the software being run.

[8-18] Keyboards are used to input data and text.

---

[6]We point out when you study modern or atomic physics you will find that, on a microscopic scale, this is not quite true. But on the scale that most engineering work is done, the permissible discrete steps are so close together as to appear continuous.

[7]Keyboard entry generally does not go immediately to the screen, but goes instead to RAM. The user and/or the software often can determine whether the keyboard input is also displayed on the screen. When it is sent to the screen, keyboard entry is generally said to be "echoed" to the screen.

Some keyboards also have a *number pad* that facilitates the entry of numbers into the computer. The number pad is a collection of keys with numerical labels arranged in a similar manner to the keys on a common calculator. These are separate from the ordinary number keys that are located across the top of the keyboard.

User-controlled input encoders other than the keyboard include the mouse, the track ball, the joystick, the touch screen, the digitizing tablet, and the light pen. All these devices are *pointing devices* for cursor control, not text entry devices.

The Mouse.   The *mouse* is a small hand-held pointing device that is moved around on a flat surface [8-19]. There is an approximate one-to-one relationship between cursor movement and mouse movement; *displacing* the mouse *displaces* the cursor. The direction of the displacement of the mouse determines the direction of displacement of the cursor. Mechanical mice usually roll on balls which activate electrical sensors that in turn send information to the computer through a connecting cord. Optical mice usually sense light reflected from a mirrored pad on which the mouse is operated. The light originates in a source attached to the mouse.

Digitizing Tablets.   *Digitizing tablets* are somewhat like mice in that there is a "pointer" that determines the position of the cursor. However, the "pointer," in the case of the digitizing tablet, is operated over a sensitive surface, not over an ordinary desk top or optically reflecting pad [8-20]. *Displacement* of the pointer over the sensitive surface produces *displacement* of the cursor on the screen.

Light Pens.   The *light pen* is typically a small pointer that is approximately the size of a pencil. It is connected to the computer through an electrical wire that provides the power for a light and/or electrical source located in the end of the pointer [8-21]. The pointer source is moved on the exterior of the screen until it is lo-

[8-19] Mouse pointing device.

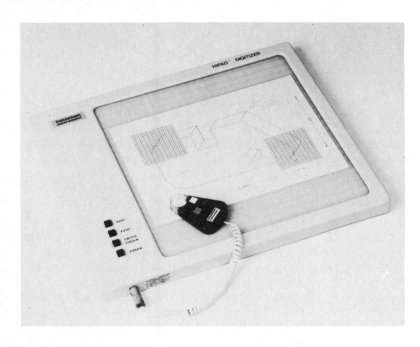

cated at the spot where the cursor is desired. When the source is activated the cursor movement is executed. The *position* of the pen is related to the *position* of the cursor.

Joysticks. The *joystick* is quickly recognized by the video game enthusiast. One end of the joystick is attached to a stationary socket, while the other end can be manipulated by the operator [8-23]. The *speed* of motion of the cursor on the screen is proportional to the

[8-21] Light pen.

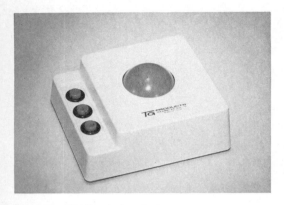

[8-22] Trackball pointing device.

displacement of the free end of the stick from the static position. Thus, it is not a displacement–displacement device like the mouse or the trackball. The direction of the cursor on the screen is related to the direction of displacement of the free end of the joystick.

Touch Screens. *Touch screens* have sensors, usually optical or acoustical, at the periphery of the screen that are sensitive to the placement of an object (your finger, perhaps) near the screen. The *location* of the objects near the screen determines the *location* of the cursor on the screen. The cursor follows the object. Due to the sensing technique and the finite size of pointing objects, the accuracy to which the cursor can be positioned is less than that obtained with most other cursor positioning devices.

Trackballs. *Trackballs* are very closely related to mice. In fact, they are essentially inverted mechanical mice [8-22]. However, instead of moving an object over a surface, the operator uses his or her fingers to rotate a ball within a socket. *Rotation* of the ball produces *displacement* of the cursor on the screen. The direction of motion of the cursor is determined by the direction of rotation of the ball. Trackballs do not require a clear area next to the computer as does the mouse. Clear areas are sometimes hard to find on a busy desk.

Thumbwheels. *Thumbwheels* are used in pairs: the rotation of one wheel determines either the vertical or the horizontal cursor displacement. The rotation of the other wheel determines the cursor displacement in the direction mutually perpendicular to the displacement caused by the first wheel. The wheels are merely variable resistors across which either the voltage or current change according to the rotation of the wheel.

Analog-to-Digital Data Converters. As pointed out previously, the physical world is almost totally an analog world. There are many instances where it is desirable to have data from analog instrumentation entered into a digital computer. Laboratory or pilot-plant monitoring, data acquisition, and data reduction are examples of situations for which direct computer entry is desirable. Conversion of such data can be carried out in devices known as *analog-to-digital (A-to-D) converters*. These devices can accept continuously varying signals and convert them to digital representations for analysis in digital computers.

Exotic Encoders. We list here several technologies that are possible now, but because of cost, are not currently in wide use. As these develop and costs decrease, they have the potential of wide acceptance. These include voice recognition and encoding devices (eagerly awaited for entering text and numerical data into the computer without typing) and image encoding devices (video cameras, optical character readers, laser scanners, etc.). Many supermarkets are already making use of the latter technologies for speeding the data entry at the checkout line.

[8-23] Joystick pointing device.

Output Decoders

Few of us can rapidly peruse binary data (strings of binary digits) and comprehend the meaning. Our senses all give us the ability to monitor and detect analog signals: shades of gray, not just black and white; awareness of amazingly broad levels of sound, not just quiet and loud; and an astonishing sense of touch. We therefore learn the most about a set of information if we can experience it in these analog ways.

We need to decode the binary output of computers to other forms of output that we can readily grasp and understand. Fortunately, there are several ways that this can be accomplished with existing equipment. We will discuss a few of these devices in the sections that follow. In our descriptions, we are including what are termed the *device drivers*. These are interfaces that help decipher the binary codes generated by the software so that they can be displayed on the final output device. These device drivers may include both hardware and additional software.

CRTs. *Cathode ray tube (CRT)* is a generic name often applied to the TV-like display screens associated with computers [8-6]. *Monitor* and *video display tube (VDT)* are other terms often used to identify the CRT. Screen displays are temporary ones, since information displayed on the screen will be lost if the screen is overwritten or the power is turned off. We are including the interface in the description of the monitor.

Computers, being digital devices, do not display continuous patterns on CRT screens. The screen is always broken into pieces or *pixels,* each of which can be controlled—turned on or off, made to blink on and off, or instructed to display any of a very limited set of colors. The number of pixels that can be addressed on the screen largely determines the quality of the display. The term *resolution* is used to describe the potential quality: 320×200 pixels (i.e., 320 horizontally by 200 vertically) on a 12-in. (diagonal) screen usually give a very grainy appearance, while 640×480 pixels make displays appear to be almost continuous. Figures [8-24] and [8-25] illustrate two different screen resolutions.

There are two important types of CRTs in use, these being distinguishable by their abilities to display information. One is the *text* monitor, on which only textual characters (letters, numbers, punctuation) can be displayed. For example, when the CPU desires to print on the screen an answer of, for example, 5, it sends one byte, an ASCII 53_{10}, 35_{16}, or 00110101_2, to the interface (see Table 8-1). The interface then takes this code into a table, usually located in a ROM on the interface, to find which of the screen dots must be illuminated in order to display the character "5." Notice that the CPU has only to deal with one byte in order to display a text character that encompasses many pixels.

The other type of display screen is that of the *graphics* monitor. It is not limited to the ROM-defined ASCII characters; it can display almost any type of design as long as it is told which pixels to illuminate. These necessary instructions must come through the CPU from the software instructions in RAM. On a monochrome graphics

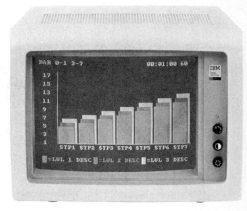

[8-24] A screen resolution of 320×200 on this 11-in. diagonal screen makes lettering and diagonal and curved lines look rather ragged.

[8-25] A screen resolution of 640×400 on this 11-in. diagonal screen makes all lettering and lines sharp and distinct.

screen (e.g., a black-and-white display), the individual bits in a binary string may be used to control the illumination—1 bit per pixel with the digit 1 being on and the digit 0 being off. This necessarily involves more information being handled by the CPU (run times increase) and more space in RAM being required for the instructions (more memory is required).

For *color graphics* displays, additional information, over and above that required by monochrome displays, must be supplied. Not only must pixels know whether or not they are to be illuminated, they must also know what color they are to display. If 256 colors are to be available, for example a byte (8-bit string) per pixel must be supplied for color coding. For high-resolution screens, for example 1024×1024 pixels, that would entail keeping track of over 1 million (1024^2) bytes of just color coding information (one byte per pixel). In this case, performance may degrade and memory requirements will increase.

Printers.   *Printers* are hard-copy devices that produce permanent records of the computer output. *Hard-copy terminals* are special types of printers that can be used for both entering information and recording output. They are particularly good for beginning programmers because they provide the user with a record of both the input and the output.

Fixed-font printers have character shapes defined by the printer hardware. They often bear names such as *impact printer, daisy-wheel printer,* and *letter-quality printer.* When the computer sends to the printer the proper binary code (e.g., an ASCII code from Table 8-1) for a particular character, the printer prints the predefined shape it has available. These printers cannot be used to print true graphics. They can only be used for *character graphics;* that is, they can only approximate the desired output using the predefined characters available to them.

In *dot-matrix printers,* printing is done by placing small dots on the paper in appropriate order or shapes such that the shapes resemble normal characters. In character or text printing mode, the printer receives a binary code (e.g., an ASCII code) from the computer, uses this to retrieve the proper character dot pattern from a stored table of dot patterns, and prints the pattern.

Most dot-matrix printers are also capable of printing in the *graphics mode,* whereby any arbitrary dot pattern can be printed. However, the computer must instruct the printer as to how the dots must be arranged, much like the computer must tell a graphics monitor which pixels to illuminate. Much more information has to be passed to the printer, resulting in degraded performance.

Laser printers use the binary output from the computer to control a small laser beam that is swept over the paper being processed. The beam sensitizes the paper, causing it to attract a fine powder, or toner, in the sensitized regions. With further processing, the toner essentially becomes permanently bonded to the paper, resulting in the printed page. The laser printer offers enhanced speed, versatility, and improved quality over that obtainable with either fixed-font or dot-matrix impact printers.

Plotters. *Plotters* [8-26], like printers, produce hard-copy records. However, unlike printers, they do not operate on binary codes that are translated into characters. Instead, the binary codes relayed to a pen plotter tell it where to take the pen, when to put the pen down on the writing surface, where to take the pen while it is down, and when to pick the pen up from the writing surface. The instructions that do this come from the software instructions in the computer's RAM.

Fast plotters do not actually make use of pens and mechanical linkages. In these, the incoming binary information defines locations to be sensitized on the final medium (usually paper) or on a drum over which the medium passes. These sensitized areas then pick up toner to give substance to the lines and areas that make up the plot.

Digital-to-Analog Converters. Digital computers are often used to control some physical process, such as changing the air/fuel ratio in a carburetor, controlling the speed of a motor, or controlling the temperature in an oven. These are uses where an analog (continuously variable) output from the computer is desired rather than a series of binary digits. Here, a *digital-to-analog (D-to-A) converter* can be used to make the conversion from the digital signals of the digital computer to the required analog signal. Such devices are commercially available.

[8-26] Plotters produce hard-copy records.

Size: Microcomputers to Mainframe Computers

Digital computers span a large range of sizes, shapes, and types. The smaller sizes include the digital hand-held calculators, some small enough to fit into your wallet or purse. At the large end of the scale, there are huge super computers that require large rooms, large amounts of electrical power, and water cooling.

Traditionally, digital computers have been grouped into the following categories: calculators, microcomputers, minicomputers, and mainframe computers. These groupings are more a result of historical development stages than of functionality. Today, only the calculator is distinct enough from the others to merit a class by itself. Considering the pace at which the technology is developing, even this distinction may soon disappear. We will discuss briefly the differences between microcomputers, minicomputers, and mainframes.

Microcomputers. Microcomputers [8-28] are the newest members of the digital computer family. They were born in the infancy of small CPUs and have matured from a hobbyist's toy to a respected tool in business and industry. They are most often thought of as small machines that conduct single tasks for one user at a time. Computational speed has typically been slower than minicomputers and, certainly, mainframes. However, recent advances in memory and CPU capabilities are changing this condition.

Microcomputers that can execute more than one task at a time (called *multitasking*) for more than one user (called *multiuser* systems) are appearing. Also, computational speed is approaching that of many minicomputers. Only two things distinguish them from

[8-27] Hand-held calculators continue to serve engineers as the prime calculating device for many computations.

[8-28] This microcomputer features a 32-bit CPU and monochrome graphics screen.

[8-29] This 32-bit minicomputer supports a multiuser environment.

[8-30] This large mainframe has a rather benign appearance that masks its speed and power.

other members of the computer family: physical size (they are somewhat smaller than minicomputers and considerably smaller than mainframes) and price. Microcomputers are available in the range of a few hundred dollars to about $10,000.

Minicomputers. Minicomputers [8-29], products of the advances made in CPU design, chronologically represent the first departure from large mainframe technology. These machines were less expensive, smaller, and easier to maintain than previous mainframes. The minicomputer allowed the placing of computing power in the physical locations where it was needed. In large companies, it was welcomed by some users as a means of breaking management's hold on centralized computer centers. For smaller companies, it offered an economical alternative to the rental of computer time from other organizations.

Today, what the minicomputer sometimes lacks in computational speed for a particular computation compared to a mainframe, it more than makes up in its ease of use and smaller queues.[8] Minicomputers typically cost in the range of tens to hundreds of thousands of dollars, considerably more than minicomputers, and substantially less than mainframes.

Mainframe Computers. Mainframes [8-30] are typically characterized by several megabytes of main memory, the capability of addressing four or more bytes, hundreds to thousands of kilobytes of fixed disk storage, and multiple magnetic tape read/write units. They

---

[8]*Queue* is a British term synonymous with "waiting line" and is a part of computer jargon. There is the "input queue" for the jobs waiting their turn to get into the computer, and the "output queue" of jobs awaiting printing or plotting.

allow *time sharing;* that is, they service many users at any one time. Mainframes are usually supported and maintained by a specialized staff employed strictly for that purpose. They are also operated in secured areas. The users are typically not co-located but access the machine through remote terminals.

Large mainframe computers are *not* a dying breed, although some of their work load has been taken over by the micro- and minicomputers. Ironically, the micro- and minicomputer revolution could, in fact, soon begin to increase the load on large machines. As more and more engineers become computer literate through the ease of use and availability of the small machines, they will continue to seek more and more computing power. All these new users may soon discover, perhaps rediscover, the power and speed of the larger machines.

There are many uses that will probably not be assumed by the micros and minis. These include the execution of large design and analysis programs that consume huge amounts of core and large computation times and the maintenance of large corporate data bases. The small computers may be used to access pieces of these larger tasks, but it seems unlikely that they will replace the mainframes.

As stated earlier, price is one of the distinguishable differences between the micro, the mini, and the mainframe. Mainframes are at the high end of the scale; they usually cost in the millions of dollars.

The progress being made in the field of computer hardware is phenomenal. From 1981 to 1984 alone, the performance of CPUs increased by a factor of 2.5, the cost of RAM decreased by more than a factor of 10, and the storage density on floppy diskettes improved by a factor of nearly 10. The price-to-performance ratio for CPUs is projected to improve continually at 15 to 20 percent per year.

These rapid advancements are driving a change that is occurring in the engineering profession. The engineer is becoming more and more inextricably entwined with the computer. Without an understanding of the machine, its role, and its use, today's engineer is obsolete.

> Man is still the best computer that we can put aboard a spacecraft—and the only one that can be mass-produced with unskilled labor.
>
> *Wernher von Braun*

GENERAL DIGITAL COMPUTER SOFTWARE

As pointed out earlier, *software* is the set of instructions that tells the computer where to get the information that it needs to process, exactly what to do with that information, and where to put the intermediate and final results. It must exist in order for the computer to do its job.

The word *algorithm* is sometimes used in referring to computational processes. In its most general use, an algorithm is a step-by-step procedure that leads to a solution of an entire class of problems in a *finite* number of steps. The mathematical formulation or mathematical model of a subject, whether in explicit or implicit form, then, is not an algorithm. But the step-by-step procedure for solving it is. Thus software, because it is a collection of step-by-step instructions, is essentially a very detailed computer-based algorithm.

Whereas the term *algorithm* can be used to describe any degree of detail of the solution steps, the term *computer program* is often used to represent the collection of the actual *lines of code* or sequential lines of computer steps that are written by a programmer in order to implement the solution on the computer. Therefore, "programs" and "software" are more nearly synonymous than are "algorithms" and "software."

This section on general software is intended to describe various types of software that can be used to carry out a user's computations. As you proceed through your academic curriculum, you will learn to recognize when the computations you need to make are within the capabilities of a hand-held calculator. Certainly, small problems with few iterations and little branching might best be solved on such devices. Since these calculators vary considerably from manufacturer to manufacturer, and since the manufacturer's instruction books provided with each machine generally cover the programming steps, we will not detail their use here.

When you need more powerful analysis tools, you must generally assess the type of computers that are available to you (micros, minis, or mainframes) and learn what software exists on these machines. There are thousands of applications software packages in existence, so do not write your own code if someone has already done it for you. Prepackaged programs range all the way from modules that can be incorporated into or used by programs you write yourself to packages that will do everything that you may need to do. We discuss some of the very general programs in the next section.

You may not always find existing routines that do what you need done. You may decide that you can improve on existing packages or you may find that your problem is unique enough to require original programming. In these cases you must invest some time and effort into writing the programs, or you must have someone (a programmer, perhaps) write them for you.

In any event, you need to apply "appropriate technology" to the solution of your problem. For example, do not use the mainframe computer just to add two numbers. Similarly, do not use your hand-held calculator to solve 20 equations with 20 unknowns. Be conscious of the resources at your disposal and use them judiciously.

In the following sections, we begin our description of various generic types of software with a discussion of operating systems, since you must always work with this system regardless of whether you want to run prepackaged programs or do original programming.

Working through an Operating System

Earlier in this chapter we discussed starting or "booting" the computer system. In that operation, certain instructions are loaded into a reserved portion of RAM so that other instructions (e.g., those contained in a user-supplied program) can be loaded and executed. The initial set of boot instructions form the working part of what is called the *operating system*.

Included in the operating system are procedures for loading a program from low-speed memory into RAM and executing it, copying a program from one form, or part, of low-speed storage to another form, or part, of low-speed storage (e.g., from floppy to floppy) and presenting a list of the files in a certain part of low-speed memory, to name only a few. In multitasking systems, the operating system keeps track of the multiple tasks that it is doing. In multiuser or time-sharing systems it must coordinate the jobs of the many users for whom programs are being executed.

Thus the operating system is the software that supervises and orchestrates the overall activity of the computer. In a well-designed operating system, users are only aware of services that are being provided (storage of programs and data, execution of programs, translation of programs, record keeping, etc.). Users are not exposed to all the details of system supervision and integration.

To accomplish a task on the computer, the user must communicate with it through operating system *commands*, commonly called *control statements* or *job control statements*. These commands do not have to be unique to a certain CPU (e.g., the command to copy files could be the same from CPU to CPU), but the actual machine instructions that are carried out when the command is issued are unique to each CPU.

There are many operating systems in use today and, unfortunately, each system seems to have its own set of commands or control statements. Examples are CP/M, MS DOS, PC DOS, UNIX, RT-11, RSTS, VM, VMS, and CMS. On mainframe and minicomputers, the operating systems are usually unique to each manufacturer. In the early stages of the microcomputer revolution, each brand of hardware also seemed to have its own operating system. However, 8-bit micros soon stabilized on CP/M as an operating system while on 16-bit machines, DOS became the de facto standard. Stabilization and standardization of an operating system makes programs much easier to move from system to system and lessens the burden on users. Stabilization leads to greater availability of software, while standardization promotes an increase in the number of people attempting to use computers.

The UNIX operating system has been implemented on nearly all sizes of computers. For this reason, it may eventually become the de facto standard for all computers. The major advantage of such an adoption is that users would then have at their disposal a set of operating commands that would be the same whether they were on a microcomputer, a minicomputer, or a mainframe.

Applications Packages

We consider applications programs to be software for which the detailed program steps are already written or coded. In using an applications program, we would not have to resort to programming languages (to be discussed later in this chapter) in order to write a program that would solve a particular problem. The use of such programs generally saves considerable time and may even do a better job than software that you would write yourself in any reasonable time.

Seemingly, there is an almost infinite number of existing software packages, with more appearing everyday. For example, in

1984, there were 12,000 software companies writing programs for microcomputers alone. At that rate, if what you need is not yet written, it probably will be some day—if you can wait.

CAD. Computer-aided drawing (CAD) packages permit the construction of engineering and architectural documentation drawings to be made on the computer. Such programs provide a precise and accurate method of drawing and permit easy duplication of all or parts of drawings already entered into the computer. They also greatly aid in the specification of dimensioning of parts, construction of lists of parts, lists of raw materials, and can even check for interference of one part of an assembly with another part of that same assembly. When actual paper copies are needed (perhaps to send to the manufacturing department), the drawings can be sent to a plotter for outputting.

Included in nearly all commercial packages is the ability to draw lines, circles, arcs, ellipses, and rectangles of various weights and structure (solid, dotted, center line, etc.). Provisions for annotating of the drawings (lettering notes, etc., on the drawing) are also provided, as is the ability to save to and reload drawings from low-speed memory. Other usual provisions include the ability to zoom (enlarge or shrink the viewing area of a drawing) and rotate.

Perhaps the single most important attribute of CAD is that it can reduce the time and effort associated with documentation drawing. Estimates that CAD can reduce these by as much as 60 percent over manual techniques have been made. This reduction is understandable since the design of products is an iterative process that typically requires many, many changes to engineering drawings.

For engineering and technology, CAD programs are, perhaps, the most useful applications programs in existence. In many companies, these packages have replaced the drawing tables, the T-squares, the triangles, the drawing instruments, and the pencils which were once the trademark of the designer/drafter. Since their first use in the design of a 1965 Cadillac automobile trunk lid, CAD programs have made steady inroads into how documentation drawings are created, stored, and updated. It is easy to understand the flurry of development work and sales efforts by so many different vendors of CAD programs since it has been conservatively estimated that less than 5 percent of the potential users have now converted from manual techniques to CAD.

We will devote Chapter 9 to a more detailed discussion of CAD and its use in engineering and technology. We present there explicit discussions of two of the leading personal computer-based CAD programs.

Spreadsheets. One of the most versatile of all applications programs is the *electronic spreadsheet* or just *spreadsheet*. Spreadsheets divide the screen into rows and columns just like an accountant's paper ledgers [8-31]. A user can enter a number or a label into any one of the boxes or "cells" formed by the intersection of a given row and a given column. The real power, however, comes in the fact that the user can enter a formula, instead of a number or label, into any of the

	A	B	C	D	E
1					
2					
3					
4					
5					
6					
7					
8					
9					
10					
11					
12					
13					
14					
15					

[8-31] Spreadsheets divide the screen of the CRT into rows and columns much like the accountant's paper ledgers are divided into rows and columns.

cells.[9] These formulas can receive inputs for the variables in the formulas from values stored in other cells. The results of the evaluation of the formulas are displayed in the cells in which the formulas reside. For example, a formula that sums the entries of the 12 cells immediately above the cell holding the formula would lead to the numerical value of the sum appearing just below the vertical list of 12 numbers. Today's most common spreadsheets bear names such as Multiplan,[10] Lotus 1-2-3,[11] and SuperCalc,[12] to mention only a few.

Spreadsheets all have commands, such as copy, move, insert row or column, delete row or column, and edit, which are used to construct and alter entries on the sheet. There are also formatting commands which allow the user to clarify and improve the appearance of the sheet. Input and output (I/O) support is always provided to allow storing, retrieving, and printing sheets.

The versatility and convenience of the spreadsheet idea is now penetrating the engineering world. The functions available for use in formulas include most of those needed for engineering calculations (e.g., sine, cosine, maximum, minimum, average, etc.). Searches for tabular data, capabilities for iterative problem solving, statistics, and table generation are some of the more useful features.

Spreadsheets now commonly support graphics. That is, the numbers displayed on the electronic sheet can also be easily displayed in graphical form for better visualization of the data.

Some spreadsheets have also been integrated with other tasks often requested by the user, such as data bases, word processing, and capabilities for communications with other computers. These enhanced capabilities have come about because of better CPUs and larger RAMs in microcomputers.

A spreadsheet can be considered to be both an applications package and a programming language. Although most of the code exists in the spreadsheet software, users have to supply or "program" their own formulas or equations. More detailed information on the use of these tools in engineering is given in Chapter 10.

Specialized Applications.   The subject of specialized application software is very broad. It extends from the spreadsheets discussed in the preceding section, to equation solvers, project schedulers, specialized design programs, and large-scale generalized design programs. The latter class includes computer-aided drawing (CAD) packages mentioned in a previous section. We discuss only two additional specific categories in this section.

In the first category, we include programs that provide the framework for a solution and require the user to supply some piece of the framework. The microcomputer revolution has produced some major contributions in this category, starting with the spread-

The Origin of Spreadsheets

The story of the invention of the spreadsheet is the epitome of all the "university-students-become-famous" stories that abound. Spreadsheets were born not in an engineering college or in a computer science department, but in a business school. As is so often the case, necessity was the mother of invention.

Daniel S. Bricklin and Robert M. Frankston, as accounting students at Harvard University, became convinced that there was a better way to handle accounting data than to enter them by hand on the standard ruled sheets divided into rows and columns. Tedious entry, complicated and repetitive calculations, and the ever-present possibility of the propagation of errors all necessitated hours—even days—of effort to complete the sheets. Then, if a professor (or executive) wanted to see what the profits would be if, for example, labor costs went up x percent and product price went up y percent, the calculations would have to be completely redone. Burn the midnight oil.

The emerging microcomputer appeared to Bricklin and Frankston as being ideally suited to simplifying this task. If fed the necessary numbers and told the necessary calculations to perform, the computer could easily produce the final results. Of course, this concept in itself was not revolutionary, for that is what computers had been doing for 30 years. What was revolutionary was the way that Bricklin and Frankston went about solving the problem.

They made the screen resemble an accountant's paper ledger (see the text) and incorporated the ability to enter numbers, labels, and formulas at various places on the screen. What was so revolutionary about the electronic spreadsheet was its interactive nature, made possible by the computer. Quick answers could be obtained for "what if?" types of questions.

Bricklin and Frankston were so successful in implementing their ideas on a 32-kilobyte RAM Apple microcomputer (which, at the time, was a new product on the market) that they are now widely credited with being largely responsible for the microcomputer infusion into the business world. Their newly developed software was marketed under the name *VisiCalc*.

---

[9] Only a number or a label or a formula could be entered in a cell, not a combination of the three.

[10] Trademark of Microsoft Corporation.

[11] Trademark of Lotus Development Corporation.

[12] SuperCalc 3 and SuperCalc 4 Trademark of Computer Associates International, Inc.

sheet discussed in the preceding section, but it is probable that the best is yet to come. Three examples, TK!Solver,[13] FORMULA/ONE,[14] and muMath/muSimp,[15] are probably early indicators of the future trend.

The first two of these, TK!Solver and FORMULA/ONE, solve simultaneous linear and nonlinear algebraic equations. These application packages, are significant because they free the user from the tasks of algorithm construction and program development. For example, if a mathematical model is developed in the engineering design process, then, assuming that the equations meet the restrictions of the programs, the designer can go directly to the solution. The designer simply types in the equations and commands the programs to solve them. The equations do not have to be entered with the unknowns on the left side of the equations, as is required by nearly all other programming forms. The programs take care of all those details.

TK!Solver and FORMULA/ONE are convenient for doing parameter analyses because they allow the use of tables of inputs and then form tables of the corresponding outputs. Graphing of the inputs and outputs can also be accomplished.

muMath/muSimp is a rational mathematics package that can do algebraic manipulation, matrix algebra, trigonometry, and differential and integral calculus.

The second category of applications package of which students need to be aware covers the mathematics and statistical *libraries* that are generally available on any large computer installation. The user provides the framework for the solution in the form of a program and the libraries are used within that framework. These packages consist of many subprograms or subroutines that can be called by high-level languages in order to perform calculations that are common to many different problems (high-level languages are discussed in the next section.

For example, if users wanted to solve a set of simultaneous equations, they might calculate the coefficients in the equations and then use an equation solver from a mathematics and statistical library to solve the equations. Similarly, users could perform integration, curve fitting, matrix inversion, statistical analysis, and other tasks by first writing programs that define, collect, or calculate preliminary data and then turning the bulk of the work over to a library routine.

Doing It from Scratch

When existing applications programs will not fill your need, there is no alternative but to write your own program. Although we call this section "Doing It from Scratch," we do not necessarily mean that you must start with the binary instructions that can be understood by your CPU. But you must start programming in some language

---

[13]Trademark of Software Arts, Inc.

[14]Trademark of Alloy Computer Products, Inc.

[15]Trademark of the Soft Warehouse.

that the machine can understand or decipher. We now take a brief look at some programming languages.

Machine Language.   The only instructions that a computer can accept *directly* are those written in *machine language*. Machine language uses strings of binary numbers to represent the individual instructions and addresses of operands. These are the instructions that must reside in RAM and be processed by the CPU in order to carry out any type of operations on the computer.

As pointed out earlier in the chapter each model or make of central processor unit has its own machine language—its own set of instructions that must match its capabilities. Knowledge of one machine language is generally not directly transferable to another machine.

Although machine language programming might be suitable for small CPU machines, it is unthinkable for programs on larger machines because the numeric code associated with each instruction must be memorized or found in a manual, and the addresses of all operands must be assigned and controlled by the programmer. This is a mind-numbing, often error-prone process.

Specialists in machine language programming must spend considerable time in training. Unfortunately, the skills acquired in this training are not highly portable. When a new CPU comes along, a new set of rules must be learned.

Fortunately, only the people who need to write machine-oriented compilers or assemblers—discussed below—are those who need to work in this language. However, these people make life possible for the rest of us.

Compilers.   Soon after computers first came into use, it became obvious that a more human-oriented form of programming than machine language had to be devised. What was needed was a way of allowing users to program in forms (characters, symbols, and syntax) that were more meaningful to them, and then providing them with a way to translate these into instructions the machine could use. Since the computer was exceptionally able in carrying out laborious and tedious tasks, it was an ideal candidate for accomplishing the intricate job of translating human-oriented forms of programming into the machine language forms it needed. That is, special programs were devised that instructed the computer to prepare a machine language program for itself.

Programs that translate other programs into machine language programs are called *compilers, translators,* or *processors*. Compilers are supplied by computer manufacturers and by companies that specialize in designing such software. Any minicomputer or mainframe computer installed to serve a number of users will have a number of compilers in internal memory (usually ROM) or stored on peripheral equipment. These can be called into use—executed, in computer terminology—by appropriate job control statements.

Languages that need to be translated into machine language are generally classified into two categories. *Interpretive* languages are those that get translated, line by line, as they are being executed. No complete, compiled version is saved; when the program is re-

FORTRAN

FORTRAN, an abbreviation for FORmula TRANslation, was the first widely accepted high-level language. It was designed so that its statements would visually resemble conventional algebraic notation, as much as the computer symbols of that time permitted. For instance, the multiplication of two variables, A and B, to yield a third variable, C, is represented by the following line of code:

$$C = A * B$$

Similarly, addition is represented in FORTRAN by

$$C = A + B$$

The FORTRAN statement

$$C = A * (B + D)$$

is an instruction to give C the value of A multiplied by the sum of B and D. These simple instructions are much easier to follow and understand than would be equivalent abstract and tedious machine language code. A FORTRAN compiler does the work of translating these instructions into machine language.

FORTRAN originated at the IBM Corporation in the mid-1950s and has been periodically upgraded to new standards. FORTRAN II was the first version widely used. FORTRAN IV, a more standard version of FORTRAN, published by the American National Standards Institute (ANSI) in the late 1960s, was the basis of most FORTRAN textbooks for the next decade. Much existing code is written in FORTRAN IV; indeed, it is still in use in many places.

FORTRAN 77, the newest ANSI standard version, came into being in 1978. This latest version contains the same basic structure as FORTRAN IV, and most programs written in FORTRAN IV will compile properly with a FORTRAN 77 compiler. However, FORTRAN 77 has some additional features that allow it to become a language much more capable of supporting structured programming principles.

FORTRAN is not necessarily restricted to scientific applications just because of its algebraic orientation. It is sufficiently powerful to handle many business problems as well and is often used in their solution.

executed, each individual program step must be recompiled again. *Compiled* languages are those in which all the program steps are compiled together when the programming is finished and prior to execution. At run time, the compiled version—often called a *load module*—is executed.

The programming language BASIC,[16] discussed below under "High-Level Languages," is the best known example of an interpretive language. Interpretive languages are convenient to use because they are usually coupled with a built-in editor that is activated automatically when a compilation error is encountered. They also go into execution more quickly than compiled languages, because the sometimes lengthy process of compiling a complete program for a compiled language is not required. However, interpretive languages execute more slowly than compiled languages, since before each step is carried out, it must first be compiled. Since scientific and engineering programs often rely heavily on looping—doing the same calculations over and over until convergence is obtained—slow execution can be a severe detriment.

Assembly Language. The first significant step above machine language programming is to teach the machine to recognize program steps written in symbolic rather than a binary machine language code. This is done by designing a special compiler called an *assembly language program, symbolic program,* or simply an *assembler*. Assembly language programs are machine oriented in the sense that, in general, each statement is compiled into one machine language statement. Because of the machine orientation, knowledge of the symbolic statements used for one computer is only of partial help in writing symbolic programs for another model.

Assembly language programs are written with mnemonic names in place of the binary codes representing operators or operand addresses. The *mnemonic* names—that is, names that bring to mind the operand—are assigned by the designers of the compiler. Accordingly, A may represent addition, S subtraction, STO store, and so on. This symbolism is much easier to remember than strings of binary numbers that might actually represent these same operations in machine language.

This book will make no attempt to describe the details of any assembler. Students should realize, however, that assemblers are an important link in the "education" of a computer. One of the first things a computer manufacturer must do when a new model is introduced is to ensure that an assembler is made available. The assembler must, of course, be written in machine language, because a computer just coming off the assembly line understands only machine language. The assembler then becomes the basic tool for *systems programmers* in writing higher-level compilers, editors, and operating systems.

Fortunately, most of us do not have to program in either machine language or assembly language. There are easier ways of using the machine that suffice. However, with these easier ways also come

---

[16] Compilers for BASIC are commercially available.

penalties. When we take the easier route, we have less freedom in deciding how we will do things, in how we will take advantage of the various parts of the computer, in how efficiently we use the available memory, and in how fast we can conduct our computations. For flexibility and speed, it is hard to beat a well-written assembly language code.

High-Level Languages. Two major developments were needed for electronic digital computation to attract the broad base of users it has today. First, it was necessary to design programming languages and their associated compilers so that the programs could be written in terms of the problem to be treated rather than in terms of an elementary instruction set hardwired into the computer. Second, it was necessary to standardize the languages to the greatest extent possible so that users would not need to learn a new language each time they had occasion to use a different computer. That is, *machine-independent* or *portable, problem-oriented* languages were needed.

Grace Murray Hopper legitimized work in the first of these two areas in 1952, with her invention of the first compiler for a language known as A-2. Work since that time has resulted in the creation of a number of high-level languages that have received wide acceptance. Such languages not only make programming easier but permit the programmer to adapt programs to most any computer simply by learning a few job control statements and ensuring that there is available a compiler for each new machine.

There are a large number of high-level languages in use today. These include FORTRAN, BASIC, Pascal, C, COBOL, ALGOL, PL/1, LISP, and Ada, to name only a few. These represent major survivors and maturing newcomers in the programming language evolution. Hundreds of other languages have been developed, but most have failed to survive the competitive environment. High-level languages, once well established, take on a life of their own—much like natural languages and systems of dimensions. They are not easily displaced even when they retain defects or are subject to theoretical criticism.

FORTRAN remains the major language for scientific and engineering computation. It has been more than adequate for most tasks, although there are doubts that it will suffice in the long term as we grow into a computer graphics world. The technical community will probably continue to court FORTRAN, since it needs to support its huge inventory of FORTRAN programs. It would be quite expensive to rewrite the hundreds of thousands of existing programs in some other language. Languages designed to make FORTRAN obsolete do appear from time to time, but widespread acceptance and controlled evolution have kept FORTRAN abreast of the latest programming concepts and computer technologies.

Checking, Debugging, and Misuse

Of course, the most important aspect of any solution technique is that it provide correct answers. However, long and complicated computer solutions often become impossible to check absolutely. All but the simplest of codes have "bugs" [8-32] that can produce erro-

BASIC

BASIC, an acronym for Beginner's All-purpose Symbolic Instruction Code, was designed primarily as a teaching language using printing or display (CRT) terminals. It was invented by Dartmouth professors John Kemeny and Thomas Kurtz in 1963. Because of its availability, ease of use, and small RAM requirements, BASIC became *the* language for microcomputers as these electronic marvels have gone from hobbyist's toys to truly sophisticated computers. Now, many beginning programming courses, particularly in high schools, are taught in BASIC. This language has many FORTRAN-like features and, in particular, retains the algebraic-like appearance.

Although BASIC has evolved well beyond the beginner's level in capability, most versions of BASIC are not as convenient as FORTRAN for solving complex analytical problems. For most versions, the major limitations are its inability to make use of generalized subprograms or subroutines, and the slow execution speeds. However, there are several versions of BASIC now available which overcome these problems. In fact, Kemeny and Kurtz's latest version, called TRUE BASIC, is intended to overcome some of the problems of the original BASIC.

[8-32] The term "bug," meaning a flaw in a program, originated in the early days of computing when the machinery included many electromechanical relays. After the Mark II computer quit one day (September 9, 1947) while Grace Murray Hopper was working on it, she found the cause to be an insect (a bug) which had been caught in one of the relays. She taped the bug into her logbook with the notation, "First actual case of bug being found." This is a photo of that page, which is now preserved in the Naval Museum at Dahlgren, Virginia.

neous, if not unexpected, results. Therefore, all computer programs must be checked and double checked.

The most often used form of checking is to have the program perform calculations for a situation (e.g., a set of data) in which you know the solution. In writing a program it is usually advisable to initially include a large number of steps that supply intermediate answers in any computerized solution. When the calculation procedure is confirmed by checking the intermediate results, these extra steps can be removed.

Programs are always subject to misuse. In any analyses, there are always assumptions that must be made in the modeling. When computer programs are used in the solution, they, too, include these assumptions. It is easy to forget about all of these original assumptions once the programming is finished and to use the program for problems that violate the assumptions that were made. This is especially true in applications programs—programs where the user does not write code but only inputs the data for a specific problem and then collects the output.

PROBLEMS

8-1. In library books that deal with the history of computing, determine what Helmut Schreyer's contribution to computing was.

8-2. In library books on computing or numbering systems, identify the person who perfected the binary numbering system.

8-3. In library books on computers or numbering systems, learn how to convert from binary numbers to decimal numbers. Convert the following binary numbers to decimal equivalents.
A. 00000001
B. 00000010
C. 00000100
D. 00001000
E. 11111111
F. 00101010
G. 01100111

8-4. What is the largest make (manufacturer and model number) of computer at your institution?
A. What CPU does it use?
B. How many bytes of RAM are installed?
C. How many bits does it hold in its internal address registers? (This is referred to as its word length.)
D. How many bits does it shuttle simultaneously over the data bus?
E. How many bytes of disk storage are available "on-line?"
F. How many tape drives does it have? What is the preferred tape density?
G. Is it operated in a time-share mode?
H. Would you call it a micro, mini, or mainframe?

8-5. If you use floppy diskettes on your computer or if floppy diskettes can be used on the computers at your institution:
A. What (physical) size are they?
B. How many bytes can be stored on these diskettes?

C. Assuming one character per byte, estimate the number of pages of single-spaced text that could be stored on one diskette.

8-6. Does your computer facility have any hard-copy terminals available for your use? What are their disadvantages?

8-7. Does your computer facility have any CRT terminals available for your use?
A. Are they capable of graphics display?
B. Do they display textual characters only?
C. How many columns of text characters can be displayed on the screen?
D. How many rows of text can be displayed on the screen?

8-8. Does your computer facility have both hard-copy terminals and CRTs available for your use? If so, which do you prefer, and why?

8-9. What input encoders are available at your computer facility?

8-10. What output encoders are available at your computer facility?

8-11. Do you own a hand-held calculator? Is it programmable? If so, how many program steps will it store in its memory?

8-12. Seek out information on the Dvorak keyboard and compare it with the Qwerty keyboard. The most often used letter in the English alphabet is the "e." Compare its placement on the two keyboards. What are the advantages and disadvantages of each?

8-13. Does your institution offer any courses in:
A. Logic simulation?
B. Microprocessor design?
C. Microprocessor applications?

8-14. What CPU is used in:
A. The IBM PC microcomputer?
B. The IBM AT microcomputer?

C. The Apple Macintosh microcomputer?

D. The Digital Equipment Company's Rainbow microcomputer?

E. The Zenith Z150 or Z100 microcomputer?

8-15. What is the name of the operating system on the computers that you use? Is it capable of handling multiple users?

8-16. For the operating system on your computer, what is the job control statement that would allow you to:

A. Obtain a listing of the files that you have stored in low-speed memory.

B. Copy a file from one part of low-speed memory to another part of low-speed memory.

C. Format a floppy diskette.

D. Execute a load module by the name of DRAW.

8-17. Does your institution offer a course in machine language programming? If so, to what computer is it specific?

8-18. What does the word *mnemonic* mean?

8-19. Does your institution offer courses in assembly language programming? If so, to what computer is it specific?

8-20. What compilers are available on the computers at your institution?

8-21. What applications programs are available on the computers at your institution?

8-22. What interpretive languages are available on the computers at your institution?

8-23. In which high-level languages does your institution offer programming courses?

8-24. Using library resources, write a one-page paper on the super language Ada. Why was it invented? For whom is it intended?

Chapter 9

Computer-Aided Drawing

A highly structured form of drawing commonly referred to by terms such as technical drafting, engineering drafting, engineering drawing, technical graphics, and engineering graphics is the type of drawing that is most often associated with engineering projects. This type of drawing, now best known as *documentation drawing,* is an essential element in manufacturing [9-1].

Documentation drawing provides necessary instructions for the manufacture or fabrication of devices and products, and has important legal as well as economic significance. These drawings are important components of the record of the designer's intentions, convey precise instructions to the manufacturing or construction components of the industry, and serve as an important base for contractual arrangements between the manufacturer, customers, and subcontractors.

Until very recently, the traditional method of producing documentation drawings has required mastery of a set of personal motor skills in using pencils, pens, and drawing instruments to ensure that finished drawings were made as accurately as possible to avoid misunderstanding or misinterpretation. Trainees usually served an apprenticeship under the eye of an experienced drafter to hone these

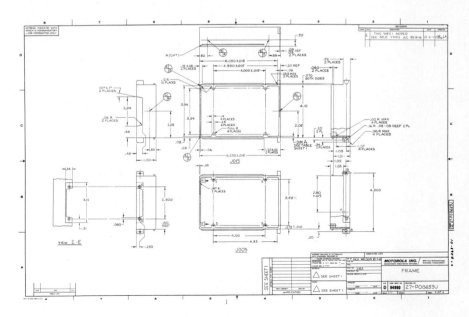

[9-1] Typical manufacturing documentation drawing.

skills and habits to acceptable levels. Notes and other symbolic information were usually added to the drawing by hand. One accepted mark of the experienced drafter was the consistency and legibility of this hand lettering.

The acquisition of these skills required both formal course work and "time on the board" at the start of a technical career. Not too many years ago, many newly graduated engineers entering the design profession could expect to spend at least a year or two on the drafting board learning how documentation drawing was done before they could expect to progress to design responsibility.

However, the situation has been changing at an accelerating rate in the past several years. In most industries, technologists have assumed responsibility for documentation drawing. At the same time, the advent of computer graphics has changed the nature of documentation drawing itself. Instead of large drafting rooms with traditional drawing boards and drafting instruments, computer terminals are taking over as the tool of choice for most documentation drawing tasks [9-2].

This modernization requires changes in the way documentation drawing is taught and learned. For this new form of drawing, the manual skill component, once so important and difficult to acquire, is now largely eliminated. Also, the computer can help overcome any lack of precision in human skill in positioning graphic elements. Computer-generated lettering and symbols enable uniform and legible notes to be added by even the most novice computer drafter.

Because of these advances, it seems apparent that documentation drawing should now be taught and learned using the computer as a drawing device without first learning the traditional hand techniques. This chapter is based on this direct approach to the teaching of CAD. CAD stands for "Computer-aided drawing" and also for "Computer-aided design." Some argument as to the distinction between them has occurred, but for our purposes here, CAD will mean computer-aided drawings which form a part of the design process.

Ideally, the student should have access to a computer graphics system. Not all systems in use today are identical in function or command structure, although most are similar. The most widely distributed CAD systems in use today are those using the IBM Personal Computer and its clones (copies by other manufacturers) as the computing hardware, and a CAD software package designed to operate on this system. There are a number of good microcomputer-based CAD software systems on the market today. The market is extremely competitive, and no one can predict with certainty which systems will endure.

Two companies currently lead the Micro-CAD market and both offer a good product at competitive prices. These products, AutoCAD and VersaCAD, will be used in this chapter to introduce the concepts of documentation drawing. Discussions will be both specific to these systems and generic so that other CAD systems may be employed with little difficulty. The special examples, where the details of AutoCAD and VersaCAD are discussed, will be called *Topics*.

Photo courtesy T&W Systems, Inc.

[9-2] CAD workstation.

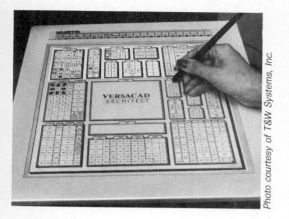

[9-3] Tablet menu in use.

COMPUTERS, PROGRAMS, AND CAD

Computers are useful in documentation drawing because they can store, retrieve, and manipulate large amounts of information. In order for the computer to act on any information, however, that information must be changed (or encoded) into a form which the computer can manipulate. This process renders the information unintelligible to most human operators. In order for people to use computers for useful work, applications software is necessary. Software is discussed in Chapters 8 and 10. The term "software" refers to a series of commands to the computer (a *program*) which cause it to act on information in desirable ways. CAD programs are quite extensive, but for the most part are invisible to the operator. The program enables the operator to interact with the computer, make choices, enter data when required, and decide on a course of action. The operator's choices are presented in a list or *menu,* from which selections may be made. There are several styles of menu presentation used in today's CAD systems.

The tablet menu [9-3], preferred by many experienced operators, displays all available commands in a chart that has been imposed over a digitizing tablet. Selections are made by the operator pointing to the command on the chart using a pointing device such as a mouse or stylus. Command locations on the chart become quite familiar with practice, so tablet selections can be made rapidly, but tablets are often confusing to the less experienced operator.

Screen menus display commands on the computer screen. Some CAD computer systems use a dual-screen configuration, reserving one screen for commands and other program control information, and the second screen for the drawing [9-4]. Other configurations use a single screen, usually displaying commands to the side or bottom of the drawing. Since screen space is limited, only a few of the available commands are displayed at one time.

The menu configuration used by many CAD systems displays a *main* menu from which categories of commands may be chosen. Upon selection of a category by the operator, the display is replaced with a submenu. Selection of a command from the submenu may result in a third display of choices. The menu structure may be compared to a tree, with the main menu as the trunk and submenus as branches. This is known as a *hierarchical menu structure* and is a common feature of large CAD systems. An inherent disadvantage of the hierarchical menu system is that only a few of the available commands are displayed at any one time. Most screen systems allow entry of commands by typing the command with the keyboard, as well as selection of the desired command by a pointing device. Operators often use both methods interchangeably.

The use of a pointing device, such as a mouse or joystick (see Chapter 8), enables the operator to move a lighted box around the menu. This lighted box is one form of *cursor* display. When the desired command choice is contained in the lighted box (cursor), it may be selected by pressing a button on the device or pressing the return (**<Enter>**) key on the computer keyboard. In the following discussion, we use the term *select* to mean the operator's choice of

[9-4] Dual-screen configuration workstation.

either input method. The text will also display all commands and keystrokes as **COMMAND** or **Y** (for "yes") to accent them.

The term "cursor" is used to refer to both the input device (mouse or joystick) and the marker on the computer screen which shows the operator the *current position* of the pointing device. As the operator moves the cursor (physical pointing device), the movements of the operator's hand are echoed by comparable movement of the cursor (screen marker). The screen cursor may take the form of a lighted box or symbol (square, caret, etc.) when moving in an area of the screen reserved for text or alphanumeric information. When moving in the graphics or drawing area of the screen, the cursor appears either as a small cross or as large cross hairs extending across the entire drawing area. Many CAD programs allow the operator to change the graphics cursor display to suit the task at hand.

Another important interactive feature of most CAD programs is the display of messages on the screen that tell the operator which actions are appropriate at any given time. These messages are shown in the *prompt line*. The prompt line message displays the input choices, command syntax, or data required. The CAD program may acknowledge a proper response to its prompt with an audible signal, or by display of a new prompt line. Improper responses to a prompt will result in an error signal (audible) or text message informing the operator of the error, or both. It is important to remember that although improper response to a CAD prompt may be embarrassing (especially to a beginning operator), it is not likely that you will damage the computer hardware or software in this way.

Additional information may be displayed on the screen of a CAD system in the form of a status line. We will discuss the information contained in a status line a little later.

One of the first things an operator must understand is the storage and retrieval of *drawing files*. A "drawing" only exists in the CAD system as a collection of electronically stored digital information. An operator usually begins a work session with the decision to edit (revise) an existing drawing or to start a new drawing. When editing, the operator must give the CAD program the *file name* of the desired drawing so that it may be retrieved and loaded into the work area. This area is known by various names in different CAD programs. *Drawing editor* (AutoCAD) and *workfile* (VersaCAD) are two such names.

When starting a new drawing, the operator must remember that in order to retain the work, a file name must be specified and the drawing must be saved to that file. There is usually no permanent record of the drawing saved unless the operator makes a deliberate choice to do so. Also, since the drawing files are saved electronically, a power outage or system "crash" (inadvertant loss of all memory registers) may irretrievably lose the work from a particular session. It is therefore essential periodically to save work in progress to backup storage media (disk or tape) to avoid frustrating (and costly!) losses of this kind.

The computer drawing topics contained in this chapter are designed to lead you through the fundamentals of documentation drawing using a popular microcomputer CAD program, such as

AutoCAD or VersaCAD. No previous experience with documentation drawing is necessary, although the ability to use freehand drawing for visualization, as discussed in Chapter 7, will be helpful.

The topics also assume that you have available a microcomputer equipped with a display screen, keyboard, input device, and a CAD program (such as AutoCAD or VersaCAD) which has been configured to your specific hardware. A plotter or dot-matrix printer will also be needed for those exercises requiring paper copies ("hard copies") of your computer drawings.

To understand this chapter, no previous computer experience is required, and no programming knowledge is necessary. When working with user-friendly applications programs such as AutoCAD or VersaCAD, you need only read the screen prompt messages and respond appropriately. Relax! This will be fun!

Topic 9-1   AutoCAD
Opening Screen

```
                    AUTOCAD
         Copyright (C) 1982,83,84,85,86 Autodesk, Inc.
         Version 2.5 (7/8/86) IBM PC
         Advanced Drafting Extensions 3
         Serial Number: 00-000000


         Main Menu

             0. Exit AutoCAD
             1. Begin a NEW drawing
             2. Edit an EXISTING drawing
             3. Plot a drawing
             4. Printer Plot a drawing

             5. Configure AutoCAD
             6. File Utilities
             7. Compile shape/font description file
             8. Convert old drawing file

         Enter selection:
```

Enter Selection: enter a number 0 through 8.

Copyright information is presented for protection of the manufacturer.

A version number and issue date identifies the particular release of the software. Most CAD programs are undergoing a continuous process of update and improvement. Some changes may result in operational differences, although the manufacturer tries to keep each new version compatible with previous versions of the product.

Advanced drafting Extensions may be purchased with the basic version of the program to meet special user needs.

Serial number identifies the specific program.

Main Menu Tasks

The operator must make a selection of one of the task numbers to initiate the following activities:

0. Exit the CAD program; control returns to the operating system.

1. Begin a new drawing. Program will prompt for the name of a file, limited to eight characters. Characters may consist of letters, digits, and "$" (dollar), "–" type the name, press ‹Enter›. Program will append the filename extension .DWG. The file will be created on the currently logged disk, unless the filename is preceded with a drive designator. Refer to the DOS (Disk Operating System) manual if you need assistance with disk and file procedures.

2. Edit an EXISTING drawing. Program will prompt for the name of the desired file, enter the drawing editor, and load the named file.

3. Plot a drawing. Program will prompt for the name of drawing file to plot, then enter plot routine. Same command is available from within drawing editor. Plotting will be discussed later.

4. Printer Plot a drawing. A plot routine for a dot-matrix printer. Similar to the routine referred to in 3. Plotting will be discussed later.

5. Configure AutoCAD. Used for installation of software, and rarely thereafter.

6. File Utilities. Same command available from within drawing editor. Will be discussed later.

7. Compile Shape/Font file. Used only when a Shape or Font file is created or modified.

8. Convert Old Drawing File. Used to convert early AutoCAD drawing files to Version 2.00.

Topic 9-1 shows the opening screens, the first screen display seen by the operator after invoking the CAD program by typing **ACAD <Enter>** for AutoCAD or **VCAD50 <Enter>** for VersaCAD in response to the DOS prompt.

THE CAD DRAWING BOARD

Topic 9-2 shows the "CAD drawing board" screen display. This is the starting position for creating a new drawing. When documentation drawing was done manually, the drafter worked in a specialized environment consisting of a drawing board, drafting equipment, reference materials, and so on. The CAD environment is also specialized, and quite different in appearance and function.

Topic 9-1   *VersaCAD*

Opening Screen

Version number and issue date identifies the particular release of the software. Most CAD programs are in a continuous process of update and improvement. Some changes may result in operational differences, although the manufacturer tries to keep each new version compatible with previous versions of the product.

V E R S A C A D   A D V A N C E D ™

Two-dimensional drafting
Version 5.0   Revision 1

copyright 1986
by T & W Systems, Inc.
all rights reserved

Duplication of this program is permitted
for archival purposes only

Press [enter] key to continue

Press ⟨Enter⟩ Key. Only Choice Possible.

Copyright information is presented for protection of the manufacturer.

The "CAD drawing board" is not really a drawing board, of course. It is the place where the graphic image is developed. In a similar way, the old drawing board was the place where a drafter developed the documentation drawing by drawing lines with a pencil or pen, using scales, templates, straightedges, triangles, and other drawing instruments. Instead of being a simple piece of furniture like a drafting table, the computer screen is really a very sophisticated electronic communication device.

Modern CAD hardware and software are said to be *interactive*. This means that the operator is able to enter into a dialogue with the computer which has been programmed to present the operator with information as it keeps track of many operating conditions and parameters. The operator makes choices, makes decisions, and supplies the computer with the data necessary to design products and produce documentation drawings.

Topic 9-2   *AutoCAD*
Cad Drawing Board

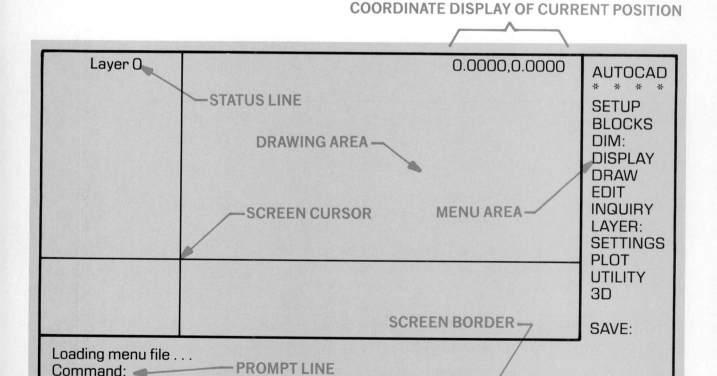

The transfer between the human operator and the electronic computer takes place at the workstation, and the screen display you see in Topic 9-2 is the medium of transfer. The display screen has been partitioned into several subareas. The drawing area can be thought of as the window through which the drawing can be viewed as it is being developed. If the drawing is simple, it can be viewed in its entirety within this window. Your first drawings will be of this type.

The operator creates a drawing by placing objects at desired locations in the drawing area. These objects may be quite simple. Simple objects, or *primitive objects* (also called just *primitives*), include points, lines, circles, arcs, polygons, and other geometric entities which may be combined to form more complex objects. These complex objects may be defined as *blocks* or *groups* and copied or added to other drawings. In fact, entire drawings, so defined as blocks or

Topic 9-2   *VersaCAD*
Cad Drawing Board

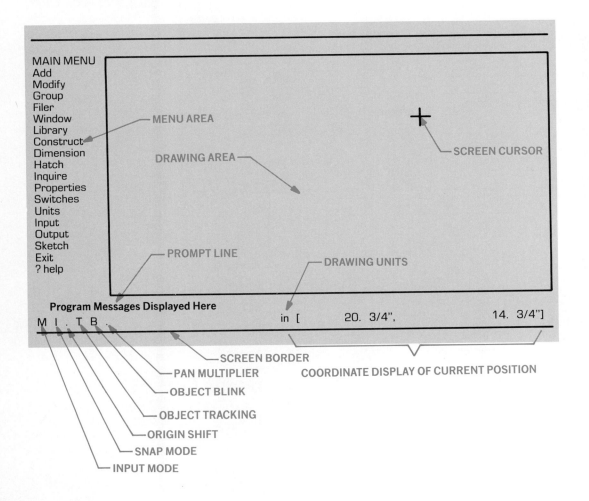

Topic 9-3

AutoCAD

Graphic Primitives

*The **DRAW** menu allows the operator to add and position primitive objects on the drawing. A list of available objects is displayed for selection, and a number of options are available. Use the on-screen help option to learn more about the available options.*

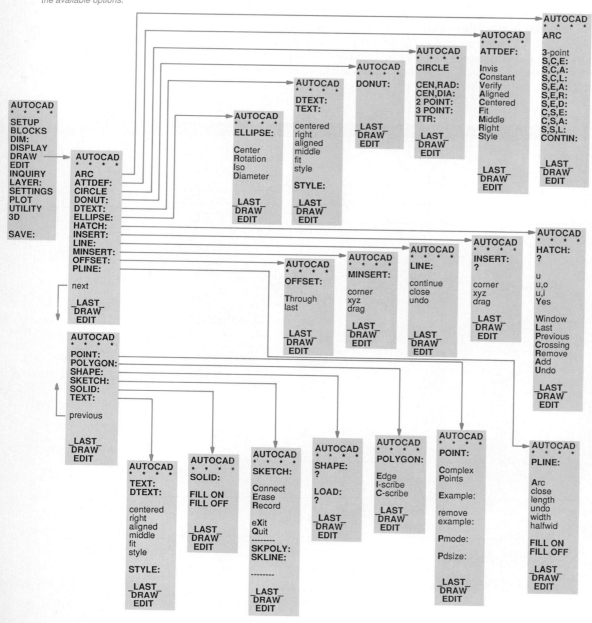

AUTOCAD
* * * * *
SETUP
BLOCKS
DIM:
DISPLAY
DRAW
EDIT
INQUIRY
LAYER:
SETTINGS
PLOT
UTILITY
3D

SAVE:

AUTOCAD
* * * * *
ARC
ATTDEF:
CIRCLE
DONUT:
DTEXT:
ELLIPSE:
HATCH:
INSERT:
LINE:
MINSERT:
OFFSET:
PLINE:

next

LAST
DRAW
EDIT

AUTOCAD
* * * * *
POINT:
POLYGON:
SHAPE:
SKETCH:
SOLID:
TEXT:

previous

LAST
DRAW
EDIT

AUTOCAD
* * * * *
ELLIPSE:

Center
Rotation
Iso
Diameter

LAST
DRAW
EDIT

AUTOCAD
* * * * *
DTEXT:
TEXT:

centered
right
aligned
middle
fit
style

STYLE:

LAST
DRAW
EDIT

AUTOCAD
* * * * *
DONUT:

LAST
DRAW
EDIT

AUTOCAD
* * * * *
CIRCLE

CEN,RAD:
CEN,DIA:
2 POINT:
3 POINT:
TTR:

LAST
DRAW
EDIT

AUTOCAD
* * * * *
ATTDEF:

Invis
Constant
Verify
Aligned
Centered
Fit
Middle
Right
Style

LAST
DRAW
EDIT

AUTOCAD
* * * * *
ARC

3-point
S,C,E:
S,C,A:
S,C,L:
S,E,A:
S,E,R:
S,E,D:
C,S,E:
C,S,A:
S,S,L:
CONTIN:

LAST
DRAW
EDIT

AUTOCAD
* * * * *
OFFSET:

Through
last

LAST
DRAW
EDIT

AUTOCAD
* * * * *
MINSERT:

corner
xyz
drag

LAST
DRAW
EDIT

AUTOCAD
* * * * *
LINE:

continue
close
undo

LAST
DRAW
EDIT

AUTOCAD
* * * * *
INSERT:
?

corner
xyz
drag

LAST
DRAW
EDIT

AUTOCAD
* * * * *
HATCH:
?

u
u,o
u,i
Yes

Window
Last
Previous
Crossing
Remove
Add
Undo

LAST
DRAW
EDIT

AUTOCAD
* * * * *
TEXT:
DTEXT:

centered
right
aligned
middle
fit
style

STYLE:

LAST
DRAW
EDIT

AUTOCAD
* * * * *
SOLID:

FILL ON
FILL OFF

LAST
DRAW
EDIT

AUTOCAD
* * * * *
SKETCH:

Connect
Erase
Record

eXit
Quit

SKPOLY:
SKLINE:

LAST
DRAW
EDIT

AUTOCAD
* * * * *
SHAPE:
?

LOAD:
?

LAST
DRAW
EDIT

AUTOCAD
* * * * *
POLYGON:

Edge
I-scribe
C-scribe

LAST
DRAW
EDIT

AUTOCAD
* * * * *
POINT:

Complex
Points

Example:

remove
example:

Pmode:

Pdsize:

LAST
DRAW
EDIT

AUTOCAD
* * * * *
PLINE:

Arc
close
length
undo
width
halfwid

FILL ON
FILL OFF

LAST
DRAW
EDIT

Topic 9-3   *VersaCAD*

Graphic Primitives

*The **ADD** menu allows the operator to add and position primitive objects on the drawing. A list of available objects is displayed for selection, and a number of options are available. Use the on-screen help option to learn more about the available options.*

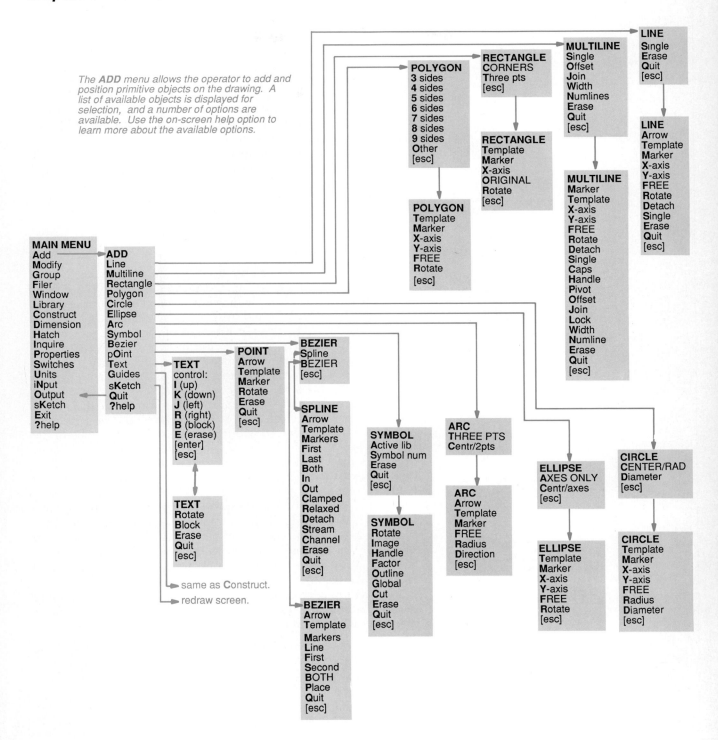

LINE
Single
Erase
Quit
[esc]

LINE
Arrow
Template
Marker
X-axis
Y-axis
FREE
Rotate
Detach
Single
Erase
Quit
[esc]

MULTILINE
Single
Offset
Join
Width
Numlines
Erase
Quit
[esc]

MULTILINE
Marker
Template
X-axis
Y-axis
FREE
Rotate
Detach
Single
Caps
Handle
Pivot
Offset
Join
Lock
Width
Numline
Erase
Quit
[esc]

RECTANGLE
CORNERS
Three pts
[esc]

RECTANGLE
Template
Marker
X-axis
ORIGINAL
Rotate
[esc]

POLYGON
3 sides
4 sides
5 sides
6 sides
7 sides
8 sides
9 sides
Other
[esc]

POLYGON
Template
Marker
X-axis
Y-axis
FREE
Rotate
[esc]

MAIN MENU
Add
Modify
Group
Filer
Window
Library
Construct
Dimension
Hatch
Inquire
Properties
Switches
Units
iNput
Output
sKetch
Exit
?help

ADD
Line
Multiline
Rectangle
Polygon
Circle
Ellipse
Arc
Symbol
Bezier
pOint
Text
Guides
sKetch
Quit
?help

TEXT
control:
I (up)
K (down)
J (left)
R (right)
B (block)
E (erase)
[enter]
[esc]

TEXT
Rotate
Block
Erase
Quit
[esc]

POINT
Arrow
Template
Marker
Rotate
Erase
Quit
[esc]

BEZIER
Spline
BEZIER
[esc]

SPLINE
Arrow
Template
Markers
First
Last
Both
In
Out
Clamped
Relaxed
Detach
Stream
Channel
Erase
Quit
[esc]

BEZIER
Arrow
Template
Markers
Line
First
Second
BOTH
Place
Quit
[esc]

SYMBOL
Active lib
Symbol num
Erase
Quit
[esc]

SYMBOL
Rotate
Image
Handle
Factor
Outline
Global
Cut
Erase
Quit
[esc]

ARC
THREE PTS
Centr/2pts

ARC
Arrow
Template
Marker
FREE
Radius
Direction
[esc]

ELLIPSE
AXES ONLY
Centr/axes
[esc]

ELLIPSE
Template
Marker
X-axis
Y-axis
FREE
Rotate
[esc]

CIRCLE
CENTER/RAD
Diameter
[esc]

CIRCLE
Template
Marker
X-axis
Y-axis
FREE
Radius
Diameter
[esc]

same as Construct.

redraw screen.

groups, may be added to other drawings. In computer graphics, the term "add" is synonymous with the term "draw".

A second major area of the computer screen is the *menu area*. This area is used to display the list of commands, or menu, with which the operator is able to interact by making selections with the cursor. The cursor changes form when moved from the drawing area (where it appears as a cross) into the menu area, where it becomes a lighted box. The menu item enclosed in the cursor box may be selected by pressing a cursor button, the space bar, or **<Enter>**. If your cursor device has more than one button, one of the buttons should "accept" screen cursor position when pressed. Your manual should clarify this. A little experimentation with pressing the buttons should also reveal their use to you.

Additional information may appear on the screen. The coordinate display of the current position of the cursor in the drawing area, status of operating modes, active layers, and similar aids to the operator may appear. These will be discussed more fully later.

Example 9-1: Menus, Prompts, and Lines

AutoCAD

To reach the screen configuration of Topic 9-2, Select 1 from the opening screen (Topic 9-1) and press **<Enter>**. You will be prompted for the name of the drawing. Type in your initials and 1, for example, **ABC1.** Wait until the screen resembles Topic 9-2.

The **AUTOCAD** menu at the right of the screen presents 14 choices (see Topic 9-2). A specific menu item may be selected by highlighting and pressing the cursor button. Each choice will cause display of a new submenu. A further choice of an item in a submenu may cause display of yet another submenu.

Notice the prompt **Command:** at the bottom of the screen. AutoCAD will wait until you (1) make a selection from the root menu shown on the screen with the cursor, or (2) type one of the menu choices from the keyboard and press **<Enter>** or the space bar. Now, make a selection. Notice that a submenu is displayed almost immediately. To cancel a command and restore the **Command:** prompt at any time, enter **CTRL C** from the keyboard.

At the bottom of each AutoCAD submenu are the choices LAST, DRAW, and EDIT. It is important to remember that whenever you become confused, or forget where you are in the menu structure, you can always return to the previous menu displayed (select LAST), or even jump back to the root menu (select **AUTOCAD**) and start the operation again. Try each of the choices in turn, returning each time to the root menu. Note that selecting the four asterisks also results in a sub-menu. You should be confident that you can always return to the familiar root menu.

1. Now, select **DRAW** from the **AUTOCAD** menu. See Topic 9-3 for the submenus under **DRAW**.

2. From the **DRAW** submenu, select **LINE:**. Immediately, a new submenu will be displayed, presenting you with specific options for drawing lines. At the bottom of this new menu will appear the familiar **LAST, DRAW,** and **EDIT**.

3. A straight line can be defined by the location of two points, the beginning endpoint and the terminal endpoint of the line. The prompt line now reads **From Point:.** AutoCAD will wait until you specify a specific point from which to start your line. Move the input device. A large screen cursor cross will appear in the blank drawing area of the screen.

4. Move the cursor to a location near the center of the screen and press the input device button to enter the location of the first endpoint on the screen.

5. Notice that the prompt line now shows **To point.** Move the cursor a short distance and enter a second point. You should observe the elastic trace line which stretches from the previous point to the current location of the cursor. This visual aid to drawing is called *rubber banding*.

6. Continue entering points to create a spiral similar to Figure [9-5]. If the drawing does not turn out the way you wish, you can always start over. To do this, return to the root menu and select **UTILITY.** From **UTILITY,** select **QUIT.** The prompt will ask: **Really want to discard all changes to the drawing?** Respond with a **Y** for "yes," and you will be returned to the main menu and can begin a new drawing. Later, we will investigate other ways to correct mistakes.

7. When finished drawing lines, press **<Enter>.** You have completed a drawing on AutoCAD!

8. Return to the root menu and select **UTILITY.** Two choices for leaving the drawing editor are presented in this menu for your selection. **QUIT** exits without saving the drawing. If you select this option, a "fail safe" prompt appears: **Really want to discard all changes to drawing?,** which requires a **Y** (yes) response in order to exit the drawing editor. Most CAD systems employ a "fail-safe" prompt to protect you against accidentally erasing a drawing file.

The other choice for terminating is **END.** This command saves (stores) the drawing to a file and exits the drawing editor. AutoCAD creates a primary drawing file which it designates with a ".DWG" file extension and a backup file designated with a ".BAK" file extension. Each time a drawing is saved, the existing backup file is overwritten, and the existing .DWG file is converted to a .BAK file.

VersaCAD

VersaCAD's main menu contains 18 items (see Topic 9-2). Menu items may be selected by highlighting and pressing the cursor button. Menu items may also be selected by typing from the keyboard. VersaCAD's menu shows one capitalized letter of each command (usually the first letter). Typing that letter alone will invoke the command.

To return to the next higher menu in the hierarchical menu structure, select **Quit,** found at the bottom of each submenu. Pressing the *escape* key (**[esc]**) will also return you to the next-highest menu but will not save work done while in the submenu. It is important to remember that whenever you become confused or forget

[9-5] Line spiral.

where you are in the menu structure, you can always return to the previous menu displayed and start the operation again. Try each of the menu choices in turn, returning each time to the main menu. You should be confident that you can always return to the familiar main menu.

1. Now, select **Add** from the main menu. See Topic 9-3 for the submenus under **ADD.**

2. From the **ADD** submenu, select **Line.** Immediately, a new submenu will be displayed, presenting you with specific commands for drawing lines. At the bottom of this new menu will appear the comforting commands **QUIT** and [esc].

3. A straight line is defined by the location of two points, the beginning endpoint and the terminal endpoint of the line. Notice the prompt at the bottom of the screen: **Place cursor and press button to define the first endpoint of the line.** Move the screen cursor to a location near the center of the drawing area and accept the location by pressing the cursor button or typing $<*>$.

4. Notice that the prompt message has changed to **Place cursor and press button to define the second endpoint of the line.** Move the cursor a short distance and enter a second point. You should observe the elastic trace line which stretches from the first point to the current location of the cursor. This visual aid to drawing is called *rubber banding.*

5. Continue entering points to create a spiral similar to Figure [9-5]. If the drawing does not turn out the way you wish, you can always start over. To do this, select **[esc].** If you wish to start over after drawing some lines and leaving the **LINE** submenu, select **Filer** from the **MAIN MENU,** and select the **New** submenu from **FILER.** The prompt will ask: **Really want to discard all changes to the drawing?** Respond with a **Y** (yes), and the drawing area will be cleared and initialized, ready for you to start over. Later, we will investigate other ways to correct mistakes.

6. When finished, select **Detach** from the screen menu to leave the line drawing mode, or select **Quit** to return to the **ADD** menu. You have now completed a drawing on VersaCAD!

7. Return to the main menu and select **Filer.** VersaCAD employs a separate menu, called "Filer," to handle drawing files and related tasks. Drawings are created in a workfile and must be transferred from this workfile to storage on a disk. Call up the Filer menu by typing **F** or by highlighting it with the cursor. To save your drawing, select **Save.** A submenu will present the choices: **All, Group, Quit, ?help.** Select **All.** The prompt **Save as what drawing?** will appear, asking you for a file name. You may use up to eight characters. Use your initials and "1" for drawing number one, for example, ABC1. A file name extension of ".2d" will be added automatically. If the drawing has been previously saved, the prompt line will ask if you wish to overfile. Overfiling means saving the current version of the drawing in

the same file as the previous version, which destroys the previous version. A **Y** (yes) response will update the saved drawing file. Upon completion of filing, the prompt line will display the message **The drawing workfile has been saved as \vcad50\draw\abc1.2d.**

To retrieve a filed drawing, call up the Filer menu and select **Get.** You will be prompted for a file name. Upon entering the name of the desired drawing, the drawing will be loaded into the workfile. If you have forgotten the name you used to file your drawing, you may type "L" for list in the filer menu for a list of all of the drawings in the file.

HELP AND TOGGLE SWITCHES

If you are learning a CAD system for the first time, you may be apprehensive about the number of things to remember in order to use the system effectively. However, many features of a CAD system are not used or needed on a regular basis. The computer itself is a powerful ally, because storage and retrieval of information is what the computer does best.

An on-screen help system is an indispensable aid for the beginning operator. Most systems are designed to accept a "help" query at any time. The operator simply selects the **help** command, types it in, or selects or types the symbol **?.** A prompt message asks the operator for the name of the command (if known) and provides information about its use. If the name of the desired command has been forgotten, a list of valid commands can be displayed by following the screen prompts.

A very effective way to speed up your acquisition of the system knowledge you will need to acquire to become a proficient CAD operator is to explore the menus, using **help** as needed to clarify the use and meaning of each unfamiliar command.

CAD systems also provide the operator with a number of operating modes which may be controlled by switching the function on or off with a *toggle switch*. Topic 9-4 shows the locations of these switches as assigned by the software to the 10 function keys of the IBM (and compatibles) keyboard. Function keys may be used at any time, even in the middle of a complex command, so they form a very powerful set of tools for the operator. Do not be concerned if you do not understand the use of all of these keys at this point. As you gain more system knowledge, you will eventually learn how to use them all.

CAD systems establish *default values* for most important operator selectable functions. This means that a preselected value (or parameter) is established by the program when it is initially loaded into computer memory. This saves the operator from having to decide upon each of these values every time a drawing is started. Often, a default choice is presented to the operator in the prompt line, usually in **[brackets]**, so that it may be selected by pressing **<Enter>.**

Exercises

The following exercises should make you more comfortable in using a CAD system. The size and position of objects you place in the drawings is not critical at this time. Later, we will investigate methods of accurately controlling size and position.

Topic 9-4 ━━━━━━━━━━   **AutoCAD** ━

Function Keys

DEFAULT VALUES ARE UNDERLINED
CONTROL KEY (CTRL) PLUS LETTER
LISTED HAS SAME EFFECT

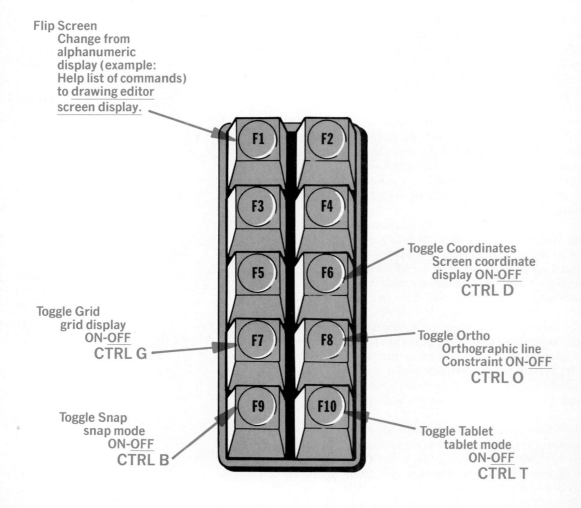

Flip Screen
   Change from
   alphanumeric
   display (example:
   Help list of commands)
   to drawing editor
   screen display.

Toggle Coordinates
   Screen coordinate
   display ON-OFF
   CTRL D

Toggle Grid
   grid display
   ON-OFF
   CTRL G

Toggle Ortho
   Orthographic line
   Constraint ON-OFF
   CTRL O

Toggle Snap
   snap mode
   ON-OFF
   CTRL B

Toggle Tablet
   tablet mode
   ON-OFF
   CTRL T

Exercise 9-1: *Title Block*

Use the rectangles command to draw the border for this and future drawings. Follow Figure [9-6] as a guide. We will discuss ways to control size later. Use the text command to place "DRAWING TITLE," "YOUR NAME," and the current date in the format "Month/Day/Year." Use the **LINE** command to finish the title block. File under the name *Title*.

Topic 9-4

Function Keys

VersaCAD

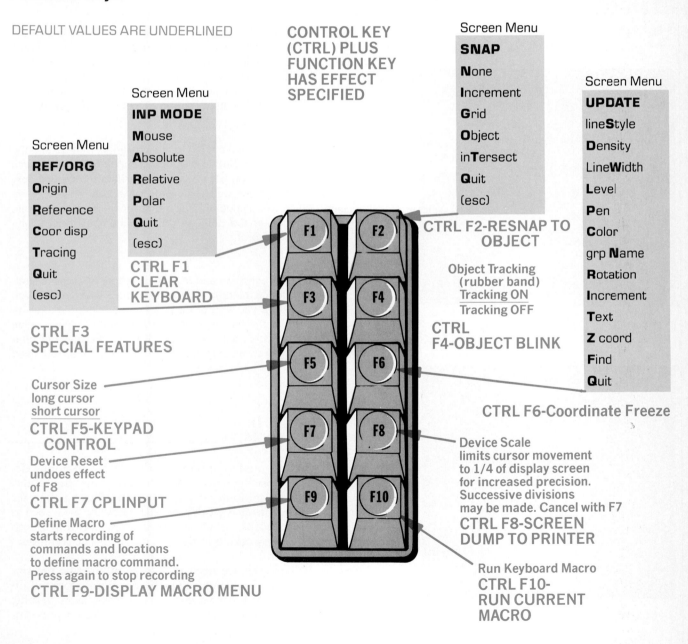

DEFAULT VALUES ARE UNDERLINED

CONTROL KEY (CTRL) PLUS FUNCTION KEY HAS EFFECT SPECIFIED

Screen Menu

SNAP

None

Increment

Grid

Object

in**T**ersect

Quit

(esc)

Screen Menu

INP MODE

Mouse

Absolute

Relative

Polar

Quit

(esc)

Screen Menu

REF/ORG

Origin

Reference

Coor disp

Tracing

Quit

(esc)

Screen Menu

UPDATE

line**S**tyle

Density

Line**W**idth

Level

Pen

Color

grp **N**ame

Rotation

Increment

Text

Z coord

Find

Quit

CTRL F2-RESNAP TO OBJECT

Object Tracking
(rubber band)
Tracking ON
Tracking OFF

CTRL
F4-OBJECT BLINK

CTRL F1
CLEAR
KEYBOARD

CTRL F3
SPECIAL FEATURES

CTRL F6-Coordinate Freeze

Cursor Size
long cursor
short cursor

CTRL F5-KEYPAD
CONTROL

Device Reset
undoes effect
of F8

CTRL F7 CPLINPUT

Device Scale
limits cursor movement
to 1/4 of display screen
for increased precision.
Successive divisions
may be made. Cancel with F7

CTRL F8-SCREEN
DUMP TO PRINTER

Define Macro
starts recording of
commands and locations
to define macro command.
Press again to stop recording

CTRL F9-DISPLAY MACRO MENU

Run Keyboard Macro
CTRL F10-
RUN CURRENT
MACRO

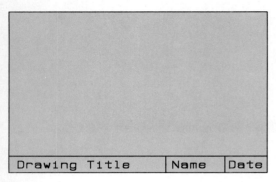

Drawing Title | Name | Date

[9-6] Title block.

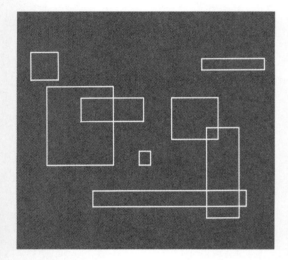

[9-7] Rectangles.

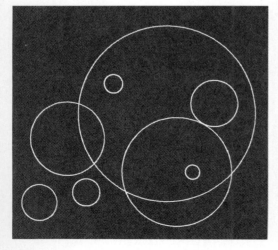

[9-8] Circles.

Exercise 9-2: *Rectangles*

Use the **RECTANGLES** subcommand to draw a series of rectangles on the screen similar to Figure [9-7]. Notice the rubber-banding feature, which allows you to see the rectangle before you accept its size and position. Explore the different methods for drawing rectangles. When finished, save your drawing to file. Name the drawing "rectang1" (short for rectangle 1—file names are limited to eight characters and letters) so that you can identify it in the file. If you wish, make several versions of the rectangle drawing, naming them "rectang2," "rectang3," and so on.

Exercise 9-3: *Circles*

Use the **CIRCLES** subcommand to draw a series of circles on the screen similar to Figure [9-8]. Notice the rubber-banding feature, which allows you to see the circle before you accept its size and position. Explore the different methods for drawing circles. When finished, save your drawing under file name circle1. If you wish, make several versions of the circle drawing, naming them "circle2," "circle3," and so on.

Exercise 9-4: *Polygons*

Use the **POLYGONS** subcommand to draw a variety of regular polygons similar to Figure [9-9]. Notice the rubber-banding feature, which allows you to see the polygon before you accept its size and position. Explore the different options under the polygon subcommand. When finished, save your drawing under the file name "polygon1." If you wish, make several versions of the polygon drawing, naming them "polygon2," "polygon3," and so on.

Exercise 9-5: *Ellipses*

Use the **ELLIPSES** subcommand to draw a variety of ellipses similar to Figure [9-10]. Notice the rubber-banding feature, which allows you to see the ellipse before you accept its size and position. Explore the different options under the ellipse subcommand. When finished, save your drawing under the file name "ellipse1." If you wish, make several versions of the ellipse drawing, naming them "ellipse2," "ellipse3," and so on.

DEFINING SPACE WITH NUMERICAL COORDINATES

The previous exercises used location indication, with a pointing device as the method of placing objects on the screen. This "free mode" of interaction with the computer is very visual and allows you to see and adjust size and location as work proceeds. Drawing aids, such as snap and grid, are available to allow very accurate control of both size and location while working in the free mode. Those techniques will be discussed later.

Many times, however, we need to numerically define the size and location of objects. Remember that the computer stores all information in the form of binary numbers. For this reason, graphic information must be encoded in number form in order for the computer to deal with it. All computer graphics images are based on a set of numerically defined locations in space.

Remember that a straight line, for example, may be defined by the location of its endpoints. Each endpoint, in turn, may be defined by an *ordered pair* of numbers (x,y) which describes the position of the point in the two-dimensional space of a plane. This ordered pair of numbers are known as the *Cartesian coordinates* of the point [9-11]. The origin (0,0) is used to reference other points in the plane. The departure from the origin in the X direction (horizontal) is conventionally given first, and the Y departure (vertical) is given second.

An ordered pair such as 3,2 describes the location of a point which is 3 units in the X direction and 2 units in the Y direction from the origin. These numbers, then, represent the *absolute coordinates* of the location. The units usually used by a CAD operator are real-world units. That is, if the CAD operator is working on a model requiring a location 3 inches in the X direction and 2 inches in the Y direction, the entire CAD program may be set to inch units. We will discuss control of units later.

An ordered pair of numbers may also specify the X and Y departures from the last point referenced, rather than from the origin. When used in this sense, the pair are said to be the *relative coordinates* of the new point.

A third method of numerical location of a point in two-dimensional space is to specify the angular displacement and distance from a reference point to the new point. These specifications are known as *polar coordinates*. Angular displacement is specified by an angle, measured in degrees with 0 at the three o'clock position, ascending in magnitude in a counterclockwise direction through 360°. Distance is specified in real-world units.

Example 9-2: Working with Numerical Coordinates

AutoCAD

1. Start a new drawing and select the **DRAW** menu. Select the submenu **Point:**.

2. Notice the prompt **Point:**. The computer will wait until you define a location. You may do so by pressing the input device button, accepting the location as described by the screen cursor, or you may type in numerical coordinates from the keyboard. First, enter a point with the input device.

3. Move the input device to the left of the center of the drawing area and press the button. You should see a small cross, which indicates the location of this point. As you move the cursor away from the point (with the input device) the cross will remain there. Notice the numbers appearing in the upper right side of the drawing area. They should change continuously as you move the input device. They represent the changing Cartesian coordinate location of the varying positions of the cursor. If the numbers do not change as you move the input device, turn them on with function key **F6** (see Topic 9-4).

4. Select **Point:** again, or press the repeat button on the pointing

[9-9] Polygons.

[9-10] Ellipses.

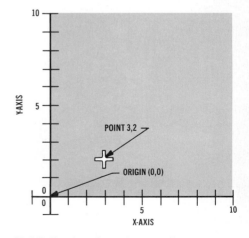

[9-11] Absolute Cartesian coordinate system.

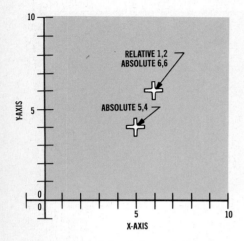

[9-12] Absolute and relative coordinates.

device to repeat the command. Now type in the ordered pair **5,4.** Another small cross should appear somewhere near the center of the drawing area. You have defined the point 5,4 with absolute coordinates.

5. Specify several more points by typing in their absolute coordinates. The computer will respond to each entry in one of three ways:
   a. A small cross (like the first one) will appear on the screen. The cross will remain there as you move the cursor to a new location. This is a successful entry.
   b. The message **Invalid point** will be displayed. This error message means that you did not follow the correct entry procedure. You must enter two numbers separated by a comma.
   c. The message ****Outside limits*** will be displayed. Move the input device to the upper right-hand corner of the drawing area. The coordinate readout of that location shows you the largest values of x and y that are valid entries. Unless you follow a special procedure, AutoCAD will not accept numerical values smaller than the current lower left-hand corner of the screen or larger than the current upper right-hand corner, if the limits switch is set to ON (it is normally OFF). Limits will be discussed later. For now, leave the limits switched OFF.

6. Now enter two or three points *relative* to another location. AutoCAD requires the symbol "@" (at) preceding the ordered pair to recognize them as relative coordinates. For example, if the last point specified was 5,4 and you want to enter a point 1 unit to the right and 2 units above point 5,4, you would enter **@1,2.** This entry would place a point at the location 6,6 in absolute coordinates [9-12]. Relative coordinates may be positive or negative. Negative specifications place the relative point to the left ($-x$) and below ($-y$) the last point specified. Positive specifications place the relative point to the right (x) and above (y) the last point specified. Positive numbers need no sign, while negative numbers must be preceded by a minus ($-$) sign. Enter several points with relative specifications.

7. Polar coordinates also reference the last point entered. AutoCAD requires the symbols @ (at) and < (less than) to be used to specify distance and angle in the format **@distance < angle.** For example, from a preceding point at 5,4, polar coordinates **@3 < 45** will define a point 3 units from point 5,4 along a 45° angle from the horizontal [9-13].

VersaCAD

1. Start a new drawing and select the **ADD** menu. Select the submenu **Point.**

2. Notice the prompt **Place cursor and press button to define the position of the point:.** The computer will wait until you define a location. You may do so by pressing the input device button, accepting the location as described by the screen cursor, or you

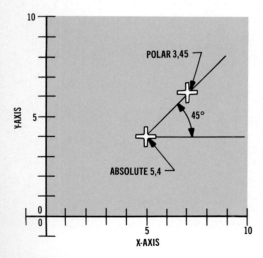

[9-13] Absolute and polar coordinates.

may type in numerical coordinates from the keyboard. First, enter a point with the input device.

3. Move the input device to the left of the center of the drawing area and press the button (or <∗>). You should see a small dot on the screen which indicates the location of the point. As you move the cursor away with the input device, the dot will remain there. Notice the numbers at the lower right side of the drawing area. They should change continuously as you move the input device. They represent the changing Cartesian coordinate locations of the varying positions of the cursor. If the numbers do not change as you move the input device, turn them on with function key **F6** (see Topic 9-4). You have been working in VersaCAD's *free mode,* using an input device specified with the **Input** menu, selected from the **MAIN MENU.** The **INPUT** menu allows a selection of one of four input devices: digitizer, mouse, plotter, or keyboard arrow keys.

4. To change from free mode to coordinate entry mode, press function key **F1.** An **INP MODE** screen menu will appear (see Topic 9-4) which presents the operator with three choices in the coordinate entry mode: Absolute, Relative, and Polar. Select Absolute. The letter "A" should appear under the prompt area and to the left, replacing the letter that indicated your input device, signifying that the current mode is absolute keyboard coordinate entry. Notice the prompt line is asking for the definition of another point with:

Absolute X:
Absolute Y:

Now type in **5 <Enter>** and **4 <Enter>.** Another small dot should appear in the drawing area. You have defined the point 5,4 with absolute coordinates.

5. Specify several more points by typing in their absolute coordinates. The computer will respond to each entry with a beep. To be accepted into the drawing data base, coordinate numbers entered should be larger than the coordinates of the lower left corner of the drawing area (usually **0,0**) and smaller than the coordinates of the upper right corner of the drawing area. Check this value by moving the cursor near the upper right-hand corner. Read the coordinates from the display at the bottom of the screen. Limits will be discussed later. For now, use only valid numbers that fall within the screen boundries.

6. Now enter two or three points *relative* to another location. To make a relative coordinate entry, press **F1,** select **Relative.** The letter **R** should appear under the prompt area and to the left, replacing the letter **A** that indicated absolute mode, signifying that the current mode is Relative keyboard coordinate entry. Notice the screen prompt:

from: 5″, 4″
Delta X:
Delta Y:

Type **1 <Enter>** and **2 <Enter>**. A point should appear 1 unit to the right of *x* = 5 and 2 units above *y* = 4, or at absolute 6,6 [9-12]. Enter two or three more points with relative coordinates. The reference point may be changed at any time by pressing function key **F3** and selecting **reference** (see Topic 9-4). If you wish to change the coordinate display to read in relative coordinates, press **F3** and select **Coor disp** (coordinate display) from the screen menu.

7. The third mode of coordinate entry, *polar coordinates,* may be activated by pressing **F1** and selecting **Polar.** The letter **"P"** should appear under the prompt area and to the left, replacing the letter **"R"** that indicated relative mode, signifying that the current mode is Polar keyboard coordinate entry. The prompt will display:

from: X",Y"

Polar angle:

Polar distance:

For example, from a preceding point at 5,4, polar coordinates **45 <Enter>** and **3 <Enter>** will define a point 3 units from point 5,4 along a 45° angle from the horizontal [9-13]. If you wish to change the coordinate display to read in polar coordinates, press **F3** and select **Coor disp** (coordinate display) from the screen menu. A selection of **Polar** will change the coordinate display to the currently active angle-distance mode. Three choices are available: decimal angle/distance, degree-minute-second/distance, and bearing/distance. The angle/distance mode may be changed through the **Units** submenu. Select **Angle** display from that menu, and then the choice desired.

Exercises

An important step in producing a CAD drawing is to plan your work carefully before you start. This is especially important if your access to a computer is limited. You can save much valuable terminal time in advance if you plan the sequence of moves you desire to make and organize any numerical data (such as coordinates) so that you will be able to find it quickly and easily.

First, make a freehand drawing of the object you wish to draw. Next, plan the sequence of steps you will need to take. For example, if you decide to draw using the line command, you will need to identify the order in which you will enter the endpoints. Label the points in the order in which you will enter them. Next, in tabular form, list each point, and its coordinates, identifying the coordinates as absolute, relative, or polar. The available information may determine which of the two end-points of a line you need to enter first. See Figure [9-14] for an example of such a table.

Exercise 9-6: *Gaskets and Plates*

Reproduce drawings from Figure [9-15] as desired or assigned. As appropriate, make use of numerical coordinates in the absolute, relative, and polar modes. Use the grid lines to determine numerical coordi-

POINT	1	2	3	4	...	n
COORDINATE MODE ABSOLUTE RELATIVE POLAR	A	R	A	P		A
X	2	1	5	—		10
Y	4	2	7	—		12
DISTANCE				5		
ANGLE				45		

[9-14] Coordinate table.

nates for those points which fall on a grid intersection by assuming that each grid space is equal to one drawing unit. Drawing units may then be related to real-world units by equating four drawing units to 1 inch, or 25 millimeters, or any other relationship you wish to establish. Each line not parallel to the grid lines has a number located adjacent to it representing the length of the line. This length plus the numerical angle measurement in degrees will allow the line to be defined in polar coordinates.

UNITS AND LIMITS

Documentation drawings must describe the size of an object as accurately as necessary in order to manufacture it. Traditionally, drawings of very large objects have been made smaller than the object itself. A good example is the common road map. Conversely, drawings of very small objects are usually made larger than the object. In either case, units of measurement on the drawing have been chosen with a specific relationship to the corresponding units on the object itself. This is known as the *scale* of the drawing, and can be specified by ratios $(1:2)$, fractions $(\frac{1}{4})$, multipliers $(3 \times)$, or equations $(\frac{1}{8}'' = 1 - 0'')$.

In traditional pencil or ink and paper documentation drawing, the first step the drafter must take is to decide on the desired scale of the drawing, considering such criteria as manufacturing processes and tolerances, paper size, and industry or government standards. The drawing proceeds from the scale selection and cannot be changed (without rework). Computer graphics has liberated the drawing from preimposed scale considerations. Although from time

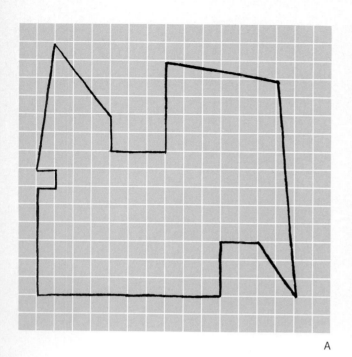

A

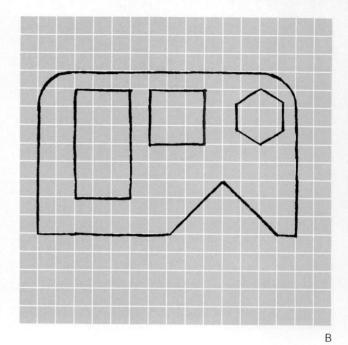

B

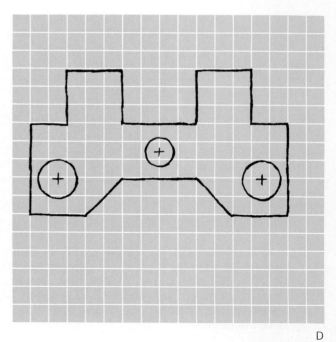

C

D

[9-15] Gaskets and plates.

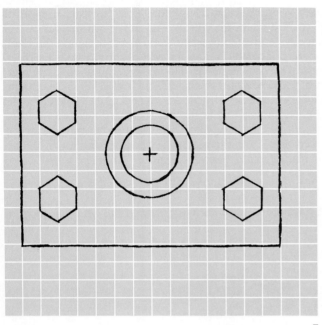

E

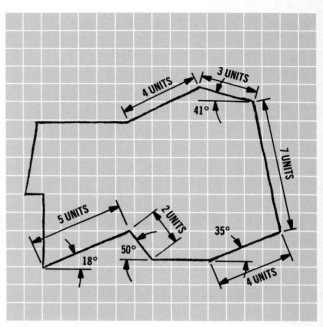

F

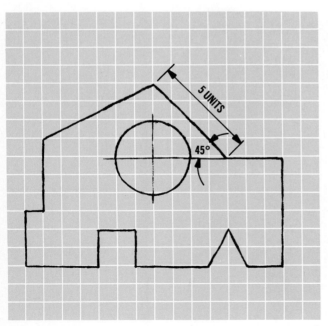

G

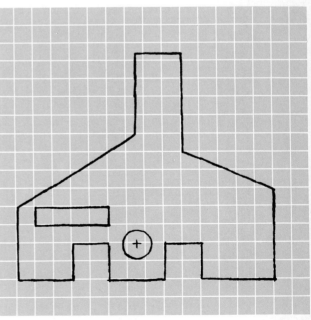

H

to time, in certain industries, making a drawing "to scale" (larger or smaller than the object it describes) may have advantages, CAD work is almost always created to *world coordinates* (drawn at full size). In fact, the operator actually creates a *model* of the real-world object, rather than making a drawing of it. The model consists of a geometric definition of the object and is created actual size. When the operator desires to see a picture of the model, the computer calculates the reduction or enlargement of the model geometry necessary to display all or a specified part of it on the desired output device—a computer screen or plotter paper. Thus, drawings of many different scales may be made from the same computer model of a real-world object.

Topic 9-5

AutoCAD

Drawing Units and Limits

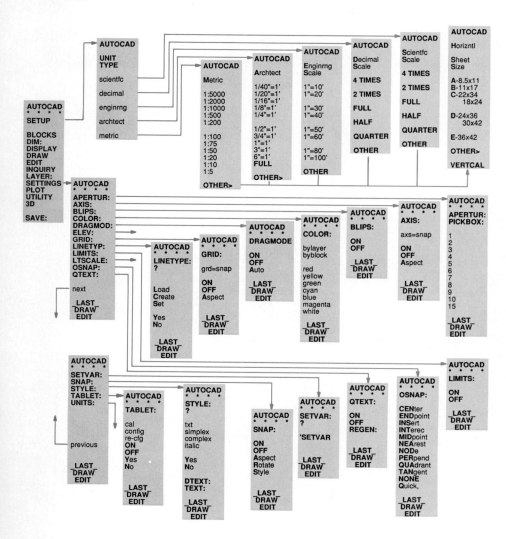

CAD systems must accommodate a variety of users with different needs. Most systems therefore allow user control of the unit specification for distance and angle measurements. The device coordinates, or drawing units of the CAD system, may be assigned appropriate units of measurement by the operator to suit the task at hand.

Example 9-3: Setting Units and Limits (Topic 9-5)

AutoCAD

1. Start a new drawing and select the **SETTINGS** menu. Select

Topic 9-5 *VersaCAD*

Drawing Units and Limits

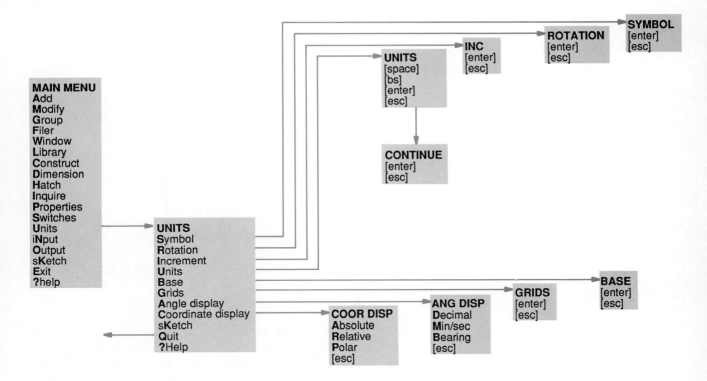

the submenu **UNITS:.** The CAD Drawing Board will be replaced with an alphanumeric (letters and numbers only) display.

2. The screen will now display the following choices:

   Systems of units:
   1. Scientific
   2. Decimal
   3. Engineering
   4. Architectural

   Enter choice, 1 to 4 <default #>:

3. Enter 3 (Engineering) and press **<Enter>.**

4. At the prompt: **Number of digits to the right of decimal point (0 to 8) <default #>:** enter 2 and press **<Enter>.**

5. Accept the default values for **Systems of angle measure:** at the next prompt by pressing **<enter>.** Also, accept the default value for the next prompt, **Number of fractional places for display of angles (0 to 8) <0>:.** Then accept the default value at the prompt, **Direction for angle 0 <0>:.** Finally, accept **NO <N>** in response to **Do you want to measure angles clockwise?**

6. Press function key **F1** to return to the CAD Drawing Board (see Topic 9-4).

7. Move the cursor around and observe the coordinate display, which now reads in feet and decimal inches. If the coordinate display does not change as you move the cursor, turn it on with function key **F6** (see Topic 9-4).

8. Repeat the sequence of steps above, each time trying a different system of units, and observe the effect on the coordinate display. Note that when the units selection is "Decimal," no units symbol is displayed. This allows the operator to assign values appropriate to the task at hand. When you are finished exploring the units options, set them to **Decimal** and two digits to the right of the decimal place.

9. Return to the **SETTINGS** submenu and select **LIMITS:.** The prompt line will now read: **ON/OFF/Lower left corner <0.00,0.00>:.**

10. This command allows the operator to set the size and location of the viewing window through which the drawing is seen. The default value for the lower left-hand corner is (0.00,0.00), or the Cartesian origin. This may be changed by typing in a new ordered pair, or accepted. Accept the default by pressing **<Enter>.**

11. The prompt **Upper right corner <12.00,9.00>** should be displayed. If it is, accept it by pressing **<Enter>.** If other numbers appear in the prompt line, type **12,9** and press **<Enter>.** If you have the units set to **Decimal** as directed in step 8 above, the drawing area now includes 12 inches horizontally by 9 inches vertically. If **LIMITS:** are "ON", AutoCAD will not accept additions of geometric primitives that extend outside those limits. The message ****Outside limits *Invalid*** will appear when

an improper entry is made. This feature may be disabled by selecting "OFF" in response to the prompt message as displayed in step 9.

VersaCAD

1. Start a new drawing and select Units from the **MAIN MENU.**

2. Select **Units** from the **UNITS** submenu (see Topic 9-5). Direct your attention to the prompt line and press the space bar. Eleven different drawing unit systems will be displayed, changing with each press of the space bar, starting with **inches.** Run through the list to get an idea of the available choices before you select **inches.** This is the default value. Escape from the **Units** submenu by pressing **<esc>.** Quit the **UNITS** menu and move the cursor about the screen, observing the coordinate display. If the coordinate display does not change as you move the cursor, turn it on with function key **F3** (see Topic 9-4).

3. Repeat the steps above, selecting a new units specification each time, and observe the changes to the coordinate display. When you are finished exploring the units options, set units back to **inches.**

4. Return to the **UNITS** submenu and select **Base.** This command allows the operator to set the size and location of the viewing window through which the drawing is seen. When selected, the program asks the operator to enter absolute coordinates for three of the four edges of the base window, with the computer calculating the fourth. The prompt area displays the current values of the base window in the following format:

left:	0.0000	**bottom:**	0.0000
right:	30.0000	**top:**	20.7941

The screen cursor will move to each numerical value in turn, and you may select the current value by pressing **<Enter>** or change it by typing in a new value and pressing **<Enter>.** Accept the current values for now.

GRIDS, SNAPS, AND INCREMENTS

Most CAD systems provide the user with a powerful set of drawing aids in the form of snaps and grids, which may be user-defined and turned on or off at will. *Grids* are a series of dots which are displayed on the screen. They can be used much like graph paper to define space. Grids are not part of the drawing file, and do not show up on a plotted drawing. Most CAD systems allow the user to define the grid increments separately in the X and Y directions, so that grid increments may be equal or unequal as needed. Some systems also allow rotation of grids to an angle other than horizontal and vertical. Grid increments may be changed as often as desired during the construction of a model.

Snap is most easily understood when used in conjunction with a grid. When snap is active (and set to coincide with the grid), the cursor is "captured" to the precise intersections of the grid, and, in effect, will not allow entry of a point into the data base which is not on a grid intersection. Snap increments may be set equal to the grid increments or to some other increments, if desired. The increments may be different in the X and Y directions. The use of snap allows great precision in placement of primitive objects while operating in the "free mode".

A special kind of snap allows the snapping of geometric primitives to other entities or objects. For example, a line may be snapped to the midpoint of another line, tangent to a circle or arc, or to the intersection of two objects.

Example 9-4: Setting Grids and Snaps

AutoCAD

1. Start a new drawing. From the **AUTOCAD** menu, select **SETTINGS.** From **SETTINGS,** select **GRID:.**

2. The prompt line will present you with the choices:
    Grid spacing (x) or ON/OFF/Snap/Aspect<0.00>:

    ON turns on (displays) the grid at default value (1). **Off** turns it off again. This "toggle switch action" may also be invoked with function key **F7** (see Topic 9-4).

    If you enter a number (e.g., 2) in response to the prompt, the X and Y increments will be set to 2. The (X) allows you to set the grid to some multiple of the current snap increment (see step 4 below). For example, an entry of **5X** would display a grid at every fifth snap increment.

    If you enter **A** (for Aspect), a new prompt appears:
    Horizontal Spacing (X)<0.00>:

    Vertical Spacing (X)<0.00>:

    You may enter the same number for both, or different numbers (e.g., **1** and **.5**) to set up a differential spacing in the horizontal and vertical directions. The (X) option works the same as for **grid spacing,** and sets the grid spacing to some multiple of the snap increment.

3. Experiment with different grid increments. If the increment is too small, the computer will display the prompt: **"Grid too dense to display"** instead of the expected grid. When you are finished, set the grid to **1** in both directions. See Figure [9-16] for an example of a screen grid.

4. Return to the **SETTINGS** submenu and select **Snap.** The following prompt will appear:
    Snap Spacing or ON/OFF/Aspect/Rotate/Style<1.00>:

    The first four choices work much the same as the grid options. Remember that grid and snap increments may be set to the same or to different increments. Also, **Snap** may be toggled on or off with function key **F9** (see Topic 9-4).

    Rotate allows you to rotate the snap grid to a new angular position and to change the base point or origin if desired. The visible grid will follow the snap grid if rotation is invoked. Also, the screen cursor will automatically align to the rotated grid.

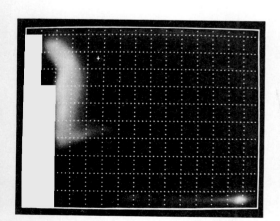

[9-16] Display of a typical grid.

Select **R** for **ROTATE.** The prompt will display:

Base point <0.00,0.00>:

Rotation angle <0>:

To rotate the grid 20° and leave the origin undisturbed, press **<enter>** and **20 <enter>.**

5. The **Style** option allows a standard grid or an isometric grid.

VersaCAD

1. Start a new drawing. Select **UNITS** from the **MAIN MENU.** Select **Grids** from the **UNITS** menu.

2. You will be prompted to enter the X spacing, or the distance between vertical lines of the grid. Enter **1.** The next prompt will ask for a "divisions" parameter. Enter **4.** This will give four divisions between each vertical grid line. See Figure [9-14] for an example of a screen grid.

3. You will be prompted to enter the Y spacing and the divisions of Y. Set these for **1** and **4,** respectively. Notice that grid spacings need not be uniform in the X and Y directions.

4. To set the snap increment, select **Increment** from the **UNITS** menu. The current increment will be displayed at the bottom of the screen. If this value is not $\frac{1}{4}$ inches type in .25 and press **<enter>.** Snap can be switched among several modes with function key **F2** (see Topic 9-4).

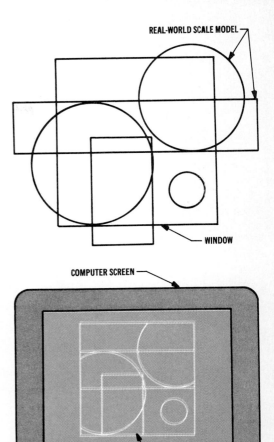

[9-17] Windows and viewports.

WINDOWS, VIEWPORTS, AND DISPLAY CONTROL ⸻

One of the most powerful features of CAD systems is the ability to change image size at will during preparation of the model. A useful analogy is to think of the computer as a synthetic camera. When using a camera, we frame a specific area of the world to capture an image. We can use lenses of varying focal lengths from wide-angle to telephoto to isolate as much or as little of the world as we wish to capture on film.

A CAD system uses computing power to perform calculations that emulate the camera lens, allowing us to "zoom in" on a model to see detail more clearly and "zoom out" to obtain an overall view. At any given moment we can define the area of the model in which we are interested with a rectangular area called the *window*. The computer must first make calculations to eliminate those portions of the data base which will fall outside the window as last defined. This process is known as *clipping* [9-17]. After the image of the model is clipped to the window, it must be mapped to some display area, which is called a *viewport*. Recall that to provide space for prompts, menus, and other information (Topic 9-2), the drawing area as defined on the computer screen is smaller than the screen itself. That drawing area is the viewport. Some CAD systems, especially those designed to operate on large mainframe computers (see pages 199–200), are capable of simultaneously displaying multiple views of the model, with each view being displayed in its own viewport on the computer screen. When working with display control on a CAD system, think of the model (which is usually created at real-world size) as being seen at all times through a window. When the window is moved closer to the model, less of the extent of the model is visible, but the parts that are visible are seen as larger. As the window is moved away from the model, more of it is displayed, but those parts displayed seem smaller. CAD operators speak of zooming or windowing IN to see enlarged detail, and zooming or windowing OUT to see more of the overall model at a smaller scale. The window may also be moved with reference to the model without changing scale or image size. This process is called *panning* in computer graphics (and also in photography). As you become more experienced as an operator, you will find that zooming or windowing and panning operations become an integral part of model building.

Example 9-5: Viewing Window Control (Topic 9-6)

AutoCAD

1. Recall an existing drawing. If you do not have a saved drawing, start a new drawing and place some geometry on the screen, utilizing techniques from previous exercises. The nature of the drawing is not important for this example.

2. Turn on a grid of 1-unit spacing in both the X and Y directions (see Topic 9-5). This will assist you in visualizing the scale changes caused by the zoom commands in the following example.

3. Select **DISPLAY** from the **AUTOCAD** menu. Next, select **ZOOM:** from the **DISPLAY** menu. Notice the seven options: All, Center, Dynamic, Extents, Left, Previous, and Window.

4. Select **All** from the screen menu or by typing **A <enter>.** **ZOOM All** resets the magnification of the screen display to show the entire drawing area. Normally, this means displaying the entire drawing area (as defined by the drawing limits) as large as possible on the screen. If, however, any portion of the drawing extends beyond the drawing limits, **ZOOM All** will adjust the magnification to include that portion in the display. It is always wise to use **ZOOM All** whenever the drawing limits are changed so that you can see the full view of your drawing area.

5. Notice the last option in the ZOOM prompt <**Scale(X)**>:. The entry of a number in response to this prompt allows you to increase or decrease the magnification of your field of view. Type **2 <Enter>.** The display will be redrawn at twice the magnification of the full view. Next, type **.5 <Enter>.** The display will appear at one-half the magnification of the full view. If you desire to use the current view as a base rather than the full view, you can enter a number followed by "X." For example, **3X** will zoom to a display which is three times the magnification of the current view.

6. The **ZOOM Previous** command allows you to step back through the five previous displays that AutoCAD has sayed.

ZOOM Previous will not recall displays beyond the limit of five. Try ZOOM Previous.

7. **ZOOM Extents** is used to show all of the objects in the model in their largest possible display. This option differs from **ZOOM All** in that it does not recognize limits in its display calculations.

8. The **ZOOM Window** option is very useful. It is one of several AutoCAD object selection techniques that uses a rectangular window to enclose objects on the screen in order to identify them for some operation. To accomplish this, select **ZOOM Window.** In response to the prompt **First point:,** enter the lower left corner of the desired window by pressing the input device button. The prompt will change to **Second point:** and you will see a rubber band rectangle appear on the screen which you can stretch to include all objects you wish to include in the new enlarged view. Upon selecting the upper right corner of the window, the display will be redrawn to show the enlarged view.

9. Two other zoom options, **ZOOM Center** and **ZOOM Left,** allow you to enter a new magnification from the current display in terms of the height of the current window, which is shown as a default value in the prompt. **ZOOM Center** centers the new display in the drawing area, while **ZOOM Left** uses the lower left-hand corner of the current display as a base point for the new display. Try each option several times until you become familiar with them.

10. The **ZOOM Dynamic** option allows the operator to zoom and pan in combination, if desired, and also displays the entire drawing extents while a selection is being made. When **ZOOM D** is chosen, the graphics viewport will be cleared of the currently displayed image and will instead display a view selection screen. This display consists of the entire drawing file displayed to the drawing extents as well as several viewport symbols and aids.

    The drawing extents are outlined in a white, solid-line box, while the current view (the last image displayed) is outlined in dotted line (and in green, if color is available). A view box for panning, with an "X" for identification, and a view box for zooming, with an arrow, are available for defining the area of interest. The operator may switch between the two boxes with the "pick" button on the pointing device. While in the zoom mode, the size of the view box may be dynamically changed by moving the pointing device. When satisfied, the operator presses **<Return>** and the newly selected area is displayed. Experiment with the **Zoom Dynamic** option until you become comfortable with it. It is a powerful option.

11. Now, select **PAN:** from the **DISPLAY** menu. The **PAN:** command allows you to view a different portion of the model without changing the magnification. Details which were clipped by the definition of the window may be restored to view by "moving the model" in relation to the fixed window. Notice that the prompt requests a displacement. You may enter a coordinate pair of numbers to indicate the desired relative displacement of the model, or two points may be selected on the screen.

AutoCAD calculates the distance between the points and uses that as a displacement value. Try several pans in different directions.

12. The **VIEW:** command allows you to save a particular screen display which you desire to return to at a later time. Select **VIEW:** from the **DISPLAY** menu. The prompt will ask you to choose from the **Delete/Restore/Save/Window** options. Upon selection of one of the options, the prompt will ask for a view name. This name may be up to 31 characters in length and should be chosen for its descriptive qualities so that it may be remembered easily. Select **Save** and enter a name of your choos-

Topic 9-6
Viewing Window Control

AutoCAD

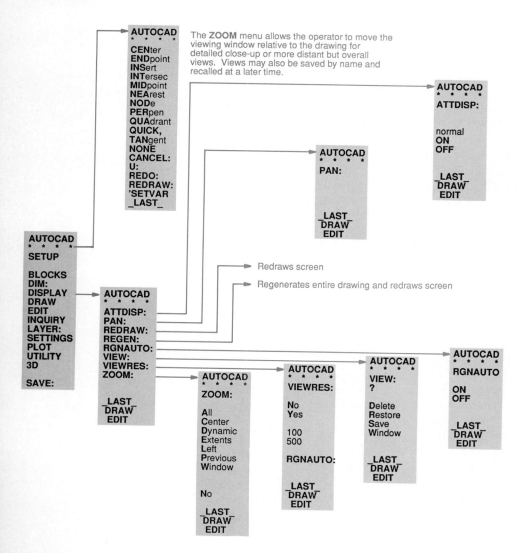

The **ZOOM** menu allows the operator to move the viewing window relative to the drawing for detailed close-up or more distant but overall views. Views may also be saved by name and recalled at a later time.

ing at the **View name:** prompt. Using one of the **ZOOM** commands, or **PAN:,** replace the screen display with another. Now, select **VIEW:** again, and the option **Restore.** Enter the view name you created, and the saved view should reappear.

VersaCAD

1. Recall an existing drawing. If you do not have a saved drawing, start a new drawing and place some geometry on the screen, utilizing techniques from previous exercises. The nature of the drawing is not important for this example.

Topic 9-6   *VersaCAD*

Viewing Window Control

The **WINDOW** menu allows the operator to move
the viewing window relative to the drawing
for detailed close-up or more distant but
overall views. Views may also be saved by
name and recalled at a later time.

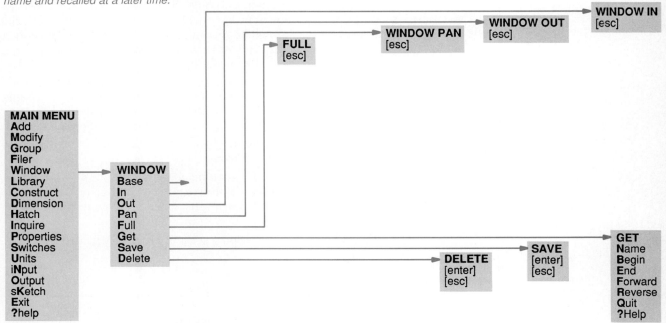

2. From the **MAIN MENU,** select **WINDOW** (see Topic 9-6). Select **Base** from the **WINDOW** menu. This option displays the model at the base window size as defined in the **UNITS** menu (see Topic 9-5).

3. Next, select **In** (for window in). You will be prompted to move the cursor to a position which will define a corner of the window defining the drawing portion to be enlarged. Follow the prompts and establish the window.

4. Repeat step 3 two more times to enlarge a detail of the drawing. Experiment with the options **Move, Scale,** and **Place. Move** allows you to move the window across the screen without changing its size. **Scale** is the default option, and allows cursor movement to change the size of the window. **Place** locks the window in position and size so that further cursor movement does not affect it.

5. Select **Get** from the **WINDOW** menu. This command allows you to retrieve previous windows, including those that have been saved by name (see step 9). You may cycle through up to 20 previous views, provided that they were established in the same direction (either in or out).

6. Select **Out** (for window out). This command allows you to specify a window smaller than the current viewport in which the existing display will be shown. The new display will be clipped to the viewport, and objects which were previously off the screen may be visible in the new display.

7. Select **Pan.** This command allows you to move the drawing one screen width or less in any direction while maintaining the same scale. Prompts will ask you to define a *handle point* by which to move the drawing. This can be any point on the drawing.

8. Select **Full.** This command will display the smallest window that contains the entire drawing.

9. Next, select **Save.** This command allows you to save a window for recall later using the **Restore** command. The **Delete** command will remove a previously named window from the drawing file.

10. The **List** command displays a list of previously named windows.

DRAWING BY EDITING OBJECTS

We have seen that CAD drawings may be created by adding primitive objects to the model. Another technique for developing the model is to modify or edit objects or parts of objects that already exist. One of the first techniques that will be useful to you is to remove or erase an unwanted object. You have seen, in Example 9-1, that you may discard the entire model and start over at any time. After several hours of productive work, however, this option will not be an attractive way to correct a minor error. All CAD systems, therefore, provide commands to erase or delete specific objects and even whole classes of objects (e.g., all points).

Systems differ in the methods provided for selection of objects to be removed, with more extensive mainframe systems providing more options for greater flexibility. Some systems remove the object from the data base immediately, whereas others suppress its display but allow it to be restored later.

Another useful technique is the ability to move an object to a new location in the model, or change its position by rotation. Such commands usually provide the operator with a choice of handling points for greater convenience. For example, a circle may be conveniently located in a model by locating its center, while a rectangle may be easier to place by specifying where one corner is located.

Where one specific shape is used several times in a drawing, it may be useful to be able to draw it one time and copy it (make a duplicate image or images). Copies may be randomly placed, or arranged in rectangular or circular arrays. Many CAD systems also allow creation of mirror-image copies and scaled copies (larger or smaller than the original).

Finally, efficient steps in the construction of a model may require that a graphic primitive be placed and then modified. For example, a line may be placed and then trimmed to match existing elements in the design.

Example 9-6: Edit and Modify (Topic 9-7)

AutoCAD

1. Start a new drawing. From the **AUTOCAD** menu, select **DRAW.** Place 8 or 10 lines, circles, rectangles, and other objects in the drawing area, in random arrangement, allowing some of them to overlap or cross.

2. Select the **EDIT** menu. Remember that you may also type **EDIT** directly from the keyboard at any time you see a **Command:** prompt. Both actions produce identical results.

3. Select **ERASE:.** The **ERASE** submenu offers a number of options (see Topic 9-7). There are two basic methods of object selection. **Last** will select the last object created. **Last** may be used to step back through the drawing file, selecting objects in the reverse order from which they were created. As each object is selected, its screen image is highlighted. The second method of object selection is **Window.** Under this option, you will be prompted for the diagonally opposite corners of a window that contains the object to be erased. Try each option. Notice that the selection prompts continue, allowing you to select as many objects as you wish. When you are satisfied with the selections, press **<Enter>** or the space bar to actually erase the selected objects. Follow the screen prompts carefully to understand which objects have actually been selected.

4. If you wish to restore an object inadvertently erased, select the **OOPS:** command. This command will restore the last object(s) selected, but not previous selections that have been erased. AutoCAD holds objects selected for an edit operation in a buffer file until the edit operation is invoked by **<Enter>.** The add, remove, and undo options refer to this buffer. Selections may be

Topic 9-7
Edit and Modify

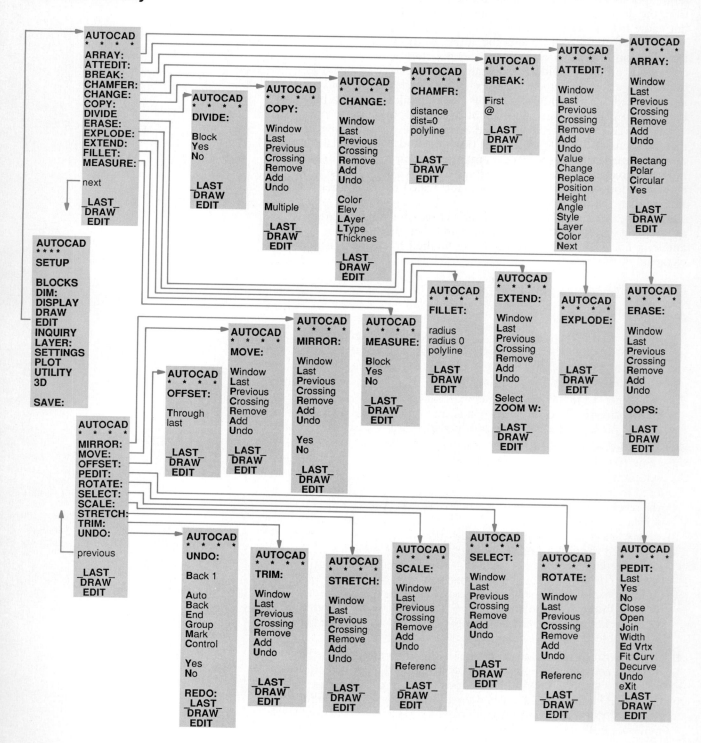

Topic 9-7
Edit and Modify

VersaCAD

The **MODIFY** menu allows the operator to make changes to the objects already added or drawn. In addition, existing objects may be copied, moved or deleted from the model.

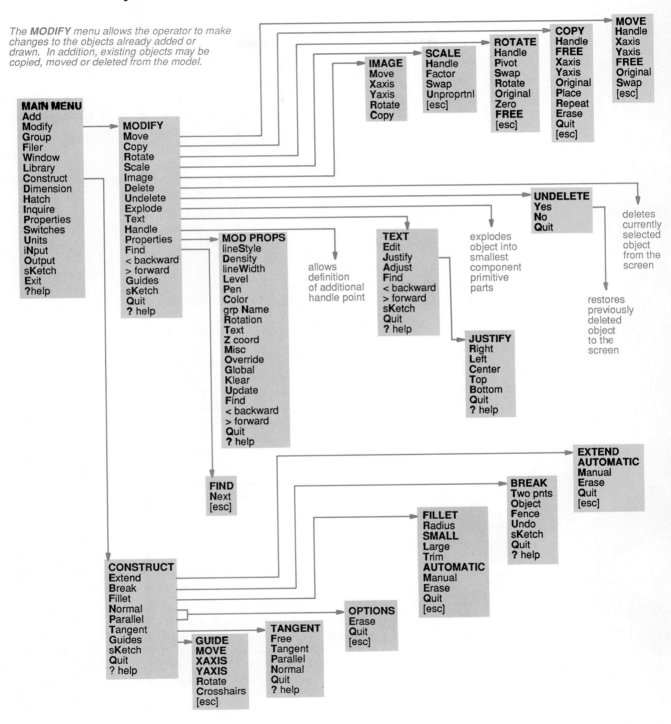

MAIN MENU
Add
Modify
Group
Filer
Window
Library
Construct
Dimension
Hatch
Inquire
Properties
Switches
Units
iNput
Output
sKetch
Exit
?help

MODIFY
Move
Copy
Rotate
Scale
Image
Delete
Undelete
Explode
Text
Handle
Properties
Find
< backward
> forward
Guides
sKetch
Quit
? help

IMAGE
Move
Xaxis
Yaxis
Rotate
Copy

SCALE
Handle
Factor
Swap
Unproprtnl
[esc]

ROTATE
Handle
Pivot
Swap
Rotate
Original
Zero
FREE
[esc]

COPY
Handle
FREE
Xaxis
Yaxis
Original
Place
Repeat
Erase
Quit
[esc]

MOVE
Handle
Xaxis
Yaxis
FREE
Original
Swap
[esc]

MOD PROPS
lineStyle
Density
lineWidth
Level
Pen
Color
grp Name
Rotation
Text
Z coord
Misc
Override
Global
Klear
Update
Find
< backward
> forward
Quit
? help

TEXT
Edit
Justify
Adjust
Find
< backward
> forward
sKetch
Quit
? help

UNDELETE
Yes
No
Quit

deletes currently selected object from the screen

restores previously deleted object to the screen

allows definition of additional handle point

explodes object into smallest component primitive parts

JUSTIFY
Right
Left
Center
Top
Bottom
Quit
? help

FIND
Next
[esc]

EXTEND
AUTOMATIC
Manual
Erase
Quit
[esc]

BREAK
Two pnts
Object
Fence
Undo
sKetch
Quit
? help

FILLET
Radius
SMALL
Large
Trim
AUTOMATIC
Manual
Erase
Quit
[esc]

CONSTRUCT
Extend
Break
Fillet
Normal
Parallel
Tangent
Guides
sKetch
Quit
? help

OPTIONS
Erase
Quit
[esc]

GUIDE
MOVE
XAXIS
YAXIS
Rotate
Crosshairs
[esc]

TANGENT
Free
Tangent
Parallel
Normal
Quit
? help

added, removed, or unselected (undo) from the buffer anytime before **<Enter>** is pressed. To back up step by step to an earlier point in the editing session, the command **Undo** or **U** will undo the results of each command in reverse order. **Redo** will undo the **"Undo"**.

5.   Select **MODE:**. Try selecting one or more objects and moving them to new positions on the screen. The options are similar to those under **ERASE.** Follow the screen prompts. **Displacement** allows you to enter relative coordinates for the new position, or to indicate the direction and distance of the desired move by selection of two points on the screen. The **Drag** option allows you to see the object as it is being moved and allows visual placement.

6.   Select **COPY:** from the **EDIT** menu. **COPY** options are quite similar to those of **MOVE**, except exact duplicate images are placed on the screen, leaving the original unchanged.

7.   Select **ARRAY:** from the **EDIT** menu. In addition to the options found in other edit submenus, **ARRAY:** contains **Rectang.** (rectangular) and **Polar** (circular). First select an object to array. Then select **Rectang..** AutoCAD will prompt for the number of horizontal rows and vertical columns desired. Negative numbers will cause rows to be added downward and columns to the left. The image selected will be duplicated in the number of rows and columns you specify. Next, you will be prompted for the space between rows and the space between columns. Try a sample rectangular array.

8.   Select **ARRAY:**. This time, choose the **Polar** (circular) option. The first prompt asks for the location of the center point of the array, while the second prompt asks for the number of items, and the image may be rotated as it is placed, or maintained in its original orientation. Try a sample polar array. See Figure [9-18] for an example of a Polar array.

9.   Select **MIRROR:** from the **EDIT** menu. This command allows you to mirror an object about a specified line of symmetry. The original image may be retained or deleted. Create a mirrored image with this command.

VersaCAD

1.   Start a new drawing. From the **MAIN MENU,** select **Add.** Place 8 or 10 lines, circles, rectangles, and other objects in the drawing area, in random fashion, allowing some of them to overlap or cross.

2.   Select **Modify** from the main menu. To modify an object, it must first be identified as the currently selected object. Any object so identified will *blink*. This can be accomplished by moving forward or backward through the list of objects (kept internally by VersaCAD in the order of their creation) with the [>] [<] keys. The list is circular. When you reach the end, the list will start over. Cause each object in the model to blink in turn until you come to the one you wish to modify.

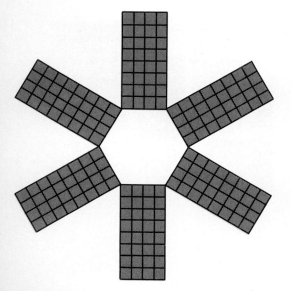

[9-18] Polar array.

A more direct way, especially with large models, is to select **Find** from the **MODIFY** menu and place the screen cursor on the desired object. Accept the location and the object should blink. If it does not, use the **[Next]** suboption under **FIND** to locate the next-closest object to the cursor position.

3. Once the desired object is blinking, select **Delete** to remove the image from the screen. If you cannot get the desired object to blink, try to turn object blink on with **CTRL F4** (see Topic 9-4).

4. Deleted objects remain in the workfile and may be restored at any time by using the **Undelete** command. Upon selection of **Undelete,** a previously deleted object will reappear in blinking mode. If you wish the object restored, answer **YES** to the undelete prompt. Try both delete and undelete.

5. To move an object, select **Move** from the **MODIFY** menu. When the desired object is selected, move the cursor to position it in the desired location. The object will follow the cursor. When you are satisfied with the new position, accept it. The object is moved by its *handle point*. The options available while moving an object are: **X axis lock** and **Y axis lock,** to restrict movement to those respective axes; **FREE,** to restore free movement from a locked option; **Original,** to return the object to its premoved position; and **Swap,** to swap the direction of an object such as an arrowhead.

6. Select **Copy** from the **MODIFY** menu. **Copy** options are quite similar to those of **Move,** except exact duplicate images are placed on the screen, while leaving the original unchanged. Options are similar to those available under **Move,** with the addition of **Repeat,** which allows you to make multiple copies of an object in one or two directions, or in a circular array [9-18]. Try these options while making copies of some of the objects in your model.

7. Select **Image** from the **MODIFY** menu. This command allows you to create a mirror image of an object about a line of symmetry. This line can be vertical, horizontal, rotated to any angle, and positioned anywhere on the screen. Options are similar to those of other modify commands. The original image of the object may be retained with the **Copy** option. Create a mirrored image using this command.

Exercises

Exercise 9-7: *Symmetrical Parts*

Reproduce drawings from Figure [9-19], as desired or assigned. As appropriate, make use of zooming or windowing, mirroring of images, symmetry, snaps, grids, moving, copying, and array techniques to facilitate construction of the models. Use the grid lines to determine numerical coordinates for those points that fall on a grid intersection by assuming that each grid space is equal to 1 drawing unit. Drawing units may then be related to real-world units by equating 4 drawing units to 1 inch, or 25 millimeters, or any other relationship you wish to establish.

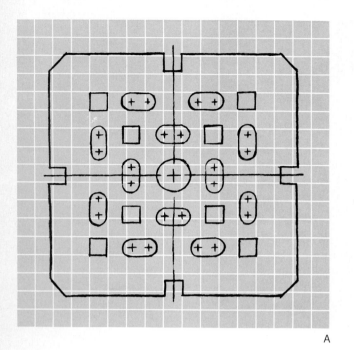

A

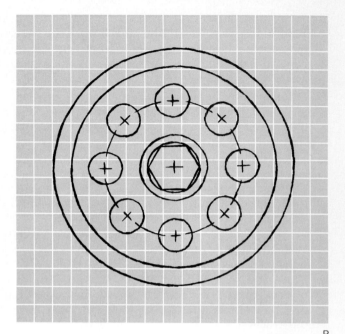

B

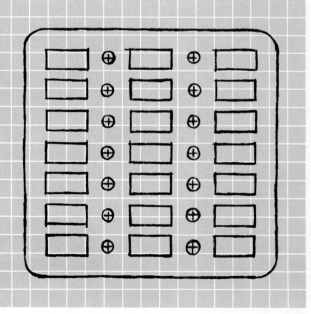

C

D

[9-19] Symmetrical parts.

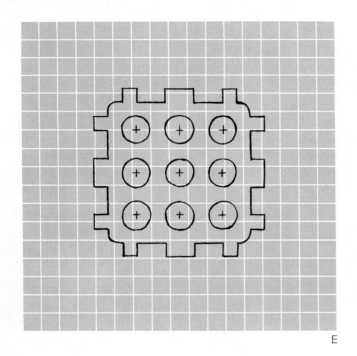

E

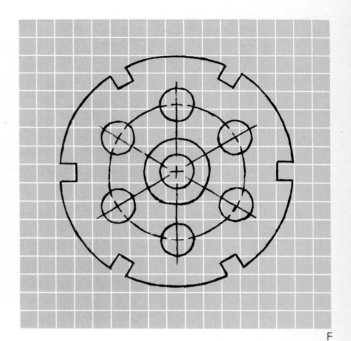

F

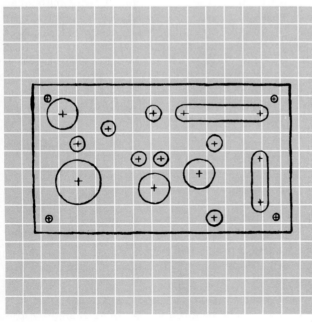

G

GROUPS, BLOCKS, AND SYMBOLS————————————————

We have seen the power of CAD at work in the form of copying and mirroring images, rather than redrawing them. In fact, if you will remember one phrase, you will understand a great deal about the appeal of CAD. The phrase is: "Never draw anything twice!"

CAD systems allow you to build up images from a set of simple primitive objects into more complex configurations. These complex configurations may be converted into a single entity and inserted into a drawing wherever needed. Whole drawings may be treated in this fashion and inserted into other drawings. The advantage of converting images into *groups* or *blocks,* as they are called, is that they may then be managed as a single entity by the CAD system, and may be selected for a change or modify operation by indicating any element of the image. Groups or blocks may be given names, may be rotated or placed in any position, may be scaled larger or smaller, and may be retained in libraries of standard parts for use by others. The potential uses of this powerful technique are practically unlimited. As you become more familiar with block or group operations, you will see more applications.

Example 9-7: Groups, Blocks, and Symbols (Topic 9-8)

AutoCAD

1. Start a new drawing. Name it "Blocks." Draw the object shown in Figure [9-20] (without dimensions) anywhere in the drawing area. Now select **BLOCKS** from the **AUTOCAD** menu.

2. Select **BLOCK:** from the **BLOCKS** menu.
   You will be prompted for the block name. A block name may have up to 31 characters. Name this one "table-1." You will next be prompted for an insertion base point. Indicate the location of the lower left corner of the block on the drawing area with your pointing device. Entry of absolute coordinates from the keyboard is also an appropriate response to this prompt.

3. The next prompt will say **Select objects:.** Select the rectangle using the method of your choice, and it will disappear from the screen (and be deleted from the drawing) but will be retained in file as a block.

4. Return to the **BLOCKS** menu and select **INSERT.**

5. The block may now be inserted into the drawing at any desired location, as many times as you wish. The first prompt will be **INSERT Block name (or ?):.** Type **table-1.** If you should forget a block name, the response **?** will produce a list of all currently named blocks in the data base of the current drawing.

6. The prompt **Insertion point:** will appear. Select a point on the screen which will be convenient. The insertion point (as you defined it) of the block will be placed here.

7. The next prompts allow you to change the X-scale factor, the Y-scale factor, the Z-scale factor, and the rotation angle of the block. Go ahead and press **<enter>** in response to all prompts.

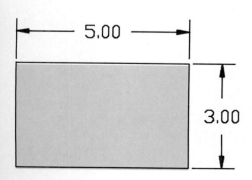

[9-20] Table-1.

The block "table-1" will be inserted into the drawing at its defined size and rotation angle.

8. Next, create a drawing of Figure [9-21] (without dimensions). Define the drawing as a block by following steps 2 and 3. Name this block "chair." Indicate the center of the arc as the base point. Now, insert the block four times to create an image similar to Figure [9-22]. To insert the inverted chair images, use a rotation angle of 180° in response to the **Rotation angle <0>:** prompt. If you wish, two **MIRROR** commands may be used in place of the last three **INSERT:** commands.

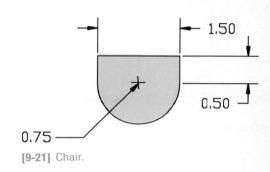

[9-21] Chair.

9. Now, create a new block named "table-1-and-chairs." You now have created a library of three parts, or *symbols:* "table-1," "chair," and "table-1-and-chairs." These could be used in combination and individually to create room furniture arrangements.

10. Increase the drawing limits to **36,27** (lower left-hand corner = 0,0; upper right corner = 36,27). We would like to produce a drawing of a rectangle with the dimensions of 24 by 36 feet, and a second rectangle concentric with the first, with overall dimensions 1.3 feet less than the first in both x and y directions to represent a wall 7.8 inches thick.

   One way to do this is first to create a "unit wall" block. This block can then be changed in scale and rotation upon insertion to create a wall of any desired length and thickness. Here's how. First, draw two horizontal lines 1 foot long and 1 foot apart, one immediately above the other. Next, create a block of the lines named "unit-wall." Select the left endpoint of the lower line as the base point. Now, insert the wall four times to create the four sides of a rectangle whose outside dimensions are 24 by 36. Start at the lower left-hand corner, proceed counterclockwise, and use the following values: (1) x-scale factor = 36, y-scale factors = .6, rotation angle = 0; (2) x = 24, y = .6, rotation = 90; (3) x = 36, y = .6, rotation = 180; (4) x = 24, y = .6, rotation = 270.

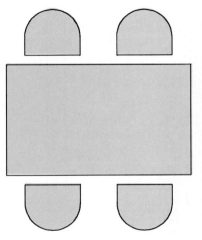

[9-22] Table-1 and chairs.

11. Insert the block "table-1-and-chairs" as required to create a model similar to Figure [9-23]. *Hint:* Use **MIRROR** creatively to decrease drawing time and make good use of **GRID** and **SNAP** to control locations. Do not dimension. Save the drawing.

12. AutoCAD allows you to insert mirror images of blocks. Simply use a negative number for the x-scale factor (step 10 above), the y-scale factor, or both.

13. AutoCAD also allows you to insert a block as a collection of individual components. Use an asterisk in front of the block name at the name prompt during the **INSERT** sequence (e.g., *chairs). This is useful if you wish to edit the block in its new position. Remember, unless you use the asterisk, blocks behave as single entities. They cannot be changed or edited by operations on their components.

14. To make later use in other drawings of a block you have defined by the procedures discussed above, you must use the command **WBLOCK** (for write block) to save the block under a drawing

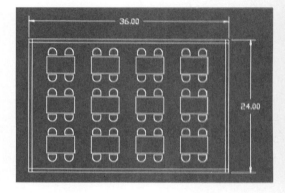

[9-23] Room with tables and chairs.

file name (eight characters, and AutoCAD supplies a .DWG extension). It may then be accessed at any time and incorporated into a new drawing with the **INSERT** command. Save the blocks "walls" and "chair" with **WBLOCK**.

Topic 9-8 ▬▬▬▬▬▬▬▬▬▬▬▬▬▬▬▬▬▬▬▬▬▬▬▬▬▬ *AutoCAD* ▬▬

Groups, Blocks, and Symbols

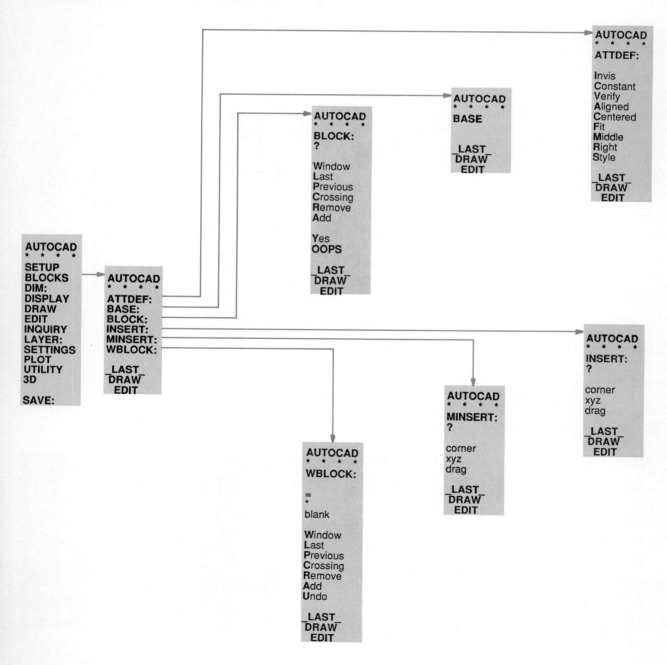

Topic 9-8

Groups, Blocks, and Symbols

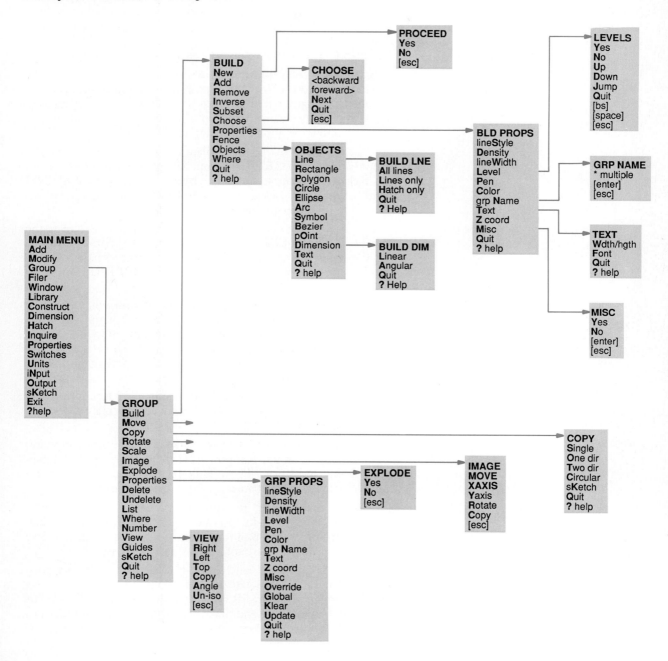

PROCEED
Yes
No
[esc]

LEVELS
Yes
No
Up
Down
Jump
Quit
[bs]
[space]
[esc]

BUILD
New
Add
Remove
Inverse
Subset
Choose
Properties
Fence
Objects
Where
Quit
? help

CHOOSE
<backward
foreward>
Next
Quit
[esc]

BLD PROPS
lineStyle
Density
lineWidth
Level
Pen
Color
grp Name
Text
Z coord
Misc
Quit
? help

GRP NAME
* multiple
[enter]
[esc]

OBJECTS
Line
Rectangle
Polygon
Circle
Ellipse
Arc
Symbol
Bezier
pOint
Dimension
Text
Quit
? help

BUILD LNE
All lines
Lines only
Hatch only
Quit
? Help

TEXT
Wdth/hgth
Font
Quit
? help

BUILD DIM
Linear
Angular
Quit
? Help

MISC
Yes
No
[enter]
[esc]

MAIN MENU
Add
Modify
Group
Filer
Window
Library
Construct
Dimension
Hatch
Inquire
Properties
Switches
Units
iNput
Output
sKetch
Exit
?help

GROUP
Build
Move
Copy
Rotate
Scale
Image
Explode
Properties
Delete
Undelete
List
Where
Number
View
Guides
sKetch
Quit
? help

COPY
Single
One dir
Two dir
Circular
sKetch
Quit
? help

EXPLODE
Yes
No
[esc]

IMAGE
MOVE
XAXIS
Yaxis
Rotate
Copy
[esc]

GRP PROPS
lineStyle
Density
lineWidth
Level
Pen
Color
grp Name
Text
Z coord
Misc
Override
Global
Klear
Update
Quit
? help

VIEW
Right
Left
Top
Copy
Angle
Un-iso
[esc]

VersaCAD

1. Start a new drawing. Press function key **F6, UPDATE** (see Topic 9-4). Select **Group Name,** type in "table-1," and press **<enter>.** Return to the main menu and select **Add.** Create the object shown in Figure [9-20] (without dimensions) anywhere in the drawing area. Now select **Group** from the **MAIN MENU.**

2. Select **Build** from the **GROUP** menu. Select **New** and answer **Yes** to the **Proceed** prompt. This clears all previous groups from memory. While still in the **BUILD** submenu, select **Properties.** Next, select **Group Name.** At the prompt, type in "table-1" and **<enter>** or just press **<enter>** if **table-1** is displayed.

3. Select **Add** and choose **Fence** to draw a window around the rectangle to add it to the group. The rectangle should now be defined as a group. You may check its status by selecting **List** from the group menu. A prompt should report that one group exists, containing four objects, and named "table-1."

4. The group may now be manipulated as a single entity using the commands of the **GROUP** menu. The group may also be saved to file. To save "table-1" to file, return to the **MAIN MENU** and select **Filer.** Select **Save** and **Group** successively. When prompted for the file name, enter "table-1."

5. Now, select **New** from **FILER** to clear the workfile. Create the image of Figure [9-21]. Follow the same procedures outlined above in steps 1 through 4. Name this group "chair". Save it under that file name and select **New** from **FILER** to initialize the workfile.

6. Next, select **Filer** and select **Merge.** The first file you will merge into the present drawing is "table-1." Enter the name and it will appear on the screen. You will want to window in to increase the size of the image for easier working. Also turn on a grid. Remember, you must select **Units** from the main menu and **Grids** from that submenu to set the grid values.

7. Now, insert the group "chair." Use **Image** and **Copy** from the **GROUP** and **IMAGE** menus, respectively, to create a drawing similar to Figure [9-22].

8. Now, create a new block named "furn" and save it to a drawing file of the same name. You now have created three parts: "table-1," "chair," and "furn." These could be used in combination and individually to create room furniture arrangements.

9. Draw a rectangle with the dimensions 24 by 36 feet. Add a concentric rectangle with overall dimensions 1.3 feet less than the first to represent a building wall thickness of 7.8 inches. File the drawing under the name "walls" for later use.

10. Insert the group "furn" and move it to the lower leftmost position of the table and chair groups of Figure [9-23]. Create a model similar to Figure [9-23]. *Hint:* Use **Image** creatively to decrease drawing time and make good use of **Grid** and **Snap** to control locations. Do not dimension.

11. VersaCAD allows you to create entire libraries of symbols similar to the groups you worked with in steps 1 through 10 above. A library may have from 1 to 1000 symbols stored for later recall. You may create several libraries as needed to categorize groups of symbols and choose from among them while preparing a model.

12. VersaCAD allows you to "explode" the group back into its component parts. Use the **Explode** submenu under **GROUP**. The objects in the group will be exploded into their component parts, while the original groups will be deleted. This is useful if you wish to edit the group. Remember, groups behave as single entities. They cannot be changed or edited by operations on their components.

LEVELS AND LAYERS

Many applications of documentation drawing require or are enhanced by preseparation of systems or components of the model onto various *levels* or *layers* (the terms mean the same thing in CAD system literature). The concept behind levels or layers is analogous to making a drawing on a series of transparent plastic sheets which are then superimposed or overlaid so that they may be viewed simultaneously. All overlays may be viewed at once, or one or more may be selectively removed from the set to display the remainder. Each individual overlay may be viewed alone or in combination with others as selected.

Many benefits are realized by this approach. In some industries (e.g., printed circuit board manufacture) physical components may actually be separated into layers during the manufacturing process. In other industries, subcontractors may be responsible for some subset of the total project. By using levels or layers, they may concentrate on that part of the project without distraction.

Clearance, interference, or fit of both moving and stationary components may need to be studied carefully in refining the design. For example, mechanism linkages and piping clearances may be reviewed through the use of levels or layers.

Large, complicated models may be simplified during their construction by turning off certain levels or layers, making the screen display easier to read and speeding up the redrawing process. Plots of certain levels or layers may be produced during design for checking purposes. Although overlay drafting was practiced to a limited extent in certain industries before the advent of CAD, the computer allows the user to extend the overlay technique to its fullest power.

Separation of objects in a CAD model usually includes text and dimensions on one level or layer, and linework and shapes on another. Linetypes are often separated, with solid (visible) lines occupying one level or layer and dashed (hidden) lines on another. Drawing title blocks and borders may be placed on a separate level or layer. If color is used, a specific color is sometimes assigned to each level or layer. Functional separations also are common: for example, a building floor plan on one level or layer and plumbing, electrical wiring, heating and ventilating, and structural framing each occupying separate levels or layers.

Example 9-8: Levels, Layers, Hatch, and Inquiry (Topic 9-9)

AutoCAD

1. Edit the existing drawing named "walls." This drawing was created with the **WBLOCK** command in Example 9-7. When the

Topic 9-9 AutoCAD

Levels and Layers Dimensions Hatch and Inquiry

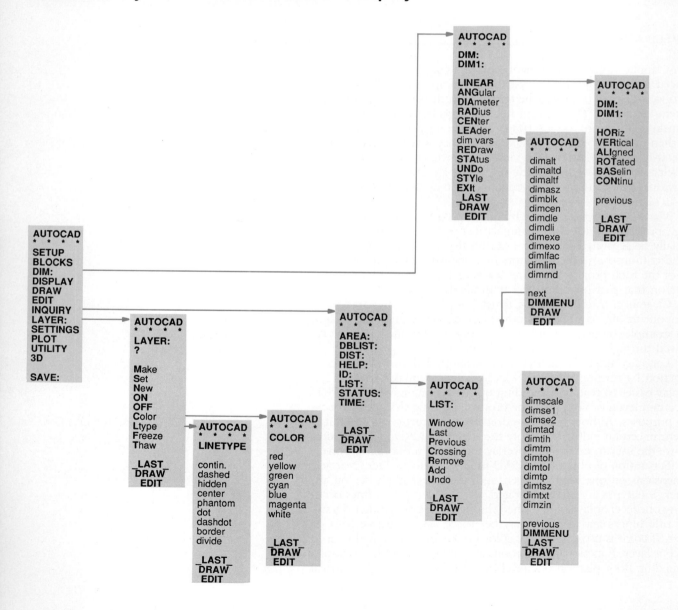

drawing is loaded into the drawing editor, it will appear on layer 0, the default layer. Select **LAYERS:** from the **AUTOCAD** menu.

2. AutoCAD allows you to work on an unlimited number of layers. Only one layer can be current at any time. A current layer receives all primitive objects as you add them, and may be modified with any of the edit commands. Layer 0, the default layer, has certain special characteristics.

Topic 9-9 ■■■■■■■   *VersaCAD* ■■

Levels and Layers Dimensions Hatch and Inquiry

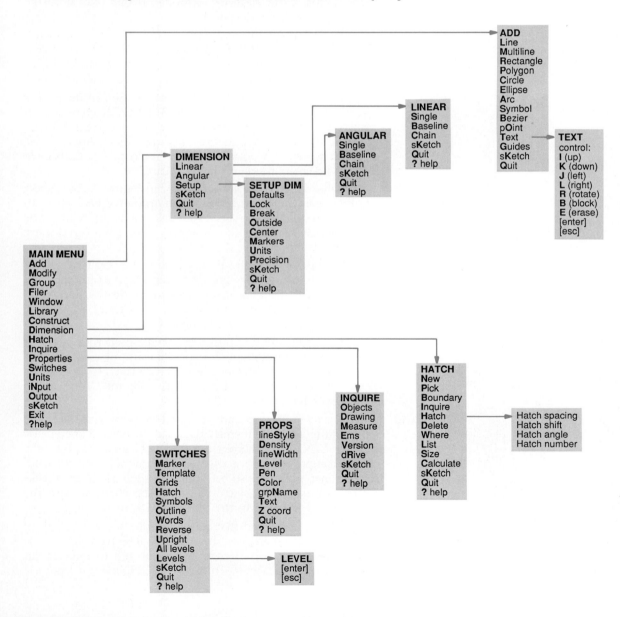

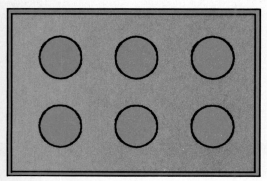

[9-24] Room with circular tables.

3. Select **?** from the **LAYER:** menu. This will provide you with a listing on a text screen of all layers currently identified and their status. Use function key **F1** to flip back to the graphics screen. Now, name two additional layers: "tables" and "chairs." To do this, select **New** from the **LAYER:** menu. At the prompt **New layer name(s):**, type in the layer names separated only by commas. You should give layers names that indicate their contents rather than abstract numbers or symbols.

4. Select **LAYER:** again and from the choices select **Set.** This will change the current layer. To the prompt **New current layer? <0>:** type "tables." Press **<enter>** twice and the status line should change from **LAYER:0** to **LAYER:TABLES.** Now all objects added to the model will reside on this layer until the current layer is changed again.

5. Enter the **DRAW** menu and draw a circle 6 feet in diameter. Use the **ARRAY:, MOVE:, COPY:,** and **MIRROR:** commands as needed to place six circular tables in a room similar to Figure [9-24]. These tables now reside on layer "tables."

6. Set a new current layer, "chairs." Insert the block "chair" that was stored as a drawing file with the **WBLOCK** command in Example 9-7. Note that inserting "chair" in the current drawing automatically enters it into the drawing's block library. Use **ARRAY:** to create a polar array of eight copies at each table [9-25]. The chairs reside on layer "chairs."

7. The drawing should now appear similar to Figure [9-26]. Select **LAYER:** again and select **OFF.** When prompted for layer(s) to turn off, type "tables". Press **<enter>** twice to redraw the screen and the tables should disappear. Don't worry, they are still in the data base!

8. **Thaw** is similar to **ON** and **Freeze** is similar to **OFF** in the fact that both **Thaw** and **ON** enable screen display of a specified layer, and **OFF** and **Freeze** suppress screen display. However, drawing image regeneration on the screen is faster if layers are frozen rather than just being turned off.

9. Color is another useful option that may be used with layers. If your hardware configuration enables the use of color, experiment with assigning different colors to each layer. **Color** may be selected by screen menu or by number, using the **Color** option in the **LAYER:** menu.

10. Linetype may be changed, and assigned to layers with the **Ltype** option in the **LAYER:** menu. First, load the desired linetype by selecting **LINETYPE:** from the **LAYER:** menu. Follow the screen prompts to display a list of available linetypes to load. Custom linetypes may also be created. If the scale of the breaks in the lines is not satisfactory, it may be changed with the **LTSCALE:** command in the **SETTINGS:** menu.

11. **HATCH** is a command that allows you to apply a variety of patterns to selected areas of the drawing. The patterns may be used for emphasis, to designate materials, and to indicate cut surfaces in sectional views. To use **HATCH,** select it from the

[9-25] Circular table and chairs.

DRAW menu. A listing of available patterns is a good place to start. Choose **?** from the **HATCH** menu. Notice the many patterns designated ANSI for the American National Standards Institute. These are important in manufacturing documentation drawing.

Choose the command **HATCH:** and select the inner and outer wall outlines of the room in the current drawing. For a hatch pattern, select **SACNCR,** the name the AutoCAD library gives to the pattern for concrete. Set the scale for the hatch pattern (following the screen prompts) at **4.** Pattern scale greatly affects the time required to build the pattern. If you make a mistake in the scale and find that an unacceptable length of time is required for generation of the pattern, press **CTRL C** to abort the hatch process at any time. Hatch patterns are created as blocks (unless you designate otherwise by preceding the name with ∗) and so may easily be erased (with **ERASE/Last**) if unsatisfactory.

12. Select **INQUIRY** from the **AUTOCAD** menu. Select **LIST:.** The prompt **Select objects:** will be displayed. Select any object with the screen cursor, press **<return>,** and a list of data will be displayed, including the layer it resides on, and other useful information, depending on the kind of object. Other options in the **INQUIRY** menu return data such as the coordinates of the endpoints of lines, their angle, and length. Area measurement is another useful option.

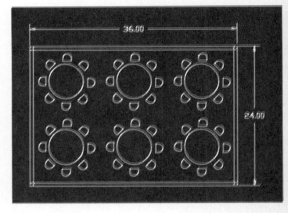

[9-26] Room with circular tables and chairs.

VersaCAD

1. Start a new drawing by clearing the workfile. Get the drawing saved in Example 9-7, step 9, "walls." When the drawing is loaded into the drawing editor, it will appear on level 1, the default level. To change levels, select function key **F6,** properties update. A screen menu will appear. Select **Level.** Type in **2.**

2. VersaCAD allows you to work on up to 250 layers. Only one layer can be active at any time. An active layer receives any primitive objects as you add them, and may be modified with any of the edit commands. Layer 1 is the default layer. The drawing you created in Example 9-7 now resides on layer 1.

3. Select **Inquire** from the **MAIN MENU.** This will allow you to determine the level on which any object resides, as well as other useful information. Select **Objects,** and **Find** the object of interest, in this case, the walls. When the selected object blinks, select **Properties.** The prompt display area will give you the current status of a number of properties of the object. You should notice **level:1.** Other useful data can be recovered with options in the **Inquire** menu, such as coordinates of points, distances between two specified points, and areas.

4. Select **Properties** again and from the submenu choices select **Level.** Type in the number, from 1 to 250, of the level that you wish to activate (receive added objects). Type **2.** Now all objects added to the model will reside on this layer until the active layer is changed again.

5. Enter the **Add** menu and add a circle 6 feet in diameter. Use the **Move, Copy,** and **Image** commands as needed to place six circular tables in the room similar to Figure [9-24]. These tables now reside on layer 2.

6. Set a new active layer 3. Merge the drawing "chair" by entering the **FILER** menu. Use **Copy** to create eight copies at each table [9-25]. *Hint:* Instead of using **Copy** for each table, **GROUP** the circular array created at the first table and copy the group. The chairs reside on layer 3.

7. The drawing should now appear similar to Figure [9-26]. All levels are now displayed. To turn them all off, select **Switches** from the **MAIN MENU.** Select **Levels** from the **SWITCHES** submenu and notice the display at the prompt line. A row of level numbers has a row of **Y** (for "yes," meaning "yes—display them) underneath." Pressing the space bar (or the backspace key) will move you through the list. For this example, select **NO** for levels 1 through 3. **Quit LEVELS** and select **All levels.** Select **No,** which has the effect of turning off any level coded "no" at the level command. Remember, the **All levels** command activates the on–off choices you made under **LEVELS.** Now, select **sKetch** and the walls, tables, and chairs should disappear. Don't worry, they are still in the data base!

8. To turn **On** one or more levels, select **Switches** again, select **Levels,** set the code **Y** (for "yes, display this level") for each level you wish to be displayed, and select **<enter>** (not [esc]!). Select **All levels,** select **No,** and select **sKetch** to redraw the display. If all occupied levels are set to **YES,** a **NO** response to **All levels** will not turn any levels off.

9. Color is another useful property that may be used with levels. If your hardware configuration enables the use of color, experiment with assigning different colors to each level. The **PROPERTIES** submenu allows you to select a new color number which will then apply to all objects being added.

10. Linestyle may be changed and assigned to levels with the **lineStyle** option under the **Properties** command. When you select **lineStyle,** an integer number input is prompted. Figure [9-27] shows the linestyles corresponding to numbers 1 through 8.

11. **HATCH** is a command that allows you to apply a variety of patterns to selected areas of the drawing. The patterns may be used for emphasis, to designate materials, and to indicate cut surfaces in sectional views. To use **Hatch,** select it from the **MAIN MENU.** Patterns can be created by selecting **New** and selecting the boundaries of the object to receive the hatch pattern.

   Select **Pick** from the **HATCH** menu and select each object in turn which forms the boundary of the area you wish to hatch. Remember, objects must blink in order to be recognized when selected. Use **Find** to make each blink in turn. Select the inner and outer wall outlines of the room in the current drawing in turn and select **All** from the **Pick** submenu to indicate that the

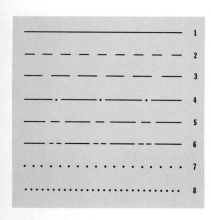

[9-27] Linestyles.

whole line and not just part of it serves as a boundary.

Next, select **Hatch** from the **HATCH** menu and specify hatch spacing, shift, angle, and number. **Spacing** is the distance, in drawing units, between hatch lines.

shift is the offset desired for double hatch lines. If single lines are desired, set **shift** to zero. **angle** defines the angle from the horizontal of the hatch pattern. **number** is the number assigned to the particular hatch pattern you develop. The pattern number is used with the **Delete, Where,** and **List** options of the **HATCH** menu. Hatch the walls with an angle of 45° and a spacing of 6 inches (.5). Remember, the spacing is in real-world units!

DIMENSIONING ――――――――――――――――――――――

Engineering documentation describes a design in a highly detailed and specialized way. All of the data and information needed to manufacture the design must be presented in an unambiguous way so that there is one and only one possible interpretation of the designer's intentions. Two complimentary forms of presentation of this information are specifications and documentation drawings.

Specifications are concerned with contractual obligations such as production quantities, delivery schedules, quality of workmanship, material properties, performance requirements, and standards of finish.

Documentation drawings describe size, shape, and extent of parts, as well as fits, tolerances, arrangement of parts, and sequence of assembly. Fabrication processes and tooling may also be reflected in documentation drawings.

Dimensioning is a very important part of size and shape description of a design. Dimensions must give all size information necessary to manufacture or build the design, and must do so in a clear and easily understood manner. Dimensions are really bits of symbolic information in the form of numbers which are attached to the graphic descriptions of the shapes that constitute the design.

Dimensioning, as well as other aspects of documentation drawing, usually must conform to one or more sets of *standards*. These are compilations of rules, restrictions, and specifications developed to ensure uniformity. Two of the best-known and most widely used of these are the ANSI (American National Standards Institute) and ISO (International Standards Organization) standards. Other, more specialized standards, such as Mil Specs (Military Specifications), also control work in certain industries. Most CAD systems have designed their dimensioning functions to conform to the ANSI standards, and some allow selection of either ANSI or ISO standards.

The practice of dimensioning involves decisions concerning which bits of size information are relevant to the manufacturing process and where to place this information on the drawing. To simplify our discussion, we will first deal with dimensioning a class of objects that can be described with a single two-dimensional drawing. These objects are primarily two-dimensional in shape, such as parts formed from sheet stock.

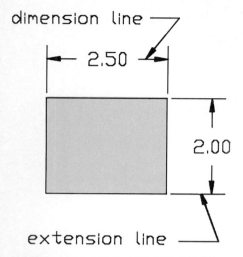

[9-28] Dimension lines and extension lines.

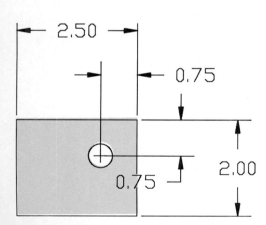

[9-29] Feature location.

Two overall dimensions are generally considered necessary: the height and width of the object. Thickness of the stock material (the third dimension) may be given with a note in drawings of this class of objects. Dimensions of this type are considered *size dimensions* because they convey the size of the object. Size dimensions are attached to the drawing with lines called *dimension lines* and *extension lines* [9-28].

The purpose of the extension lines is to remove the dimension from the outline or contour so that the dimension does not interfere with clear perception of the shape description provided by the drawing. The dimension line is always parallel to the line whose length it is defining and is drawn to the same length as that line. The dimension itself is the number associated with the dimension line. The dimension can be inset in a broken dimension line, as in Figure [9-28], or placed above (or beside) and parallel to the line. The dimension may be stated as a theoretical length or may indicate acceptable deviations or *tolerances* from this length.

The next step in the process of dimensioning is to determine if one or more *features* exist in the object. A feature is a collection of geometric primitives that relate to each other because of proximity or because of function. In Figure [9-29] the circle (representing a hole) constitutes a feature. Each feature that can be identified must also be given a size dimension, in this case, the diameter of the hole. Another type of dimension is required for each feature, a *location dimension*. This dimension defines the position of the feature relative to other, more general geometry of the part. In Figure [9-29] the hole is located by giving the position of its center.

Overall height, width, and the size and location of each feature are the normal dimensions required to manufacture a two-dimensional part. Superfluous or duplicate dimensions should never be given, but all information needed to manufacture the part must be explicitly provided. No assumptions, guesses, or calculations should be needed to completely understand the size and location description of a part.

Placement of necessary dimensions is partially a matter of judgment, keeping in mind that clarity, lack of clutter, and common sense are good standards to apply to placement of dimensioning. Keep crossing of lines to a minimum, and especially avoid crossing a dimension line with anything. Keep dimensions away from the object lines (the shape description). It is always a good idea to make a freehand drawing to plan dimension placement before going to the CAD workstation. This saves valuable terminal time and helps you to anticipate layout problems. Some CAD systems allow you to "rubber band" dimension placement, a real help in making the visual judgments needed to produce a balanced and coherent drawing.

Dimensioning of drawings of three-dimensional objects follows all of the rules and strategies we have discussed so far, but must consider the conventional method of describing three-dimensional objects on a two-dimensional surface: *multiview drawing*. The most important rule to remember for dimensioning is to always display the dimension adjacent to a *true length* view of the line whose length is being defined.

A true length view of a line occurs in any view where the line is perpendicular to the observer's line of sight. Hold a pencil in front of your eyes so that you are looking directly at the point and cannot see the length of the pencil, as in Figure [9-30]. Now, rotate the pencil slowly until you are looking directly at (perpendicular to) the side of the pencil and you see its full length. This position is equivalent to a true length view of a line in a drawing. Now, slowly rotate the pencil back to the first position. Notice that, as you slowly rotate the pencil, its image becomes shorter and shorter. This is called *foreshortening*. The extreme position when you cannot see the length at all is called a *point view*. Notice that when the pencil is in the true length position, it is perpendicular (at right angles) to your line of sight [9-31].

Multiview drawings present several views of the same object, arranged in a conventional manner, with each view obtained by looking at the object from a line of sight rotated 90° from that of the adjacent view [9-32]. The purpose of this method of presentation is to show each line of the part in its *true length* in one or more of the views. When dimensioning a multiview drawing, always place the dimension adjacent to the true length view of the line, never adjacent to a view where it shows as a foreshortened line or as a point.

A corollary to this true length rule is to always dimension a feature in the view which shows its characteristic contour in *true shape* (all lines of a true shape view of a surface are seen in true length).

Example 9-9: Dimensioning

AutoCAD

1. Start a new drawing. Create a line drawing similar to Figure [9-33].

2. Next, select **DIM:** from the **AUTOCAD** menu. Notice two choices in the **DIM:** menu: **LINEAR** and **angular:.** An example of an angular dimension is shown in Figure [9-34]. Select **LINEAR.**

3. Select **horiz:.** The prompt **First extension line origin or RETURN to select:** will appear. You can select the origins manually or press **<return>** if you want AutoCAD to determine them automatically. Press **<Enter>.** The prompt **Select line, arc, or circle:** will appear. Select one of the horizontal lines to dimension. You will be prompted for **Dimension line location:.** Indicate a point on the screen that you wish the dimension line to pass through. The last prompt will be: **Dimension text <calculated length>:.** Press **<Enter>** and AutoCAD will calculate the dimension based on the endpoint location of the line you selected.

4. The command **Undo:** will delete the last dimensioning entry if you wish. If you are not satisfied with text size or other variables associated with dimensioning, the command **dim vars** will allow you to change them. Select **dim vars.** The list of variables is shown in Topic 9-9. The variable **dimscale** is the overall scale factor applied to all dimension variables. If the text height of

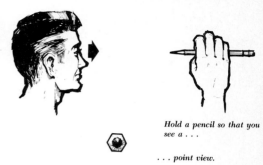

Hold a pencil so that you see a . . .

. . . point view.

[9-30] Point view.

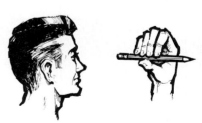

[9-31] True length view.

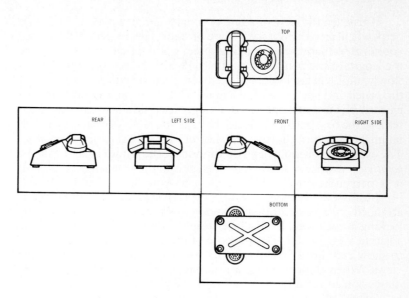

[9-32] The six principal views.

dimensioning notes is not satisfactory, try changing the value of **dimscale.** To see the current variable settings, select **status:** from the **LINEAR** submenu. Note that a dimensioning variable must be set *before* placing a dimension that you want it to affect, and this setting will remain in effect for additional dimensions until you change it.

5. Continue dimensioning all of the horizontal and vertical sizes and locations.

6. Return to the **DIM:** menu and select **diametr:.** Follow the prompts to place the leader and dimension the hole. **diametr:** will automatically precede the size of the hole with the ANSI symbol (Ø) for diameter.

7. If the part shown in Figure [9-33] had depth in addition to height and width, it would be represented by a multiview drawing. Although six principal views of a part are possible in multiview drawing, only enough views to describe the part completely are actually used. The part shown first in Figure [9-33] would require two views. Either a top view or a side (profile) view would be needed. Figure [9-35] shows the two views in proper position, with all needed dimensions shown. Notice that only the depth dimension of 1.00 is attached to the top view. The depth dimension cannot be shown in the front view. It must be shown in either the top or the side view, where depth appears true length. Add the top view to your drawing and place the appropriate dimensions.

VersaCAD

1. Start a new drawing. Create a line drawing similar to Figure [9-33].

2. Next, select **Dimension** from the **MAIN MENU.** Notice the choices: **Linear, Angular,** and **Setup.** An example of an angular dimension is shown in Figure [9-34]. Select **Setup.**

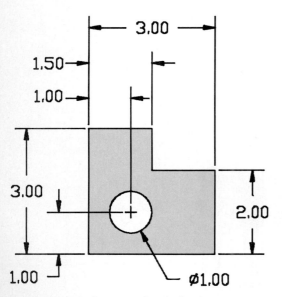

[9-33] Dimensioning example, single view.

3. **Setup** allows you to change dimensioning options. Select **Break** and set **Yes** for linear. This will cause dimension lines to be broken to receive the text. Also, select **Precision** and set for a two-digit decimal fraction. Return to the **DIMENSION** menu.

4. Select **Linear.** Select **Single.** Now, select **Two points** as the entry mode, following the screen prompts. Next, place the cursor at one end of one of the horizontal lines to be dimensioned and accept the position. Now, move the cursor to the other end of the line and enter that point as well. You will be prompted for **Dimension line position.** First, set the **Y axis** suboption to restrain the dimension line to the vertical position. Indicate a point on the screen that you wish the dimension line to pass through. The last thing you will need to do is to place the dimension text in the break in the dimension line. You may have to rotate the text to achieve the desired orientation. Use the **Rotate** option for this. Each time you select rotate, the text will be rotated by an increment which can be set by pressing **F6** and selecting **ROTATION** (see Topic 9-4). The default value is 90°. You may also want to adjust the text height or width. To do this, press **F6** and select **Text.** Follow the screen prompts.

5. Continue dimensioning all of the horizontal and vertical sizes and locations.

6. To dimension the hole, you will need to use the text command and the line command to draw the leader. Both are in the Add menu. Select **Line** first and start to draw toward the circle. Before you place the last endpoint, select the Arrow option by typing **A.** Next add the text. The ANSI standard symbol for diameter is "Ø" [9-33].

7. If the part shown in Figure [9-33] had depth in addition to height and width, it would be represented by a multiview drawing. Although six principal views of a part are possible in multiview drawing, only enough views to describe the part completely are actually used. The part shown first in Figure [9-33] would require two views, and either a top view or a side or profile view would be needed. Figure [9-35] shows the two views in proper position, with all needed dimensions shown. Notice that only the depth dimension of 1.00 is attached to the top view. The depth dimension cannot be shown in the front view. It must be shown in either the top or the side views, where depth appears true length. Add the top view to your drawing and place the appropriate dimensions.

EXERCISES

Exercise 9-8

Reproduce drawings from Figure [9-36] as desired or assigned. As appropriate, make use of zooming or windowing, mirroring of images, symmetry, snaps, grids, moving, copying, and array techniques to facilitate construction of the models. Use the grid lines to determine numerical coordinates for those points that fall on a grid intersection by as-

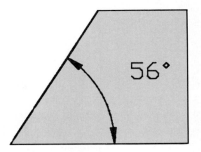

[9-34] Angular dimension.

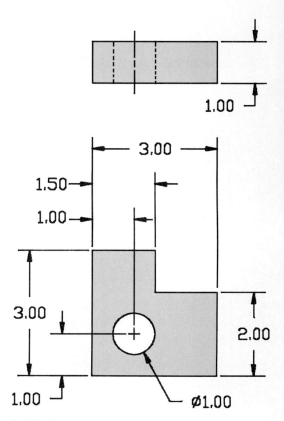

[9-35] Dimensioning example, multiview.

suming that each grid space is equal to 1 drawing unit. Drawing units may then be related to real-world coordinates by equating four drawing units to 1 inch, or 25 millimeters, or any other relationship you wish to establish. Dimension each drawing, using those dimensions needed to describe completely the size and location of each feature of the part and overall dimensions of the part.

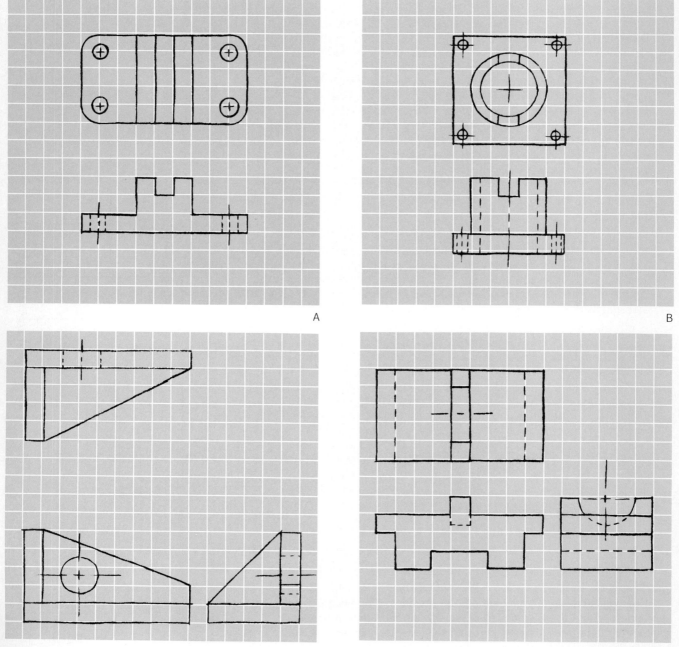

A

B

C

D

[9-36] Dimensioning exercises.

ISOMETRIC DRAWING ON CAD SYSTEMS————————

The drawings we have been considering so far have been derived by selecting a viewing direction perpendicular to major surfaces of the part in order to show lines and surfaces in true length and shape. This is necessary for dimensioning, as we have seen. A major shortcoming of these drawings is that they make it difficult to visualize the shape of the object being depicted without considerable study.

Pictorial drawings are easier to understand, as they look more like the views of the part we would expect to see under ordinary circumstances. Figure [9-37] compares a multiview drawing with a type of pictorial drawing known as an *isometric* drawing. The term "isometric" means equal measure and is derived from the fact that the three spatial coordinates of height, width, and depth are foreshortened equally in an isometric drawing. This type of drawing was comparatively easy to produce with standard manual drafting equipment, and has become the type of pictorial drawing most often seen in engineering environments. Most CAD systems provide features to facilitate isometric drawing.

Figure [9-38] shows the orientation of the isometric drawing related to the two-dimensional surface of the computer screen. It is important to remember that the methods of producing isometric drawings discussed here will create the illusion of a three-dimensional part but actually only a two-dimensional data base is created. True three-dimensional models generated on a CAD system allow the user to perform many useful operations, such as rotation and analysis, which cannot be done with an isometric drawing, which is only a two-dimensional illusion of a three-dimensional object. These techniques are beyond the scope of this chapter.

Example 9-10: Isometric Drawing

AutoCAD

1. Start a new drawing. Select **UNITS:** from the **SETTINGS** menu. Set the units to decimal, with one digit displayed to the right of the decimal place. Select **LIMITS:**. Set the upper left limits to 12,9.

2. Select **GRID:**. Type in the value **.5** and press **<enter>**. Select **SNAP.** At the prompt **Snap spacing or ON/OFF/Aspect/ Rotate/Style<1.0>:**, select **Style.**

3. At the prompt **Standard/Isometric:**, select **Isometric.** The grid will change to an isometric style [9-39] and may be toggled on and off with the F7 function key. At the prompt, **Vertical Spacing <1.0000>:**, type in **.5**.

4. The command **Isoplane** allows you to change the screen cursor style to conform to one of the three isometric planes. At the prompt **Left/Top/Right (Toggle):**, select **Left.** The screen cursor changes to conform to the isometric grid, left plane. The key sequence **CTRL E** will also allow you to toggle through the cursor styles in sequence.

5. Select **DRAW.** Next, select **LINE.**

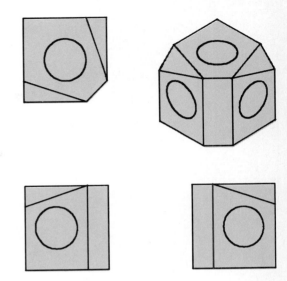

[9-37] Multiview and isometric drawing.

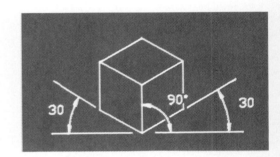

[9-38] Isometric drawing orientation.

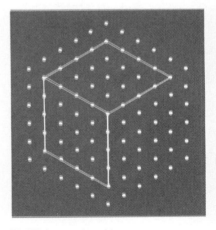

[9-39] Isometric grid.

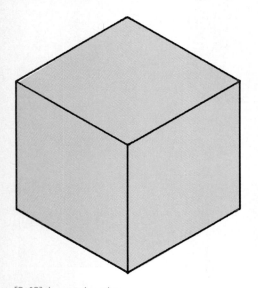

[9-40] Isometric cube.

6. Draw from point to point and create a 2 × 2 isometric cube, as shown in Figure [9-40].

7. Continue to use **LINE** to modify the cube by creating the additional lines that bound the inclined and oblique surfaces, as illustrated in Figure [9-41]. When working in the top plane of the cube, select **Isoplanetop.** When working in the right surface, select **Isoplaneright.** When **Isoplane** is selected, the isometric cursor can be toggled through right, left, and top by pressing **<Enter>.** The cursor can also be toggled in the midst of a command by pressing **Ctrl E.**

8. Next, select **EDIT,** and select **BREAK.** Break and erase unwanted line segments using break and erase [9-42]. You will want to zoom in, for greater control, as you work.

9. Now, draw an isometric circle (ellipse) on each face of the cube. To accomplish this, draw a circle, make it into a block, and stretch it when inserting. Select **CIRCLE** from the **DRAW** menu. Draw a circle with a diameter of 1.0 somewhere to the side of the cube. Select **BLOCKS** and **BLOCK:.** Name the block "B." Choose the center of the circle for the insertion base point. Select the circle in response to the prompt and it will disappear.

10. Next, select **INSERT:.** Give the block name (B) in response to the prompt, and position the cursor in the center of the top isoplane (horizontal surface of the cube). Accept the position and type in the value **1.2** in response to the prompt for the X-scale factor. Type in **0.7** for the Y-scale factor. Use **0** for the rotation angle. These factors will give a correctly proportioned isometric ellipse when the block is inserted.

11. Continue to use **INSERT** to add an ellipse to the left- and right-side surfaces of the object, as shown in Figure [9-43]. Notice that the rotation angle for the ellipse on the right isoplane must be 60°, and the rotation angle for the left isoplane must be 120°.

VersaCAD

1. Start a new drawing by selecting **Filer** from the **MAIN MENU** and **New** to clear the workfile. Select **UNITS,** set **INCHES,** and set the base for 12 inches wide.

2. Set the long screen cursor with function key **F5.** Set **SNAP** to **GRID** with function key **F2.**

3. Create the multiview drawing of the 2-inch cube shown in Figure [9-40].

4. Now, to create the isometric drawing, window out so that the multiview drawing occupies the lower left quarter of the screen. The procedure is to create a group of each of the three faces in turn and convert them into isometric form. First, select **GROUP, Build, New,** and **Yes (proceed with New).** Fence in the right-side view to form a group. Now select **View** from the **GROUP** menu. Select **Right** (the default), and **Copy,** since we want the multiview drawing to remain as a reference. When

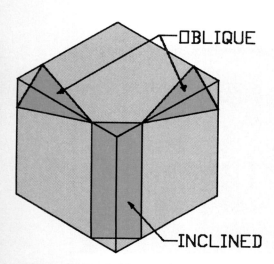

[9-41] Inclined and oblique surfaces.

prompted for a handle point, indicate point 1 in Figure [9-44]. Move the screen cursor into the upper right corner of the screen (clear area) and indicate where you want to draw the isometric view. The right isometric face will be drawn.

5. Continue to repeat the steps with the other two faces. Figure [9-44] shows the three faces before editing, with numbered dots showing the position of the handle points.

6. Use the **MODIFY** menu to edit the drawing by finding and deleting unwanted lines. Use **sKetch** to redraw the image.

7. Select the **Add** menu to add necessary lines to complete the isometric drawing as shown in Figure [9-43]. You will find the snap mode function key **F2** handy for this, especially for **SNAP TO OBJECT** and **SNAP TO INTERSECTION** (see Topic 9-4). Follow the screen prompts.

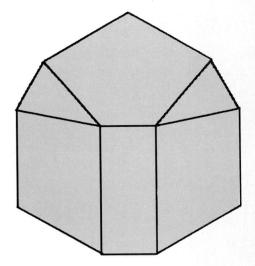

[9-42] Trimmed and cleaned surfaces.

PLOTTING

Plotting allows you to make a paper copy, also known as a "hard" copy, of part of a drawing file or of the entire file. Plotting includes copies of graphic images produced by a *printer,* which is ordinarily used for text output.

Plotters vary widely in their features and must be matched to the particular system you are using, or configured for the specific hardware and software in use. CAD programs will allow you to specify plotter pen sizes, colors, plotting speeds, and plotter-provided linetypes if these options are available on your plotter.

In general, CAD software allows you to select the portion of your drawing that you wish to plot and specify the size and proportions of the area you wish it "mapped" to. The computer will perform the necessary calculations to size the image to fit the area specified. Usually, the default plotting option plots the current screen display as large as possible in the standard plotting area as set on the plotter you are using. Most plotters used in engineering documentation applications are set to plot to one or more standard paper sizes. For example, the plotting area on "A"-sized paper (8.5 by 11 inches) is commonly 8 by 10.5 inches wide.

It is often desirable, or necessary, to plot to a standard scale, such as one-half of full size, meaning $\frac{1}{2}$ inch on the drawing represents 1 inch on the real-world model. This scale can be written as $\frac{1}{2}$, .5 = 1, or 1 = 2.

As stated earlier, one of the most useful characteristics of CAD systems is that plot scale is entirely independent of model size. The same data base can be used to produce plots at many different scales to fit different requirements.

Example 9-11: Plotting

AutoCAD

1. Enter the **MAIN MENU** and select **3, Plot a drawing.** The same plot routine may be entered from within the drawing editor by selecting **PLOT** in the **AUTOCAD** menu. The only

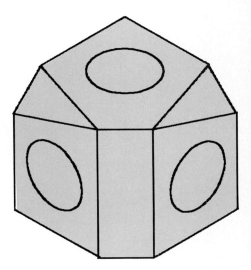

[9-43] Isometric ellipses.

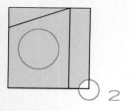

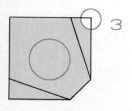

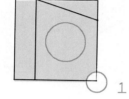

[9-44] Handle points.

difference is that plotting from within the drawing editor only plots the drawing you are currently working on.

2. You will be prompted for the file name. Enter the name of a drawing you have worked on and press <Enter>.

3. You will be presented with five choices for the part of the drawing you wish to be plotted. The choices and their effects are:
   a. *D (Display):* plots the view that was displayed on the screen just before the **SAVE** or **END** command was used.
   b. *E (Extents):* plots the portion of the drawing that contains primitive objects.
   c. *L (Limits):* plots the entire drawing area defined by the **LIMITS** command.
   d. *V (View):* plots a view that you previously saved using the **VIEW** command in the **DISPLAY** menu (see Topic 9-7).
   e. *W (Window):* plots the portion of your drawing that you capture in a window specified by lower left and upper right corners.

4. After following the screen prompts for your choice, the current plot setup specifications are presented. A typical example is shown below:

   Plot will NOT be written to a selected file.

   Sizes are in inches

   Plot origin is at (0.00,0.00)

   Plotting area is 10.50 wide by 8.00 high (MAX size)

   Plot is NOT rotated 90 degrees

   Hidden lines will NOT be removed

   Plot will be scaled to fit available area

   Do you wish to change anything? <N>:

   If these specifications are correct, press <enter> to continue the plotting process.

5. If you wish to change any of the specifications, type **Y** and press <Enter>. Follow the screen prompts to change any of the specifications, or simply press <Enter> to accept the default values.

   When you reach the last specification, plot scale, the following prompt will be displayed:

   Plotted inches = Drawing Units or Fit or ?: <F>

   Selecting **FIT** will instruct the computer to calculate the largest display of the area previously selected to be plotted. You may also enter an equation to specify scale. An entry of $1 = 1$ will produce a plot at the same size as the model. An entry of $.5 = 1$ will produce a half-size plot, while the entry $2 = 1$ will produce a twice-size plot.

6. If you wish to stop a plot in progress at any time, type **Ctrl C** to abort.

7. A printer plot capability is also available with AutoCAD. This is convenient for quick, inexpensive check prints of drawings in progress. Two ways are provided for entering this portion of the program. From the main menu, enter **4** for **Printer Plot a drawing.** You will be prompted for a drawing name. The same

plot routine may be entered from the drawing editor. In this case, only the drawing currently in the drawing editor can be printer-plotted. From the ROOT MENU select **PLOT.** The submenu will display two choices, **PLOTTER** and **PRINTER.** If you choose **PRINTER,** the prompt **What to plot—Display, Extents, Limits, View, or Window <D>:** will be displayed. AutoCAD printer-plots only the drawing area of the screen, not the menu or prompt areas. Follow additional screen prompts to set up your printer-plot.

VersaCAD

1. From the **MAIN MENU,** select **Output.** If you are currently working on a drawing file, you may print it. If not, first select **Filer** and **Get** a drawing file.

2. The **OUTPUT** menu gives you the choices of **Plotter** and **pRinter.** Either choice uses default specifications and no further decisions are necessary. The **PRINTER** option sends a "screen dump" to the printer, printing out everything on the screen, including menus and prompts. The **Display** option allows customization of the color display, if you have a computer with a high-resolution graphics board.

3. The **Specs** option of the output menu allows you to modify plotting specifications. Select **Specs.** The first option on the **SPECS** submenu is **Window.** This enables you to plot the portion of your drawing that you capture in a window specified by lower left and upper right corners.

4. The **Boundary** option allows you to place the area to be plotted on the paper relative to a 0,0 reference point.

5. The **Factor** option allows you to input a scaling factor. Select this option. The program will prompt you for a ratio of plotted size to real-world coordinates. To plot $\frac{1}{2}$ inch on paper as representing 1 foot on the model, enter **.5 <Enter>** and just **<Enter>.**

6. VersaCAD allows you to name and save plot specifications for future use. To do this, select **Save** from the **Specs** menu. Enter a name of up to seven characters, and the plot spec will be saved to the workfile and eventually to your drawing file. A plot spec may be recalled with the **Get** option from the **Specs** menu. **List** from the same menu provides a list of all plot specs saved to the workfile.

7. If you wish to stop a plot at any time, press **Esc.**

EXERCISES

Exercise 9-9

The following objects are designed to give you some additional practice in using a CAD program. Complete the drawings you select or are assigned. Dimension and plot each drawing.

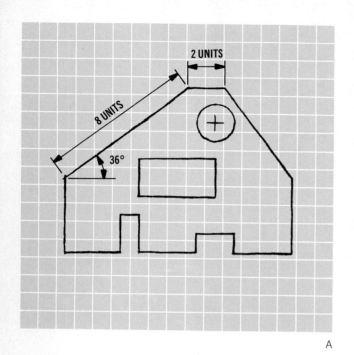

2 UNITS

8 UNITS

36°

A

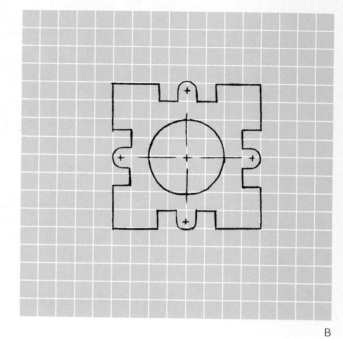

B

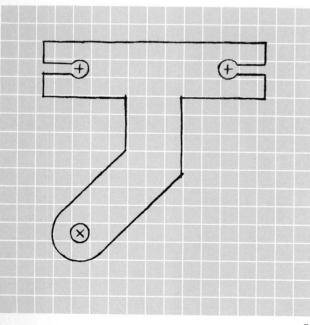

C

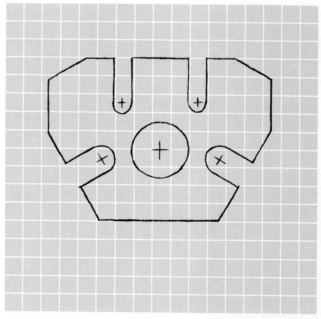

D

[9-45] Additional exercises.

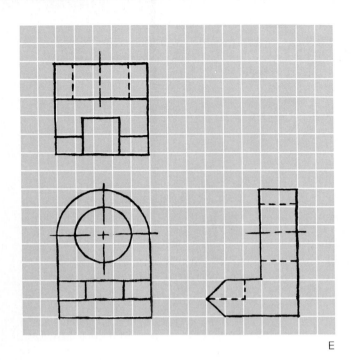

E

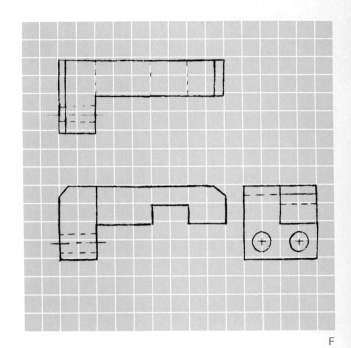

F

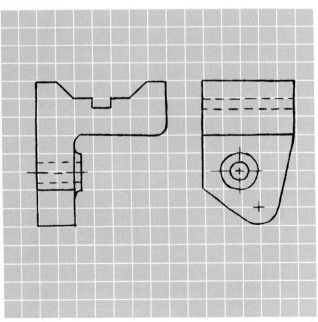

G

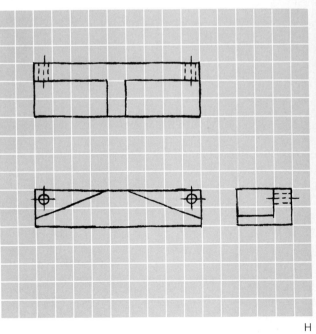

H

Chapter 10

Problem Solving Using Spreadsheets

The origin of spreadsheets was discussed in Chapter 8, where it was projected that they have just begun to have an impact on the engineering community. As pointed out there, spreadsheets are applications programs that can be programmed to solve specific problems. The use of the term "programmed" in this sense means that the user must adhere to the commands and rules of syntax that the spreadsheet program establishes in order to solve any problem. This sounds very much like the process of following the necessary rules and syntax required by high-level languages such as BASIC or FORTRAN when writing computer programs. The primary difference, however, is the framework within which users can solve problems and view results when using a spreadsheet. Although not as flexible as programming from "scratch," the spreadsheet can often provide quicker, more thorough, and much easier solutions to interpret (e.g., it can graphically display results) than specialized computer programs.

The small trigonometric Table 10-1 was constructed on a spreadsheet in about 15 minutes. What is even more impressive is

TABLE 10-1

Angle (deg)	Sin	Cos	Sin*Cos
0	.000	1.000	.000
5	.087	.996	.087
10	.174	.985	.171
15	.259	.966	.250
20	.342	.940	.321
25	.423	.906	.383
30	.500	.866	.433
35	.574	.819	.470
40	.643	.766	.492
45	.707	.707	.500
50	.766	.643	.492
55	.819	.574	.470
60	.866	.500	.433
65	.906	.423	.383
70	.940	.342	.321
75	.966	.259	.250
80	.985	.174	.171
85	.996	.087	.087
90	1.000	.000	.000

that within another 5 minutes, Figure [10-1] was plotted to graphically show the behavior of the three functions tabulated in Table 10-1 (the functions are the sine, the cosine, and the product of sine and cosine). We include this table and figure not just to show the versatility of the spreadsheet, but to whet your appetite for its use. Indeed, the spreadsheet is much more versatile than can be displayed here. More examples are developed later. But first we need to explain the concept of the spreadsheet and how it functions.

We will give a generic discussion of spreadsheet programming but we will supplement it with details on both *SuperCalc4*[1] and *Lotus 1-2-3*.[2] In addition to the normal capabilities found in almost every spreadsheet, both of these have the ability to accommodate iterative solutions and both include the added enhancement of graphics. Iterative capability is important in many engineering problems of the implicit type—problems that will not yield to direct solution techniques. Also, graphing capability can add a whole new dimension to the understanding of data and results. The special examples, where the details of SuperCalc4 and Lotus 1-2-3 are discussed, will be called *Topics*.

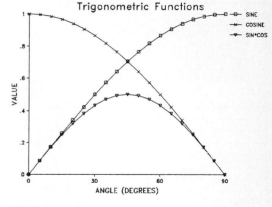

[10-1]

THE CONCEPT OF THE CELL

Spreadsheets are workspaces that consist of a matrix of many rows and columns. They are much like ruled paper accounting forms, although the actual rulings are not visible on the computer screen.

The intersection of each row and each column is the location of a *cell*, the smallest unit that contains information in a spreadsheet. This information can consist of text, numeric data, or formulas. As might be expected, there must be a way to specify the type of information that is to go into a cell and this is covered in detail in the following pages. In most spreadsheets the columns are distinguished by alphabetic letters, starting in the leftmost column with A and progressing with A, . . . ,Z, AA, . . . , AZ, BA, . . . , BZ, and so on. The rows are generally designated by sequential numbers, with row 1 being at the top of the sheet. Cells are then *addressed* by their [column letter] [row number] as in A1 or D73 [10-2].[3]

A cell address can be thought of as a variable name representing a location containing a value much like a variable name in BASIC or FORTRAN represents an address in memory (or RAM) where the current value of the variable is stored. Whenever a variable name is used in a BASIC or FORTRAN formula or expression, the value of the variable is fetched from the address in memory that has been

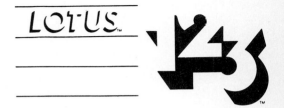

---

[1] Trademark of Computer Associates.

[2] Trademark of Lotus Development Corporation.

[3] The major exception to this is Multiplan (trademark of Microsoft), which uses numbers to designate both rows and columns. Absolute cell addresses are specified with an R before the row number and a C before the column number, as in R3C7 for the cell at the intersection of the third row and the seventh column. Relative cell addresses are specified with relative locations placed in square brackets, such as in R[-1]C[+2] denoting the cell located one row up (-) and two columns to the right (+) of the cell that contains this relative address.

[10-2] Rows and columns of a spreadsheet. Cell C5 is highlighted by a colored rectangle.

	A	B	C	D	E
1					
2					
3					
4					
5					
6					
7					
8					
9					
10					
11					
12					
13					
14					
15					

[10-3] Three different ranges of cells. The largest colored rectangle is the range A1.C5. The colored partial column is the range E2.E14. The colored partial row is the range B13.C13.

assigned to that variable name by the compiler or interpreter. So cell addresses, such as A1 or D73, can be used in expressions to represent the value that each cell contains, much like variable names are used in FORTRAN or BASIC! Couple this capability with that of being able to store expressions or formulas, not just values, in cells and we have touched on the real power of the spreadsheet. This should become clearer after we discuss formulas below.

Cell addresses are also required in spreadsheet commands to carry out those commands on a particular cell or group of cells. Spreadsheet commands are actions that let you take control of the spreadsheet; they are discussed in a separate section below.

Range of Cells

There is often a need to describe a number of contiguous cells in a shorthand notation. In practice, the most useful collection of contiguous cells is one in which the cells form a rectangle, the rectangle generally being called a *range*. A range of cells is generally denoted by specifying the address of the upper left-hand corner and the address of the lower right-hand corner of the range, separated by the *range delimiter* or *operator*. For example, the range A1.C5[4] refers to cells A1, A2, A3, A4, A5, B1, B2, B3, B4, B5, C1, C2, C3, C4, C5 [10-3]. Both A1 and A1.A1 refer to the single cell A1.

Example 10-1
State the range of cells that are visible in the worksheet screens (Screens 2) of Topic 10-1.

Solution
The range is A1.H12. On most screens, newly loaded blank worksheets will actually display about 20 working rows and eight columns. We have shown only 12 rows in Topic 10-1, for brevity.

Size Limitations and the Working Area

Current spreadsheet manufacturers commonly advertise that they can address a rather large number of cells (see Topic 10-1). However, the actual number of cells that can be used is determined by the size of RAM and the complexity of a particular sheet and is usually much smaller than the potential maximum.

Most spreadsheet problems require fewer cells than the maximum permitted by reasonable size RAMs (e.g., 256K). *However, for efficient use of memory it is important when you use spreadsheets that you confine your work to the top left corner of the sheet, working in the smallest rectangle possible.* We will refer to this rectangle as the *working area* of the sheet.

---

[4] Some spreadsheets, including Lotus 1-2-3, use a period,".", as the **range delimiter** or *operator* and some, including Multiplan by Microsoft, use a colon, ":". Others, including SuperCalc4, can use either. We will use the period as the range delimiter in the range specification.

The Screen as a Window

It is not unusual for the working area of a spreadsheet to contain more cells than can be displayed at one time on the limited size screen of most CRTs. Most programs use the screen as a *window* through which you can see only part of the worksheet. The screen windows in Topic 10-1 are viewing columns A through H and rows 1 through 12. Figure [10-4] graphically illustrates the potential spreadsheet, the working area, and the screen window.

The screen window can be moved around on the sheet to view different areas. We will describe how this is done in the next section. It is also possible to split the CRT screen in order to view two[5] different parts of a sheet at the same time by opening a second window simultaneously with the first one.

THE CURRENT CELL AND ENTERING DATA

Through the window afforded by the screen, the user can enter any one of the following three types of data into the current cell:

1. An alphanumeric string.
2. A number.
3. A formula.

The current or active cell is that cell in which the cursor is currently located, the cursor being the large highlighted rectangle that always appears in one of the cells located in the screen window. The cursor can be moved to any desired cell in order to make that cell active. This is accomplished by using the cursor control keys (either the keys on the keyboard with the arrows on them, or the S-E-D-X cursor diamond control keys actuated by holding down the Control key and touching S for left, E for up, D for right, or X for down). In Topic 10-1 the cell B2 is the active cell because it contains the cursor (the highlighted box in the active worksheet). That same cell also contains the text or label (an alphanumeric string) "Cell B2".

If you desired to make the cell M20 the active one, you would use the cursor control keys to move the cursor to the right and down on the sheet. When the edge of the window is reached, the sheet will begin to scroll by the window, taking the window where ever you desire. There is another way to reach a cell quickly and that is with the aid of the GoTo command or the GoTo key. It is discussed briefly with the other commands below.

Most spreadsheets reserve several lines in the screen display for aiding users in programming and using the sheet efficiently. We have identified these areas of the screen in Topic 10-1, by using notes shown in colored lettering (the colored lettering is our addition and does not appear in actual use). The first of these is the *current cell status line,* which always gives the address of the current cell (in Topic 10-1 note that it reads "B2" since the cursor is in cell B2) and its contents (note the "Cell B2" in the status lines of Topic 10-1). It

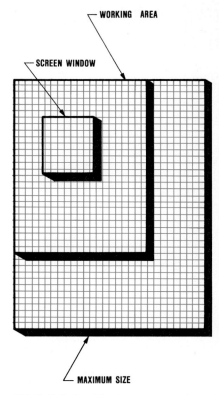

[10-4] Relationship between maximum-size sheet, the working area, and the screen window. The working area should be in the upper left corner of sheet. The screen window can be taken anywhere on the sheet to view the cells.

---

[5] Some spreadsheets, such as Multiplan by Microsoft, allow more than two windows to be open.

Topic 10-1 ━━━━━━━━━━━━━━━━━━━━   Lotus 1-2-3 ━━

Opening Screens and Ultimate Size

Shown below are the opening screen and a sample spreadsheet screen for Lotus 1-2-3. In Lotus 1-2-3, the opening screen represents the Lotus Access System and contains a menu of items that can be selected. The menu items are listed in the horizontal line beginning with 1-2-3 and ending with Exit. When the term 1-2-3 is highlighted in the menu (1-2-3 is the default item), an **<Enter>** will begin the 1-2-3 spreadsheet program.

1-2-3's spreadsheets can be as large as 256 columns by 8192 rows. Potentially, 1-2-3 can have up to 2,097,152 cells.

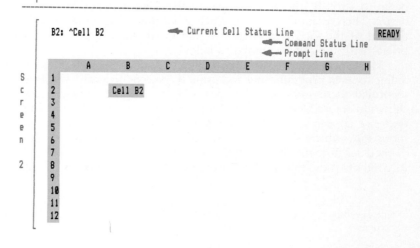

Topic 10-1 ▆▆▆▆▆▆▆▆▆▆▆▆▆▆▆▆▆▆▆▆▆▆▆▆▆▆▆▆▆▆▆▆▆ **SuperCalc4** ▆

Opening Screens and Ultimate Size

Shown below are the opening screen and a sample spreadsheet screen for SuperCalc4. An **<Enter>** takes you from the opening screen to the spreadsheet.

SuperCalc4 spreadsheets can occupy up to 255 columns and 9999 rows, providing, potentially, up to (255 × 9999 = 2,549,745) cells.

S
c
r
e
e
n
1

```
┌──────────────────────┐
│   SuperCalc4 (tm)    │
└──────────────────────┘

Version  1.00
S/N 0000-000000, IBM DOS
(WITH 8087 NDP)


Copyright 1986
Computer Associates International, Inc.
```

Software superior by design.

Press F1 for information about SuperCalc4 or other Computer Associates products.
Press any key to start

S
c
r
e
e
n
2

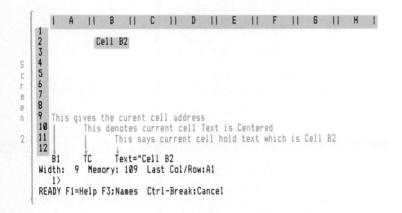

```
    | A  || B   ||  C  ||  D  || E  ||  F  ||  G  || H  |
 1
 2        Cell B2
 3
 4
 5
 6
 7
 8
 9  This gives the curent cell address
10        This denotes current cell Text is Centered
11            This says current cell hold text which is Cell B2
12
   B1      TC      Text="Cell B2
Width:  9  Memory: 109  Last Col/Row:A1
   1>
READY F1=Help F3:Names  Ctrl-Break:Cancel
```

```
 |    A    ||    B    |  C   ||     D     |
1 123-45-6789 ←SS # Entered as Label
2       -6711 ←SS # Entered as Number
3
4
```

```
 A1               Text:"123-45-6789
Width: 11   Memory:220 Last Col/Row:B2
   1)
```

[10-5] Notice that the entry on the current cell status line has the " symbol preceding it. This is the text definition or prefix character in SuperCalc4. This symbol was entered before the SS # for cell A1. It was not included when the same SS # was entered in cell A2, resulting in the SS # being interpreted as a formula and the result being displayed in the cell. If the cursor were located in cell A2, the current cell status line would show 123-45-6789, without any " preceding it.

may also contain some information that is peculiar to the particular spreadsheet you are using. We shall discuss the other two special lines on the screen later in their specific context.

Entering Text or Labels

In most cases, you will enter text or labels into the current cell simply by typing the desired text and pressing the Enter key.[6] As you are typing, the text is displayed on the current cell status line. You may use the backspace key to correct typing mistakes.

Text entered into the cell in this manner will generally be left justified as a default. That is, the text will begin at the left edge of the active cell and proceed to the right.

If the text overfills the cell (i.e., if the text string has more characters than there is space in the active cell), some spreadsheets will display the overflowing text in the next cell(s) to the right if that (those) cell(s) is (are) empty.[7] An example of this is shown in Figure [10-5], where the entry in cell B1, which is "←SS # Entered as Label," flows over into cells C1 and D1. If the adjacent cells to the right are not blank, only the part of the text that will fit into the active cell will be displayed.

In some spreadsheets, the text or label string will always be truncated to fit the length of the cell. Of course, there is usually more than one way to solve a problem. In a later section we discuss the command that allows you to change the width of individual columns.

Text can also be displayed centered in the active cell or right justified. But the method by which this is accomplished depends on the actual spreadsheet being used. See Topic 10-2 for the details in SuperCalc4 or Lotus 1-2-3.

---

[6] The Enter key will vary depending on the computer you are using. On IBM PCs and compatibles, this is the key marked ↵.

[7] On some spreadsheets, such as SuperCalc4, the adjacent cells to the right must be truly empty (we discuss the blank or erase commands later), which includes not being formatted.

Topic 10-2 ■■■■■   Lotus 1-2-3 ■■■

Current Cell Status Line Editing

In 1-2-3, the cursor control keys behave somewhat like the enter key, entering the text into the active cell, but then moving the cursor on to the next cell.

Beginning the label or text string with a ' will cause the text to be left justified, ^ (shift 6) centers the string, and " right justifies the string. To change the alignment of previously entered text in 1-2-3, you may either retype and reenter the text, or you may edit (we discuss editing the cell contents later) the ', ^, or " that is already there (these symbols are not visible when text is displayed in a cell). To enter strings that otherwise would appear as formulas or numbers, simply precede the string with any of (', ^, ").

In cases where the alphanumeric string being entered appears to the program as a formula or a number (the latter is a very specific formula), the string is usually not entered as a text, but as a formula or number. For example, if you desired to enter a social security number 123-45-6789 as a label, but simply typed 123-45-6789 **<Enter>,** the number -6711 would appear in the active cell for many spreadsheets [10-5]. That is because these programs would believe you were typing in a formula that consisted of 123 minus 45 minus 6789, the result of which is -6711. To enter special cases such as these as text or labels, precede the text with the appropriate text definition or prefix character (see Topic 10-2).

Example 10-2

Using an actual spreadsheet, enter the following labels into various cells in the window of the screen:

January	Energy	J7
122nd Airbourne	COUNT	@modular

Solution

Take the cursor to the cell where you want the label entered, type in the label, and execute the entry with the **<Enter>** key. The label should appear in the cursor cell on the sheet. Special problems you may encounter here are:

SuperCalc4: You should encounter no problems here except with J7, which is a valid cell address and makes the program think that a formula is intended. Since cell J7 is probably empty, a 0 (zero) appears in the cell where you entered J7. If you enter "J7," you should be able to enter the label J7. There is a command toggle / G " which when on will make SuperCalc4 think that every entry, unless preceded by a ", is a formula.

Lotus 1-2-3: The labels that cause problems here are J7, 122nd Airbourne and @modular. Lotus 1-2-3 thinks all of these are formulas, because they start with a cell address, a number, and a function symbol (the @; see Topic 10-3), respectively. Preceding each of these three with one of the text prefix characters will cure the problem (see Topic 10-2).

Topic 10-2 ▰▰▰▰▰▰▰▰▰▰▰▰▰▰▰▰▰▰▰▰▰▰▰▰▰▰ *SuperCalc4* ▰

Current Cell Status Line Editing

In SuperCalc4, you may use the backspace key in order to reach an earlier mistake. You may also use the Control-Break (hold down the control key and push the break key and then release both) to erase your entry on the status line.

Text formatting is done with the /**F**ormat command (discussed later). To enter strings that otherwise would appear as formulas or numbers, just precede the string with a double quote (") for SuperCalc4.

If errors are made in formula entry and the formula gets entered as a label or text, you must either repeat the entry or edit the cell entry to both correct the mistake and to remove the " that precedes the entry. Edit is discussed under Command Mode.

Entering Numbers

On nearly all spreadsheets, numbers are entered simply by typing the number into the current cell status line and entering. Usually, no number definition character need precede the number itself. Any-time the first character typed on the current cell status line is a number (0 through 9), most programs assume that what will follow is either more numbers or a formula. The numbers can be integers with no decimal point or decimal part, fixed point (i.e., have a decimal point and a decimal part), or be entered in scientific notation using an E to denote the start of an exponent of 10 (e.g., 7.5E13 for 7.5×10^{13}). In sophisticated spreadsheets, these numbers can have as many as 15 significant figures and be as large as 10^{63} or as small as 10^{-64}, although this will vary from spreadsheet to spreadsheet.

Numbers are generally right justified in the current cell, although some programs allow you to reformat to left justified. The display format of a number [i.e., whether it is preceded by a dollar sign ($), is displayed in exponential notation, or has a fixed number of digits after the decimal point] is controlled by the cell format command (discussed further below).

There is a distinct difference between a series of digits entered as a number and a series of digits entered as text (i.e., preceded by the appropriate text definition or prefix character). Numbers entered into a cell can be used by formulas that are entered into any other cell and their true value will be used. That is, if a formula in some cell takes a 2 that has been entered in another cell as a number 2, and adds 1 to it, the result will be 3. On the other hand, all text or label strings, including strings of digits entered as text, default to zero when inadvertently used by formulas that reside in other cells. Thus, if the formula mentioned above takes a 2 that has been entered in another cell as text or a label 2, and adds 1 to it, the result will be 1 [10-6].

Example 10-3

Enter the following numbers into cells A1, B2, C3, respectively, of a spreadsheet:

2   -4.7   9.666666E17

Solution

Take the cursor to cell A1, type 2 into the current cell status line, and push the **<Enter>** key.

Take the cursor to cell B2, type -4.7 into the current cell status line, and push the **<Enter>** Depending on the formatting of the cell and your equipment, this negative number may appear in (), or it may display in a different color than positive numbers.

Take the cursor to cell C3, type 9.666666E17 into the current cell status line, and push the **<Enter>** key. Since the default cell width is typically nine characters on most spreadsheets, this number probably got shortened in the display (we cover the column-width command later, which will allow you to change the width of a column). Was it rounded or truncated? Although the number may have been shortened in the display, the full number is retained in memory and used in any calculations.

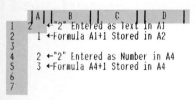

```
   |A||   B   ||   C   ||   D   |
1  2 ←"2" Entered as Text in A1
2    1 ←Formula A1+1 Stored in A2
3
4    2 ←"2" Entered as Number in A4
5    3 ←Formula A4+1 Stored in A4
6
7

A1              Text:="2
Width:  3  Memory:220 Last Col/Row:B5
  1)
```

[10-6] Notice that on the current cell status line, the 2 in cell A1 was entered as text or label. (How can you tell?) Notice that it does not work in the formula of cell A2. Textual entries default to 0 when used in formulas on most spreadsheets (SuperCalc4).

Entering Formulas

The ability to enter formulas into the cells of a spreadsheet is probably the most powerful attribute of spreadsheet programming. A formula is typed into the current cell status line and then entered into the current cell by actuating the **<Enter>** key. Once a formula is entered in a cell, only the number that results from the evaluation of the formula is displayed on the sheet. In addition, when that cell is referenced in other formulas (by using the cell address, e.g., A1 of D78), the result of the formula is used rather than the formula itself.

Unfortunately, this trait of displaying only the results means that by looking at the sheet, you cannot tell whether the numbers displayed are results that have been calculated from formulas or are simply numbers that have been entered.

The current cell status line plays an important role here, for if a formula has been entered in the current cell, the formula will be displayed on the status line. It is often helpful to remember that the current cell status line always shows what is actually stored in the current cell, not just what is displayed in that cell on the sheet.

Operations Possible. Nearly all common spreadsheets allow for the following arithmetic operations to be performed within any formula:

■ Exponentiation (signified by either ^ or **).

■ Multiplication (* and division (/).

■ Addition (+) and subtraction (−).

Most spreadsheets (including both SuperCalc4 and Lotus 1-2-3) follow the order preference rules adopted by most programming languages, which specify:

■ Exponentiation is done first.

■ Multiplication and division are next.

■ Addition and subtraction are last.

■ Within any of the groups above, apparent conflicts are resolved in left-to-right order (e.g., c/d*e multiplies the quotient of c/d by e).

However, there are a few spreadsheets that use as a preference rule that *all expressions* are evaluated from left to right, so be careful. In any case, parentheses can be used to override the rules above, because expressions within parentheses are always evaluated first. Therefore, the liberal use of parentheses is always wise.

Constants and Variables. Both constants and variables can be entered in formulas. Similar to BASIC and FORTRAN expressions, constants are values that do not change from one evaluation of the expression to the next. In Equation (10-1), the constant values are 3 and 4.6; these values will not change with each evaluation of the expression. The size and number of significant figures permitted are the same as discussed in the preceding section on numbers.

Variables, on the other hand, are quantities that you expect to change during the use of the spreadsheet. They are represented by a cell address, as described earlier. Some spreadsheets also enable the user to give a specific name to a cell and these user-defined names can also be used in formulas in lieu of the actual cell address. For example, if you wanted to perform calculations with the formula

$$y = 3x^2 + 4.6 \qquad (10\text{-}1)$$

you might enter this formula into cell A2. But before doing this, you must decide where the formula will find the value of x to use in the calculation. You cannot enter the value of x into cell A2 without destroying the formula you wish to put there. Assuming that we decide to use 5 as the desired value of x, we might choose to enter this number into cell A1 on the sheet. The formula we enter in cell A2 is then

$$3*A1\hat{\ }2 + 4.6 \qquad (10\text{-}2)$$

If we have not made an error in entry and if we have already entered the 5 in cell A1, the result of 79.6 will appear in cell A2 almost immediately [10-7].

Of course, the spreadsheet formula represented by Equation (10-2) looks a little strange since the variable x in Equation (10-2) now has a nonmnemonic name. The use of cell addresses in place of more recognizable variable names contributes to the difficulty usually encountered in debugging (i.e., in finding errors in) spreadsheet models. Some spreadsheets allow you to assign an arbitrary name to a cell or group of cells and then to reference that cell by the chosen name instead of by the more cumbersome cell address. We will discuss naming of cells and groups of cells later in this chapter.

Except for the name of the variable x, Equation (10-2) looks very much the same as the right-hand side of Equation (10-1). Notice that we have not specified an equal sign in the spreadsheet formula, only the right hand side of Equation (10-1). Once a formula is stored in a cell and the sheet is recalculated, the result is returned to the screen in the location of the cell where the formula is entered. That is, in this example the spreadsheet returns the value of the calculation to the screen location of cell A2—cell A2 is, in essense, the variable y.

In the building of formulas, you may expect a spreadsheet to do any calculations that can be described with the available algebraic operations and the available functions (built-in functions are described later). Single-cell formulas are limited to 240 characters in both Lotus 1-2-3 and SuperCalc4.

Fortunately, like most programming languages such as BASIC or FORTRAN, formulas look much like their algebraic counterparts, making them easy to construct and "read." However, equations involving multiple levels (e.g., fractions) must be written by enclosing both the numerator and denominator in parentheses and separating them with a division sign (/) with the numerator preceding the denominator. Parentheses are the only delimiter or separator available for groups of terms; the use of brackets ({ } or [ ]) is not permitted. However, parentheses can be nested, as in

$$((A1+A2)-2*(A3\text{-}A4))/(A5\text{-}A6)$$

The liberal use of parentheses is advised.

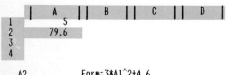

A2   Form=3*A1^2+4.6
Width: 9 Memory:220 Last Col/Row:A2
    1}

[10-7] Notice on the current cell status line that a formula has been entered in cell A2, but the result is displayed on the spreadsheet (SuperCalc4).

Once formula (10-2) was entered in cell A2, the spreadsheet "recalculated" the current sheet. In this case, the sheet consisted of the cell A1, which contained the number 5, and the cell A2, which contained the formula given in Equation (10-1). By specifying the cell address A1 in the formula contained in A2, we have effectively instructed the spreadsheet to "fetch" the contents of A1 each time the spreadsheet is recalculated and Equation (10-1) is evaluated. The actual format of the numbers (i.e., how many decimal places, etc.) will depend on how the cells are formatted (formatting is discussed later).

On the sheet, only two numbers will be showing, a 5 in cell A1 and 79.6 in cell A2. With the cursor in cell A1, the entry on the current cell status line will read 5. With the cursor in cell A2, the entry on the current cell status line will be the formula, 3*A1^2 + 4.6. Thus, by looking at the current cell status line, you can distinguish between what is actually stored in each cell and what is displayed in the cell location on the sheet. *Remember, a formula can be stored but only the result is displayed on the sheet.*

Using Results from Other Formulas. Formulas, of course, can depend on the results of other formulas. Spreadsheets accommodate this simply by allowing the results of other formulas to be used by referencing the cell address of the results. For example, assume that we were interested in calculating

$$\alpha = 2y - 25 \qquad (10\text{-}3)$$

where y is the result of Equation (10-1) for which we have already prepared a spreadsheet method of solution [cell A2 contains y, and A1 contains x, the variable that appears in Equation (10-1)]. In cell A3 of our existing sheet, we would enter the formula

$$2*A2 - 25 \qquad (10\text{-}4)$$

As soon as it is entered, we should see 134.2 appear in cell A3 on the sheet (see Figure [10-8]). Again, the formula shows on the current cell status line, as long as the cursor is located in cell A3.

Example 10-4

Express the equation

$$\left(1 + \frac{3 + z^2}{y^5}\right)\frac{z}{x}$$

in spreadsheet formula form, where x will be placed in cell A1, y in B1, z in C1. Enter the formula into D1.

Solution

Move the cursor to cell D1 and type into the current cell status line the formula

$$(1 + (3 + C1^2)/B1^5)*C1/A1$$

followed by an **<Enter>**. If you have not made an entry mistake and if nonzero values have been placed in cells A1 and B1, you should see the results in cell D1. If A1 and/or B1 are empty or 0 (zero), you will get an error appearing on the sheet in cell D1, telling you that something is

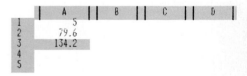

```
        |    A    ||   B   ||   C   ||   D   |
  1  |        5
  2  |      79.6
  3  |     134.2
  4  |
  5  |

    A3              Form=2*A2-25.
  Width:  9  Memory:220 Last Col/Row:A3
      1)
```

[10-8] The formula in cell A3 references the result of the formula that resides in cell A2. See Figure [10-7] to verify that a formula which references the value in cell A1 is stored in cell A2 (SuperCalc4).

wrong. (You are dividing by zero.) Just type in some nonzero values for A1 and B1, and the error message should go away. If you made an entry error, you will have to retype the equation without error or wait until we discuss the edit command to fix it.

Relative Versus Absolute Cell Addresses. Depending on the spreadsheet you are using, cell addresses embedded in formulas [e.g., in Equation (10-4)] may be "relative" addresses or they may be "absolute" addresses. Relative addresses used in formulas are interpreted strictly in terms of the relative location of the cited address with respect to the cell holding the citation.

In the example of the preceding section, cell A3 cited cell A2, which is straightforward. Should either SuperCalc4 or Lotus 1-2-3 be the spreadsheet in use, the citation in A3 is really interpreted as meaning the following.

> Bring to cell A3 the result (not the formula, but the result) found in the cell located zero columns over and one row up (the relative distance between A3 and A2), multiply it by 2, subtract the number 25 from that product, and place the result in cell A3.

The tricky part comes if we want to replicate or copy the formula into other cells on the sheet. Once you begin to understand spreadsheets, you will probably find copying quite useful, but it has to be done with some awareness if the newly copied formula are to give correct results.

If we were to copy[8] the contents of cell A3 to cell B3, we might be surprised to see that the number 134.20 did not appear in B3, but that (−25.) did appear (assuming that cell B2 is blank). If we check what is actually stored in cell B3 by taking the cursor to that cell and observing what appears on the current cell status line, we find that the formula has changed slightly from what it was in cell A1 before replicating or copying [10-9]. It now reads

$$2*B2 - 25. \tag{10-5}$$

not

$$2*A2 - 25. \tag{10-6}$$

as we might have expected. What has happened is that the *relative* address to cell A2 stored in A3 was copied as a *relative* address to cell B2 (the cell located zero columns over and one row up) when placed in cell B3.

In practice, it is relative addresses that are most useful due to the way replicating or copying are often used in setting up a worksheet. Relative addresses can cause problems at times, especially when you want a number of copied formulas to all reference a single cell for some common data input. For this you need to use an absolute cell reference—or you need to be very careful.

Most spreadsheets provide a special notation for entering absolute cell addresses into formulas. For example, Lotus 1-2-3 allows you to embed a "$" sign ahead of the column letter and/or the row

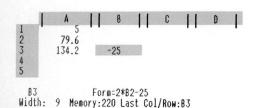

[10-9] The formula in cell B3 was replicated or copied from cell A3 using relative cell addresses. Cell A3 has the same formula as it did in Figure [10-7] (SuperCalc4).

---

[8] Replication or copying are spreadsheet commands and are discussed in more detail later.

number in an address to signify that the column and/or row reference is absolute and should always be treated as such when copying the reference to another cell. A2 is an absolute address to cell A2; $A2 is an absolute reference to column A and a relative address to row 2 (the latter depends on the relative row displacement from the citing cell). A$2 is a relative reference to column A and an absolute reference to row 2.

SuperCalc4 provides two alternatives. First, the "$" can be used just as it is in Lotus 1-2-3. Second, it also provides for absolute cell addressing in a somewhat indirect but quite adequate manner during copying of existing references. In the latter method, cell references in any formula are entered using the typical (column, row) specification. Then when copying a formula to any other cell, there is an option in the copy command that allows you to specify whether the references in the formula being copied are to be treated as relative or absolute.

When replicating or copying formula on any spreadsheet, you should always ask yourself: *"Are the relative cell addresses implicit in the formula being copied appropriate to the new location to which it is being copied, or are absolute cell references required?"*

Relative versus absolute cell addressing is a difficult concept to understand. As a result, it is one of the concepts that most often leads to spreadsheet errors. To master the concept, you should practice with some simple problems.

Calculation and Recalculation of Formulas. It is typical for most spreadsheets to *calculate* or *recalculate* all formulas in all cells every time the entry in any one cell is changed, regardless of whether or not all the formulas depend on the changed cell. If there are a large number of formulas on a spreadsheet, the recalculations could take an annoying amount of time and slow the data entry. However, this *automatic recalculation* mode usually can be overridden in order to improve data-entry speed. That is, a *manual recalculation* mode can be invoked, whereby numbers can be entered without recalculating new values for the formulas. However, the user must remember to force a recalculation of the sheet at the completion of the data entry task; otherwise, the results of any formula will not reflect the new data.

In the example displayed in Figure [10-7], assuming that the 5 for x is already entered in A1, the result of the formula for y was calculated as soon as it was entered. That is why the value of 79.6 appeared in cell A2 so quickly. If you were now to take the cursor back to cell A1 and enter a 4 in place of the 5 (just enter a 4 on the current cell status line and push the **<Enter>** key—the 5 will be overwritten), a new value of 52.6 appears in cell A2 almost immediately if the automatic recalculation mode is in effect. What a delight!

When a sheet is *calculated* or *recalculated,* the calculations generally start from the upper left corner (cell A1) and proceed downward and to the right. That is, the formula in cell A1 will generally be calculated before the formula in cell Z50. The more advanced spreadsheets allow several recalculation modes to be instituted. Most allow at least a choice of either recalculation by rows or recalculation by columns.

Of course, the calculation process, when initiated, may encounter a cell formula that references values from cells that are below or to the right of the current cell. Row-wise or column-wise recalculation procedures may not be able to handle these situations.

Some spreadsheets[9] provide a *natural* recalculation pattern which proceeds through the computations keeping all the dependent cells uncalculated until the cells on which they depend have been calculated. This allows users to be more flexible in their placement of formulas throughout the sheet. It particularly gives the technical user more freedom.

Circular References. With the natural mode of recalculation, the capability for any cell to reference any other cell, regardless of its position on the sheet, is assured. However, the special case of two cells referencing each other requires special attention. For example, if cell A1 contains a formula that references cell B1, and cell B1 contains a formula that references cell A1, then it would appear that the calculations may be "going in circles."

Such situations are referred to as *circular references* and may be intentional or accidental. Most sheets give a warning to the user when a circular reference exists so that if the circular reference is unintentional, it can be corrected. If unintentional circular references exist, they usually lead to significant errors if uncorrected.

Problems of an *iterative* nature make use of circular references, as will be demonstrated in the sample problems at the end of this chapter. Therefore, circular references can be intentional. In iterative problems, equations or formulas must be solved over and over again, with the solution being updated during each iteration until the final result *converges* or stops changing.

Intentional circular references are resolved by forcing the sheet to be recalculated a sufficient number of times so that the formula results in the circular cells converge or stabilize. Some spreadsheets allow the user either to set the number of sheet recalculations to execute in an attempt to establish *convergence,* or to set a *convergence tolerance* to strive for in resolving the circular references. This should become clearer in Example 10-9.

The Functions Available. One of the features that make formula construction so convenient is the availability of a large number of established *functions* that can be used for constructing formulas. For example, suppose you want to enter into cell A3 a formula that represents the equation

$$z = \sin \theta + \exp x \qquad (10\text{-}7)$$

with ($\theta = \pi/4 =$) 3.14159/4 radians being located in cell A1[10] and ($x =$) 0.1 located in cell A2. With the cursor in cell A3, you would

---

[9]This includes Lotus 1-2-3, SuperCalc4, and Multiplan.

[10]It is important to remember that in almost all computer work, the trigonometric functions require their arguments to be in *radians,* not degrees. In Equation (10-8) the argument of the sine could have been (A1*3.14159/180.) if the θ had been entered in cell A1 in degrees.

type into the current cell status line the formula[11] (see [10-10])

$$SIN(A1) + EXP(A2) \qquad (10\text{-}8)$$

and follow it with an **<Enter>.** The result of the calculation, 1.812277, will be displayed almost immediately in cell A3 of the sheet, but again, the format of the answer will depend on how the cell has been formatted.

Thus the mathematical formula for the trigonometric sine function and the exponential function do not have to be entered into cells that require their computation. These mathematical formulas are available, under specific[12] and reserved names, as a part of the spreadsheet package. Common functions on most spreadsheets include

Trigonometric functions:
    Cosine of a number.
    Sine of a number.
    Tangent of a number.

Inverse trigonometric functions:
    Arccosine of a number.
    Arcsine of a number.
    Two-quadrant arctangent.
    Four-quadrant arctangent of a number.

Exponential and logarithms:
    Exponential of a number.
    Log_e of a number.
    Log_{10} of a number.

Constants: Pi.

Arithmetic functions:
    Absolute value of a number.
    Division remainder.
    Maximum of values in a range.
    Minimum of values in a range.
    Number of nonblank values in a range.
    Square root of a number.
    Sum of values in a range.

Logical functions:
    If-then-else with operators $=, \leq, \geq, <, >, \neq$, NOT, AND, OR.

Statistical functions:
    Average of nonblank values in a range.
    Standard deviation of nonblank values in a range.
    Variance of nonblank values in a range.

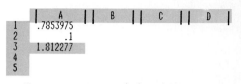

[10-10] From the current cell status line you can see that we have entered into cell A3 a formula that requires sines and exponentials. The result magically appears in the cell on the spreadsheet. The spreadsheet recognizes the names SIN and EXP, among others, and supplies the means to calculate them (Super-Calc4).

---

[11] Some spreadsheets require the "@" sign to be the first character in any function name.

[12] The specific names used by most spreadsheets are very close to those used by most programming languages, such as FORTRAN and BASIC. Some spreadsheets do require that the symbol "@" precede the more common function name.

Financial functions:
    Net present value.
    Internal rate of return.
    Present worth.

Miscellaneous functions:
    Integer part of a decimal number.
    Rounded-value part of a decimal number.
    Random number generator.
    Table lookup.

Specific but limited examples for both SuperCalc4 and Lotus 1-2-3 are given in Topic 10-3. Help screens in both spreadsheets enumerate all of the available functions.

Topic 10-3

Intrinsic Functions

	SuperCalc4[a]	Lotus 1-2-3[a]
Trigonometric functions		
cosine (*c/v* in radians)	COS(*c/v*)	@COS(*c/v*)
sin (*c/v* in radians)	SIN(*c/v*)	@SIN(*c/v*)
tangent (*c/v* in radians)	TAN(*c/v*)	@TAN(*c/v*)
Inverse Trigonometric functions		
arc cosine (returns radians)	ACOS(*c/v*)	@ACOS(*c/v*)
arc sine (returns radians)	ASIN(*c/v*)	@ASIN(*c/v*)
arc tangent (returns radians)	ATAN(*c/v*)	@ATAN(*c/v*)
4-quadrant arc tangent (returns radians)	ATAN2(*c/v$_1$,c/v$_2$*)	@ATAN2(*c/v$_1$,c/v$_2$*)
Exponentiation and Logarithms		
exponentiation, e$^{(c/v)}$	EXP(*c/v*)	@EXP(*c/v*)
log$_e$	LN(*c/v*)	@LN(*c/v*)
log$_{10}$	LOG(*c/v*) or LOG10(*c/v*)	@LOG(*c/v*)
Constants		
pi (3.14159+)	PI	@PI
Arithmetic functions		
absolute value	ABS(*c/v*)	@ABS(*c/v*)
count non-blank cells	COUNT(*range*)	@COUNT(*range*)
maximum value in a range	MAX(*range*)	@MAX(*range*)
minimum value in a range	MIN(*range*)	@MIN(*range*)
division remainder	MOD(*c/v$_1$,c/v$_2$*)	@MOD(*c/v$_1$,c/v$_2$*)
square root	SQRT(*c/v*)	@SQRT(*c/v*)
sum of values in range	SUM(*range*)	@SUM(*range*)
Statistical functions		
average of values in range	AV(*range*) or AVERAGE(*range*)	@AVG(*range*)
standard deviation of values in range	STD(*range*)	@STD(*range*)
variance of values in range	VAR(*range*)	@VAR(*range*)
Financial functions		
internal rate of return	IRR(*[guess,]range*)	@IRR(*guess,range*)
future value	FV(*pmt,int,per*)	@FV(*pmt,int,per*)
net present value	NPV(*disc,range*)	@NPV(*c/v,range*)
payment	PMT(*prin,int,per*)	@PMT(*prin,int,per*)
present value	PV(*pmt,int,per*)	@PV(*pmt,int,per*)

Example 10-5

Investigate the ability of your spreadsheet to do natural-order recalculation by placing the sine of x in cell A1 and the value of x in cell B2. x is to be entered in degrees.

Solution

Move the cursor to cell B2 and enter, for example, 45. Then move the cursor to cell A1 and enter SIN(B2*PI/180), remembering that you need the argument in radians, not degrees (in some spreadsheets, you will need to use @SIN instead of SIN; see Topic 10-3). Did an answer of 0.707 appear in cell A1? The recalculation command that allows you to choose the recalculation mode is discussed in the next section.

SuperCalc4 and Lotus 1-2-3

	SuperCalc4[a]	Lotus 1-2-3[a]
Logical functions		
if-then-else	IF(cond,c/v₁,c/v₂)[b]	@IF(cond,c/v₁,c/v₂)[c]
and	AND(cond_a,cond_b)[d]	#AND#
or	OR(cond_a,cond_b)[e]	#OR#
not	NOT(cond)[f]	#NOT#
true (returns 1)	TRUE	@TRUE
false (returns 0)	FALSE	@FALSE
not available	NA	@NA
error	ERROR	@ERR
Miscellaneous functions		
integer part (truncates)	INT(c/v)	@INT(c/v)
random number generator	RAN or RANDOM or RAND	@RAND
rounded value	ROUND(c/v,places)	@ROUND(c/v,places)
table lookup	LOOKUP(c/v,range)	@HLOOKUP(c/v,range,row#)
	(also HLOOKUP and VLOOKUP)	@VLOOKUP(c/v,range,column#)
Arithmetic and relational operators		
add	+	+
subtract	−	−
multiply	*	*
divide	/	/
raise to power	^ (or) **	^
percent	%	
equal to	=	=
not equal to	<>	<>
less than	<	<
less than or equal to	<=	<=
greater than	>	>
greater than or equal to	>=	>=

[a]c - reference to a single Cell; c/v - a Cell reference or a Value; *cond* - some algebraic Condition (e.g., A5 > 5.); *disc* - Discount rate; *guess* - a Guess; *int* - Interest; *per* - Period or term; *places* - number of Places after the decimal; *pmt* - Payment; *prin* - Principal; *range* - a Range of cells (e.g., B3.C6); *v* - a Value entered here.
[b]The function IF takes the value of c/v₁ if *cond* is satisfied or the value of c/v₂ if *cond* is not satisfied.
[c]The function @IF takes the value of c/v₁ if *cond* is satisfied or the value of c/v₂ if *cond* is not satisfied.
[d]The function OR takes the value of 1 if *either cond_a or cond_b* are satisfied. Otherwise, the value of OR is 0 (zero).
[e]The function AND takes the value of 1 if *both cond^a and cond^b* are satisfied. Otherwise, the value of AND is 0 (zero).
[f]The function NOT takes the value of 1 if *cond* is not satisfied. Otherwise, the value is 0 (zero).

TAKING CHARGE OF THE SHEET
COMMAND MODE OPERATION _____

So far we have discussed how to move around on spreadsheets, how to enter labels and data, and how to enter equations or formulas. These constitute what may be called the *data entry* mode of operation. However, nearly all spreadsheets have a *command mode* of operation which is used to make alterations to the sheet itself or to the entries on the sheet. The command mode puts you in charge of the sheet.

The most common way of entering the command mode is to actuate the "/" or slash key.[13] This key activates a command mode status line which generally shows a menu of items that can be carried out (see Topic 10-1 for the location of the command status line). In some spreadsheets, this menu is inhospitable to new users in that it consists of only single-letter representations of the commands that can be chosen. You select the command you want by typing the single-letter representation. To ease this unfriendly environment, these spreadsheets provide easy access to a help screen (help screens are discussed later), which gives a more user-friendly description of the commands available.

In some spreadsheets, the menus consist of whole-word representations of the commands available. SuperCalc4 for example, gives two command status lines, or menu lines, and a prompt line. The lines appear in Figure [10-11]. The first two lines show the menu items that are immediately accessible. The last line displays a short description of the highlighted item in the menu lines. Different items in the first menu can be chosen by using the Cursor Control keys to space over to the desired item and then actuating the **<Enter>** key (the cursor has been moved to the item Copy in Figure [10-11]). Once you become familiar with the menus and submenus, all you have to do is actuate the key corresponding to the first letter of the menu item you wish to select and the program responds. This single-keystroke technique has become a commonly used one. Figure [10-12] displays the main menu help screen available in SuperCalc4.

Figure [10-13] shows the main command menu line in Lotus 1-2-3. The second line displays the submenu items that will be available if the highlighted item in the first line menu were to be chosen. In keystroke sequences that end with a user input (such as a cell reference or range reference), an **<Enter>** is usually required to terminate the entry and execute the command. We will demonstrate the use of many of the commands in the example problems later in this section.

As important as the command keystrokes is the *clear* or *cancel key*. This key either cancels the entire command sequence that you

_____

[13]The major exception to this is Multiplan by Microsoft, which uses the escape key (generally labeled *Esc*) to access the command mode menu. In Multiplan this menu is always visible at the bottom of the screen, on what is the equivalent of a command mode status line.

```
Arrange  Blank  Copy  Delete  Edit   Format  Global   Insert  Load  Move  Name
Output  Protect  Quit  Save  Title  Unprotect  View  Window  Zap  /more
      2>/
MENU   Duplicate or replicate a cell range or graph range
```

[10-11] The initial command menu for SuperCalc4.

```
Slash Commands                              SuperCalc4 AnswerScreen
  /Arrange    >  Sorts cells in ascending or descending order.
  /Blank      >  Removes or empties contents of cells or graphs.
  /Copy       >  Duplicates graphs or cell contents and display formats.
  /Delete     >  Deletes a range.
  /Edit       >  Allows editing of cell contents.
  /Format     >  Sets display format at Entry, Row, Column, or Global levels.
  /Global     >  Changes global display or calculation options.
  /Insert     >  Adds empty rows or columns.
  /Load       >  Reads spreadsheet or portion from disk into memory.
  /Move       >  Relocates existing rows or columns.
  /Name       >  Defines Named Ranges.
  /Output     >  Sends display or cell contents to printer, screen or disk.
  /Protect    >  Prevents future alteration of cells.
  /Quit       >  Ends the SuperCalc4 program.
  /Save       >  Stores the current spreadsheet on disk.
  /Title      >  Locks upper rows or left-hand columns from scrolling.
  /Unprotect  >  Allows alteration of protected cells.
  /View       >  Displays data as Pie, Bar, Line, Area, X-Y or Hi-Lo graph.
  /Window     >  Splits the screen display.
  /Zap        >  Erases spreadsheet and format settings from workspace.
  //          >  Data  Export  Import  Macro
```

[10-12] The help screen for the main command menu of SuperCalc4 (see Figure [10-11] for the main menu).

have started (e.g., SuperCalc4) or backs you up one menu with each press (as in Lotus 1-2-3 and SuperCalc4).

In SuperCalc4, there are three cancel keys. The Control-Break (hold down the key labeled Ctrl, push the key labeled Break, and then release both keys) cancels the entire entry line. Both the Backspace key and the Esc key back you up either one item or one keystroke in the current entry sequence.

In Lotus 1-2-3, the Escape key (labeled Esc) is the cancel key on IBM PC and compatibles. It cancels the current menu and returns you to the previous menu or submenu that you were in.

In the following sections we give you a brief introduction to commands found in most spreadsheet programs. The list is far from exhaustive; we cover only those that you will need to get started. Others you will have to pick up from built-in help files (discussed later) or from more extensive sources such as the program manuals.

Note: Most of the commands discussed below carry out actions in ranges or groups of cells. To make your job easier, many spreadsheets are written to anticipate where you want the action initiated. You can help make this noble feature work for you by taking the cursor to the cell that will be first involved in the action, prior to invoking the desired command. We reiterate this in a few of the command descriptions below.

Global Settings (Topic 10-4)

On most spreadsheets there is a series of commands that allow you to set features (formats, column widths, protection, recalculation mode, etc.) for the entire sheet. There is always a default set of

```
Worksheet  Range  Copy  Move  File  Print  Graph  Data  Quit
Copy a cell or range of cells
```

[10-13] Initial command menu for Lotus 1-2-3.

features active when you first load a new, blank sheet, but the global commands allow you to override these defaults. The number and type of features that can be set in this manner varies widely, however.

We have given some of the global settings special attention below. These include erasing the entire sheet, setting all column widths, formatting all cells, and setting the recalculation mode.

With the global settings as defaults, features for individual cells, columns, rows, or ranges can be changed with the commands that follow.

Arranging or Sorting the Entries (Topic 10-5)

There is often a need to sort the rows of a spreadsheet based on the entries in a certain column, or to sort the columns based on the entries in a certain row. This is typically done with the *arrange* or *sort commands*. Sorting can usually be done in ascending or descending order, depending on your preference. Alphabetical ordering of textual entries is usually truly alphabetical, but on some spreadsheets it is actually done according to the ASCII codes for the characters. In the latter case, all the uppercase letters are placed ahead of lowercase

Topic 10-4 ▰▰▰▰▰▰▰▰▰▰▰▰▰▰▰▰▰▰▰▰▰ *Lotus 1-2-3*

Global Settings Commands

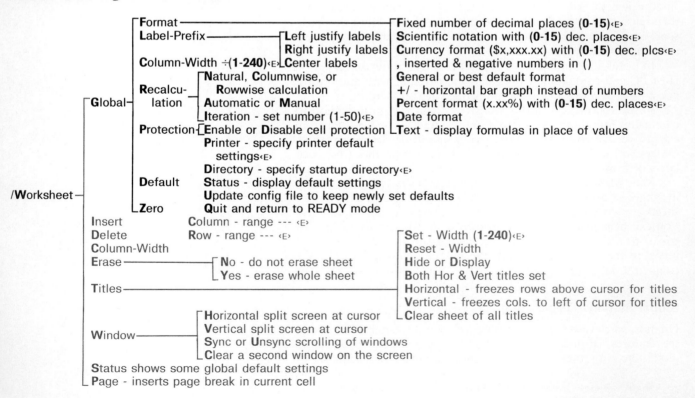

letters, permitting, for example, "Zebra" to be listed ahead of "deAnza."

It is always good practice to save or file a copy of your worksheet (see "Saving and Reloading Worksheets" below) before arranging or sorting, because it may be difficult to get back to where you started once you have invoked the arrange or sort command. Also, formulas that rely in a complicated way on other cells for inputs may give strange results after arranging or sorting.

Blanking or Erasing Cell Entries (Topic 10-6)

When a spreadsheet is first loaded into a computer, none of the cell locations have anything stored in them. Thus, there is no memory (RAM) required in order to store the contents of the empty cells.

As soon as something (e.g., text, numbers, or formulas) is placed in any of the cells, memory (RAM) is then required to retain it. If it is later desired to remove the contents of the cell, the cell can be returned to its original truly empty state with the *blank* or *erase command*. Once blanked or erased, the cell will no longer require memory storage locations for its contents.

Topic 10-4 ▬▬▬▬▬▬▬▬▬▬▬▬▬▬▬▬▬▬▬▬ *SuperCalc4* ▬

Global Settings Commands

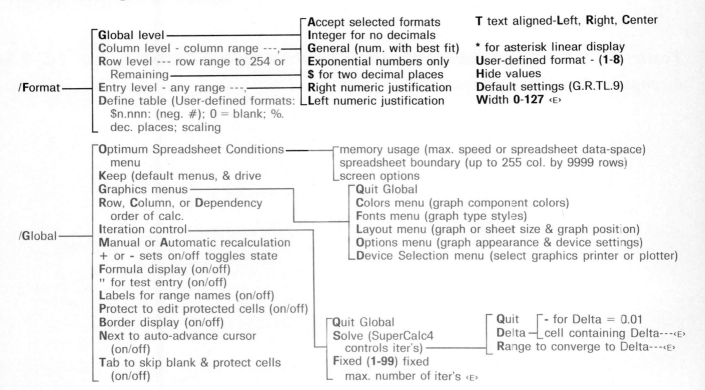

You may think that overwriting the contents of a cell with spaces would return the cell to its truly empty state. However, spaces are one-byte binary characters (ASCII 32_{10}; see Table 8-1) which require memory just like any other character. Therefore, you should always blank or erase any cells you no longer need. Be careful, for once cells are blanked or erased, their previous contents are irretrievable.

Changing the Column Widths (Topic 10-7)

All spreadsheets display a default column width when the program is first loaded into memory. Most often this width is nine characters. However, nearly all spreadsheets allow you to set the width of the columns to the value you desire. Usually, you can do this on a global basis (i.e., change all the columns on the sheet), or you can do this on a column-by-column basis. You cannot change the width of individual cells in a column without changing the width of all the cells in that column.

We narrowed column A in Figure (10-5) to three spaces in order to avoid a lot of blank spaces between the left-justified entry in cell A1 and the right-justified entry in cell A4. Do you remember why A1 is left justified and A4 is right justified?[14]

---

[14]The contents of cell A1 was entered as text which is left justified by default. The contents of cell A4 was entered as a number; numbers are right justified by default. The numbers displayed as results from formula (cells A2 and A5) are also right justified by default.

Topic 10-5 ▬▬▬▬▬▬▬▬▬▬▬▬▬▬▬▬▬▬▬▬▬▬▬ Lotus 1-2-3 ▬

Arrange or Sort Data Command Keystrokes

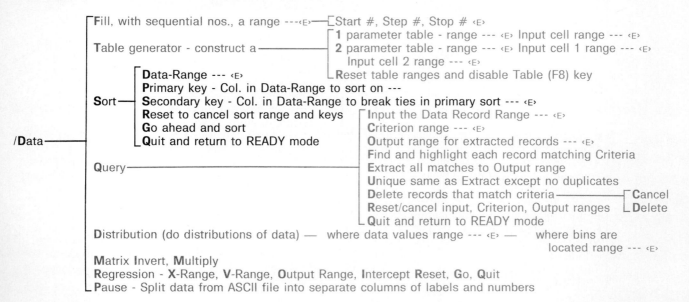

Copying Cell Contents (Topic 10-8)

Of great benefit for making lists of numbers and formulas are the *copy* or *replicate commands*. With these commands you can place in other cells something that you have entered in a single cell. This saves a tremendous amount of time over that required to enter similar contents in many cells individually.

These commands often are programmed to anticipate what cell or cells are to be replicated or copied. To make best use of this anticipation, you should generally place the cursor in the cell you wish to copy before initiating the command.

As we have mentioned previously, you must know whether the cell addresses that exist in the formula to be copied should be copied as a relative cell address or as an absolute cell address. Unless you take specific action, they generally will be copied as relative addresses by default. Since most copying is done in situations where relative cell addresses are desired, this works ideally. For example, assume that we use cell A1 as an input cell for a quantity that was referenced in a formula in cell B1. Now try to visualize copying the formula in B1, down column B into cells B2 through B10, in order to see the output from 10 different inputs we will enter into cells A1 through A10. If the formula $A1^2$ resided in the original cell B1, then after replicating, the formula in cell B2 will automatically be $A2^2$, that in B3 will be $A3^2$, and so on. This is, of course, exactly what we needed, relative cell addresses and not absolute addresses.

Topic 10-5 ▬▬▬▬▬▬▬▬▬▬▬▬▬▬▬▬▬▬▬ *SuperCalc4* ▬

Arrange or Sort Data Command Keystrokes

		row number ---	‹E› for entire row; ascending sort; no adjust							
	Row	‹E› current row	, col. range ---,	**A**scend	**A**djust	**G**o	‹E› primary	row number,	**A**scend
/**Arrange**		col. number	, row range --- ,	**D**escend	**N**o adjust	**O**ptions	, secondary	col. letter,	**D**escend
	Col	‹E› current col.	‹E› for entire column; ascending sort; no adjust						

Editing the Cell Contents (Topic 10-9)

There are two ways to change what is stored in a particular cell. One is to reenter the contents by typing the new value into the current cell status line and actuating the **<Enter>** key. This works fine for short entries but becomes error prone for long entries.

When the entries become long, it is desirable to be able to edit the contents with out retyping the whole entry. This can be done with the *edit command*.

The edit command generally brings the cell contents (label,

Topic 10-6 ▪▪▪▪▪▪▪▪▪▪▪▪▪▪▪▪▪▪▪▪▪▪▪▪▪▪ Lotus 1-2-3 ▪▪▪

Blanking or Erasing (Entire Sheet) Command Keystrokes

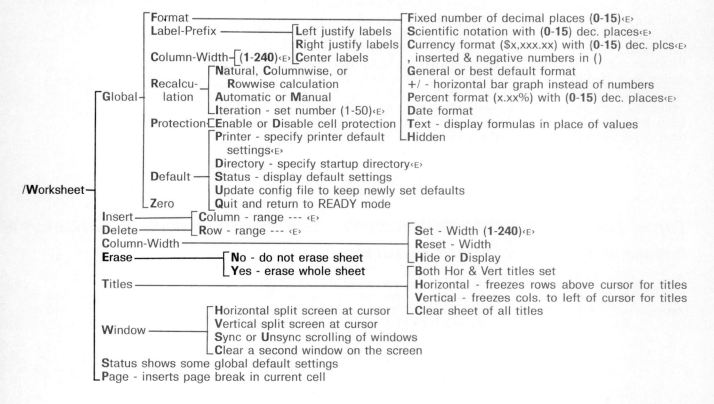

number, or formula) into the command status line so that edit keys can be used to alter them. These edit keys typically include the Cursor Control keys, the Backspace key, the Delete key, and the Insert key.

Formatting Cells (Topic 10-10)

We have mentioned several times previously that worksheet cells can be formatted to alter the manner in which data and labels can be

Topic 10-6 ▰▰▰▰▰▰▰▰▰▰▰▰▰▰▰▰▰▰▰▰▰▰ *SuperCalc4* ▰▰

Blanking or Erasing (Entire Sheet) Command Keystrokes

/**Z**ap ──────┬ **Y**es to delete current spreadsheet; retains settings of Global menus, Output Setup, & Directory
　　　　　　　├ **N**o to cancel this command
　　　　　　　└ **C**ontents, same as Yes, but also retains User-defined format table settings

Blanking or Erasing (Ranges) Command Keystrokes

/**B**lank ──────┬ range --- ‹E›
　　　　　　　├ ‹E› for current cell
　　　　　　　└ graph range

Column Width (All Columns) Command Keystrokes

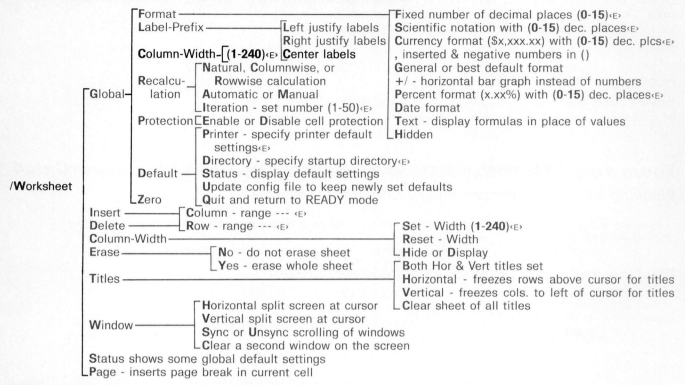

Column Width (Individual Columns) Command Keystrokes

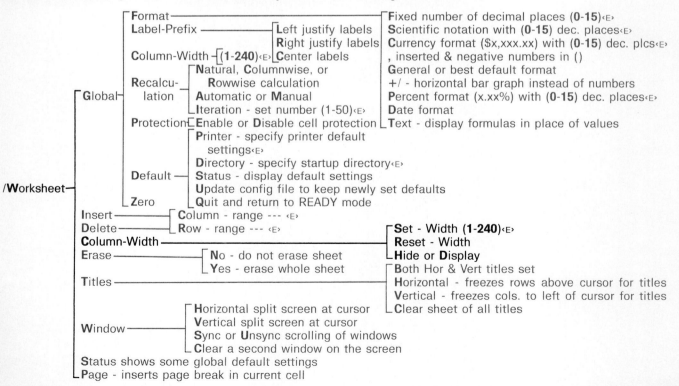

Column Width (All Columns) Command Keystrokes

/Format
- Global level
- Column level - column range ---,
- Row level --- row range to 254 or
  - Remaining
- Entry level - any range ---,
- Define table (User-defined formats: $n,nnn; (neg. #); 0=blank; %; dec. places; scaling

- **A**ccept selected formats
- **I**nteger for no decimals
- **G**eneral (num. with best fit)
- **E**xponential numbers only
- **$** for two decimal places
- **R**ight numeric justification
- **L**eft numeric justification

- Text aligned - **L**eft, **R**ight, **C**enter
- * for asterisk linear display
- **U**ser-defined format - **(1-8)**
- **H**ide values
- **D**efault settings (G,R,TL,9)
- **W**idth of columns - **(0-127)** column width

Column Width (Individual Columns) Command Keystrokes

/Format
- Global level
- Column level - column range ---,
- Row level --- row range to 254 or
  - Remaining
- Entry level - any range ---,
- Define table (User-defined formats: $n,nnn; (neg. #); 0=blank; %; dec. places; scaling

- **A**ccept selected formats
- **I**nteger for no decimals
- **G**eneral (num. with best fit)
- **E**xponential numbers only
- **$** for two decimal places
- **R**ight numeric justification
- **L**eft numeric justification

- Text aligned - **L**eft, **R**ight, **C**enter
- * for asterisk linear display
- **U**ser-defined format - **(1-8)**
- **H**ide values
- **D**efault settings (G,R,TL,9)
- **W**idth of columns - **(0-127)** column width

displayed on the sheet. Nearly all spreadsheets include a *format command* that provides the means for doing this. Formatting can generally be done on a global basis (i.e., to all the cells in the working area[15]), to a group or range of cells, or to any individual cell. Common choices include:

1. Text or labels:
    a. Left justified (usually the default).
    b. Right justified.
    c. Centered.
2. Numbers and formula results:
    a. Right justified (usually the default).
    b. Left justified.
    c. Fixed point rounded to a specified number of places after the decimal point.
    d. Integer (truncated decimal parts).
    e. Floating point (e.g., 3.1E06).
    f. With or without "$" sign.
    g. With or without embedded commas (e.g., 1,000 or 1000).

---

[15] On nearly all spreadsheets it is usually unwise to format all the cells that can be accommodated by the memory, for formatting generally takes up memory space and reduces the maximum number of cells available. It is also unwise to format all the cells that are in the working area, even though the working area is often substantially smaller than the maximum possible sheet. It is good practice to format no more cells than is necessary for clearness and appearance.

Topic 10-8 ▬▬▬   *Lotus 1-2-3* ▬▬

Replicating or Copying Command Keystrokes

/**C**opy   ⎡FROM range --- ‹E› TO range --- ‹E›

Topic 10-9 ▬▬▬   *Lotus 1-2-3* ▬▬

Cell Editing Command Keystrokes

The **F2** function key on IBM PCs and compatibles invokes editing of the current cell.
The Cursor Control keys take you any place in the editing line.
The Backspace key deletes the character to the left of the cursor.
The Delete key deletes the character over the cursor.
Typing any character inserts that character where the cursor is located.

h. With negative numbers enclosed in parentheses.

i. String of *'s, +'s, or −'s replacing the number.

GoTo Command or Key (Topic 10-11)

You may occasionally have a need to jump quickly to a given cell, especially when the given cell is not currently in the window of the screen. This can be done with either the *GoTo* command or the *GoTo key*. Most spreadsheets have one or the other available to aid in your efficient use of large sheets.

Inserting or Deleting Rows and Columns (Topic 10-12)

For problems that are solved repeatedly, it is typical for worksheets to grow from a modest beginning to a very elaborate solution sheet. Quite often it is necessary to add a row or a column in order to better maintain an orderly work space.

Generally an *insert command* is used to add rows or columns within the working area in order to make more cells available. The *delete command* is used to remove rows or columns from the sheet. Most spreadsheets are programmed to anticipate that you will want the row or column added or deleted where the cursor is located. Thus things are made much easier if you place the cursor in the row or column where you want the new one placed before issuing the insert or delete command.

Topic 10-8 ■■■■■■■■■ *SuperCalc4* ■■

Copying Command Keystrokes

```
                   ┌ from range ---, to upper/left cell of destination range ┐ ┌ ‹E› adjust   ┌ No adjust
/Copy ─────────────┤                                                          ┤ ┤              │ Ask for adjust
                   └ from * graph number (1-9), to graph number (1-9) ‹E›     ┘ └ , options ─── │ Values only
                                                                                                └ + - * /
```

Topic 10-9 ■■■■■■■■■ *SuperCalc4* ■

Cell Editing Command Keystrokes

```
/ Edit ────────┬ any cell --- ‹E›
               └ ‹E› for current cell
```

Also,

The **F2** function key on IBM PCs and compatibles invokes editing of the current cell.

The Cursor Control keys take you any place in the editing line.

The Backspace key deletes the character to the left of the cursor.

The Delete key deletes the character over the cursor.

The Insert key acts as a toggle. You must first toggle the Insert key on before you can insert new characters. The new characters are inserted at the cursor location. When the Insert key is toggled off you do not insert, but you *overtype* the characters where the cursor is located.

Topic 10-10 ■■■■■■■■■■■■■■■■■■■■   *Lotus 1-2-3* ■■

Formatting (All Cells) Command Keystrokes

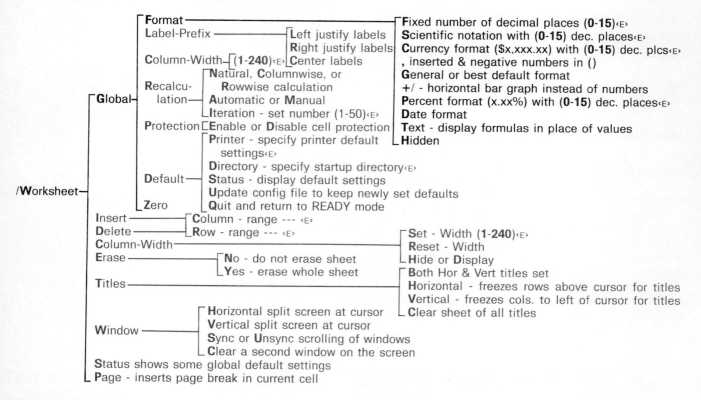

```
                 ┌Format─────────────────────────      ┌Fixed number of decimal places (0-15)‹E›
                 │ Label-Prefix ─────────┌Left justify labels  │ Scientific notation with (0-15) dec. places‹E›
                 │                        ├Right justify labels │ Currency format ($x,xxx.xx) with (0-15) dec. plcs‹E›
                 │ Column-Width─┌(1-240)‹E›└Center labels       │ , inserted & negative numbers in ()
                 │              ┌Natural, Columnwise, or        │ General or best default format
         ┌Global─┤ Recalcu-    │  Rowwise calculation           │ +/ - horizontal bar graph instead of numbers
         │       │ lation──────┤ Automatic or Manual            │ Percent format (x.xx%) with (0-15) dec. places‹E›
         │       │             └Iteration - set number (1-50)‹E›│ Date format
         │       │ Protection─┌Enable or Disable cell protection│ Text - display formulas in place of values
         │       │            ┌Printer - specify printer default└Hidden
         │       │            │  settings‹E›
         │       │            │ Directory - specify startup directory‹E›
         │       │ Default─────┤ Status - display default settings
         │       │            │ Update config file to keep newly set defaults
         │       └Zero────────┤ Quit and return to READY mode
/Worksheet─┤ Insert ─────────┌Column - range --- ‹E›
         │ Delete ─────────└Row - range --- ‹E›             ┌Set - Width (1-240)‹E›
         │ Column-Width                                     │ Reset - Width
         │ Erase ──────────┌No - do not erase sheet          └Hide or Display
         │                 └Yes - erase whole sheet          ┌Both Hor & Vert titles set
         │ Titles ─────────────────────────────────────────┤ Horizontal - freezes rows above cursor for titles
         │                                                  │ Vertical - freezes cols. to left of cursor for titles
         │                 ┌Horizontal split screen at cursor └Clear sheet of all titles
         │ Window ─────────│ Vertical split screen at cursor
         │                 │ Sync or Unsync scrolling of windows
         │                 └Clear a second window on the screen
         │ Status shows some global default settings
         └Page - inserts page break in current cell
```

Formatting (Range of Cells) Command Keystrokes

```
         ┌Format - [see choices under WGF above]; range --- ‹E›
         │ Label-Prefix - Left justify, Right Justify, or Center labels in range --- ‹E›
         │ Erase range --- ‹E›
/Range ──┤ Name────────────┌Create or modify range name┌name or select from list ‹E› range --- ‹E›
         │ Justify range --- ‹E›  │ Delete a range name────┘
         │ Protect range --- ‹E›  │ Labels - create names for   ┌Right. Down.┌range ‹E›
         │                        │  range of labels───────    └Left. Up───┘
         │                        └Reset delete all range names
         │ Unprotect range --- ‹E›
         │ Input data to unprotected cells in range --- ‹E›
         │ Value - allows values in a range to be copied without formulas
         └Transpose - copy a range, switching columns and rows
```

Topic 10-11 ■■■■■■■■■■■■■■■■■■■■   *Lotus 1-2-3* ■■

GoTo Command Keystrokes

On IBM PCs and compatibles
The **F5** function key [you supply the cell address to go to]

Topic 10-10 ▰▰▰▰▰ *SuperCalc4*

Formatting (All Cells) Command Keystrokes

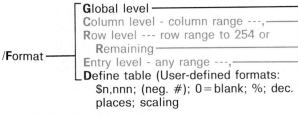

/Format —
- Global level
- Column level - column range ---,
- Row level --- row range to 254 or
- Remaining
- Entry level - any range ---,
- Define table (User-defined formats: $n,nnn; (neg. #); 0=blank; %; dec. places; scaling

- Accept selected formats
- Integer for no decimals
- General (num. with best fit)
- Exponential numbers only
- $ for two decimal places
- Right numeric justification
- Left numeric justification

- Text aligned - Left, Right, Center
- * for asterisk linear display
- User-defined format - (1-8)
- Hide values
- Default settings (G,R,TL,9)
- Width of columns - (0-127) column width

Formatting (Range of Cells) Command Keystrokes

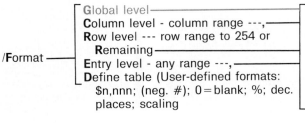

/Format —
- Global level
- Column level - column range ---,
- Row level --- row range to 254 or
- Remaining
- Entry level - any range ---,
- Define table (User-defined formats: $n,nnn; (neg. #); 0=blank; %; dec. places; scaling

- Accept selected formats
- Integer for no decimals
- General (num. with best fit)
- Exponential numbers only
- $ for two decimal places
- Right numeric justification
- Left numeric justification

- Text aligned - Left, Right, Center
- * for asterisk linear display
- User-defined format - (1-8)
- Hide values
- Default settings (G,R,TL,9)
- Width of columns - (0-127) column width

Topic 10-11 ▰▰▰▰▰ *SuperCalc4*

GoTo Command Keystrokes

= [you supply the cell address to go to]

Insert Row or Column Command Keystrokes

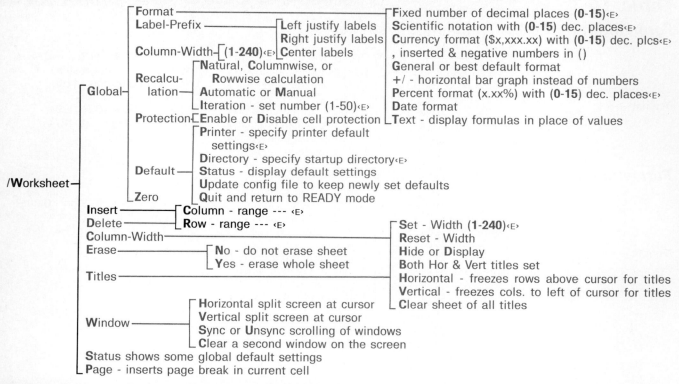

```
                    ┌─Format──────────────────────────────┌─Fixed number of decimal places (0-15)‹E›
                    │ Label-Prefix───────────┌─Left justify labels   Scientific notation with (0-15) dec. places‹E›
                    │                         │ Right justify labels  Currency format ($x,xxx.xx) with (0-15) dec. plcs‹E›
                    │ Column-Width─┌(1-240)‹E›│ Center labels         , inserted & negative numbers in ()
                    │              ┌─Natural, Columnwise, or          General or best default format
                    │ Recalcu-     │   Rowwise calculation            +/ - horizontal bar graph instead of numbers
             ┌─Global│ lation───────Automatic or Manual              Percent format (x.xx%) with (0-15) dec. places‹E›
             │       │              └─Iteration - set number (1-50)‹E› Date format
             │       │ Protection─┌Enable or Disable cell protection └Text - display formulas in place of values
             │       │            ┌─Printer - specify printer default
             │       │            │    settings‹E›
             │       │ Default────│ Directory - specify startup directory‹E›
             │       │            │ Status - display default settings
             │       │            │ Update config file to keep newly set defaults
             │       └─Zero       └ Quit and return to READY mode
/Worksheet───┤ Insert─────────┌Column - range --- ‹E›
             │ Delete──────────Row - range --- ‹E›
             │ Column-Width──────────────────────────────┌─Set - Width (1-240)‹E›
             │ Erase──────────┌─No - do not erase sheet    Reset - Width
             │                └ Yes - erase whole sheet    Hide or Display
             │ Titles─────────────────────────────────────Both Hor & Vert titles set
             │                                             Horizontal - freezes rows above cursor for titles
             │                ┌─Horizontal split screen at cursor  Vertical - freezes cols. to left of cursor for titles
             │ Window─────────│ Vertical split screen at cursor    └Clear sheet of all titles
             │                │ Sync or Unsync scrolling of windows
             │                └ Clear a second window on the screen
             │ Status shows some global default settings
             └ Page - inserts page break in current cell
```

Delete Row or Column Command Keystrokes

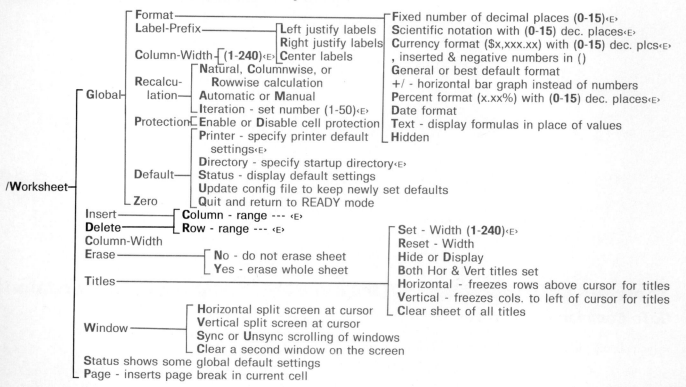

```
                    ┌─Format──────────────────────────────┌─Fixed number of decimal places (0-15)‹E›
                    │ Label-Prefix───────────┌─Left justify labels   Scientific notation with (0-15) dec. places‹E›
                    │                         │ Right justify labels  Currency format ($x,xxx.xx) with (0-15) dec. plcs‹E›
                    │ Column-Width─┌(1-240)‹E›│ Center labels         , inserted & negative numbers in ()
                    │              ┌─Natural, Columnwise, or          General or best default format
                    │ Recalcu-     │   Rowwise calculation            +/ - horizontal bar graph instead of numbers
             ┌─Global│ lation───────Automatic or Manual              Percent format (x.xx%) with (0-15) dec. places‹E›
             │       │              └─Iteration - set number (1-50)‹E› Date format
             │       │ Protection─┌Enable or Disable cell protection  Text - display formulas in place of values
             │       │            ┌─Printer - specify printer default └Hidden
             │       │            │    settings‹E›
             │       │ Default────│ Directory - specify startup directory‹E›
             │       │            │ Status - display default settings
             │       └─Zero       │ Update config file to keep newly set defaults
             │                    └ Quit and return to READY mode
/Worksheet───┤ Insert─────────┌Column - range --- ‹E›
             │ Delete──────────Row - range --- ‹E›
             │ Column-Width                               ┌─Set - Width (1-240)‹E›
             │ Erase──────────┌─No - do not erase sheet    Reset - Width
             │                └ Yes - erase whole sheet    Hide or Display
             │ Titles─────────────────────────────────────Both Hor & Vert titles set
             │                                             Horizontal - freezes rows above cursor for titles
             │                ┌─Horizontal split screen at cursor  Vertical - freezes cols. to left of cursor for titles
             │ Window─────────│ Vertical split screen at cursor    └Clear sheet of all titles
             │                │ Sync or Unsync scrolling of windows
             │                └ Clear a second window on the screen
             │ Status shows some global default settings
             └ Page - inserts page break in current cell
```

Insert Row or Column Command Keystrokes

```
        ┌ Row of empty cells──── row range (one or more empty rows) ---‹E›
/Insert─┤ Column of empty cells─column range (one or more empty columns) ---‹E›  ┌ Right (dislocated data moves right)
        └ Block of empty cells──block range (an empty rectangle) ---‹E› ───────  └ Down (dislocated data moves down)
```

Delete Row or Column Command Keystrokes

```
        ┌ Row ──row range (to delete one or more rows of data) --- ‹E›
/Delete─┤ Column─column range (to delete one or more columns of data) --- ‹E›
        ├ Block──block range --- , ──────────────────────────────  ┌ Left (data to right of deletion moves left)
        └ File──filename (to delete a file from disk) ‹E› or F3 for Directory  └ Up (data below deletion moves up)
```

On most spreadsheets, inserting and deleting rows or columns will automatically alter cell addresses used in formulas so that they reflect the changes in the location of the references caused by the new rows or columns.

Moving Cells (Topic 10-13)

For the same reasons that you may need to add or delete rows or columns to your sheet, you may also want to move existing cell contents to different locations. This is accomplished with the *move command*. Some spreadsheets can move only rows or columns, so check the one you are using.

This command, like the insert and delete command, will generally alter cell addresses used in formulas so that they reflect the changes in the location of these references caused by the move. The command also anticipates what range of cells you will want to move. It helps to have the cursor in the upper left-hand corner of the range to be moved when the command is invoked.

Naming Cell Ranges (Topic 10-14)

Back when we discussed the solution of Equation (10-1) on a spreadsheet, we pointed out the desirability of being able to assign arbitrary names to cells or ranges of cells. Suppose that we were able to give the cell A1 a name of X and then reference that cell in other formulas as X instead of A1. For example, if we were able to assign such a name, Equation (10-2), to be entered into cell A2, could then

---

Topic 10-13   Lotus 1-2-3

Moving Cells Command Keystrokes

/**M**ove ———⎡FROM range --- ‹E› TO range --- ‹E›

Topic 10-14   Lotus 1-2-3

Naming Cell Ranges Command Keystrokes

```
              ⎡Format - (see choices under /WGF above): range --- ‹E›
               Label-Prefix - Left justify, Right Justify, or Center labels in range --- ‹E›
               Erase range --- ‹E›          ⎡Create or modify range name⎤
/Range ——————  Name ——————————  Delete a range name _____⎦—⎡name or select from list ‹E› range --- ‹E›
               Justify range --- ‹E›         Labels - create names for range of labels —⎡Right. Down⎤
               Protect range --- ‹E›        ⎣Reset delete all range names               ⎣Left. Up  ⎦⎡range ‹E›
               Unprotect range --- ‹E›
               Input data to unprotected cells in range --- ‹E›
               Value - allows values in a range to be copied without formulas
              ⎣Transpose - copy a range, switching columns and rows
```

be entered as

$$3*X^2 + 4.6 \qquad\qquad (10\text{-}9)$$

which looks much more like its algebraic counterpart than does Equation (10-2).

As another example, suppose that we wanted to enter into cell B13, the sum of all the numbers in the columnar range B1.B12. In the absence of the ability to name cells or ranges of cells, the formula in cell B13 would have to be

SuperCalc4: SUM(B1.B12)   (10-10)

or

Lotus 1-2-3: @SUM(B1.B12)

Were we able to assign the range B1.B12 a name that might be representative of the contents of those cells, perhaps ENERGY, the formula could then be written as

SuperCalc4: SUM(ENERGY)   (10-11)

or

Lotus 1-2-3: @SUM(ENERGY)

Some spreadsheets (Lotus 1-2-3 and SuperCalc4, in particular) allow the user to give names to individual cells or to cell ranges when desired. Whenever a defined name is used in a formula, the spreadsheet automatically makes the substitution of cells addresses corresponding to that defined name so that data can be found and the necessary computations made.

Topic 10-13 ▬▬▬▬▬▬▬▬▬▬▬▬▬▬▬▬▬▬▬▬▬▬▬▬ *SuperCalc4* ▬
Moving Cells Command Keystrokes

```
             ┌ Block from block range - - -, to range - - -‹E›
Move ────────┤ Row from row range ---, to row number ---‹E›
             └ Column - from col. range ---, to col. letter ---‹E›
```

Topic 10-14 ▬▬▬▬▬▬▬▬▬▬▬▬▬▬▬▬▬▬▬▬▬▬▬▬ *SuperCalc4* ▬
Naming Cell Ranges Command Keystrokes

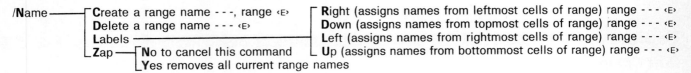

```
          ┌ Create a range name - - -, range ‹E›      ┌ Right (assigns names from leftmost cells of range) range - - - ‹E›
/Name ────┤ Delete a range name - - - ‹E›             │ Down (assigns names from topmost cells of range) range - - - ‹E›
          │ Labels ──────────────────────────────────┤ Left (assigns names from rightmost cells of range) range - - - ‹E›
          └ Zap ─┬ No to cancel this command          └ Up (assigns names from bottommost cells of range) range - - - ‹E›
                 └ Yes removes all current range names
```

Printing Worksheets (Topic 10-15)

All useful spreadsheets provide a means of making hard-copy output from the sheet. This is typically done with a *print command*. The more sophisticated sheets allow you to print headers and footers, choose column and row borders, turn on or off the column and row identifiers, and send control sequences to the printer for special effects.

Often, you may also elect to print to a file instead of directly to the printer. The file can be printed later if desired.

Quitting the Session (Topic 10-16)

Essentially all spreadsheets have a *quit* command that allows you to exit from the spreadsheet and return to the operating system of the computer on which you are operating. These commands nearly always quiz you about your true intentions of quitting as a reminder to you to save your spreadsheet if you have not already done so. The reminders are also welcome when you have accidentally invoked the quit command, for they usually offer you a chance to resume your spreadsheet work instead of quitting.

Topic 10-15   *Lotus 1-2-3*
Printing Command Keystrokes

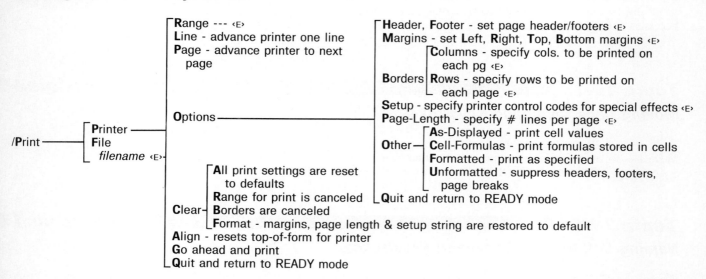

Topic 10-16   *Lotus 1-2-3*
Quit Command Keystrokes

/**Q**uit ———— **N**o - do not end session
                Yes - end session

(Be sure that you have saved your work)

Recalculating the Sheet (Topic 10-17)

As mentioned previously, spreadsheets can calculate all the formulas on the sheet in several ways. You can generally decide when to recalculate, although the default is generally to recalculate automatically any time that one of the entries has changed. You may want to change this default in order to speed up data and formula entry on larger sheets. It can be changed from "automatic" to "manual" or back, with the *recalculation command*.

In the manual mode of recalculation, there is a designated *recalculation key* which, when depressed, will force an immediate recalculation of the sheet. It is very easy to forget to force recalculation when you are using the manual mode. Since your formulas never get updated until you do so, you need to stay alert to this potential problem.

On those spreadsheets that allow the "natural" recalculation mode, this is generally the default scheme. It can generally be changed to row-wise or column-wise recalculation order, although there is little reason ever to do this. With the natural recalculation order, it is often desirable in technical problems that require iteration, to change the number of calculation iterations to be performed

---

Topic 10-15 ▰▰▰   *SuperCalc4* ▰

Printing Command Keystrokes

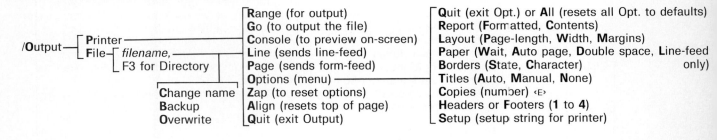

```
                                      ┌ Range (for output)            ┌ Quit (exit Opt.) or All (resets all Opt. to defaults)
                                      │ Go (to output the file)       │ Report (Formatted, Contents)
               ┌ Printer ─────────────┤ Console (to preview on-screen)│ Layout (Page-length, Width, Margins)
/Output ───────┤ File ─┬ filename, ───┤ Line (sends line-feed)        │ Paper (Wait, Auto page, Double space, Line-feed
               │       └ F3 for Directory Page (sends form-feed)       │ Borders (State, Character)              only)
               │            ┌ Change name Options (menu) ─────────────┤ Titles (Auto, Manual, None)
               │            │ Backup    Zap (to reset options)         │ Copies (number) ‹E›
               │            └ Overwrite Align (resets top of page)     │ Headers or Footers (1 to 4)
                                      └ Quit (exit Output)             └ Setup (setup string for printer)
```

Topic 10-16 ▰▰▰   *SuperCalc4* ▰

Quit Command Keystrokes

```
          ┌ No to cancel this command
/Quit ────┤ Yes to exit from SuperCalc4 (does not save current work) ┌ program filename ‹E›
          └ To quit & or load program specified ─────────────────────┴ F3 for Directory options
```

each time the sheet is calculated. This number can be changed (it is usually set to 1 as a default) with the recalculation command.

Saving or Reloading the Sheet (Topic 10-18)

Spreadsheets would not be so widely used if they could not be saved for use at a later time. If all data and formulas had to be reentered each time you wanted to rework the data, spreadsheets would never take the place of programming languages. However, spreadsheet programs generally have *file saving* and *file retrieving commands* for storing and reloading worksheets.

As with most file storage work, you must give a worksheet a name when you store it. This name must be consistent with the naming rules of the operating system you are using.

Topic 10-17   Lotus 1-2-3

Recalculation Mode Command Keystrokes

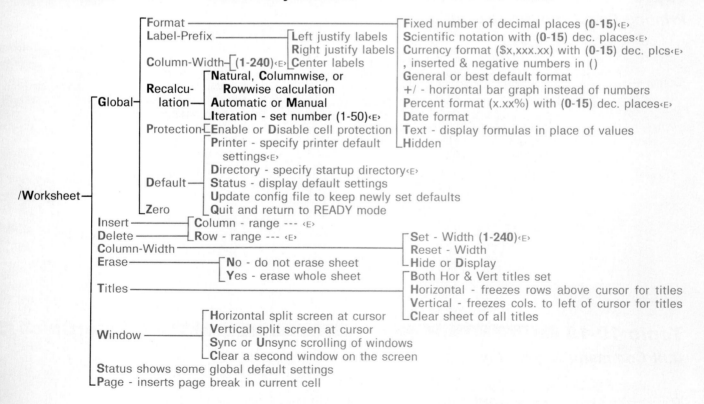

F9 FUNCTION KEY IS THE FORCE RECALCULATION KEY (IBM PC AND COMPATIBLES)

The more sophisticated spreadsheets allow you to store your entire worksheet sheet or only some range or part of the sheet. The default storage format is generally binary and unique to the product you are using. If the binary format is used, it is improbable that you will be able to read or load these files into any other brand of spreadsheet or other computer program. However, most programs allow you to choose other storage formats in case you need to access the data from other programs.

When you retrieve or load previously saved files into the program, you must, of course, know the name under which the file was stored. If it was stored in a format other than binary, you may be required to take special action, depending on the product you are using. We will say more later about loading ASCII files into spreadsheets.

Topic 10-17   *SuperCalc4*

Recalculation Mode Command Keystrokes

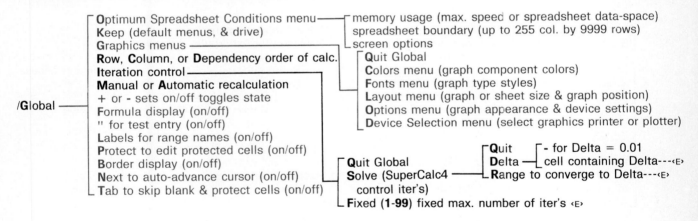

Windows: Opening and Closing (Topic 10-19)

The *window command* allows the user to open a screen window to view two portions of the worksheet at the same time on the screen. This is convenient for large sheets where the input cells are more than a screen away from the output or results cells. With a window open, both the input and output cells can be viewed on the same screen, for example. The window command can also be used to close the window. Additional options under the window command often allow you to link together the scrolling of the windows.

A designated *window change key* allows you to change the location of the cursor from one window to the other.

Topic 10-18   Lotus 1-2-3

Saving and Reloading File Command Keystrokes

File Saving

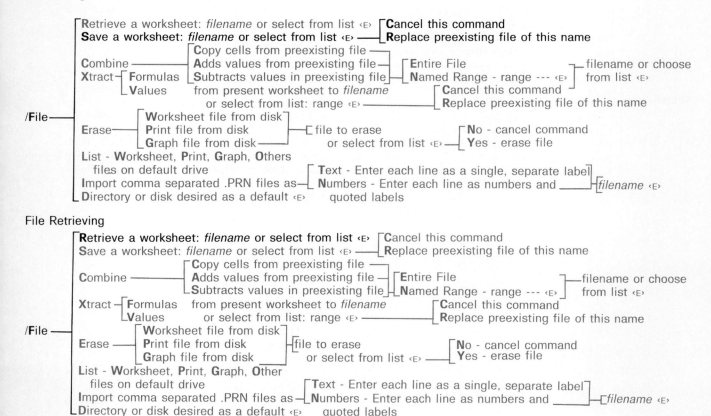

File Retrieving

FINDING HELP ON-SCREEN

When actuated, the *help key* generally brings you a screen full of useful information. It is generally one of the function keys, often the F1 key or the "?" key. The help level is usually between what the terse menu items bring to mind and what you will find in the instruction manual for the program.

Actuated when using one of the commands, it brings information concerning that command. Actuated when in the data entry mode, the help key generally brings you information about entering labels, data, and formulas.

Topic 10-18 █ ─ ─ ─ *SuperCalc4* █

Saving and Reloading File Command Keystrokes

File Saving

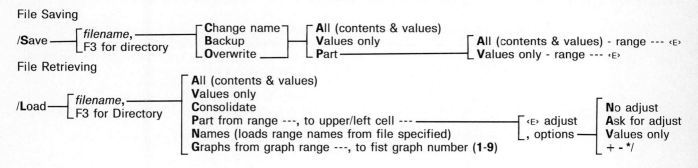

File Retrieving

GRAPHING

Graphing of data from a spreadsheet is, perhaps, second in importance only to the ability to enter formulas in the cells. Graphing allows you to put your eyes to work in interpreting your data. You quickly gain new insights and understanding that are often not possible by just looking at the maze of numbers on the sheet.

Only a few spreadsheets include graphing capabilities as an integral part of the spreadsheet package. Although there are stand-alone graphing packages that can do plotting of spreadsheet files that have been stored, the integral packages make it possible to take quick looks at the numbers displayed on the sheet and to permit quick changes to be made in cells. We will concentrate here on the integral packages. Therefore, this section is less generic than the previous topics. We will give a very concise overview of how the graphing is carried out and then continue the more detailed Topics sections covering graphing in SuperCalc4 and Lotus 1-2-3.

Topic 10-19   Lotus 1-2-3

Window: Open and Close Command Keystrokes

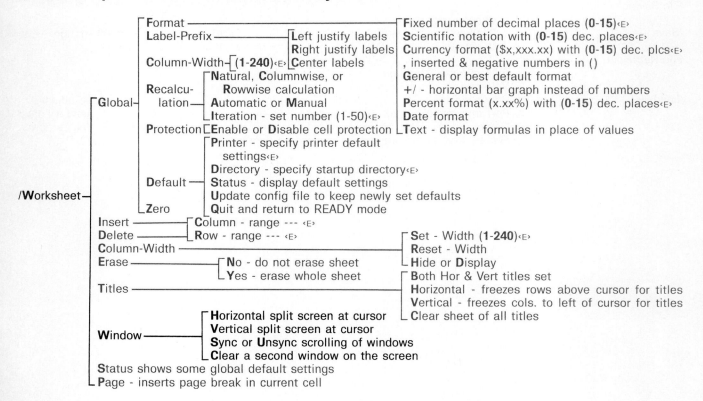

F6 FUNCTION KEY MOVES CURSOR TO NEXT WINDOW (IBM PC AND COMPATIBLES)

The *graphing commands* are typically a part of the Command Mode of operation and can be initiated from the menu accessed with the "/" key. In SuperCalc4, the proper menu item is **View**; in Lotus 1-2-3 it is **G**raph (see Topic 10-20).

Choosing the Type of Graph

The type of graph you wish to plot is typically specified by selection of the proper menu choice under the graph command (the / **V**iew in SuperCalc4 or / **G**raph in Lotus 1-2-3). Most spreadsheets that support graphics include the following types of graphs:

■ *Pie charts:* Each data point of a set is displayed as a slice of pie whose size is proportional to the ratio of that point to the sum of all the data.

■ *X-Y plots:* This is the most common engineering data plot. Each data pair is plotted as a point on the graph. One limitation is

---

Topic 10-19 ▉▉▉▉▉▉▉▉▉▉▉▉▉▉▉▉▉▉▉▉▉▉▉▉ *SuperCalc4* ▉

Window Open and Close Command Keystrokes

```
           ┌ Horizontal split
           │ Vertical split
/Window ───┤ Clear to right or below split
           │ Synchronize split-wise scroll
           └ Unsynchronize split-wise scroll
```

F6 FUNCTION OR ; KEY MOVES CURSOR TO NEXT WINDOW

that multiple data sets must generally all share the same set of independent variables.

■ *Bar charts or line plots:* These two plots are related. They both plot data points at a height above the horizontal axis proportional to the size of the data. Points from the same data set are typically spaced equally along the horizontal axis. The horizontal axis is sometimes referred to as the time line (it is assumed that all data sets share the same intervals along the time line). The bar chart draws bars vertically upward from the horizontal axis to each data point. The line plot connects adjacent data points with straight lines.

■ *Stacked bar charts or area plots:* These are similar to the bar and line charts above, except that common time-line data points from each data set are stacked vertically on top of the previous data set, instead of being plotted adjacent to one another.

■ *Hi-lo plots:* These graphs are similar to bar charts and line plots except that vertical lines are drawn at each time-line data point, but between the maximum data point and the minimum data point—thus the name "hi-low."

Topic 10-20 ██████████████   ██ Lotus 1-2-3 ██

Graphics Command Keystrokes

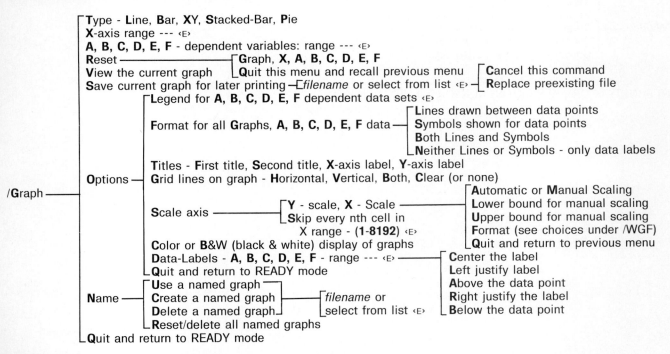

X Is the independent variable for Xy plots specified by range
A, B, C, D, E, F are six sets of dependent variables specified by range.
    Not all are necessary, but you must at least have one set. Pie charts
    are plotted from A; you must have only A defined for pie charts.

Of course, missing from this collection are semilog, log-log, and polar plots (graphing is dicussed in Chapter 11). The log plotting can be accommodated by using the log function to compute the logarithms of the data points you wish to plot. The axis of a graph made with these new logarithms will not be labeled in the conventional manner,[16] but the plot will be a logarithmic one. Polar plots cannot be accommodated.

Choosing the Data to Be Graphed

Most integrated spreadsheet/graphics packages permit one or more sets of data to be displayed in graphical form, depending on the type of graph desired. Pie charts, for example, can display only one set of data, whereas line, bar, and *x-y* plots may accommodate several sets (up to ten in SuperCalc4 and six in Lotus 1-2-3). Each set of data

---

[16]The conventional axes on logarithmic scales are labeled in terms of the numbers whose logarithms are plotted. In the approach to logarithmic plots suggested here for spreadsheets, the logarithmic axes will be labeled in terms of the logarithms of the original numbers.

Topic 10-20 ▰▰▰▰▰▰▰▰▰▰▰▰▰▰▰▰▰▰ *SuperCalc4*

Graphics Command Keystrokes

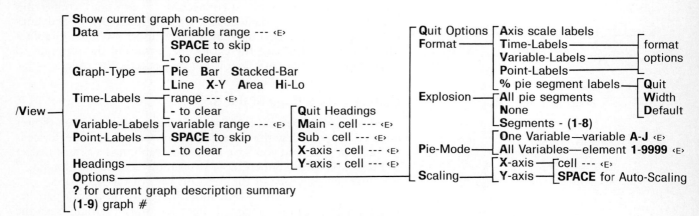

Up to nine graphs can be defined (graph #), with each having (except for pie graphs) up to 10 data sets [A through J, changed by the space bar under/**V**iew **D**ata]. Set A is the independent variable for *xy* plots.

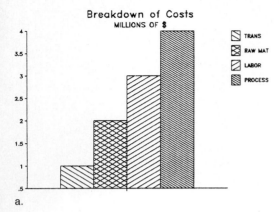

a.

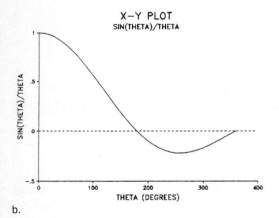

b.

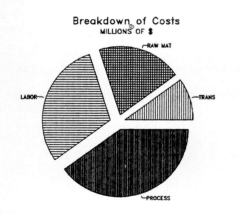

c.

[10-14] Example spreadsheet plots (SuperCalc4).

must reside in contiguous cells that form a rectangle, because they are identified to the program by their range (e.g., C3.C20, or C3.M3, or C3.F21).

Viewing the Graph

Once you have identified the data sets to be plotted and have chosen the type of graph, you simply actuate the *graphing key* to view the graph on the CRT (assuming that you are using a graphics monitor). In both SuperCalc4 and Lotus 1-2-3, the graphing key is a function key[17] on machines that have function keys [in SuperCalc4 it can also be a Ctrl T (i.e., hold down the key labeled Ctrl, push the T key, and then release both keys)]. With or without function keys, the selection of the **S**how item of the **V**iew menu in SuperCalc4 will draw the graph on the CRT. The / **G**raph **V**iew will execute a similar action in Lotus 1-2-3.

Topic 10-20 covers the graphing commands. Example plots are shown in Figure [10-14].

Formatting the Graph

Commonly buried in the graphing commands are the user choices for several formatting features for graphs. These features allow you to add titles, axis labels, legends, and data point labels to your graph. It also allows you to select, for appropriate types of graphs, whether you want data points without connecting straight lines, connecting straight lines without data points, or both data points and lines [these options in SuperCalc4 are chosen under / **G**lobal **G**raphics **O**ptions]. You also have the choice of permitting the program to scale your axes automatically or you may specify the scales.

Getting a Hard Copy of the Graph

Hard copy of graphs can be obtained on any printer or pen plotter which is supported by the package you are using. Once the device is properly in place and you have defined a graph, it is a simple matter to make a plot from SuperCalc4. You simply actuate the plot key,[18] which is generally one of the function keys of the keyboard.

Lotus 1-2-3 requires a little more preparation on your part before hard copy can be obtained. First you must / **G**raph **S**ave the graph, giving it a name when requested. This writes a file with the chosen name and a .PIC extension to storage. You must then **Q**uit the graph menu and / **Q**uit **Y**es the 1-2-3 spreadsheet. This will then put you back into the Lotus Access Program, where you must **P**rint**G**raph **S**elect the .PIC file to be plotted (you will select the one you just saved), followed by **O**ptions **F**onts 1 and selection of the graph name font, followed by **O**ptions **F**onts 2 and selection of the fonts for the balance of the labels on the graph, followed by a **Q**uit of

---

[17] For IBM PCs and compatibles, the F10 function key initiates display of the latest graph on the screen for either SuperCalc4 or Lotus 1-2-3.

[18] On IBM PCs and compatibles, the plot key is the F9 function key or Ctrl Y (i.e., hold down the key labeled Ctrl, push the "Y" key, and then release both keys).

the option menu, followed by a selection of **G**o in the PrintGraph menu. When the graph has plotted, you must **Q**uit the PrintGraph menu to reenter the Lotus Access Program.

LOADING OTHER DATA INTO THE SPREADSHEET

When Is It Useful?

We have just shown that sophisticated spreadsheets can handle large amounts of data. They can do data reduction, statistical analysis, and plotting. Spreadsheets are also fairly fast and are extremely easy to use (we have not yet spent much time on the latter point, but it will be illustrated in the sample problems that follow).

All of the foregoing attributes make spreadsheets desirable for analyzing many kinds of data, whether they be generated and recorded in the field, in the laboratory, or from analysis (possibly on the computer).

How to Do It

A few spreadsheets do not provide for any means of "foreign" data entry except by hand. Fortunately, some of the better spreadsheet programs give you several options for loading "foreign" data, most of which are too lengthy to explain here. However, several spreadsheets, including SuperCalc4 and Lotus 1-2-3, provide an easy way to import ASCII files[19] into their sheets.

The ASCII files should consist of rows of data, with each row ending in a carriage return and line feed (ASCII 13_{10} and ASCII 10_{10}, respectively). Data within each row that are intended for individual cells must be separated from each other by a comma (referred to as "comma delimited"). Data that are intended to be text or label entries must be enclosed in quotation marks "like this".

SuperCalc4 can load CSV files directly into a spreadsheet under the //Import menu item. However, there are several other types of files that can be loaded under this command, including saved Lotus 1-2-3 worksheet files. It is possible to load the desired files anywhere within the present spreadsheet through the //Import main menu item. See the SuperCalc4 Quick Reference guide in the front endpapers of this book for menu choices.

Lotus 1-2-3 makes loading these ASCII files simple if the ASCII CSV file has a .PRN extension. While in the 1-2-3 spreadsheet, you take the cursor to where you would like the upper left-hand corner of the ASCII file data to be loaded and invoke the / File Import commands. You will then be asked to supply the name of the ASCII file to be loaded. If all goes well, you will see your data appear

---

[19] ASCII files consist of all ASCII characters between ASCII codes 32_{10}, and 127_{10}, and includes 8_{10}, 10_{10}, 12_{10}, and 13_{10}, which are the backspace, line feed, form feed, and carriage return, respectively. The absence of escape characters (ASCII 27_{10}) and other control characters makes these files rather inert when transferring them around from program to program or computer to computer.

on the worksheet. Like SuperCalc4, the file is loaded row for row, with each entry in each row being given its own column.

SOME FURTHER SPREADSHEET PROBLEMS

We will cover four additional spreadsheet problems to demonstrate the versatility of the spreadsheet as an engineering tool. The first is intended to demonstrate several commands that we have talked only briefly about in the preceding text. The example shows the construction of tables much like that in Figure [10-1]. We will partially duplicate the table in Appendix III.

Example 10-6

Construct a multicolumn table with the following attributes:

Table title: "DECIMAL INTEREST RATE = ". *Interest rate:* i = ".02".

Column 1: titled "PERIOD", the independent variable, having values starting with 1, ending with 5, and incrementing by 1.

Column 2: titled "F/P" and listing dependent variable $(1 + i)^n$, where n is the period and i is the interest rate.

Column 3: titled "P/F" and listing the reciprocal of column 2.

Column 4: titled "F/A" and listing dependent variable $((1 + i)^n - 1)/i$.

Column 5: titled "A/F" and listing the reciprocal of column 4.

Column 6: titled "P/A" and listing dependent variable $((1 + i)^n - 1)/(i(1 + i)^n)$.

Column 7: titled "A/P" and listing the reciprocal of column 6.

Solution

To give a little more space between the numbers in the table we are about to create, let's expand the column widths to 10 characters for at least columns A through G (seven columns). For SuperCalc4 and Lotus 1-2-3, the keystrokes are (Topic 10-7):

SuperCalc4: / **F C A.G,10 <Enter>**

Lotus 1-2-3: / **W G C 10 <Enter>**

Place the cursor in cell C2 and enter the table header. Place the cursor in cell F2 and enter the number .02. Choose row 4 for entering the column titles, entering PERIOD into cell A4, F/P into cell B4, and so on. If we decide that these labels would look better right justified, we can invoke the following keystrokes:[20]

SuperCalc4: / **F E A4.G4 <Enter> T R A**

Lotus 1-2-3: / **R L R A4.G4 <Enter>**

Now start the table, beginning with the first column. Enter the number 1

---

[20] In Lotus, the " label prefix character could have been used when entering the label. This would have automatically right justified the label in the cell.

in cell A6. Enter the formula A6 + 1 into cell A7. Replicate the formula in A7 into cells A8.A10 by placing the cursor in cell A7 and executing the following keystrokes[21] (Topic 10-16):

SuperCalc4: / **C A7,A8.A10 <Enter>**

Lotus 1-2-3: / **C <Enter> A8.A10 <Enter>**

In either case, note that the formulas are copied using relative (or "adjusted," in SuperCalc3 language) cell addresses. Each cell references the cell immediate above itself. Column A should now contain the numbers 1, 2, 3, 4, 5.

Enter the proper formulas in row 6 of the columns. For example, into cell B6, enter the formula $(1 + F2)$ $^A6$; into cell C6 enter the formula 1/B6; and so on. We have to be extremely careful here because the cells will be referencing a single cell (F2) for the value of the interest rate, i. If we replicate the formulas in cells B6.G6 down their respective columns (that is the easy way to complete the sheet), we will want absolute cell references to cell F2 where the interest rate, i, resides. Cell F2 is not separated by the same relative displacement from all cells. Therefore, relative cell addresses to cell F2 are not appropriate.

SuperCalc4: replicate the formula in B6 to cells B7.B10 by taking the cursor to cell B6 and executing the following commands (Topic 10-8):

/ **B6,B7.B10,A**

Because we added the **A** (for Ask) at the end, the prompt line will ask us which cell references in the formula are to be "adjusted" (considered relative) and which are *not* to be "adjusted" (these will be considered absolute). Answer **N** for reference F2 and **Y** for Reference A6. The column should fill with numbers nearly matching the table in Appendix III.

Lotus 1-2-3: In order to make F2 an absolute reference in cells B6.G6, it should be entered as F2 in the formulas found there. The formula in cell B6 should read $(1 + \$F\$2)$ $^A6$. The formula can be copied to cells B7.B10 by taking the cursor to cell B6 and invoking the following keystrokes (Topic 10-8)

/ **C <Enter> B7.B10 <Enter>**

After copying all the formulas down their respective columns, the table should nearly match the table in Appendix III. However, the appendix table is given to four decimal places, so we should perhaps format cells B6.G10 (Topic 10-10).

SuperCalc4 keystrokes: First we should set up our format selection for general use:

/ **F D**

and change User-defined format #1 to **N**o dollar sign and four decimal places (use the cursor control keys and editing keys to make the changes) and exit that screen with the Control-Break key (or Ctrl Z).

---

[21] In Lotus, a better alternative is provided by the Data Fill commands invoked by the keystrokes

/ **D**ata Fill A6.A10<**Enter**> 1<**Enter**> <**Enter**> <**Enter**>

There is no comparable command in SuperCalc4.

Then initiate

/ F E B6.G10,U 1A

Lotus 1-2-3 keystrokes:

/ R F F 4 <Enter> B6.G10 <Enter>

The resulting table appears as follows (from SuperCalc4):

	A		B		C		D		E		F		G	
1														
2				DECIMAL INTEREST RATE =						.02			
3														
4	PERIOD		F/P		P/F		F/A		A/F		P/A		A/P	
5														
6	1		1.0200		.9804		1.0000		1.0000		.9804		1.0200	
7	2		1.0404		.9612		2.0200		.4950		1.9416		.5150	
8	3		1.0612		.9423		3.0604		.3268		2.8839		.3468	
9	4		1.0824		.9238		4.1216		.2426		3.8077		.2626	
10	5		1.1041		.9057		5.2040		.1922		4.7135		.2122	

In case you did not get a table that matched the above, we have shown the same spreadsheet with formulas being displayed instead of the formula results.[22] Check these formulas with what you have. This is for SuperCalc4. Lotus 1-2-3 will differ only slightly because all references to F2 should read F2 to be absolute cell references, which also applies to SuperCalc4.

	A		B		C		D		E		F
1											
2				DECIMAL INTEREST RATE =				0.02			
3											
4	PERIOD		F/P		P/F		F/A		A/F		P/A	A/P
5											
6	1		(1+F2)^A6		1/B6		(B6-1.)/F2		1/D6		D6/B6	1/F6
7	A6+1		(1+F2)^A7		1/B7		(B7-1.)/F2		1/D7		D7/B7	1/F7
8	A7+1		(1+F2)^A8		1/B8		(B8-1.)/F2		1/D8		D8/B8	1/F8
9	A8+1		(1+F2)^A9		1/B9		(B9-1.)/F2		1/D9		D9/B9	1/F9
10	A9+1		(1+F2)^A10		1/B10		(B10-1.)/F2	1/D10		D10/B10	1/F10

Before we leave this problem, let us explore the graphing of some of the data on the spreadsheet we have just made. We will choose the

---

[22] In SuperCalc4, you can display the formulas stored in the cells by toggling the cell display with the keystrokes **/ G F** (toggling means that if you issue these same keystrokes a second time, you will revert back to results being shown in the cells). A somewhat similar feature is available in Lotus. It is accessed with the commands: **/ Worksheet Global Format Text**. However, regaining the original worksheet is not as convenient as it is in SuperCalc4, if more than one numeric formula is used. A simple solution is to save the sheet, command **/ WGFT**, and then reload the saved sheet after viewing the formulas. Formulas in individual cells or a range of cells in Lotus 1-2-3 can be displayed with the **/RFT** command and reset to values with the **/RFR** command.

data in the last column (G), labeled A/P, to plot as a function of the first column (A), labeled PERIOD.

In SuperCalc4, we set up the plot by defining the data on the sheet to plot and by choosing the type of plot. The command keystrokes are

/ **V D A6.A10 <Enter> D** space bar

G6.G10 <Enter> X F10

The keystrokes /**VD** allow you to enter sets of data, the default of which is set A (this shows right above the prompt line). Set A is the independent variable if an XY plot is requested. The keystrokes above have just defined the data in cells **A6.A10** to be the independent variable set A. In entering the second set of data (the second D in the keystroke sequence), the [space bar] toggles the line above the prompt line to define data set B, which we assign the data in cells **G6.G10.** We then define the **G**raph-Type to be an **X**-Y plot and request it to be displayed with the **F10** function key. The graph should magically appear. Remember what you formerly did to plot a data set?

We can define labels for this plot if we place the labels we wish to use in cells on the spreadsheet. The /**VT,** /**VV,** /**VP,** and /**VH** commands are then used to make the labels in these cells usable on the graph. The /**VO** command also allows the choice of many options. The /**GGO** command also allows more options to choose from, in order to customize the screen to your liking.

In Lotus 1-2-3, graphing is done just as easily, perhaps even more so. The keystrokes are

/ **G X A6.A10 <Enter> A G6.G10 <Enter> V**(iew)

and, like magic, the plot should appear on your screen if you have a graphics monitor and capability. The **X** selects for definition the independent variable from your sheet (we have assigned X to be the data in cells A6.A10). The **A B C D E F** options under the main menu /**G** command allow you to define six sets of dependent variables (in our use we have defined data set **A** to be the data in cells **G6.G10**). The **V**iew command produces the plot on your screen. You can add labels under the /**G**raph **O**ptions command. When you want to quit the **G**raph commands, just select the **Q**uit option.

The next example problem is intended to highlight the use of the IF-THEN-ELSE function that is available on most good spreadsheets.

Example 10-7

Set up a spreadsheet to solve for the roots of a quadratic equation using the binomial theorem.

Solution:

Let the quadratic equation be designated by

$$ax^2 + bx + c = 0$$

The binomial theorem, given by

$$x = \frac{-b \pm (b^2 - 4ac)^{1/2}}{2a}$$

can be used to solve for the roots (the values of x that satisfy the quadratic equation). The number and type (real or imaginary) of roots, of course, depend on the value of the discriminant, $(b^2 - 4ac)$.

Given below is a spreadsheet, with formulas showing in the cells, that solves for the roots. (This is a SuperCalc4 sheet that has been toggled to display the formula.) You should enter the text, numbers, and formulas into the cells in the manner we have covered previously. Only after switching the

/ **G F**

toggle will your sheet look like the following one.

```
!       A      !!    B    !!
1   This sheet solves the quadratic equation:
2     aX^2+bX+c=0   Text
3
4   Enter here      a              1
5   Enter here      b             12
6   Enter here      c              3
7
8   The Discriminant      B5^2-4.*B4*B6
9
10      See below for formula
11    stored here in A10.B11
12
13
14
```

Stored in cell A10 is the formula
    IF(B8 < 0 , ("REAL__PART") , ("X1 = "))
Stored in cell A11 is the formula
    IF(B4 < 1.E−10 , ("1__ROOT") ,)
            IF(B8 < 0 , ("IMAGINARY"),)
                IF(B8 > 0 , ("X2 = "), ("1__ROOT"))
Stored in cell B10 is the formula
    IF(B4 < 1.E − 10 , B6/B5 ,)
            IF(B8 < 0 ,-B5/(2*B4) ,)
        IF(B8 > 0 , (−B5+SQRT(B8))/(2*B4) , −B5/(2*B4))
Stored in cell B11 is the formula
    IF(B4 < 1.E−10 , (" ") ,)
        IF(B8 < 0 , SQRT(−B8)/(2*B4) ,)
    IF(B8 > 0 , B5 ˆ2−4.*B4*B6(−B5−SQRT(B8))/(2*B4) , (" "))

The IF-THEN-ELSE statements in cells A10.B11 get a little complicated, but basically they all have the form

IF (*condition is satisfied, give the function IF this value, else give the function IF this value*)

where *condition is satisfied* is any mathematical expression that in-

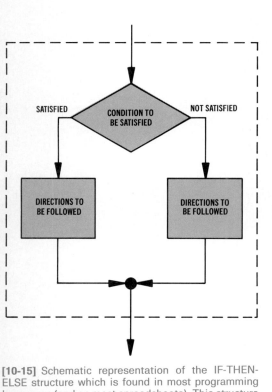

[10-15] Schematic representation of the IF-THEN-ELSE structure which is found in most programming languages (and on most spreadsheets). This structure always makes some test specified by the user (a condition to be satisfied) and, based on the results of the test, executes the directions in the left-hand route if the condition is satisfied *or* executes the directions found in the right-hand route if the condition is not satisfied. In the Lotus 1-2-3 and SuperCalc4 IF statement, the condition to be satisfied and the two sets of directions are placed within the single set of parentheses that follows the IF. The condition to be satisfied is specified ahead of the first comma, the left-hand directions after the first comma and before the second comma, and the right-hand directions after the second comma.

cludes a comparison ($=,<,>,<=,>=,<>$) and *give the function IF this value* and *else give the function IF this value* are formulas or numbers. In SuperCalc4, these can also be text strings that are written to the cell if the strings are enclosed in double quotes and parentheses ("like this")[23] and if they are short ($\leq$ nine characters). In Lotus 1-2-3 (version 2.0), text strings must be enclosed in double quotes; there is no reasonable limit on the length of the string.[24]

In both SuperCalc4 and Lotus, the *give the function IF this value* and *else give the function IF this value* can be another IF-THEN-ELSE. We have spread apart the IFs above for clarity to show the use of such *nested* structures. In the actual cells the statements are all on one line of code and do not need embedded spaces.

The IF-THEN-ELSE statements in column A cause messages to be displayed that tell the user the number of roots and whether the roots are real or imaginary. The statements in column B are used to display the numerical value of the root(s). The pseudocode for the statements in columns B11 and B12 are given below to aid you in understanding what they do. We will let you write the pseudocode for cells A11 and A12.

Cell B10:

If B4 < a very small number $\approx$ 0
   display the single root B6/B5
Else
   If B8 < 0
   Display the real part of complex root
Else
      If B8 > 0
      Display one of two real roots
      Else
      Display the only real root
End of formula

Cell B11:

If B4 < is a very small number $\approx$ 0
   there is no second root - print nothing
Else
   If B8 < 0
   Display the imaginary part of complex
      root
   Else
      If B8 > 0
      Display second of two real roots
      Else
      There is no second root - print nothing
End of formula

The actual sheet in operation looks like the following screen (from SuperCalc4):

Pseudocode is simply a way of organization of one's thoughts on what steps are necessary to program the solution of the problem at hand. It uses English-like statements that have counterparts in most programming languages (and most spreadsheets). The pseudocode for the IF-THEN-ELSE structure, which is available in the IF statement of SuperCalc4 or the @IF statement of Lotus 1-2-3, is given by:

If the appropriate *condition is satisfied* then do something

Else
   do something else

End the if

The *do something* and the *do something else* parts, in the case of SuperCalc4 or Lotus 1-2-3, compute values to be given to the function IF or @IF. Most modern textbooks on programming languages discuss pseudocode.

---

[23] The ()s may be omitted here.

[24] Lotus (version 1A) users would have to replace the text strings that appear as ("text") with a numeric code of some sort and, perhaps, a legend for this code in another cell (e.g., cell A12).

```
|        A        ||   B   ||
1  This sheet solves the quadratic equation:
2     aX^2+bX+c=0
3
4  Enter here    a            1
5  Enter here    b            12
6  Enter here    c            3
7
8  The Discriminant           132
9
10                   X1= -.2554374
11                   X2= -11.74456
12
```

Whenever the value of a, b, or c is changed in cell B4, B5, or B6, respectively, the new roots appear almost immediately in cells B10 and B11.

The following example demonstrates another useful feature of spreadsheets, the table-lookup capability. This feature has many possibilities for use with tabular data that are not easily expressed in equation form.

Example 10-8
Assume that we want to easily determine the minimum-size copper wire that would be needed to carry a known direct current over a given distance with a maximum possible voltage drop. We want to pick a standard-size wire for economical reasons, but standard sizes increase in discrete steps. This means that the resistance of these standard wires will also vary in discrete steps. Expressing the relationship between standard size and resistance in a mathematical relationship is not satisfactory since permissible resistances are not continuous. Therefore, we need to incorporate a table lookup into our solution.

Solution
We use Ohm's law to calculate the maximum resistance that will be required. This relationship is

$$R = \frac{V}{I}$$

where R is the resistance (in ohms), V is the voltage (in volts), and I is the current (in amperes). The resistance per foot of length for wire can be found in standard handbooks, so we will calculate the resistance per length that we will need to match our conditions. That is, we will calculate

$$\frac{R}{L} = \frac{V}{IL}$$

where L is the given length in appropriate units. We then look up this value in a table and let it determine the appropriate wire size.

A possible spreadsheet solution to this problem is shown below.

The spreadsheet is Supercalc4 and the / **G F** toggle has been switched to show the formulas in the cells rather than the results of the formulas.

	A	B	C	D	E
1	This sheet uses Ohm's law to determine			Ohms/Mft	WireGauge
2	the size of wire to use in order to			.09827	0
3	not exceed a given voltage drop over			.1239	1
4	a specified distance, at a given current.			.1563	2
5	Inputs:			.197	3
6	Current (amps)	.5		.2485	4
7	Max Voltage drop	40		.3133	5
8	Distance (specify units below)	20		.3951	6
9	Units,enter 1-ft,2-mi,3-m,4-km	2		.4982	7
10	Outputs:			.6282	8
11	Resistance- Ohms/1000ft *see below for formula*			.7921	9
12				.9989	10
13	Wire gauge needed LOOKUP(B11,D2.D18)			1.588	12
14				2.525	14
15				4.016	16
16				6.385	18
17				10.15	20
18				16.14	ERR
19					

We are using column A to display labels that prompt the user as to what is needed and where the inputs go. Column B is the working column. Notice that columns D and E contain a table, with column D containing the resistance (in ohms) per unit length (per 1000 ft) for standard copper wire sizes, and column E containing the corresponding wire gauge sizes.

There are only two formulas stored in cells on this sheet. One can be found in cell B11 and one in B13. The formula in cell B11 calculates the resistance per foot of wire from the input information on current, length (including units), and maximum possible voltage drop, all of which must be supplied by the user. The formula is given by

IF(OR(B9 < 1,B9 > 4) = 1,ERROR,1000.*B7/B6/(B8*IF(B9 = 1,1,

IF(B9 = 2,5280,IF(B9 = 3,1/.3048,1000/.3048)))))

Notice the use of the nested IF functions (@IF in Lotus 1-2-3) within the formula, just as if they were another constant or variable (see Example 10-7) if you have not done so already.

Pseudocode for this formula is given by the following:

If B9 < 1 or B9 > 4 then
    display ERROR for resistance and wire size
        (inappropriate units have been specified)
Else
    Calculate R/L from V/I/(L*const) where the constant is determined from the input units:
        If B9 = 1
            the const = 1
        Else
            If B9 = 2
                the const = 5280

```
                    Else
                       If B9 = 3
                          the const = 1/.3048
                       Else
                          the const = 1000/.3048
              End of formula
```

If you are using Lotus 1-2-3, you will need to replace OR(B9 < 1,B9 > 4) = 1 with the Lotus equivalent B9 < 1#OR#B9 > 4, which is more conventional in its use of the logical OR function. You will also have to add the @ to the IF function and replace ERROR with @ERR (see Topic 10-3). The built-in help facility in Lotus 1-2-3 is very good (it is reached with the F1 function key on IBM PCs and compatibles). Do not hesitate to use it.

In SuperCalc4, the OR function returns a 1 if either expression in the argument list is met; otherwise, it returns a 0. The other logical functions in SuperCalc4 (see Topic 10-3) behave in a similar fashion. Use the program's built-in help screens to verify the details on each one.

The other formula, located in cell B13, is much simpler looking than the nested IFs. In SuperCalc4, its syntax is LOOKUP *(cell reference, table range),* where *cell reference* is the cell address of the independent variable to be used in entering the table (in our case, the value from cell B11 for the maximum resistance per unit length that can be tolerated). The *table range* second parameter of LOOKUP is the range of cells which defines where the table's independent variables are located. If this is a column range (e.g., D1.D12), the dependent variables are assumed to be located in the next column over (here, E1.E12). If *table range* is a row reference (e.g., A20.L20), the dependent variables are assumed to be in the next row down (correspondingly, A21.L21).

LOOKUP searches for the last value in the range of numbers that is less than or equal to the independent variable given (the *cell reference*) and returns the adjacent value to the right of the search column or below the search row. This is exactly what we want in this example problem, since we need the wire with the next smallest resistance.

The comparable LOOKUP function is not available in Lotus 1-2-3. However, both SuperCalc4 and Lotus 1-2-3 allow a more sophisticated table structure than the SuperCalc4 LOOKUP function permits. In both programs there are two table-lookup functions: VLOOKUP and HLOOKUP (precede these with the @ sign for 1-2-3). They both have arguments of *(cell reference, range, offset).* As implied in the names, or @VLOOKUP in 1-2-3 is used for vertical tables where the dependent and independent variables are both in columns. For HLOOKUP (or @ HLOOKUP in 1-2-3, both variables must be horizontal in rows.

The argument *cell reference* gives the location of the unique independent variable with which you want to enter the table in order to find the corresponding dependent variable (for our example it would be cell B11). The argument *range* encompasses at least the independent and dependent variables that constitute the table (our example would need a *range* of D2.E18). The *offset* is the number of columns (or rows) to move over (or down) from the independent variable column (or row) to find the dependent variable (in our problem it would be one).

Lotus 1-2-3 will permit text strings to be used for either the independent variables or the dependent variables. There is no need to enclose the text strings in quotes or parentheses. SuperCalc4 will also permit the use of textual values for either variable as long as the textual values are enclosed in both () and "", like ("this") and the textual strings contain no more than nine characters [excluding the () and ""].

The sheet below shows the working version with values showing, not formulas. Any time one of the four input variables is changed (cells B6 through B9), the resistance value (cell B11) and wire size output (cell B13) are immediately updated.

	A	B	C	D	E
1	This sheet uses Ohm's law to determine			Ohms/Mft	WireGauge
2	the size of wire to use in order to			.09827	0
3	not exceed a given voltage drop over			.1239	1
4	a specified distance, at a given current.			.1563	2
5	Inputs:			.197	3
6	Current (amps)	.5		.2485	4
7	Max Voltage drop	40		.3133	5
8	Distance (specify units below)	20		.3951	6
9	Units,enter 1-ft,2-mi,3-m,4-km	2		.4982	7
10	Outputs:			.6282	8
11	Resistance- Ohms/1000ft	.757575757576		.7921	9
12				.9989	10
13	Wire gauge needed	8		1.588	12
14				2.525	14
15				4.016	16
16				6.385	18
17				10.15	20
18				16.14	ERROR
19					

You should experiment with the table limits. That is, input values for current, voltage, and distance that yield values for the resistance per length to exceed the table limits. Also, note what happens when numbers less than 1 or greater than 4 are entered for the units key.

The table columns do not have to be located close to the cells that use the LOOKUP command. In our example we could have moved them to columns that were not in the window of the screen. There are two other ways of hiding a table in either SuperCalc4 or Lotus 1-2-3 (version 2.0). We could specify the width of the table columns (here, D and E) to be 0 (zero) in SuperCalc4 or the column-width to be hidden in Lotus 1-2-3. We could also format them with the Hide option in either program.

The last spreadsheet problem involves the solution of an iterative problem. Such problems are often encountered in engineering and it is important to understand how they can be solved on a spreadsheet. The key is in initializing the problem to get the iteration going and then determining when it has converged to an answer sufficiently close to the correct one.

Example 10-9
    Set up a spreadsheet to solve

$$x^2 - 4 = 0$$

Solution

This equation can, of course, be solved by factoring, the solution being $x = \pm 2$. However, we will use this simple example to demonstrate iteration on the spreadsheet using Newton's method to find the roots of equations.

Newton's method tells us that if x_{old} is an existing approximation to the root of the equation

$$f(x) = 0$$

then a new and better approximation for x can be obtained from the relationship[25]

$$x_{new} = x_{old} - \frac{f(x_{old})}{f'(x_{old})}$$

For our $f(x)$ [where $f'(x) = 2x$]

$$x_{new} = x_{old} - \frac{x^2 - 4}{2x}$$

If we can make a first guess for x_{old}, this relationship allows us to calculate a better value for x. Then using this for x_{old}, we can get an even better value for x by applying the equation for x_{new} again.

The trick in doing iteration problems on the spreadsheet is to work in the initial guesses to start the calculations. Once an iteration has started, spreadsheets with iterative capability can keep it going, but there must be a way to initiate the process. This initiation can be done with the IF function statement. We show below a SuperCalc4 iteration sheet for this problem. Formulas are displayed in the cells where they are stored.

We have decided to use named ranges in coding this problem. That is, we have assigned names to various cells and have used these names in formulas that reference the named cells. For example, we have assigned the name "GUESS" to cell B2, the cell that contains our first guess. This naming of cell B2 is done by first taking the cursor to that cell and initiating the following keystrokes:

SuperCalc4: / **N C GUESS,**‹E›

Lotus 1-2-3: / **R N C GUESS,**‹E›

Had we not named cells B1, B2, and B7, the formula in cell B5 would have been IF(B1=0,B2,B7) instead of IF(FLAG=0,GUESS,XNEW).

```
CIRC|     A       ||              B              |  Cell    Name
 1  |  Iteration flag 0                          |  ← B1    FLAG
 2  |     First guess 100                         |  ← B2    GUESS
 3  |Converg Tolerance .0001                      |
 4  |                                             |
 5  |         xold IF(FLAG=0,GUESS,XNEW)          |  ← B5    XOLD
 6  |         xnew +XOLD-(XOLD^2-4)/(2*XOLD)      |  ← B6    XNEW
 7  |      counter IF(FLAG=0,0,COUNTER+1)         |  ← B7    COUNTER
 8  |                                             |
 9  |                                             |
10  |                                             |
11  |                                             |
```

Note here that the following range names have been defined

Cell	Name
← B1	FLAG
← B2	GUESS
← B5	XOLD
← B6	XNEW
← B7	COUNTER

Note also that the range names have been used in the formulas instead of the cell addresses

---

[25] Here $f'(x)$ is the derivative of $f(x)$ with respect to x. If you have not had a course in calculus, please accept this on faith.

Column A of this SuperCalc4 sheet is being used to prompt the user and contains various labels or text. Column B is where the work gets done.

There are three user-supplied numbers in column B. (On a Lotus 1-2-3 worksheet, the convergence tolerance that is requested in cell B3 cannot be used effectively. Thus, this cell would be left blank in Lotus 1-2-3.) Notice that the first guess for x is entered into cell B2. This value of x is then used in cells B6 to calculate a better value of x.

Cell B1 contains what we will call the *iteration flag,* which the user gives the value of either 0 or any other number. If this flag is 0, the iterative problem is initialized and made ready for the iterative solution.

Cell B13 contains an IF function (@IF in Lotus 1-2-3) that examines the value of the iteration flag (B1). If the flag is 0, the value of cell B5 is set to the first guess of x (from cell B2). After the flag in B1 gets set to 1 or any value other than zero, cell B5 takes its value from cell B6, which, as we shall see, holds the result of the last updated x.

Of course, this reference of cell B6 by cell B5 is a circular one that must be resolvable by the spreadsheet you are using. SuperCalc4 informs you that there is a circular reference on the sheet by displaying the CIRC in the upper left-hand corner (on Lotus 1-2-3 the warning appears in the lower right-hand corner). SuperCalc4 will allow you to resolve circular references, as will Lotus 1-2-3. Many spreadsheets will not.

During each calculation of the sheet in the natural calculation mode, the calculations proceed downward and to the right. When a circular reference is encountered (cell B5 references cell B6 and cell B6 references cell B5), the cell nearest the upper left-hand corner of the sheet is calculated first. Therefore, when the FLAG is set to 1, cell B5 takes its value from B6, which holds the best approximation that was computed during the last sheet calculation. Then cell B6 is computed and returns a value that is even better than the value in cell B5 (which holds the value that B6 produced during the previous calculation.) If the values in B5 and B6 continue to change with each calculation, the problem has not converged and more iterations must be done.

Now, in operation, you initialize the sheet by setting the iteration flag to 0 (you may have to push the Recalc key, which is the F7 function key in SuperCalc4 or the F9 function key in Lotus 1-2-3, depending on whether the recalculation mode is set to manual or automatic). This places the guess for x (from cell B2) in cell B5 and calculates a better estimate of x for display in cell B6.

When the iteration flag is set to 1 (or any value other than zero), the sheet has a guess and is free to iterate. How the iteration proceeds, however, is determined by you, the user. Both SuperCalc4 and Lotus 1-2-3 programs can be set to either automatic or manual mode of recalculation with a fixed number of iterations—experiment with this by trying different numbers of iteration, starting with a small number and increasing it if needed. The command keystrokes for setting this are (Topic 10-15):

SuperCalc4: / **G [M** or **A]** / **G IF** [supply number] **<Enter>**
Lotus 1-2-3: / **W G R [M** or **A]** / **W G R I** [no.] **<Enter>**

To operate the sheet, enter a guess for the root in cell B2 and enter a 0 in cell B1. If you are on SuperCalc4 and want the sheet to stop iterating automatically when your specified convergence tolerance is reached, enter the convergence tolerance in cell B3 (we will say more about this later). The sheet should reset itself (you will need to push the Recalc key if you are in the manual mode of recalculation). If the sheet does not reset itself, you probably have a bug in the program which you will have to fix. Carefully examine all formulas.

If the sheet did reset itself, enter a 1 in B1. The numbers in B5 and B6 should change[26] (you will need to push the Recalc key if you are in the manual mode of recalculation). If the numbers do not change, you have an error to debug. If they did change, push the Recalc key to watch them converge even closer to the true answer (if possible). The sheet will undergo another set of iterations every time you push the Recalc key, even when you are in the automatic mode.

SuperCalc4 will automatically iterate until the convergence criteria you specify has been met if you carry out the following command keystrokes:

/G I S D B3, R B5 <Enter> Q

With this "solve" mode on, you should be able to reset the sheet (B1 = 0), initiate the iteration (B1 = 1), and watch as the iteration occurs. The sheet will automatically stop iterating when the value in cell B5 changes by less than the tolerance set in cell B3. We hope you are not in an infinite loop—if you are, use the Control-Break key to abort.

Correct sheets for guesses of +100 and −50 are given below for your reference. The first sheet shows after initialization, while the second sheet shows the results after iteration. Set the number of iterations to 1, and watch this converge iteration by iteration.

The sheet after initializing and before iterating with a guess of +100:

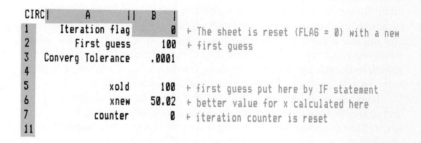

```
CIRC|       A      ||   B   |
 1    Iteration flag        0    ⊢ The sheet is reset (FLAG = 0) with a new
 2       First guess      100    ⊢ first guess
 3    Converg Tolerance   .0001
 4
 5              xold      100    ⊢ first guess put here by IF statement
 6              xnew    50.02    ⊢ better value for x calculated here
 7            counter        0    ⊢ iteration counter is reset
11
```

The sheet after converging:

```
CIRC|       A      ||   B   |
 1    Iteration flag        1    ⊢ The iteration is started (FLAG = 1)
 2       First guess      100    ⊢ first guess is no longer used but
 3    Converg Tolerance   .0001    ⊢ the tolerance is used to check convergence
 4
 5              xold        2    ⊢ previous XNEW is put here by IF statement
 6              xnew        2    ⊢ better value for x calculated here
 7            counter       10    ⊢ solution converges in 10 iterations
```

---

[26] SuperCalc4 will display the numbers as they are being recalculated. Lotus 1-2-3 will only display the numbers at the end of the fixed number of iterations you have told it to complete. It displays a WAIT sign in the upper right-hand corner of the sheet while calculating.

The sheet after converging from an initial guess of −50:

```
CIRC|      A       ||  B   |
1        Iteration flag        1
2            First guess      -50
3      Converg Tolerance     .0001
4
5                  xold       -2  ← another root is found by resetting the
6                  xnew       -2  ← sheet with another guess and reiterating
7               counter        9  ← solution converges in 9 iterations
```

Note the solution has converged to another root (−2) than that obtained with a guess of +100.

Cell B8 contains a counter that records the number of iterations completed. The IF statement (@IF in Lotus 1-2-3) causes this cell to display a zero if the sheet is being reset (B1=0). However, if the sheet is undergoing iteration (B1 not equal to 0), this cell displays one more than it did during the last iteration and thus counts the iterations. Since it references itself, it alone constitutes a circular reference. Notice from the spreadsheets displayed above that it took 10 iterations to converge to a root of 2 from an initial guess of 100 and 9 iterations to converge to a root of −2 from an initial guess of −50.

Our purpose in presenting this material on spreadsheets has been to demonstrate their versatility and to give you some working knowledge of their capability. We have stressed here only small and relatively simple problems in order not to cloud the learning process. However, once you learn the fundamentals, you will be able to solve much more complicated problems.

PROBLEMS

10-1. Duplicate Table 10-1 by programming a spreadsheet.

10-2. Duplicate Figure [10-1], using the spreadsheet that you used in Problem 10-1 if it has graphics capability. If your spreadsheet does not have graphics capability, do you have access to any plotting programs that can load and plot data stored in a spreadsheet file? If so, plot a graph that is as similar to Figure [10-1] as possible.

10-3. Enter your social security number on a spreadsheet in cell B2 (R2C2 for Multiplan).

10-4. Enter the following numbers into a columnar range of cells.

3.0

3

1,238,876

0.0000586

10.568952

Copy this column of cells into the next five columns to the right. Then format each column for one of the following formats.

a.   Integer.
b.   Decimal with zero places after the decimal.
c.   Decimal with three places after the decimal.
d.   With embedded commas.
e.   As dollar values to the penny and with dollar signs.
f.   In scientific notation.

10-5. Enter the following formulas into the cells of a spreadsheet.

a. $y = mx + b$.
b. $y = 3(x + 2)^2$
c. $y = 3x^2 + 2z$
d. $x_1 = x_2 + vt + at^2/2$
e. $E = mc^2$
f. $F = ma$
g. $E = IR$
h. $y = \sin x$
i. $z = e^{y-2}$
j. $y\text{-}\tan x/\cos x$

10-6. Explore the behavior of your spreadsheet.
a. What happens to the value of $\tan x$ when x approaches 90°?
b. What happens to the value of $\arctan x$ when x becomes large and positive? when x becomes large but negative?
c. Does the square root function give the correct answers for all sizes of numbers, both large and small?

10-7. Enter a string of numbers in a column on your spreadsheet but leave one cell in the string blank. In the cell immediately below the last number in the string, enter the built-in function that averages this string of numbers (include the blank cell in the range to be averaged). Has the formula included the blank cell into the average? That is, was the blank cell averaged in as a zero, or was it ignored?

Enter into the cell below the one holding the average, the built-in function for counting the number of entries in a range of cells. With this function, count the number of entries in the string of numbers you have entered, including the blank cell. Was the blank cell counted or was it ignored?

10-8. Enter the formulas below into a spreadsheet and evaluate them for a few choices of independent variables.

a. $f = \dfrac{c}{2}\sqrt{\left(\dfrac{p}{l}\right)^2 + \left(\dfrac{q}{w}\right)^2 + \left(\dfrac{r}{h}\right)^2}$

[This formula is used in acoustics for finding the normal modes of resonance (f) in a rectangular room ($l \times w \times h$). c is the velocity of sound. p, q, and r are nonnegative integers representing different modes of resonances.]

b. $p = P + \frac{1}{2}\rho V^2\left[1 - \left(\dfrac{v}{V}\right)^2\right]$

(This formula is used in aerodynamics.)

c. $t_1 = \dfrac{R_t - R_0}{R_{100} - R_0}(100) + \delta\left(\dfrac{t_2}{100} - 1\right)\dfrac{t_2}{100}$

(This formula is used in thermometry.)

10-9. Write expressions for the *condition* of an IF function that will check for the items listed below.

a. $i = 0$
b. i greater than or equal to 1
c. i greater than 0 (If i is INTEGER, does this differ from part *b*?)
d. $z < 17.3$
e. $(a + b)/2 \geq \sqrt{ab}$
f. $z_{min} \leq z \leq z_{max}$
g. x and y are within 0.005 of each other
h. k_6 is not more than seven counts larger than k.
i. k_6 and k are not more than seven counts apart
j. The relative difference between e^2 and $\cos(3a + b)$ does not exceed 1 per cent of e^2.
k. $0 \leq a \leq a_{max}$ and $0 \leq b \leq b_{max}$
l. $x > x'$ or $y > y'$
m. a within 0.05 of a_{max} but a_{max}
n. q does not have a fractional part

10-10. Enter some numbers in cells B2 and B3 of your spreadsheet. In cell B4 enter a formula that will display twice the value of the number in B2 if B2 is greater than B3. Otherwise, the value displayed in B4 should be 0.5 times the value of B3.

10-11. Obtain all real roots of the following equations on a spreadsheet using Newton's method. Use a convergence criteria of 0.0001. (The derivatives are shown in [ ].)

a. $f(x) = 7.3x^3 - 20x^2 + 14x - 3 = 0$
    $[f'(x) = 21.9x^2 - 40x + 14]$
b. $f(x) = 6x^4 - 2x - 3.6 = 0$
    $[f'(x) = 24x^3 - 2]$
c. $[f'(x) = 4 + 6.3x^{-1} + 4.32x^{-2} - 7.91 \times 10^2 x^{-3} = 0$
    $[f'(x) = -6.3x^{-2} - 8.64x^{-3} + 23.73 \times 10^2 x^{-4}]$

10-12. Use the data in Table 11-1 to construct a table on a spreadsheet. Program the sheet so that when you enter a wire gauge in the range of the first column into some cell on the sheet, an appropriate value of cross-sectional area is returned in another cell.

10-13. (For spreadsheets that can load comma-delimited ASCII files with graphing capability.) Using an editor or word processor, create a comma-delimited file that has the following two lines of data:

14.5, 16., 17.5, 19., 20.5, 22., 23.5, 25.
1, 2, 3, 4, 5, 6, 7, 8

Store this file as an ASCII file (editors typically do this; most word processors can do this if specifically instructed), and load this into your spreadsheet. Assuming that the first row contains the dependent variable and the second row contains the independent variable, plot these data by pairing up the points using the first point in the first line with the first point in the second line, the second point in the first line with the second point in the second line, and so on.

Chapter 11

Principles Used in Engineering Analyses

THE NEED TO DOCUMENT ANALYSES _____

There is a subset of the infamous Murphy's "laws"[1] that seemingly governs bureaucracies. Two members of this subset are cited here in order to give you an idea of their "tongue-in-cheek" flavor: (1) massive expenditures obscure the evidence of bad judgments (known as the First Bureaucratic Bylaw), and (2) a system that performs a certain function or operates in a certain way will continue to operate in that way, regardless of need or changed conditions (you may recognize this as Newton's law of systems inertia).

Another important one of these "laws" that relates more particularly to the material in this chapter states: "A memorandum is written not to inform the reader, but to *protect* the writer." Called Acheson's rule of the bureaucracy, this statement implies that effective communication is less important than protecting the job of the bureaucrat.

Although this may seem to violate all engineering codes of ethics, engineers occasionally do need to obey this "law." They often write for protection reasons. This statement is made without hesitation and without tongue-in-cheek. We will provide some clarification below.

The documentation in writing of ideas, analyses, and results has a number of purposes other than personal protection. Foremost among these purposes is communication—informing others of the author's thoughts, decisions, progress, and conclusions. In our rapidly changing and complex technology, seldom will an entire project be carried out from conception to completion by a single person. The team approach is crucial for much of the nation's industries. There may be bankers to be persuaded, fellow workers (including,

[11-1] Why me?

---

[1]There is some evidence to suggest that Murphy was a development engineer at Wright Field (Ohio) Aircraft Lab. In 1949, being frustrated with a measuring device that was not working because it had been incorrectly wired by a technician, he was reportedly heard to say, "If there is any way to do it wrong, he will." Over the years, this rule has been generalized to: "If anything can go wrong, it will." It has also been expanded into a large set of corollaries. As Dickson has written, "It has been suggested that Murphy . . . helped more people get through crises, deadlines, bad days, the final phases of projects, and attacks by inanimate objects than either pep talks, uplifting epigrams, or the invocation of traditional rules. It is true that if your paperboy throws your paper into the bushes for five straight days it can be explained by Newton's law of gravity. But it takes Murphy to explain why it is happening to you" (see Paul Dickson, "The Official Rules of Engineering," *New Engineer,* December 1978, p. 56).

perhaps, your boss) to be informed, patent attorneys to be educated, and perhaps even licensing boards to be convinced. Lack of communication among any of the team members who are working on the project is almost always catastrophic.

More and more industrial communication is being handled through computerized electronic mail or through the use of digital data bases (rather than by exchanging "hardcopy"). However, the form of the documentation is not the important feature. The content and ease of assimilation (reading and understanding) are the important features. There must be an *effective* transfer of information!

Documentation is also necessary for protection, as has been pointed out previously. However, unlike the bureaucrat, who may write to protect his job, engineers write mainly to protect their credibility and their ideas. Patents, for example, are never granted for ideas that exist only in an inventor's mind. Ideas must be documented in hard copy; otherwise, legal protection for original ideas cannot be guaranteed. This does not mean that formal patent applications must be filed on every idea, but at the very least, the inventor's personal written log should describe each idea, its date of conception, and the identities of witnesses.

For internal documentation and communication, many companies (and engineering professors) require engineers to keep a written log or engineering notebook. When this log is properly used, all ideas, notes, derivations, progress, and meeting minutes are recorded in it. Neatness and conciseness are not required, but completeness, in the sense that the record conveys enough information so that a knowledgeable reader can understand the intent, is paramount.

An occasional review of this information often restimulates old ideas and or rejuvenates a stagnant project. Companies often require that each filled workbook be placed on file within the company archives.

Often, engineers must be able to document particular information in order to establish and/or protect their professional credibility. For example, if you were requested to serve as an expert technical witness in a court case, you would generally be asked to document your credentials. Your testimony, either given in the courtroom or through depositions, would be scrutinized as much for its ability to establish your competence and credibility, as for its ability to inform the court on matters pertinent to the case.

Engineering students are no different than practicing engineers in their need to communicate effectively. Their success or failure in their educational pursuit depends on their ability to express themselves through the written word. Faculty, like clients of practicing engineers, have great difficulty separating those students who do not understand a subject from those who understand it but cannot properly express themselves.

THE DETAILS OF DOCUMENTATION: PRESENTATION, UNCERTAINTIES, RESOLUTION, AND MISTAKES

Presentation

When shopping in your favorite grocery store or drugstore, did you ever notice the emphasis placed on product packaging? It is a fact that marketing success often depends on how attractive the product appears, rather than on just functionality alone. The failure of the Edsel and the success of the Mustang automobiles [11-2 and 11-3] cannot be explained by the fact that the Edsel was a more comfortable car and offered much more interior space than the Mustang. The difference in marketing appeal was focused on the Mustang's exterior "packaging" and sporty appeal, not on the relative merits of the two interiors.

Unfortunately, packaging (i.e., the method of presentation) plays a strong role in the acceptance of technical documentation. But fortunately, this desirable attribute of attractiveness in packaging usually manifests itself somewhat differently in the case of technical analyses than for commercial products. The readers of technical literature are not looking for the racing stripe down the side or the color that just matches their stockings. If they were looking for fantasy and excitement they would be reading Ian Fleming's books about James Bond.

The astute technical reader is looking for *conciseness, clarity, and accuracy*. These attractive features "sell" technical literature. It should not always be necessary for the reader to rederive the formulae, reperform the calculations, or replot the data in order to fill in missing pieces and to understand the work. On the other hand, the reader should not have to be carried through every step of the problem in order to comprehend the concept.

There must be a middle ground. The key to finding it is for engineers to understand their purpose in writing and to know the level of their readers.

There are many documentation formats that can provide the structure for a technical document. Most large companies require a standard format for their reports; many smaller companies do not. However, all successful formats have some similar features, some of which are:

1. An abstract (a very short, less than one page, description of the full report).
2. A statement of the problem or subject being addressed.
3. An explanation of the problem solution or conclusions reached.
4. A description of the significant features of the solution or findings.
5. The supporting data.

Next we wish to explore some of the ways of presenting data and some of the problems that often arise in the manipulation of numbers and the performance of calculations.

[11-2] The Edsel—a marketing failure.

[11-3] The Mustang—a marketing success.

A single photographic reproduction is said to have the same intrinsic value as one thousand eloquent articulations.

Translation: A picture is worth a thousand words.

A rotating fragment of mineral collects no bryophytic plants.

Translation: A rolling stone gathers no moss.

If it is necessary to undertake to complete a task, one should strive to complete the job in an exemplary manner.

Translation?

It is impossible to form an authoritative opinion concerning a hardbound compilation of literary information by mere observation of its protective bindings.

Translation?

Tables. A compact and efficient method of presenting data is in the form of *tables* of numbers. As with equations, tabular data have independent and dependent variables. In Table 11-1 gauge number is the *independent variable* since the other quantities listed in the table are commonly determined (calculated or measured) for each gauge. The other quantities (cross-sectional area and ohms per kilometer) are known as *dependent variables*.

Tables with only one independent variable (such as Table 11-1 below) are sometimes referred to as *single-parameter* tables. Similarly, tables with two independent variables are referred to as *two-parameter* tables. Tables in which there are three independent variables are usually represented as a series of two-parameter tables. Each table in such a series might give data for two of the variables, but each table would represent a different value of the third variable. Tables for data with four or more independent variables are rarely encountered.

Tables are particularly useful for data in which no interpolation is required (where only discrete values of the independent variable are possible) or for data in which the precision of the numbers would be compromised by plotting in graphical form (discussed in the next section). Table 11-1 contains physical and electrical data for copper

TABLE 11-1 WIRE TABLE FOR STANDARD ANNEALED COPPER [AMERICAN WIRE GAUGE (BROWN & SHARPE) SOLID CONDUCTOR; TEMPERATURE = 20°C]

Wire Gauge	Diameter (mm)	Cross-Sectional Area (mm^2)	Resistance (ohms/km)
0000	11.6840	107.219	0.1608
000	10.4038	85.011	0.2028
00	9.2659	67.432	0.2557
0	8.2525	53.488	0.3224
1	7.3482	42.409	0.4065
2	6.5430	33.624	0.5128
3	5.8268	26.665	0.6463
4	5.1892	21.149	0.8153
5	4.6203	16.766	1.0279
6	4.1148	13.298	1.2963
7	3.6652	10.551	1.6345
8	3.2639	8.367	2.0610
9	2.9058	6.631	2.5988
10	2.5883	5.261	3.2772
12	2.0526	3.309	5.2100
14	1.6276	2.081	8.2841
16	1.2908	1.309	13.1759
18	1.0236	0.823	20.9482
20	0.8118	0.518	33.3005

wire in the standard wire sizes in the range from 0000 to 20 gauge. Since, for economic reasons, wire of standard size is almost always used, there is usually no interpolation required when using this table.

Tables should always be organized for the ease of use of the reader. This means, above all, that only fundamental data should be presented—data that can be useful in many different ways, not just in an apparatus similar to the one that you have chosen for your experiment. For example, if you were measuring the viscosity of a lubricating oil as a function of temperature by measuring the time it takes for the oil to flow out of a container (the Saybolt technique for viscosity determination), the measurements you might make are temperature and time intervals. However, no one except you is really interested in the time intervals you measure; what is much more useful are viscosities that are implicit in the time intervals measured. You need to use a "data reduction" scheme to deduce viscosity from your measurements so that you could tabulate the fundamental quantity viscosity versus temperature.

Other rules that aid the user are:

1. Align the decimal points in each column of similar numbers.
2. Use scientific notation for large numbers.
3. If users will commonly interpolate from the table, choose convenient intervals for tabulating the independent variables. Intervals of 1, 2, 5, 10, or multiples thereof, are preferred.
4. Label all data with the names of the quantities being tabulated and their units.
5. Title the table with enough information to make clear its contents and limitations.
6. Place extended information about such things as limitations and sources of nonoriginal data in footnotes to the table.

Graphs.   The use of tables to display data is only one way of presenting information to prospective users. Often, graphical presentation is a superior method of showing data that might otherwise be placed in a table. Sometimes, graphical and tabular methods complement each other and the inclusion of both is useful.

Graphs are presentation forms that allow the data to be expressed as "pictures" in which lengths or areas are used to depict the size of the numbers. When displayed in this way, lengths and areas are much more quickly assimilated than are collections of numbers. The old adage "A picture is worth a thousand words" is certainly true in this case.

Types of Graphs.   A graphical display of information may take any of several forms, depending on the type and use of the information to be presented. Often, *pictographs* provide for the most rapid interpretation to the widest possible audience. They are used routinely in newspapers and on television. An example is shown in Figure [11-4].

Pie charts are ideal for presenting data that represent various parts of a whole or total entity. Such charts allow the user to understand quickly the contribution of each part to the total. In Figure

North Carolina

Georgia

South Carolina

California

Virginia

Wyoming

Leading States in
the Quarrying of Granite
Each symbol represents
1,000,000 tons

[11-4] Pictograph.

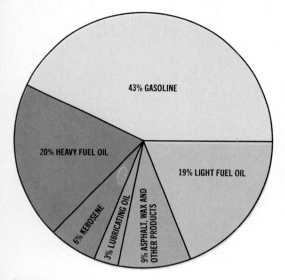

[11-5] Pie chart (data expressed as parts of a whole).

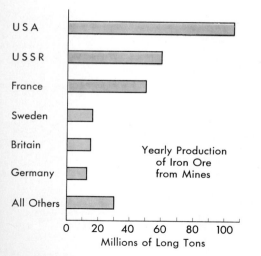

[11-6] Horizontal bar graph.

[11-5], you can readily see from the sizes of the slices of the "pie" that gasoline is by far the largest product in the refining of crude oil.

Bar charts are graphs in which data for the dependent variable are shown as bars whose lengths are proportional to the size of the data. In Figure [11-6], for example, *country* is the independent variable while the *amount of iron ore* is the dependent variable. We can see that the United States produces almost twice as much iron ore as does its next closest competitor.

A *histogram* is a special type of bar chart in which the dependent variable is usually the *number of "things"* that fall into "ranges" (often called *bins* or *cells*) shown on the independent variable axis. In the histogram of Figure [11-7] the vertical bars represent the *number of light bulbs* (expressed as a percentage of those tested) that failed in various ranges of operating hours. The largest fraction, here 24 percent, failed within 800 to 900 hours of operation although some, 1 to 2 percent, lasted more than 1600 hours. Figure [11-7] readily gives you information on how long you might expect a light bulb from the particular group tested, to last. You can see that histograms are convenient for showing statistical data.

The bulk of engineering data are presented using *line graphs* or *x-y* plots. Such graphs are usually more exact and better suited for use in interpolation, extrapolation, and investigation of data trends. In *linear* axis or *linear* scale line graphs, the distance along each axis is proportional to the numbers representing the data [11-8 and 11-9]. In *semilogarithmic* or *semilog* plots, one axis of the plot (the linear one) is *proportional to the numbers* representing one of the variables, while the other axis (the logarithmic one) is *proportional to the logarithm of the numbers* representing the other variable [11-10]. *Log-log* plots are graphs in which both axes are *proportional to the logarithms of the numbers* representing the respective variables [11-11]. On any logarithmic axis, it is *not* conventional to label the axis with the value of the logarithm of the numbers plotted, but rather to label the axis in terms of the data numbers being plotted (which are the antilogs of the logarithm numbers). This fact always makes the logarithmic axes appear nonlinear [11-10 and 11-11] and more difficult to read; that is, equal increments on logarithmic axes do not correspond to equal intervals for the variables being plotted. For example, Figure [11-10] shows that the first major division along the horizontal logarithmic axis represents a change from 10^4 to 10^5 loading cycles while the next major horizontal division represents a change from 10^5 to 10^6, considerably more.

Semilog plotting, where only one variable is given a log scale, is convenient in cases where

1. the variation of the data is such that it may be desirable to compress the larger values of one variable (for example when one variable spans a large range of numbers), or

2. there is an exponential behavior of the data.

To illustrate the latter phenomenon, let us consider data that satisfy the following form of equation:[2]

$$y = e^x$$

---

[2]Similar analyses apply to the equation $y = 10^x$.

Distribution of failures of 75-watt lamps
operated at rated voltage

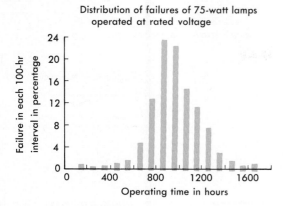

[11-7] Vertical bar graph.

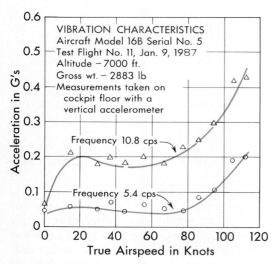

[11-9] Linear scale graph.

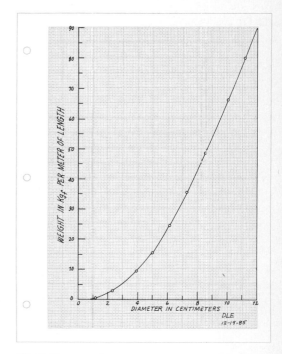

[11-8] Linear scale graph.

Taking logarithms (to the base 10) of both sides yields

$$\log_{10} y = \log_{10} (e^x)$$

or

$$\log_{10} y = x \log_{10} e = 0.4343x$$

A plot of (log y) versus (x) would be a straight line. *Thus exponential relationships plot as straight lines on semilog graphs and, conversely, straight lines on semilog graphs signify exponential relationships.*

Log-log plotting, where both variables are assigned log scales, is convenient in the following types of cases:

1. The variation of the data is such that it may be desirable to compress values of both variables (e.g., when both variables span large ranges of numbers).

2. The data satisfy algebraic equations containing power relationships or roots.

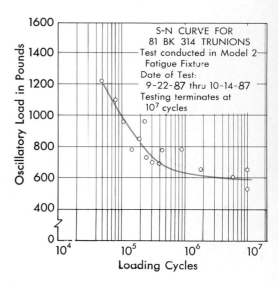

[11-10] Semilog plot.

[11-11] Log-log plot.

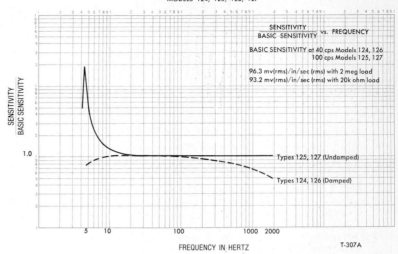

To illustrate the latter phenomenon, consider data that satisfy the following form of equation:

$$y = \alpha\, x^n$$

where α and n are constants. Taking logarithms of both sides yields

$$\log y = \log \alpha + n \log x$$

The latter equation represents a straight-line relationship between $(\log y)$ and $(\log x)$. The line would have a slope of (n) and a y intercept of $(\log \alpha)$ in the $(\log y)$ versus $(\log x)$ plane. *Thus power relationships between variables plot as straight lines on log-log graphs and, conversely, straight lines on log-log graphs demonstrate that there are power relationships between the variables.*

Each power of 10 that can be plotted on a logarithmic axis is said to be a *cycle*. For example, Figure [11-10] is a three-cycle semilog plot since three powers of 10 (10^4 through 10^7 loading cycles) or three orders of magnitude in the number of loading cycles can be accommodated on the horizontal axis.

Polar graphs are often used where a variable quantity is to be examined with respect to various angular positions. Examples are the light output of luminous sources, the response of microphone pickups, and the behavior of rotating objects at various angular positions. An example of a graph plotted on polar coordinate paper is shown in Figure [11-12].

Graphing Practice.   Much graphing is now done using application programs on computers. For example, most spreadsheets (discussed in Chapter 10) have graphing capabilities. In addition, there are many special applications graphing programs for micros, minis, and mainframes. Depending on the software package being used, the user may or may not be able to exercise much control over the plots. However, there are some important rules that need to be followed to promote ease of use. In some cases, this may mean that a new set of

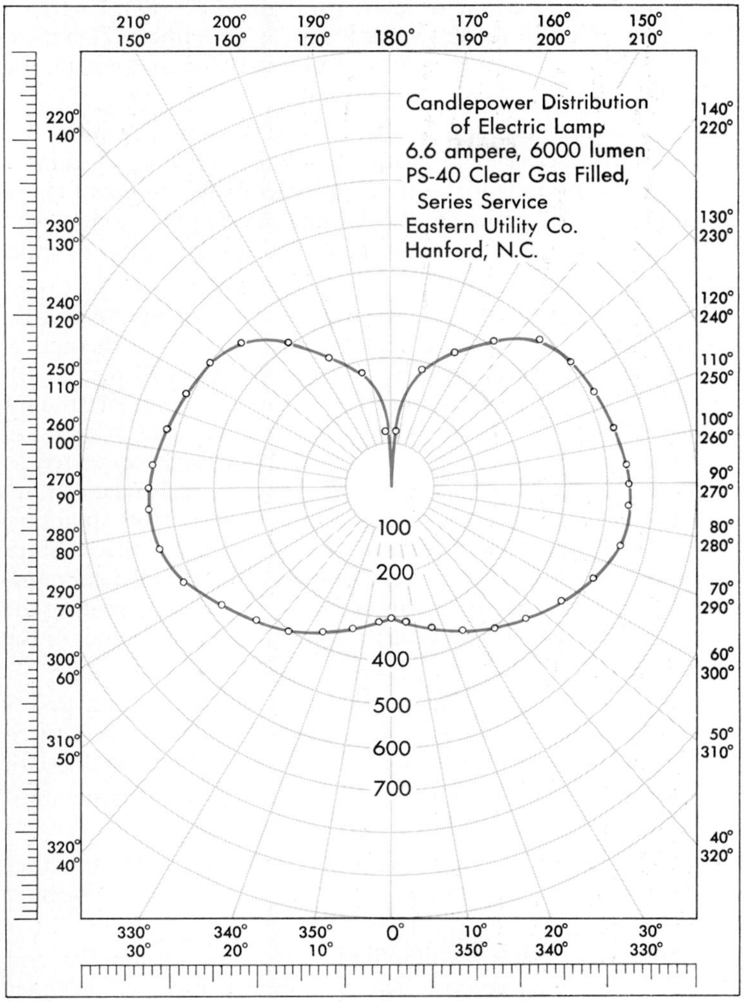
Candlepower Distribution
of Electric Lamp
6.6 ampere, 6000 lumen
PS-40 Clear Gas Filled,
Series Service
Eastern Utility Co.
Hanford, N.C.
100
200
400
500
600
700
180°
0°

[11-12] Polar graph.

axis labels will have to be added to a machine-made graph. The important rules are as follows:

1. It is customary to plot the independent variable along the abscissa (the horizontal axis) and the dependent variable along the ordinate (the vertical axis).
2. The scale for each axis must be suitable for the paper and the data being used. On linear plots, scale divisions of 1, 2, 5, 10, or a multiple of these numbers, are *mandatory* for ease of use. Never use a scale that requires awkward interpolation. For example, labeling points along an axis as 21.07, 23.27, 25.47, and so on, does not promote ease of use. For logarithmic axes, label the powers of 10 [11-10 and 11-11], and perhaps some intermediate values if space permits.
3. On linear axes, it is desirable to show zero as the beginning of the ordinate and abscissa axes unless this would compress the data unnecessarily. If compression is severe, show zero at the origin and use a broken axis line to alert the reader that the data

are compressed (see the ordinate on Figure [11-10]). Logarithmic axes never go to zero since log(0) is undefined. Therefore, choose the origin to be the lowest power of 10 that will permit plotting of the data.

4. The points for each data set being plotted should be identified by using distinctive identification symbols such as small squares, triangles, diamonds, or other simple geometric figures [11-9]. Distinctive linework such as solid line, dashed line, long dash–short dash line also may be used to aid identification of independent data sets.

5. Graphs may be drawn for theoretical relationships, empirical relationships, or measured relationships. Curves of theoretical relationships will not normally have point designations [11-8]. Empirical relationships should form smooth curves or straight lines, depending on the form of the mathematical expression used [11-9]. Measured data points that can*not* be expected to have an underlying mathematical or empirical relationship, should be connected by straight lines drawn from point to point. Otherwise, data obtained from measured relationships should be drawn to average the plotted points (called curve fitting) [11-9 and 11-10]. In this situation, curves representing the measured data do not necessarily go from center to center of the points.

6. Curves used to approximate the behavior of the data points *should not* be drawn through the symbols that enclose the plotted points, but rather stop at the perimeter of the symbols [11-10].

7. Exponential notation should be used when labeling large numbers on the abscissa and ordinate.

8. The abscissa and ordinate variables should be labeled together with their respective units of measure: for example, *Weight in Newtons*.

9. The title of the graph should refer to the ordinate values first and the abscissa values second. The title should also include other descriptive information, such as sizes, weights, names of equipment, date that the data were obtained, location where the data were obtained, serial numbers of the apparatus, name of manufacturer of apparatus, and any other information that would help describe the graph.

10. The title should be placed on the sheet where it will not interfere with the curve.

11. The name (or the initials) of the person preparing the graph and the date the graph was plotted should be placed on the sheet [11-8].

Uncertainties

It should be obvious that the diameter of a saucepan and the diameter of a diesel engine piston, although each may measure about 15 cm, will usually be measured with different accuracies. Similarly, a measurement of the area of a large ranch that is valued at $0.05 per square meter would not be made as accurately as a measurement of a piece of commercial property that is valued at $10,000 per square meter. Given enough resources (equipment and money), we could increase the accuracy of almost any measurement, but there are many situations that simply do not merit such effort. There is almost always, however, a need to know the approximate accuracy in any measurement we make. The concepts of uncertainty and significant figures are often used to express the accuracy of single measurements.

Uncertainties are the differences between observed measurements (or calculated numbers) and the true values of those measurements (or numbers). Since physical quantities cannot be measured exactly, the true values cannot be known. The engineer's idea of uncertainty, then, is that of an inaccuracy, where complete accuracy is impossible. The word "error" is often used as a synonym for uncertainty in science and engineering literature, but since it is often used as a synonym for mistake or blunder in layperson's conversation, it is preferable not to use the term at all.

Realizing that inaccuracies are a fact of life, we attempt to assess the amount of uncertainty we might have in any measurement or in any calculation that makes use of measurements. We try to establish *bounds* on the uncertainty, a *lower bound* and an *upper bound* between which the true value is presumed to lie. For example, we might say that the measurement is 4.173 ± 0.002 meters to indicate an uncertainty of 0.002 meter. In this case, the true value is presumed to be bracketed between the upper bound of 4.175 meters (4.173 + 0.002) and a lower bound of 4.171 meters (4.173 − 0.002). Bounds are sometimes given as a percentage of the quantity: 57 ± 3 percent liters. Here the bounds are said to be *relative bounds*. The uncertainty in this case is 3 percent of 57, or about 1.7 liters and implies an upper bound of about 58.7 and a lower bound of about 55.3.

More commonly in engineering calculations, the uncertainty is implied by the manner in which the value itself is stated (with no $\pm$ quantities attached). The usual practice is to retain no more than one questionable digit for any measured or calculated quantity. The uncertainty then is taken to be one of the units which the questionable digit represents (e.g., 1, 10, or 100), or plus or minus one-half of the unit. If the 6 in 4026 is the questionable digit, we normally would say that the uncertainty is 1, or ± 0.5, since the 6 represents the number of 1's in 4026. We will amplify on this technique in the discussion of significant figures below.

The uncertainties or inaccuracies that appear in results arise from three causes. One cause is the uncertainties that exist in the original data used in the calculation. These always cause uncertainties in the final output and, for most scientific and engineering work, are the most troublesome.

Another cause of uncertainties in the output of any calculation is the numerical technique used. Finally, the third cause is related to the limit on the number of digits that can be retained during any computation. We discuss the latter two sources of error briefly in a later section on numerical and rounding errors.

Significant Figures.   A *significant figure,* or *significant digit,* is any *digit* within a number that may be considered to be reliable as a result of measurements or mathematical computations. For example, if the digit 3 within the number 143,580 representing some measurement, truly represents the number of thousands in that measurement, then 3 is said to be a significant figure. Owing to the way in which measurements are made, if the digit 3 in the number 143,580 is a significant figure, then the 1 and the 4 are also significant figures. As you proceed from left to right in a number (e.g., from the 3 to the 5 to the 8, etc.), you will eventually encounter a digit which is doubtful due to the way that it was measured or determined (due to the accuracy of an apparatus or an instrument or of human precision).

All the digits in a number, from left to right, up to and including the doubtful one, are said to be significant figures. Therefore, if the 8 in 143,580 is the doubtful digit, then this number is said to contain, or be accurate to, five significant figures. What this means is that the actual value, if it could be determined, should be somewhere between 143,575 and 143,585. Note that the 8 is really in doubt. Note also that once we are certain of the doubtful digit, we will have just found another way of expressing the uncertainty in the number. That is, 143,580, with the knowledge that 8 is the doubtful digit, is equivalent to writing the absolute bounds as $143,580 \pm 5$, or the relative bounds as $143,580 \pm 0.0035\%$.

In the number 12.43, there are four significant figures if 3 is the doubtful digit, while in 0.123456, there are six significant figures if 6 is the doubtful digit.

For numbers in which the doubtful digit is located to the right of the decimal point, it is misleading to add further digits (including zeros) to the right of the doubtful digit. To do so violates a standing tradition of using the doubtful digit as the last digit and hides the true accuracy of the number. Hence 12.43, by tradition, implies that the number is accurate up to the doubtful digit of 3, whereas 12.430 implies that the 3 is a reliable number and the 0 is the doubtful digit.

For numbers in which the doubtful digit is located to the left of the decimal point, we cannot omit the zeros that may lie between the doubtful digit and the decimal point. Thus the number 987,600. does not make clear which is the doubtful digit. It could be the 6 or it could be either of the two zeros. In such a case we can write the number in scientific notation with the doubtful digit shown to the right of the decimal point as the last digit in the number (as in 9.876×10^5). This same condition can be requested on spreadsheets by formatting the appropriate cells for scientific notation.

To avoid decreasing the accuracy of a calculated set of numbers, a programmer usually assigns the outputs more digits than are actually significant based on the uncertainties of the inputs. Thus the outputs often look more accurate than they really are. We shall return to this problem in our discussion on resolution.

Resolution

Calculators and computers have the capability of retaining many digits of *resolution* in the numbers they handle. Most calculators will accept the entry of numbers with up to 10 digits. The resolution of numbers accepted by computers depends on the word size and on whether the number is integer or decimal. For decimal numbers, the smallest number of digits accepted by computers is usually 7 and the largest, 15.

It is often very easy to be lulled into believing that all the digits that can be displayed are always significant. But there is a very important distinction between resolution (the ability to display many digits) and accuracy (the requirement that all digits used are accurate or meaningful).

For example, if we were to multiply 2.2×1.1 on most hand-held calculators, we might observe results that vary from 2.42 to 2.42000000. Using the discussion above as a reference in determining the doubtful digit, these results would imply anything from three significant figures to nine significant figures. Let us explore the accuracy of this calculation for a moment.

Based on the lower bounds or the uncertainty implied in the numbers 2.2 and 1.1, the smallest product that we could expect would be 2.15×1.05 or 2.2575. Similarly, using the upper bounds, the largest product that we might expect would be 2.25×1.15 or 2.5875. From this it is apparent that the doubtful digit in the product is the one immediately following the decimal point. Thus the original product of 2.4200 . . . has a doubtful digit of 4, giving the number only two significant figures. The product written as 2.4 implies the correct uncertainty, since the 2 is accurate and the 4 is doubtful.

Products and Quotients.   Products and quotients have no more significant figures than those contained in the least accurate number used as an input. For example, in the multiplication of two numbers, one having five significant figures and one having only two, the product will have only two significant figures.

Sums and Differences.   Sums and differences are no more accurate than the doubtful digit among the numbers being added or subtracted that represents the largest unit (1's, 10's, 100's, etc).

Example 11-1

If 301×10^3 and 4028 are being added, the number representing the thousands (1000's) in the final result will be doubtful since the number of thousands in the first number is in doubt. Therefore, the sum, 305,028, has 5 as its first doubtful digit going from left to right. This sum should, by convention, be written as 305×10^3 in order to imply no more than its true accuracy.

Example 11-2

If you owned 5.01×10^3 acres of land and purchase 0.1 acre more to obtain frontage on a highway, you are implying a false accuracy to say that your total acreage is now 5010.1. Your holdings after the acquisition are still *somewhere* between 5005 and 5015 acres, just as they were previously. Of course, you could pay to have your land surveyed more accurately so that you can make legitimate use of more resolution in the numbers.

So it is in engineering measurements, there is an economic trade-off between increased accuracy (which may require more expensive instruments, better test rigs, and so on) and the *need* for the accuracy.

Subtraction of Nearly Equal Numbers. Great care must be exercised when taking the difference of two numbers that are close to each other in size.

Example 11-3

Subtracting 1110 ± 5 from 1160 ± 5 gives 50 ± 10, which shows that the 5 is doubtful; the difference in this case has only one significant figure. The uncertainty, especially the relative uncertainty has greatly increased, because the true answer lies somewhere between 40 $(1155 - 1115)$ and 60 $(1165 - 1105)$. In the notation introduced in this text, this would be represented by 50 ± 10 percent or 50 ± 20 percent. The original numbers were represented by relative bounds of 1110 ± 0.45 percent and 1160 ± 0.43 percent.

The point of the examples above is that you should never let all the numbers that a calculator or computer can display lull you (or your client or your instructor) into a false sense of accuracy. Calculating machines do not increase the accuracy of the input. If imprecise data are entered, then imprecise numbers will be generated, regardless of the number of digits of readout possible.

A similar problem exists in making laboratory measurements. Improved designs in digital equipment have made possible data acquisition systems capable of high resolution (displaying many digits). Connected to low-accuracy measuring instruments, these devices are fantastic in providing many insignificant digits in their output. If these outputs are connected directly to a computer for data reduction and manipulation, it is exceedingly easy (but ***naive***) to believe the final displayed result of, say, 30.0034801°C in a situation where the thermocouple (a temperature-measuring device) being used is accurate to only ±0.1°C.

Numerical and Rounding Errors. Nearly all numerical calculations involve some degree of approximation. This is due mainly to the fact that we have to deal with finite limits on the numbers of digits storable and on the numbers of operations performable. For example, representing the reciprocal of three (1/3) with a finite number of terms is never truly accurate. Similarly, including only a finite number of terms in an infinite series for a trigonometric function or an exponentiation always involves some error.

Errors that we encounter or introduce because of our finite mathematical procedures are called *numerical errors*. These errors are introduced when, for example, we use a finite number of terms in an infinite series or when we stop an iterative calculation before we reach the truly correct answer.

The fact that all numbers must be represented on a computer or calculator in a finite number of digits can cause *rounding errors*. Combining (adding or subtracting) large and small real numbers, or subtracting two nearly equal numbers, often requires very large resolution (many digits) to maintain numerical accuracy.[3] When more digits are required to represent a number than a machine can physically handle, the numbers must be truncated or rounded. In *truncation,* all the digits beyond the maximum number permitted by the machine are discarded. In *rounding,* all the digits beyond the maximum number are discarded, but the last retained digit is increased by one unit if the closest dropped digit is greater than 5.

Numerical and rounding errors are closely related. Once generated, these errors may propagate and increase significantly under certain conditions. This is particularly true in large "number-crunching" programs, where 10^6 or even 10^9 calculations are often performed. The study of numerical and rounding errors and their interplay forms a large part of the important field of *numerical analysis.* The subject is beyond the scope of this text.

Fortunately, the resolution and capacity of most calculators and computers keep these numerical and rounding errors to negligible proportions in many cases.

Mistakes

When computations are performed by calculators or computers, there is a great tendency to believe the results. The attitude is: "If the calculator did it, it must be correct." The results are often accepted without question, even when a little reflection would indicate that something is seriously wrong with the answer. Indeed, if computations do turn out to be wrong but cause no harm to life or property, the computer usually gets the blame. Nearly all of us have seen it happen. However, when incorrect calculations lead to serious problems, such as a building collapse or a chemical plant explosion, the situation ceases to be a humorous inconvenience. In these serious circumstances you should note that computer reputations do not suffer, calculators are not sued, and computing machines are never incarcerated. It is the people responsible for the calculations who are held responsible.

A *mistake* can be thought of as an error in action, calculation, or judgment that has been caused by poor reasoning, inadvertence, or carelessness. Even a seemingly minor mistake, such as leaving out one keystroke in programming, can produce results that are totally invalid. Let us examine three common sources of mistakes.

1. *Poor reasoning that results from using an inaccurate mathematical description.* The mathematical relationships used in engineering calculations must be valid or none of the calculations that result from their use will be valid. Calculations that result from the use of incorrect mathematical descriptions may appear to proceed logically and even give reproducible results. But this is of no consequence. If the mathematical model is inaccurate or unsuitable, the results obtained will be equally unsuitable.

---

[3]We use the term *numerical accuracy* here to represent the accuracy in computing with numbers in which all the digits are assumed to be significant. As we discussed previously, all the digits may or may not be significant or meaningful based on the accuracy of the data entered.

Identification and Correction of Mistakes

How might you recognize that you have a questionable result, and, if so, how do you clear up the problem so that you can rely on the results obtained? There are several things you can do, but all methods are not applicable to all situations. Several of the more common procedures are listed below:

- *Evaluate the reasonableness of results.* In most real applications of computation you will be dealing with physical quantities that have some meaning to you. In these cases your intuition is often a guide to the numerical range of results that might be appropriate.

- *Make mental estimates of the range of the expected result before you begin the calculation process.* In this way, you will avoid letting the outcome of the calculation influence your judgment of the expected results. If the result varies significantly from your expectations, check your procedures as well as your calculation. Example: In solving a problem involving parallel resistors, your understanding of the problem should tell you two things: first, that the equivalent resistance will be less than that of the smallest resistor in the circuit, and second, that the outcome cannot be negative. Any results that violate these bounds will be incorrect. Try to apply this type of reasoning in advance for every problem that you solve.

- *Check dimensional consistency.* Verify dimensional consistency for all of your proposed calculations before beginning the calculations. Recheck the dimensions and units if the results are widely different from your expectations.

- *Step-by-step checking.* Another technique is to look closely at intermediate results. If computer calculations are being performed, print out the intermediate results in which you might be otherwise uninterested. If the detailed result for any step differs significantly from an estimate you might make, or better still, from a hand calculation you can perform, stop immediately and identify the reason.

- *Double checking.* Repeat the entire calculation. If possible, use another sequential problem solving procedure for the second attempt. Forcing yourself to use a different technique may keep you from repeating a simple mistake.

- *Printing.* Examine printouts of the programs to ensure that you have entered the coding correctly.
- *Test data.* Run test or fictitious data through your procedure. This should be data for which you know the answer. If it works correctly for the test data, then you can treat your data with greater, but not absolute, confidence.
- *Practice.* Build your proficiency in devising and executing computational procedures by working many problems and exercises. Work problems found in other books. Those with answers given can be particularly helpful, but avoid looking at the answers until you have made your own mental estimate of the result and have actually calculated it. If your result does not match the answer given, practice the troubleshooting techniques above to identify and correct the difficulty. By following this procedure you will develop a facility and flexibility in using your calculator and give yourself confidence in your ability to organize and execute computations.

You must also devote attention to guarding against computational blunders. All of us make a mistake from time to time. However, it is important that you are able to recognize mistakes and correct them when they occur.

2. *Poor reasoning that results from using invalid computational procedures.* Fortunately, many improper computational procedures are discovered by error trapping routines in calculators and computers. Some of these mistakes are mathematical, such as trying to take roots of negative numbers, performing division by zero, or attempting to find the arccosine of numbers with magnitudes larger than 1. Others result from improper or incorrect coding. Although it is annoying to have the computer call these improper procedures to your attention, it is a far better condition than the alternative: invalid results masquerading as valid ones. Many of these self-identifying mistakes can be corrected as soon as they are encountered. Some may be more difficult to trace to their unique cause. But at least you have been made aware that a mistake exists.

Much more serious are computational procedures that are accepted by the computer as being valid, but are not really the proper steps for the solution of the problem at hand. Calculators and computers give you no warning that such logical errors exist. The machines cannot decide whether the sequence is appropriate for your problem.

It is possible, with totally invalid procedures, to get answers that are close enough to the true or expected results that they look valid *purely by accident.* Even a nonworking clock is correct twice a day! At these two times of day you might not even recognize that something is amiss.

3. *Inadvertence and carelessness.* Avoiding the two sources of mistakes described above will not ensure valid results. The problem-solving procedure must be executed properly, and the results must be recorded accurately before you are finished. Carelessness and lack of attention to detail are probably the biggest obstacles on the path to achieving correct results. Thus, consistent vigilance is the price to obtain correct calculations. Remember, you must execute the computations as carefully as you create them.

PROBABILITY AND STATISTICS

We have now spent some time discussing the difference between accuracy and resolution of numbers. We have tried to make clear that all numbers have some limit as to their accuracy even though they may be displayed with considerable resolution. Let us now turn to the science of inexact numbers, better known as probability and statistics.

Probability

No doubt, you already have some idea of what is meant when someone says that an event is either highly probable or highly improbable. For example, we expect that you will accept the statement that even though the famous racehorse Man o' War won many races in the past (20 wins in 21 races), it is highly *improbable* that he will do so again (the horse died in 1947). Similarly, we suspect you would readily agree that it is highly *probable* that the sun will emerge over the eastern horizon tomorrow morning.

However, there are a large number of occurrences or events which exhibit much less certainty than the two examples in the last paragraph. Examples such as the result of a toss of a coin (excluding an edge landing, there are two possible outcomes), the toss of a die (here there are six possible outcomes), and the drawing of a playing card from a full deck (now there are 52 possible outcomes) are only a few of the nearly infinite number of events in the world in which the outcome is by no means certain.

Of course, you may have noticed that we have named only those events that are important in certain games of chance, when, indeed, there are many situations or events that are not certain but which are important in engineering and technology. We mention games of chance primarily because it was the wagering on such games by the well-to-do in the eighteenth and nineteenth centuries that fostered the development of the field now known as probability theory.

To give the word *probability* a technical and quantitative definition, we say that the *probability is the ratio of the number of favorable (or desired) outcomes from an event to the total number of possible outcomes, when the event is repeated a large number of times.* We will use the letter P to denote probability. Note that the probability is zero ($P = 0$) if a favorable (or desired) outcome never occurs (for example, $P = 0$ for Man o' War winning a race in the future). Also, the probability is 1 ($P = 1$) if a favorable (or desired) outcome occurs in every event ($P = 1$ for the sun rising above the eastern horizon tomorrow morning[4]).

If we were to observe, on the average, a favorable outcome half the time when observing a large number of events, the probability of observing a favorable outcome would be $P = \frac{1}{2}$ or $P = 0.5$. Thus the probability of finding a head when flipping a legitimate coin is P (head) $= \frac{1}{2}$; similarly, the probability of finding a tail is P (tail) $= \frac{1}{2}$. Note that the probability of finding either a head or a tail is P (head or tail) $= 1$, since the only possible outcomes are favorable or desired ones. This is a specific result of the fact that the sum of the probabilities of all possible outcomes for an event must equal 1. The algebraic expression of this fact,

$$\sum_i P_i = 1$$

is often useful to remember while working problems involving probability.

There are two other principles that are useful in the calculation and use of probabilities. They are:

1.   If $P(A)$ and $P(B)$ are the probabilities that favorable outcomes A

---

[4]There is apparently some evidence that in the distant past, major and sudden changes have occurred in the orientation of the earth's axis and perhaps even its direction of rotation about that axis. Such happenings would reduce the probability for the sun rising in the east to a value ever-so-slightly less than 1.

and B, respectively, will result from an event, the probability of either A or B occurring will be

$$P(A \text{ or } B) = P(A) + P(B)$$

That is, the individual probabilities add.

2. If $P(A)$ and $P(B)$ are the probabilities that favorable outcomes A and B, respectively, will result from an event, the probability that *both* A and B will occur in two sequential events will be

$$P(A \text{ and } B) = P(A) \cdot P(B)$$

That is, the individual probabilities multiply in this situation.

Returning to games of chance, we wish to give some brief examples of probability.

Example 11-4
Calculate the probability of rolling a 2 with a die.

Solution

$$P(2) = \tfrac{1}{6} = 0.1666+$$

Interpretation: One out of six times (on the average) a 2 will be rolled. You are seeking one favorable outcome (the rolling of a 2) when there are six possible outcomes (you could roll a 1, a 2, a 3, a 4, a 5, or a 6) for the event of rolling a die.

Example 11-5
Calculate the probability of obtaining either a 2 or a 3 when rolling a die.

Solution

$$P(2 \text{ or } 3) = P(2) + P(3)$$
$$= \tfrac{1}{6} + \tfrac{1}{6} = \tfrac{1}{3} = 0.3333+$$

Interpretation: One out of three times (on the average) a 2 or a 3 will be rolled.

Example 11-6
Calculate the probability of obtaining a 1 followed by a 6 on two successive rolls of a die.

Solution

$$P(1 \text{ and } 6) = P(1) \cdot P(6)$$
$$= \tfrac{1}{6} \cdot \tfrac{1}{6} = \tfrac{1}{36} = 0.02777+$$

Interpretation: One out of every 36 pairs of rolls, on the average, will yield a 1 followed by a 6. This could also be interpreted as the probability of obtaining a 1 and a 6 when two dies (often referred to as a pair of dice) are thrown simultaneously.

Example 11-7
Calculate the probability of drawing any single card you wish to identify [say, the queen of hearts (QH)].

Solution

$$P(\text{QH}) = \tfrac{1}{52} = 0.01923+$$

Interpretation: Once out of every 52 times you remove a single card from a full, well-shuffled deck (52 cards), on the average, you will draw the queen of hearts.

Before we dwell too long on games of chance, let us proceed to a discussion of statistics.

Statistics

Figure [11-13] shows an example set of data that have been assembled by taking measurements of some phenomena at various points in time. As is typical of all data, the measurements do not all yield the same value. The variations might be due to several possible causes, including inaccuracy of the actual measuring instruments, inaccuracy in reading and/or recording the instruments, and variations in the actual quantities being measured.

In their efforts to be brief but explicit, engineers and technologists often use two or three calculated numbers to express the "best value" of all the data and the variation that exists in the data set. Two of these calculated numbers relating to the "best value" are the arithmetic mean (or average) and the median.

Arithmetic Mean. The *arithmetic mean* is the value about which the data tend to cluster. It is often referred to as the *average*. It is calculated by obtaining the sum of the individual measurements and dividing this quantity by the number of measurements made. This process may be represented mathematically by

$$\overline{X} = \frac{1}{n} \sum_i X_i = \frac{X_1 + X_2 + X_3 + \cdots + X_n}{n}$$

where $\overline{X}$ = arithmetic mean
$\quad X_i$ = an individual measurement (the ith measurement)
$\quad n$ = total number of measurements

Table 11-2 displays the data displayed in Figure 11-13 and shows a hand method of calculation of the arithmetic mean, or average. Nearly all spreadsheets have a standard function which will return the average of a range (or list) of numbers. (See Topic 10-3 for the SuperCalc4 and Lotus 1-2-3 function names.)

The horizontal line in Figure [11-13] represents the arithmetic mean or average of that data set. The average *may* have some relationship to a true average of the phenomenon being measured, but it also may not. For example, it may be that the instruments used for the measurements had some inherent errors (usually referred to as *systematic errors*) that biased all of the readings or it may be that the phenomenon being measured actually had a time-dependent average which is masked by the scatter and lack of long-time data.

Assuming that the values represented in Figure [11-13] have no systemic errors and that the data possess no time-varying arithmetic mean, the calculated arithmetic mean does represent significant in-

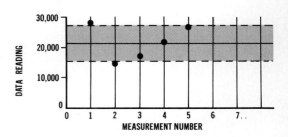

[11-13] Sample data set.

TABLE 11-2 STATISTICS OF THE SAMPLE DATA SET SHOWN IN FIGURE [11-13]

Data Reading X_i	$X_i - \overline{X}$	$(X_i - \overline{X})^2$
28,000	6,800	46,240,000
14,800	−6,400	40,960,000
16,300	−4,900	24,010,000
20,200	−1,000	1,000,000
26,700	5,500	30,250,000

$\Sigma X_i = 106{,}000$ $\qquad\qquad \Sigma(X_i - \overline{X})^2 = 1.4246 \times 10^8$

$$\overline{X} = \frac{106{,}000}{5} \qquad\qquad \sigma = \sqrt{\frac{1.4246 \times 10^8}{5-1}}$$

$$= 21{,}200 \qquad\qquad = 5{,}968$$

formation about the true average value to be associated with the actual phenomenon.

Median.　The *median* is similar to the arithmetic mean in that it is also a value about which the data tend to cluster. The median is the midpoint (not average) of a group of data, determined in the way we describe. Its determination requires that the data be arranged in increasing or decreasing order. When the total number of data is odd, the median is then the middle number of the ordered set of numbers. If the total number of data is even, the median is the arithmetic mean of the two middlemost numbers in the set.

For the data set displayed in Figure [11-13] and tabulated in Table 11-2, the median is 21,200. Of the two measures of the central point of a data set, the arithmetic mean or average is the more common. The median is not often used.

Standard Deviation.　It is desirable and instructive to have a means of expressing the uncertainty, or the scatter of a series of measurements about the arithmetic mean of that data. To do this, we often make use of what is known as the *standard deviation* (usually designated as σ), defined as

$$\sigma = \sqrt{\frac{\sum_i (X_i - \overline{X})^2}{n-1}}$$

where $(X_i - \overline{X})$ represents the deviation of a single observation from the mean, and n is the number of observations or data points. Table 11-2 displays a hand method for computing the standard deviation of the data displayed in Figure [11-13].

On most spreadsheets, there is a standard function which will automatically return the standard deviation of a range (or list) of numbers. (See Topic 10-3 for the SuperCalc4 and Lotus 1-2-3 function names.)

In Figure [11-13] we have shown two horizontal dashed lines, one drawn at a value of the ordinate equal to the average plus the standard deviation $(\overline{X} + \sigma)$, and one drawn at the mean minus the standard deviation $(\overline{X} - \sigma)$. The standard deviation is large when the scatter of the data about the mean is large. It is small when the scatter is small. Hence, in one number, the standard deviation provides information about the scatter of the data.

HISTOGRAMS AND PROBABILITY DISTRIBUTIONS _____

If a large glass jar were filled to the top with marbles and placed in view of a large class of students and each student was asked to write down an estimate of the number of marbles in the jar, it is extremely unlikely that every student would estimate the same number and that this number would be the exact number of marbles in the jar. Rather it is likely that, if the answers were compiled, a pattern of distribution of estimates would focus upon a certain estimated number of marbles.

If, for simplicity in plotting, the estimates are grouped into blocks to the nearest 100 marbles, a graph of this distributor might look like Figure [11-14]. This figure is plotted so that the width of a column is equal to the interval, in this case 100 marbles, and the height is equal to the frequency, which is the number of persons making any given block of estimates.

If the number of persons making estimates of the marbles were increased and the blocks within which the estimates fall were made smaller, the histogram probably would take on an appearance similar to Figure [11-15]. When a fairly large number of measurements have been used to construct a histogram, we can confidently say that for any bin value on the horizontal axis, the probability of obtaining that value is equal to the number of occurrences measured for that bin value (the ordinate on the histogram) divided by the total number of occurrences. Thus by dividing the number of occurrences for each bin value by the total number of occurrences for all bin values, a *probability distribution* can be obtained. Such a probability distribution gives the probability of finding each bin value.

For example, if we were to roll a legitimate die a large number of times and record the number on top of the die each roll, we could construct a histogram of this experiment. With a legitimate die being

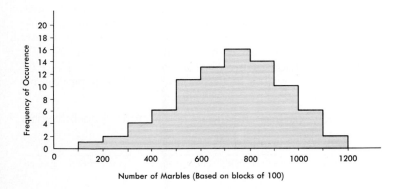

[11-14] Histogram of estimates of marbles in a jar.

Histogram of a large number of estimates.

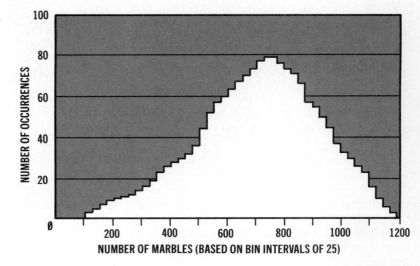

rolled a large number of times, we should find that each number on the die would come up the same number of times as any other number. The histogram for 600 rolls should resemble Figure [11-16], which shows that the number of occurrences of each number is very close to 100. The probability distribution constructed from this histogram is shown in Figure [11-17], where it can be seen that the probability of rolling any one number is $100/600 = \frac{1}{6} = 0.16666+$. Indeed, these are the probabilities that we used for the die-tossing experiments discussed in the preceding section on probabilities.

The probability diagram constructed from Figure [11-15] is shown in Figure [11-18]. If we were to continue to increase the number of bins (and thus decrease the interval for each bin) and increase the number of students participating, we might expect to see the probability distribution assume the shape of a smooth curve. For many types of problems, the shape of the probability distribution approaches a bell-shaped curve known as the *normal* or *Gaussian*[5] *distribution,* such as shown in Figure [11-19]. The general mathematical expression for the probability curve is an exponential function of the form

$$y = \frac{1}{\sqrt{2\pi}\sigma} e^{-\frac{1}{2}\left(\frac{X_i - \bar{X}}{\sigma}\right)^2}$$

It is important to note that not all phenomena exhibit normal distributions, but there are enough in science, engineering, and technology to make it of great value. It should be clear that the distribution for tossing of a die does not form a normal distribution [11-17].

From an inspection of the normal probability curve [11-19], determined either by trial or by derivation, several principles can be observed:

1. Small deviations from the mean occur more frequently than large ones.

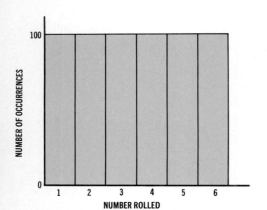

[11-16] Histogram of die rolls.

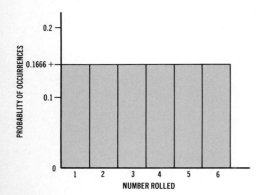

[11-17] Probability of die rolls.

<hr>

[5] Karl Gauss derived the equation for this curve from a study of errors in repeated measurements of the same quantity.

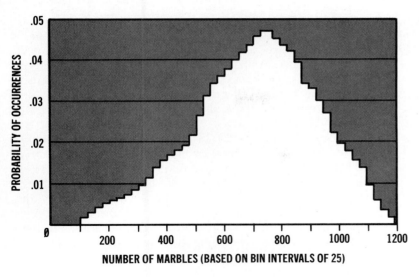

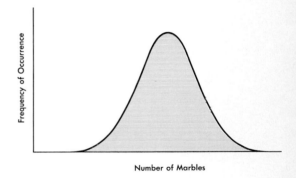

[11-18] Probability of estimates of number of marbles.

[11-19] Normal probability (Gaussian) curve.

2. Deviations of any given size are as likely to be positive as they are to be negative.

3. Very large random deviations from the mean seldom occur.

We have labeled that arithmetic mean (calculated as described earlier) on Figure [11-20]. We have also shown with dashed lines the region defined by plus or minus one standard deviation ($\pm\sigma$, as σ was defined before) from the arithmetic mean. It can be shown that 68.3 percent of all occurrences or observations fall within this shaded area.

Using shading, we have shown an area falling within $\pm2\sigma$ of the arithmetic mean of the normal distribution in Figure [11-21]. It can be shown that 95.5 percent of all occurrences or observations fall within this shaded area defined by $\pm2\sigma$ from the geometrical mean.

Let us now explore one possible way of using the probability distribution.

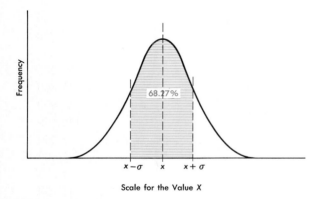

[11-20] Normal probability curve showing the 1σ area.

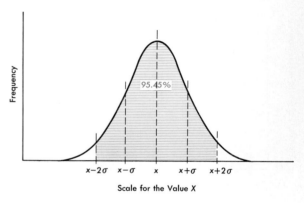

[11-21] Normal probability curve showing the 2σ area.

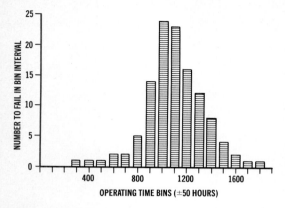

[11-22] Histogram of light-bulb operating time before failure.

Example 11-8

Suppose that tests were conducted on a particular brand and size of light bulb. The purpose of the tests was to measure the lifetimes of these bulbs. Since we probably would not expect the life of all bulbs to be exactly the same, it would not be unreasonable to construct a histogram using lifetime along the bin axis with the number of bulbs that fail at each bin value plotted on the vertical axis (see Table 11-3 and Figure [11-22]).

It can be seen in Figure [11-23] that the probability distribution constructed from the histogram of Figure [11-22] reveals a nearly normal behavior. The probability of finding a bulb with a particular bin lifetime is simply the number of bulbs found in each bin divided by the total number of bulbs (117). From the initial data the arithmetic mean or average lifetime is 997 hours and the standard deviation from this average is 244 hours. Notice that the arithmetic mean cannot be calculated from averaging the number of light bulbs in the bins. Instead, one must compute this average by dividing the total number of hours of operation for all the bulbs (116,700) by the total number of bulbs (117). Since we do not have data on each bulb that was tested, we can approximate the total hours associated with the bulbs in a particular bin by multiplying the mean bin value by the

TABLE 11-3

Mean Bin Value (hours) ±50 hrs	Number of Light Bulbs That Failed in Bin (hrs)	Number of Hours Associated with Bulbs in This Bin
0	0	
100	0	
200	1	200
300	1	300
400	1	400
500	2	1,000
600	2	1,200
700	5	3,500
800	14	11,200
900	24	21,600
1000	23	23,000
1100	16	17,600
1200	12	114,400
1300	8	110,400
1400	4	5,600
1500	2	3,000
1600	1	1,600
1700	1	1,700
	$\Sigma = 117$	$\Sigma = 116,700$

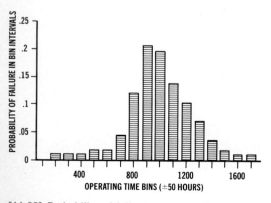

[11-23] Probability of failure of light bulbs.

number of bulbs in that bin (the product appears in the third column of Table 11-3). The total number of hours for all bulbs is then approximated by summing these individual products over all bins (116,700).

The standard deviation is computed by first squaring the difference between each bin lifetime and the arithmetic mean, multiplying by the number of bulbs in that bin and summing these over all the bins. Extracting the square root of the number obtained by dividing the sum above by the number of bulbs minus 1 yields the standard deviation.

From the data on normal distributions just discussed in the text, we can predict that 68.3 percent of a large lot of these light bulbs would fail while operating between 753 and 1241 hours $(\overline{X} \pm \sigma)$ and 95.5 percent would fail between 509 and 1485 hours $(\overline{X} \pm 2\sigma)$. We can also calculate that $(100 - 95.5)/2 = 2.25$ percent of the bulbs would last more than 1241 hours.

There are other standard probability distributions in use in science and engineering. These include the Weibull distribution, the gamma distribution, and the exponential distribution.

The Weibull probability distribution is often used in engineering because it has exceptional flexibility. The mathematical expression for the distribution is given by

$$y = \frac{k}{c}\left(\frac{x}{c}\right)^{k-1} e^{-(x/c)^k}$$

where k and c are constants that can be chosen to best fit the data at hand. It is through the choice of these constants that the flexibility of the distribution is realized.

One interesting use of the Weibull distribution is to represent the probability distribution for wind speeds that occur at some particular geographic site. When wind speed data are collected at a site over a long period of time, a probability distribution of the data often resembles the Weibull distribution with k assuming the value of 2 and c assuming a value of about 1.12 times the long-term mean wind speed. Figure [11-24] shows a typical wind speed histogram along with a superimposed Weibull distribution whose ordinate has been multiplied by 8760, the number of hours in a year, in order to compare directly with the histogram wind data.

Notice that the normalized Weibull distribution in Figure [11-24] is somewhat skewed from a normal or Gaussian distribution. The conclusion that we wish to make here is that not all data are normally distributed. In the case of wind data, they are not randomly generated. The winds are related in a complicated way to the physics of the atmosphere. Thus it is not reasonable to expect them to be Gaussian in their distribution. Thus you should be cautious when interpreting data in terms of the Gaussian distribution.

Standard Error of the Mean

It usually is desirable to evaluate the uncertainty of the arithmetic mean. We know that the uncertainty of the mean, $\overline{X}$, is considerably less than the uncertainty of any single observation, X_i. The uncer-

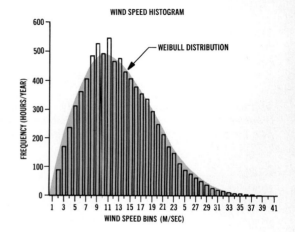

[11-24] Wind speed histogram.

tainty of the mean can be expressed in the following form:

$$\sigma_m = \frac{\sigma}{\sqrt{n}}$$

and this is usually approximated by

$$s_m = \frac{s}{\sqrt{n}}$$

where σ_m is standard error of the mean, s_m is the estimated standard deviation of the mean, s is the standard deviation of a sample, and n is the number of observations in the sample.

Since the true value is never known, an estimate based on mathematical processes can be made as to the confidence that can be placed in the sample mean as an estimate of the true mean.

Example 11-9

The mean of 25 measurements of an angle gives a value of 32° 17.1′; s is 1.2′. What is the probable range of the true value?

Solution

The sample mean of 32° 17.1° is the best estimate to use for the mean.

$$s_m = \frac{1.2'}{\sqrt{25}} = 0.24'$$

There is a 68.27 percent certainty that the true mean lies between 32° 16.9′ and 32° 17.3′.

There is a 95.45 percent certainty that the true mean lies between 32° 16.6′ and 32° 17.6′.

Since the true value is never known, an estimate based on mathematical processes can be made as to the confidence that can be placed in the mean of the sample ($\overline{X}$) as an estimate of the true mean.

PROBLEMS

11-1. Plot a graph showing the relation of normal barometric pressure of air vs. altitude. Plot values up to and including 15,000 ft.

Altitude (Feet above Sea Level)	Normal Barometric Pressure (in. Hg)	Altitude (Feet Above Sea Level)	Normal Barometric Pressure (in. Hg)	Altitude (Feet above Sea Level)	Normal Barometric Pressure (in. Hg)	Altitude (Feet Above Sea Level)	Normal Barometric Pressure (in. Hg)
0	29.95	3000	26.82	9,000	21.4	15,000	16.9
500	29.39	4000	25.84	10,000	20.6		
1000	28.86	5,000	24.9				
1500	28.34	6,000	24.0				
2000	27.82	7,000	23.1				
2500	27.32	8,000	22.2				

11-2. Plot a graph of maximum horsepower transmittable by cold-drawn steel shafting vs. diameter for a speed of 72 rpm based on the formula

$$hp = \frac{D^3 R}{50}$$

where hp = horsepower

D = diameter of shaft, in.

R = rpm of shaft

Calculate and plot values for every inch diameter up to and including 8 in.

11-3. Plot a graph on the variation of the boiling point of water vs. pressure.

Boiling Point (°C)	Pressure (mm Hg)	Boiling Point (°C)	Pressure (mm Hg)
33	38	98	707
44	68	102	816
63	171	105	907
79	341	107	971
87	469	110	1075
94	611		

11-4. Plot the variation of pressure vs. volume, using data as obtained from a Boyle's law apparatus.

Pressure (cm Hg)	Volume (cm³)	Pressure (cm Hg)	Volume (cm³)
50.3	23.2	76.8	15.1
52.5	22.4	79.7	14.7
54.5	21.5	82.7	14.1
56.9	20.9	84.2	13.6
59.4	19.6	87.9	13.2
63.0	18.5	90.6	12.8
65.3	17.8	93.5	12.5
67.2	17.3	95.7	12.3
72.6	16.1	101.9	11.4
74.5	15.6		

11-5. Using data in Problem 11-4, plot a graph pressure vs. the reciprocal of the volume.

11-6. Plot the relation between magnetic flux density (B, tesla) and magnetizing force (H, ampere-turns/meter) for a specimen of tool steel. This graph will form what is customarily called a B–H curve.

H (ampere-turns/m)	B (tesla)	H (ampere-turns/m)	B (tesla)
0	0	10,840	1.375
2,710	0.900	13,550	1.409
5,420	1.180	16,260	1.422
8,130	1.302	18,970	1.466
21,680	1.486	29,810	1.535
24,390	1.498	35,220	1.557
27,100	1.523		

11-7. The formula for converting temperatures in degrees Fahrenheit to the equivalent reading in degrees Celsius is

$$°C = \frac{5}{9}(°F - 32°)$$

Plot a graph so that by taking any given Fahrenheit reading between 0 and 220° and using the graph, the corresponding Celsius reading can be determined.

11-8. Plot the variations of efficiency with load for a ¼-hp, 110-V, direct-current electric motor, using the following data taken in the laboratory.

Load Output (hp)	Efficiency (percent)	Load Output (hp)	Efficiency (percent)
0	0	0.175	56.5
0.019	24.0	0.195	58.0
0.050	42.0	0.248	59.1
0.084	44.9	0.306	58.0
0.135	50.7	0.326	56.2

11-9. Plot the values given in Problem 11-1 on semilog paper [log(pressure) versus altitude].

11-10. Plot the values given in Problem 11-4 on semilog paper [log(pressure) versus volume].

11-11. Plot the values given in Problem 11-1 on log-log paper.

11-12. Plot the values given in Problem 11-4 on log-log paper.

11-13. The following data were taken from an acoustical and electrical calibration curve for a type 1126 microphone. The test was run with an incident sound level of 85 dB perpendicular to the face of the microphone.

Frequency (Hz)	Relative Response (dB)
20	−40
50	−29
100	−19
400	−5
1,000	+1
2,000	+1
3,000	0
6,000	−4
10,000	−11

Plot a graph on semilog paper showing the decibel response with frequency.

11-14. A Weather Bureau report gives the following data on the temperature over a 24-hr period for October 12.

Midnight 47°	10 am 55°	6 am 63°
2 am 46°	Noon 68°	8 pm 58°
4 am 44°	2 pm 73°	10 pm 57°
6 am 43°	4 pm 75°	Midnight 57°
8 am 49°		

Plot the data.

11-15. A test on an acorn-type street lighting unit shows the mean vertical luminous intensity distribution to be as shown below.

Midzone Angle (deg)	Candlepower at 10 ft	Midzone Angle (deg)	Candlepower at 10 ft
180	0	85	156
175	0	75	1110
165	0	65	1050
155	1.5	55	710
145	3.5	45	575
135	5.5	35	500
125	8.5	25	520
115	13.5	15	470
105	22.0	5	370
95	40.0	0	370

Plot the data. (While data for only half the plot are given, the other half of the plot can be made from symmetry of the light pattern.)

11-16. From data determined by you, draw a pie chart to show one of the following.
  a. Consumption of sulfur by various industries in the United States.
  b. Budget allocation of the tax dollar in your state.
  c. Chemical composition of bituminous coal.
  d. Production of aluminum ingots by various countries.

11-17. A series of test specimens of a crank arm, part 466-1, was tested for the number of cycles needed to produce fatigue failure at various loadings. The results of the tests are shown below. Plot a graph of load against operating cycles (S–N curve) on semilog paper [log(cycles) versus load].

Specimen Number	Oscillatory Load (lb)	Operating Cycles to Produce Failure
1	960	1.1×10^5
2	960	2.2×10^5
3	850	2.4×10^5
4	800	4.2×10^5
5	800	6.0×10^5
6	700	2.4×10^5
7	700	5.1×10^5
8	650	1.8×10^6
9	600	7.7×10^6
10	550	1.0×10^7

(Note that the proper number of significant figures may not be given in the reading.)

11-18. Compute the percent error.
  a. Reading of 9.306 ± 0.003
  b. Reading of 19165 ± 2.
  c. Reading of 756.3 ± 0.7
  d. Reading of 2.596 ± 0.006

11-19. Compute the numerical error.
  a. Reading of 35.219 ± 0.03 percent
  b. Reading of 651.79 ± 0.01 percent
  c. Reading of 11.391 ± 0.05 percent
  d. Reading of 0.00365 ± 2 percent

11-20. A surveyor measures a property line and records it as being 3207.7 ft long. The distance is probably correct to the nearest 0.3 ft. What is the percent error in the distance?

11-21. The thickness of a spur gear is specified as 0.875 in., with an allowable variation of 0.3 percent. Several gears that have been received in an inspection room are gaged, and the thickness measurements are as follows: 0.877, 0.881, 0.874, 0.871, 0.880. Which ones should be rejected as not meeting dimensional specifications?

11-22. A rectangular aluminum pattern is laid out using a steel scale which is thought to be exactly 3 ft long. The pattern was laid out to be 7.42 ft by 1.88 ft, but it was subsequently found that the scale was incorrect and was actually 3.02 ft long. What were the actual pattern dimensions and by what percent were they in error?

11-23. Add and then express the answer to the proper number of significant figures.

a.	b.	c.
11.565	858.7	1.39395
4.900	404.3	8.7755
226.55	54.42	10.6050
82.824	19.8	49.201
17.668	8.775	88.870
108.77	12.04	108.887

11-24. Subtract and then express the answer to the proper number of significant figures.

a. 6508.	b. 8.104	c. 0.04642
3379.	7.891	0.0199

d. 731.16	e. 7.114	f. 10276.
189.28	16.075	61581.

11-25. Multiply and then express the answer to the proper number of significant figures.

    a. 5167.　　　b. 32105.　　　c. 535.58
       238.　　　　　 5.28　　　　 0.2759

    d. 84.636　　　e. 1.03975　　f. 0.0548
      30869.　　　　 54682.　　　 0.00376

11-26. Divide and then express the answer to the proper number of significant figures.

$$\text{a. } \frac{3928.}{5636.} \qquad \text{b. } \frac{216.75}{53.83} \qquad \text{c. } \frac{7.549}{3.069}$$

11-27. A series of weighings of a sample of metal powder are made with the following results:

Weight of a sample, grams

2.020	2.021	2.021	2.019	2.019
2.018	2.021	2.018	2.021	2.017
2.017	2.020	2.016	2.019	2.020

Compute $\overline{X}$ and σ for the weighings.

11-28. A series of measurements of the length of a concrete runway is made using a steel tape. The results (in meters) are tabulated below:

1363.7	1364.5	1364.0	1363.8	1364.0
1364.1	1363.9	1364.1	1363.9	

Compute $\overline{X}$ and σ of these measurements.

11-29. A series of readings was taken, using an electronic interval timer, for one complete swing of a pendulum to occur, [11-P29]. The data are tabulated as follows:

Time (seconds)	Number of Occurrences	Time (seconds)	Number of Occurrences
1.851	1	1.859	18
1.852	3	1.860	15
1.853	6	1.861	12
1.854	9	1.862	10
1.855	12	1.863	5
1.856	14	1.864	4
1.857	18	1.865	2
1.858	19	1.866	1

    What is the mean time of a swing, and what would be the standard deviation for the data?

11-30. Take 10 coins and toss them at least 25 times, keeping count of the number of heads and tails for each toss. Plot a histogram.

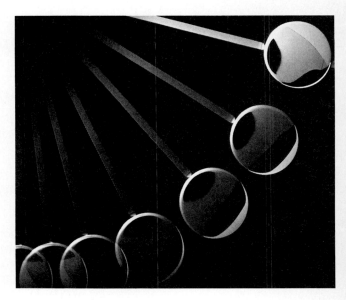

[11-P29] Pendulum.

11-31. The distribution of ages of a group of recruits at an Army camp is given in the accompanying table. Plot a histogram and sketch a probability curve for the ages. Show the σ and 2σ locations. Does this graph show any unusual departures from a standard probability curve? Find the mean and median.

Age (years-months)	Number of Persons	Age (years-months)	Number of Persons
18-1	1	19-7	9
18-2	0	19-8	5
18-3	1	19-9	3
18-4	3	19-10	3
18-5	8	19-11	0
18-6	5	19-12	2
18-7	8	20-1	5
18-8	10	20-2	1
18-9	14	20-3	0
18-10	7	20-4	2
18-11	12	20-5	6
18-12	11	20-6	0
19-1	11	20-7	1
19-2	6	20-8	0
19-3	10	20-9	2
19-4	8	20-10	0
19-5	7	20-11	0
19-6	6	20-12	0

Chapter 12

Dimension and
Unit Systems

THE LAWS OF NATURE AND THEIR
MEASURABLE QUANTITIES: DIMENSIONS_____

> Nature will tell you a direct lie if she can.
> *Charles Darwin, 1809–1882*

Perceptions. We interpret the universe by evaluating our perceptions. More than any other group, those individuals who are generally called *scientists* try to make sense out of their perceptions. They look for reoccurrences of observed phenomena and develop concepts to predict the happenings around them. The *physical scientists* who have preceded us have distilled the predictive schemes and inventions of man down to what we now call the *physical laws of nature*. Some examples are:

Conservation of mass.

Newton's second law.

Conservation of energy.

Law of universal gravitation.

Second law of thermodynamics.

Maxwell's equations of electrodynamics.

A thorough discussion of these, and others, must be left for a more in-depth study in science and engineering. These physical laws relate observable or measurable quantities. Examples of common quantities are:[1] distance (L), time (t), force (F), mass (M), electrical charge (Q), area (A), volume (V), speed (S), work (W), power (P), temperature (T); the list goes on. Notice here that we are referring to the measurable *phenomena* themselves and *not* to the numbers that represent the sizes or magnitudes of the measurements. These measurable phenomena are called *dimensions*.

You may realize that some of these dimensions can be expressed in terms of other dimensions. For example, speed, a dimension, is defined to be distance divided by time, both of which are dimensions. In referring to speed, we may argue about whether we actually mean distance traveled divided by total elapsed time (usually called average speed) or speed at some instant of time (usually called instantaneous speed, a quantity that you will become more familiar with as you study calculus). However, such discussions are not important here. What is important is to understand the concept of measurable quantities or dimensions.

Dimensions can arbitrarily be assigned scales of measure. How-

[12-1] Length, force, and time are measurable quantities.

---

[1] Letters set in colored type signify dimensions, whereas quantities set in black type represent variables that have magnitudes.

ever, since dimensions are usually related to one another through the fundamental laws of nature and the definitions of man, we usually do not assign an arbitrary measuring scale to every dimension but only to a select set. We arbitrarily select some dimensions and then let the physical laws tell us what the resulting dimensions of the other measurable quantities will be. The selected dimensions are called *primary* or *fundamental dimensions*. The remaining dimensions, those that can be expressed as combinations of the primary ones, are called *secondary* or *derived dimensions*.

For example, if we were to choose length and time as primary dimensions, we could derive speed as a secondary dimension since it is defined to be distance per unit time. On the other hand, we could just as easily take length and speed as primary dimensions and use time as a secondary dimension derived from speed and distance.

If we were to assume that length, time, *and* speed are all primary dimensions, we would have to alter our relationship between the three to get meaningful results. In the latter case our definition is over specified. We will have trouble if we want to assign values arbitrarily to all three variables in an equation that involves only three variables. We have to introduce a dimensional constant (an invariant number, often experimentally determined, that has dimensions associated with it) into the equation to make it correct and useful. This process should become clearer as we discuss k_n and k_g, constants in Newton's second law and the classical law of gravity, respectively.

The quantities that are used in the equations of science and engineering nearly always have some physical meaning or interpretation: a force, a distance, a voltage, and so on. Thus, essentially all of the equations we use should be dimensionally homogeneous, since they relate observed quantities. "Dimensionally homogeneous" means that when the dimensions of each term in the equation are substituted for the term and algebraic reduction is carried out, both sides of the equation will contain the same collection of dimensions ordered in the same arrangements.

At first glance, some equations may not appear to be dimensionally homogeneous, but that is probably only because the dimensions of a constant within the equation have been neglected.

[12-2] Voltage is a measurable quantity.

Example 12-1

A, an area, and *S*, a speed, are involved in the following equation relating physical quantities or variables:

$$Q = S(A - P)$$

Determine the dimensions on both P and Q.

Solution

Since the equation relates physical phenomena, it should be dimensionally homogeneous. P must have the same dimensions as A since they are added together (you cannot add apples and oranges). Also, Q must have the same dimensions as the product of SP. We know[2]

$$A \overset{\mathrm{d}}{=} L^2 \quad \text{and} \quad S \overset{\mathrm{d}}{=} L/t$$

---

[2]The symbol d above the = denotes "dimensionally equivalent to," meaning that the left-hand side of the equation has the same dimensions as the right-hand side of the equation.

since areas, being the products of two lengths, have dimensions of L^2, and speeds have traditional dimensions of L/t. Using this information

$$P \stackrel{d}{=} L^2$$

and

$$Q \stackrel{d}{=} \left\{\frac{L}{t}\right\} L^2 \stackrel{d}{=} L^3/t^2$$

Example 12-2

Solve for the dimensions of the conversion factor k in the expression

$$\frac{L^2 t^3 T}{F^4} \stackrel{d}{=} k\left(\frac{L^5 t F^2}{Q^2}\right)$$

Solution:

$$k \stackrel{d}{=} \frac{t^2 T Q^2}{F^6 L^3}$$

Checking:

$$\frac{L^2 t^3 T}{F^4} \stackrel{d}{=} \overbrace{\frac{t^2 T Q^2}{F^6 L^3}}^{k} \left(\frac{L^5 t F^2}{Q^2}\right)$$

$$\frac{L^2 t^3 T}{F^4} \stackrel{d}{=} \frac{L^2 t^3 T}{F^4} \stackrel{d}{=} L^2 t^3 T F^{-4}$$

Problems

Solve for the dimensions of the conversion factor k.

12-1. $\dfrac{FTL^2}{tM^3} \stackrel{d}{=} k\dfrac{t^5 M}{T^2}$

12-2. $k\left(\dfrac{QM}{TF^2}\right) \stackrel{d}{=} \sqrt{L^4 I t Q^8}$

12-3. $t^2 \sqrt{LM^5} \stackrel{d}{=} k\left(\dfrac{FT^2}{M^3}\right)$

12-4. $k(Ft^2 TL^{-2} M^{-3}) \stackrel{d}{=} M^5 L t F^{-3}$

12-5. $M^2 FT^{-5} L^{-2} \stackrel{d}{=} k\sqrt{MTt}$

12-6. $\sqrt{LT^3 F^{-2} M} \stackrel{d}{=} k\sqrt{TF^3 M^6}$

12-7. $k\dfrac{\sqrt{T^3 Q}}{L^2 F^{-2}} \stackrel{d}{=} MTLF$

12-8. $k(F^2 T \sqrt{Lt^{-2}}) \stackrel{d}{=} t^{-3} T^{-2}$

12-9. $FL^3 Q^{-1} M^{-3} \stackrel{d}{=} k\sqrt{L^2 Q^{-1}}$

UNITS

The scales or size subdivisions that we use in expressing the magnitudes of the various dimensions are referred to as *units*. Dimensions are measured in terms of units. Micrometers (μm), millimeters (mm), feet, yards, miles, and light-years are all examples of units used to express magnitudes of the dimension of length (L). Table 12-1 gives some typical units used to express the magnitude of common dimensions.

We pointed out in the preceding section that the equations of science and engineering are nearly always dimensionally homogeneous. Dimensionally homogeneous equations must also be homoge-

TABLE 12-1 TYPICAL UNITS FOR COMMON DIMENSIONS

Common Fundamental Dimensions	Typical Units
Length, L	Micrometers, millimeters, feet, yards, meters, rods, chains, miles, light-years
Time, t	Microseconds, seconds, minutes, hours, days, months, years, decades, centuries
Force, F	Dynes, newtons, ounces, pound-force, tons
Mass, M	Grams, pound-mass, kilograms, slugs

neous or balanced in terms of the units. Thus, substitution of the applicable units for each term in the equations, and reduction of the algebraic expressions that result, should yield left- and right-side values that are identical. This salient fact can be quite useful in solving problems, although it is seldom exercised by beginning students.

If the left-hand side of a dimensionally homogeneous equation has the dimensions of FL/t (force · length/time), then the right-hand side must reduce to dimensions of FL/t. If the left-hand side of a dimensionally homogeneous equation has *units* of (pound · meters)/year, then the right-hand side must reduce to the same units of (pound · meters)/year. In the latter case if the right-hand side reduced to (newton · meters)/second while the left-hand side was still in units of (pound · meters)/year, then the equation would contain a units error, although it would still be dimensionally correct. Results derived in the presence of such an error should never be trusted; however, such equations may yield to correction, barring a more complicated error, by the introduction of simple unit conversion factors. If the equation cannot be balanced dimensionally, this most often indicates that more serious problems exist.

Example 12-3

The stress in a certain column may be calculated by the following relationship:

$$\sigma = \frac{F}{A}\left[1 + \left\{\frac{\ell}{k}\right\}\frac{R}{\pi^2 nE}\right]$$

where[3] σ = induced stress, lb-force/in.^{2}

F = applied force, lb-force

A = cross-sectional area of member, in.^{2}

ℓ = length of bar, in.

k = radius of gyration, in.

R = elastic limit, lb-force/in.^{2}

E = modulus of elasticity, lb-force/in.^{2}

n = dimensionless coefficient for different end conditions

---

[3]lb-force (lb$_f$) means pounds of force. We discuss this expression and its meaning later in this chapter.

(a) Is this equation dimensionally homogeneous?
(b) Do the units balance?

Solution:

Collecting all the dimensions on both sides of the equation results in

$$\frac{F}{L^2} \stackrel{?}{=} \frac{F}{L^2}\left[1 + \left\{\frac{L}{L}\right\}^2 \frac{F/L^2}{F/L^2}\right] \stackrel{d}{=} \frac{F}{L^2}$$

which shows dimensional homogeneity. Collecting all of the units on both sides of the equation:

$$\frac{\text{lb-force}}{\text{in.}^2} \stackrel{?}{=} \frac{\text{lb-force}}{\text{in.}^2}\left[1 + \left\{\frac{\text{in.}}{\text{in.}}\right\}^2 \frac{\text{lb-force/in.}^2}{\text{lb-force/in.}^2}\right]$$

$$= \frac{\text{lb-force}}{\text{in.}^2}$$

This demonstrates that the equation is homogeneous in units.

Problems

12-10. Is the equation $a = (2S/t^2) - (2V_1/t)$ dimensionally homogeneous if a is an acceleration, V_1 is a velocity, t is a time, and S is a distance? Prove your answer by writing the equation with fundamental dimensions.

12-11. Is the equation $V_2^2 = V_1^2 + 2as$ dimensionally correct if V_1 and V_2 are velocities, a is an acceleration, and s is a distance? Prove your answer by rewriting the equation in fundamental dimensions.

12-12. In the homogeneous equation $R = B + \frac{1}{2}CX$, what are the fundamental dimensions of R and B if C is an acceleration and X is a time?

12-13. Determine the fundamental dimensions of the expression $B/g\sqrt{D} - m^2$, where B is a force, m is a length, D is an area, and g is the acceleration of gravity at a particular location.

12-14. The relationship $M = \sigma I/c$ pertains to the bending moment for a beam under compressive stress. σ is a stress in F/L^2, C is a length L, and I is a moment of inertia L^4. What are the fundamental dimensions of M?

12-15. The expression $V/K = (B - \frac{7}{4}A)A^{5/3}$ is dimensionally homogeneous. A is a length and V is a volume of flow per unit of time. Solve for the fundamental dimensions of K and B.

12-16. Is the expression $S = 0.031V^2/fB$ dimensionally homogeneous if S is a distance, V is a velocity, f is the coefficient of friction, and B is a ratio of two weights? Is it possible that the numerical value 0.031 has fundamental dimensions? Verify your solution.

12-17. If the following heat transfer equation is dimensionally homogeneous, what are the units of k?

$$Q = \frac{-kA(T_1 - T_2)}{L}$$

A is a cross-sectional area in square feet, L is a length in feet, T_1

and T_2 are temperatures (°F), and Q is the amount of heat (energy) conducted in Btu per hour.

12-18. In a swimming pool manufacturer's design handbook, for a pool whose surface area is triangular, you find the following formula: $V = 3.74Rt\theta$, where V = volume of pool in gallons, R = length of base of triangular-shaped pool in feet, t = altitude of triangular-shaped pool in feet if t is measured perpendicular to R, and θ = average depth of pool in feet. Prove that the equation is valid or invalid.

12-19. You are asked to check the engineering design calculations for a sphere-shaped satellite. At one place in the engineer's calculations you find the expression $A = 0.0872\Delta^2$, where A is the surface area of the satellite measured in square feet, and Δ is the diameter of the satellite measured in inches. Prove that the equation is valid or invalid.

12-20. The U.S. Navy is interested in your torus-shaped lifebelt design and you have been asked to supply some additional calculations. Among these is the request to supply the formula for the volume of the belt in cubic feet if the average diameter of the belt is measured in feet and the diameter of a typical cross-sectional area of the belt is measured in inches. Develop the formula.

12-21. From the window of their spacecraft two astronauts see a satellite with foreign insignia markings. They maneuver for a closer examination. Apparently the satellite has been designed in the shape of an ellipsoid. One of the astronauts quickly estimates its volume in gallons from the relationship $V = 33.8ACE$, where the major radius (A) is measured in meters, the minor radius (E) is measured in feet, and the end-view depth to the center of the ellipsoid (C) is measured in centimeters. Verify the correctness of the mathematical relationship used for the calculations.

12-22. An engineer and his family are visiting in Egypt. The tour guide describes in great detail the preciseness of the mathematical relationships used by the early Egyptians in their construction projects. As an example he points out some peculiar indentations in a large stone block. He explains that these particular markings are the resultant calculations of "early day" Egyptians pertaining to the volume of the pyramids. He says that the mathematical relationship used by these engineers was $\odot = \square \uparrow$, where $\odot$ was the volume of pyramid in cubic furlongs, $\square$ was the area of the pyramid base in square leagues, and $\uparrow$ was the height of the pyramid in hectometers. The product of the area and the height equals the volume. The engineer argued that the guide was incorrect in his interpretation. Prove which was correct.

12-23. Develop the mathematical relationship for finding the weight in drams of a truncated cylinder of gold if the diameter of the circular base is measured in centimeters and the height of the piece of precious metal is measured in decimeters.

12-24. Is the equation $F = WV^2/2g$ a homogeneous expression if W is a weight, V is a velocity, F is a force, and g is the acceleration of gravity? Prove your answer.

THE PHYSICAL LAWS OF NATURE AND DIMENSION/UNIT SYSTEMS

We concentrate first on "mechanical" dimensions (those playing an important role in mechanics: distance, speed, acceleration, force, mass, time) and their relationship to the physical laws. Later we discuss "electrical and magnetic" and "thermal" dimensions.

Newton's Second Law

As an introduction to *dimension/unit systems* consider the dimensions involved in Newton's second law. This law relates the forces acting on an object to both its mass and its acceleration (how fast the object acquires or loses velocity). For a constant-mass object it is given by[4]

$$\mathbf{F} = k_n M \mathbf{a} \qquad (12\text{-}1)$$

where M is the mass of the object, $\mathbf{a}$ is the acceleration, k_n is a constant of proportionality, and $\mathbf{F}$ is the resultant force acting on the object.

If we consider Newton's second law to be dimensionally homogeneous (experience tells us it must be), we have some flexibility in choosing what we will consider to be the primary dimensions. If you apply the principles of algebra, you will note that we cannot arbitrarily choose primary dimensions for all the quantities in the equation ($\mathbf{F}$, k_n, M, and $\mathbf{a}$). If we did, how can we expect both the left- and right-hand sides to reduce to the same set of dimensions?

Also, if we choose any two of the quantities to have arbitrary primary dimensions, only the *combination* of the remaining two is determined. That is, the secondary dimensions of either of the remaining two quantities are not uniquely specified.

We conclude from the discussion above that the dimensions of three of the four quantities in Equation (12-1) can be specified arbi-

One of the confusing things to students in the application of this law has always been the distinction between mass and weight. Mass is a property of all matter that does not depend on the environment or on the surroundings in which the matter finds itself.[5] This book has a mass, you have a mass, and each star has a mass. Each of these masses is independent of whether it is in the vicinity of some other object, is isolated in outer space, or is darting through the universe.

However, masses attract each other according to the law of universal gravitation (discussed in the next section). That is, two objects exert forces on each other, forces that attempt to pull the objects together. In many situations we call this force of attraction the *weight*. For example, the earth applies a gravitation attraction to you (and you to it). The force on you, or your weight, though, varies depending on how physically separated you are from the earth and how close you are to other objects that attract you. Thus your *weight is a force* and that force depends on your environment or surroundings.

Weight changes as the surroundings change. In the vicinity of the moon, for example, your weight is considerably reduced. This is because the moon, when you are close to it, does not attract you with the same force as does the earth, when you are near it. Remember, however, that your mass, which does not depend on your surroundings, would be independent of whether you are on the moon or on the earth.

A physicist named Eotvos, over a 25-year period starting in about 1890, showed that the masses that appear in Equation (12-2) are the same masses that appear in Newton's second law. His work is just one more example of the detailed devotion and effort that has made possible our present understanding of nature and our advanced state of technology.

---

[4]The boldface symbols that are used here and in some other equations in this chapter represent *vector* quantities . . . quantities that have both direction and magnitude.

[5]The mass that we are discussing here is known in the world of physics as the "nonrelativistic or rest mass." This label applies to objects that travel at speeds considerably less than the speed of light (3×10^{10} m/s). The mass of an object that travels at speeds approaching the speed of light is different from the mass the object has at much slower speeds. The theory of relativity deals with very fast moving objects.

[12-3] America's "number one" sport depends on the athletes' ability to use a force to strike a mass and impart to it an acceleration. The force from the bat acts over a very short time period. It produces an acceleration that takes the ball velocity from about 90 mph toward home plate, to perhaps 140 mph toward the outfield fence. This picture shows the legendary Babe Ruth imparting an acceleration to one of many balls which he hit over the outfield fence.

trarily. The dimensions of the fourth and last quantity are then uniquely determined by the mathematical form of the equation. Table 12-2 shows all of the primary and secondary possibilities available to us (we have listed the dimensions of **a** as L/t^2, using the definition of **a**).

The Law of Universal Gravitation

Newton was also the first person to formulate what is now called the (classical) law of universal gravitation. This principle expresses the force (**F**) with which any two bodies in the universe attract each other as

$$\mathbf{F} = \frac{k_g M_1 M_2}{r^2} i \qquad (12\text{-}2)$$

where M_1 and M_2 are the masses of two gravitating bodies 1 and 2, respectively, r is the distance of separation, i is a unit vector (magnitude 1, no dimensions, and a direction along the line between bodies 1 and 2), and k_g is a constant of proportionality.

In establishing dimension/unit systems, we have to make sure that our choices are consistent with this equation, as well as with Newton's second law. The dimensions and units of any three of the four dimensional quantities in Equation (12-2) (the four are **F**, M, r, and k_g) can be arbitrarily specified; however, when we do this, the fourth is then determined automatically. In nearly all common dimension/unit systems, the dimensions of **F**, M, and r are determined before any consideration is given to the law of gravitation. This means that k_g will then be uniquely determined; it cannot be arbitrarily specified.

Example 12-4

Suppose that you carried out an experiment to measure the gravitational attraction between two objects. Object A has a mass of 4.4×10^{20} heavies (abbreviated hev), while object B has a mass of 7.4×10^5 heavies (you have chosen these nonstandard units because you have not yet read the next section). When you separate A and B by 2.01×10^{-2} strides (st), the magnitude of the force of attraction you measure is 4.68×10^{-23} pulls (pu). Determine the magnitude and units of k_g in the law of universal gravitation. (Although Newton formulated the law of universal gravitation, he could never determine the value of k_g.)

Solution

Solving Equation (12-2) for k_g yields (here F means the magnitude of **F**)

$$k_g = \frac{F r^2}{M_1 M_2}$$

Substituting the appropriate values yields

$$k_g = 4.68 \times 10^{-23} \cdot (2.01 \times 10^{-2})^2 / (4.4 \times 10^{20} \cdot 7.4 \times 10^5)$$

$$\underbrace{\qquad}_{\text{pu}} \cdot (\underbrace{\qquad}_{\text{st}})^2 / (\underbrace{\qquad}_{\text{hevs}} \cdot \underbrace{\qquad}_{\text{hevs}})$$

or

$$k_g = 5.8 \times 10^{-53} \text{ pu} \cdot \text{st}^2/\text{hev}^2$$

Now in these units, we can always use the relationship given by

$$F = 5.8 \times 10^{-53}(\text{pu} \cdot \text{st}^2/\text{hev}^2)M_1 M_2/r^2$$

to relate force, masses, and distance of separation. That is, we have a general law that holds good in these nonstandard units.

THE NEED FOR STANDARDS

As the world continues to grow effectively smaller through transportation and communications marvels, the advantages of having all people use the same dimensional systems and units should be obvious. Many of the unique dimension/unit systems used by earlier cultures have disappeared under attempts to standardize these important features from culture to culture.

Unfortunately, we still do not have one standard dimension/unit system in use throughout the world. The common ones that persist today can be classified into three types based on what are chosen as primary dimensions and what result as secondary dimensions. These are exhibited in Table 12-3. Notice that length (L) and time (t) are primary dimensions in all three categories, but mass (M) and force (F) are not.

Electrical quantities are more involved than those listed in Table 12-3 and will be discussed later.

The SI System

The SI system (from Le Système International d'Unités or International System of Units) is the most internationally accepted dimension/unit system. Of all the industrial nations of the world, only the United States has not converted to the SI system in an extensive way. After several years of investigation by Congress, legislation was enacted in 1974 to implement a changeover to the SI system. However, two factors have worked against this implementation. First, no specific date was set for a mandatory changeover, and second, the conversion process is extremely expensive because it requires much physical retooling of machinery and equipment. The mental conversion that is required to allow people to begin to "think" in terms of SI quantities (kilograms, meters, etc.) is also a deterrent.

Of course one of the attractive features of the metric system, on which the SI system draws, is that prefixes such as *milli*, *centi*, and *kilo* that are commonly added to unit names are all related by powers of 10 (Table 12-4); this makes it a decimal-based system.

Mass. In the SI system, the standard unit for the fundamental dimension of mass is the kilogram (kg) [12-5]. The standard for this unit has been internationally accepted since 1889. It is defined to be the mass (not the volume, weight, or composition) of a certain platinum–iridium cylinder that is maintained under carefully controlled

TABLE 12-2 SOME POSSIBLE CHOICES FOR PRIMARY DIMENSIONS IN NEWTON'S SECOND LAW

Primary Quantity	Primary Dimension	Secondary Quantity	Secondary Dimension
Option 1			
F	F		
a	L/t^2	M	Ft^2/L
k_n	$(1)^a$		
Option 2			
M	M		
a	L/t^2	F	ML/t^2
k_n	$(1)^a$		
Option 3			
F	F		
M	M	k_n	Ft^2/ML
a	L/t^2		
Option 4			
F	F		
M	M	a	F/M
k_n	$(1)^a$		

[a](1) is intended to imply unity or no dimensions.

[12-4] The law of universal gravitation expresses the force with which any two bodies in the universe attract each other.

TABLE 12-3   MOST COMMON CHOICES FOR DIMENSIONING SYSTEMS

Category 1	Category 2	Category 3
Measurable quantities having primary dimensions		
—	Force	Force
Length	Length	Length
Time	Time	Time
Mass	—	Mass
k_n[a]	k_n[a]	—
Measurable quantities having secondary or derived dimensions		
Force	—	—
—	Mass	—
—	—	k_n[a]
k_g[a]	k_g[a]	k_g[a]
Velocity	Velocity	Velocity
Acceleration	Acceleration	Acceleration
Area	Area	Area
Volume	Volume	Volume
Energy	Energy	Energy
Power	Power	Power
Pressure	Pressure	Pressure
Density	Density	Density
Specific weight	Specific weight	Specific weight

[a]The dimensions of k_n and k_g are discussed on pages 393, 395, and 397.

The adoption of a standard based on some very fundamental phenomenon, such as the orange-red light of krypton, allows a unit to be precisely reproduced anywhere in the world, indeed the universe. For example, if scientists and engineers throughout the world need to compare lengths accurately, each could obtain some krypton 86 and set up his or her own local standards lab. This would ensure that person A is talking about the same length as person B. If the standard meter were based on the length of some agreed-upon bar (which it was prior to 1960), kept in a single place at specified conditions, each worker would have to produce copies of that bar, keep them at the same conditions, and at some time make direct comparisons against the standard by traveling to the home of the original.

You may have noticed that the accepted standard for the kilogram mass does not yet have this same type of fundamental definition that leads to independent reproducibility. In the case of kilograms, "traceability of a measurement to the standard," a term often used in standards work, implies going to France and making a direct comparison against the original.

conditions at the International Bureau of Weights and Measures in Sèvres, France.

The gram (g), another unit that is often used for mass, is, by definition, 1/1000 of a kilogram.

Length.   The standard unit for length is the meter (m). As our technology has advanced over the years, the definition of the meter has been refined to the point where it now personifies, perhaps more than the other dimensions, the high-tech world in which we live. A meter, as redefined in 1960, is that distance in space that is equivalent to 1,650,763.73 wavelengths (in a vacuum) of the orange-red light emitted by krypton 86 (a specific isotope of that element). For most of us this is an abstract definition.

It is, of course, much easier to develop an intuitive feeling for this unit after seeing a "meter" stick in a high school or college freshman physics lab (a meter stick is a few inches longer than a yardstick). But because all materials expand and contract with temperature and pressure, any definition based on the length of a certain material has its limitations. (Similarly, the wavelength of light varies with the index of refraction of the material through which it passes; thus, the standard krypton 86 radiation must be measured in a vacuum.)

Although the meter is the standard length in the SI system, the units of millimeter (mm), centimeter (cm), and kilometer (km) are often used to simplify the writing of digits. If a device is only 1 mm long, it is often inconvenient to list it at 0.001 m.

Time.   Once you get the wavelengths of krypton 86 counted, you might want to get started counting the vibrations of a cesium 133 atom, for it holds the key to the definition of the standard time unit, the second. In the time required to count 9,192,631,770 vibrations of this atom, 1 standard second will have elapsed. Obviously, this international standard, set in 1967, also reflects our modern era.

However, the technique of counting atomic vibrations is quite analogous to counting the number of swings of the pendulum in a wall clock and equating a given number of these swings to a certain period of time. Only the accuracy, reproducibility, and preciseness are different.

k_n.   In the SI system, the constant k_n in Newton's second law [Equation (12-1)] is arbitrarily taken to be unity (a magnitude of 1 with no dimensions). We can make this assumption as long as the dimensions and units of force (the only variable left in Newton's law that is not yet defined) are set by the equation and not by us. We will ensure this in the next paragraph.

Force.   With the choice of 1 for k_n, Newton's second law [Equation (12-1)] reduces to

$$\mathbf{F} = M\mathbf{a} \tag{12-3}$$

in the SI system. It is through this law that the (secondary) dimensions of force are established. Thus force must have the dimensions of the product of $M\mathbf{a}$ or ML/t^2.

When we begin to explore forces in the SI system, Newton's law tells us that a force of $1 \text{ kg} \cdot \text{m/s}^2$ is required to accelerate a mass of 1 kg at 1 m/s^2. This comes from direct substitution

$$\mathbf{F} = 1 \text{ kg} \cdot 1 \text{ m/s}^2 = 1 \text{ kg} \cdot \text{m/s}^2 \qquad (12\text{-}4)$$

This gives us a meaningful way of assigning numbers to forces. If a force produces an acceleration of 2 m/s^2 when applied to 1 kg of mass, the force must have a magnitude of $2 \text{ kg} \cdot \text{m/s}^2$. Two important features should be noted from this exercise. The first is that we can now compare one force with another; the second is that we have ensured that Newton's law works in this dimension/unit system.

Since the strange set of units for the dimension of force seems awkward (have you ever seen a spring scale that was graduated in units of $\text{kg} \cdot \text{m/s}^2$?), it is customary to rename any force of $1 \text{ kg} \cdot \text{m/s}^2$ to be a *newton*. This is the standard, but *derived*, unit in the SI system. Note that it is intimately related to having a standard kilogram, a standard meter, and a standard second. Note also that the term *newton* applied to forces is an alias [12-7].

TABLE 12-4   METRIC PREFIXES

Prefix	Multiplication Factor	SI Symbol
exa	10^{18}	E
peta	10^{15}	P
tera	10^{12}	T
giga	10^{9}	G
mega	10^{6}	M
kilo	10^{3}	k
hecto	10^{2}	h
deca	10^{1}	da
deci	10^{-1}	d
centi	10^{-2}	c
milli	10^{-3}	m
micro	10^{-6}	μ
nano	10^{-9}	n
pico	10^{-12}	p
femto	10^{-15}	f
atto	10^{-18}	a

Example 12-5

Consider a mass of 68.4 kg acted on by a force of 667.2 N. What is the acceleration of this mass under these conditions?

Solution

From Newton's second law, the magnitude of the acceleration **a** is

$$a = \frac{F}{M} = \frac{667.2 \text{ N}}{68.4 \text{ kg}} = 9.75 \text{ N/kg}$$

Substituting the collection of units for which newton is an alias results in

$$a = 9.75 \frac{\text{kg} \cdot \text{m}}{\text{s}^2} \cdot \frac{1}{\text{kg}} = 9.75 \text{ m/s}^2$$

k_g. In the SI system, we have seen that M and L are primary dimensions, whereas F is a secondary dimension. We can also see that Equation (12-2) demonstrates that if F, M, and L are all chosen, then k_g cannot be arbitrarily specified. It must have dimensions of [6]

$$k_g \overset{\text{d}}{=} \frac{F \cdot L^2}{M^2} \overset{\text{d}}{=} \frac{\overset{F}{\overbrace{M \cdot L}}}{t^2} \frac{L^2}{M^2} \overset{\text{d}}{=} \frac{L^3}{M \cdot t^2} \qquad (12\text{-}5)$$

In terms of the standard units in the SI system (kg, m, N), k_g is an experimentally determined constant. The accepted value is

$$k_g = 6.673(\pm 0.003) \times 10^{-11} \text{ m}^3/(\text{kg} \cdot \text{s}^2) \qquad (12\text{-}6)$$

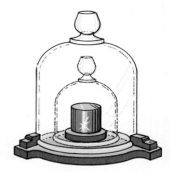

INTERNATIONAL
PROTOTYPE KILOGRAM

[12-5] The standard for the unit of mass, the *kilogram*, is a cylinder of platinum–iridium alloy kept by the International Bureau of Weights and Measures near Paris. A duplicate in the custody of the National Bureau of Standards serves as the mass standard for the United States. This is the only base unit still defined by an artifact.

---

[6]Remember, the symbol d above the equal sign is intended to imply "dimensional equivalence," whereby the terms on the left and right sides of the equation have the same dimensions. Numerical equality is not intended.

[12-6] A balance can be used to compare the mass of one object with another (usually known) mass. However, the balance really compares the weight of one object with the weight of the other. Since both masses are located in the same gravitational field, the masses are equal when the weights are equal.

The only problem with tradition is that it sometimes lasts too long.

Source unknown

The English Engineering System

The English Engineering system has been used in the English-speaking countries of the world for decades. However, most of these countries have now effectively changed to the SI system in order to better participate and compete in the international marketplace and to take advantage of the multiples of 10 that separate various sub-units within the SI system. Ironically, even England now shuns the English Engineering system in favor of the SI system.

At the present time, the English Engineering system's chief user, but not necessarily a proponent, is the United States, where there are two common forms in use. These are the *FMLt* system and the *FLt* system. The latter is sometimes called the British Gravitational system.[7] The two systems differ in the choices of primary dimensions.

The English Engineering system is so deeply embedded in the U.S. engineering profession that it will probably be at least a decade or two before the adoption of the SI system can be considered successful. Most of the engineering journals where important work can be published and disseminated now require that SI units be used in all papers. American journals, however, will also accept the English Engineering equivalents if they are entered in parentheses alongside the SI values.

Most machinery that is being shipped abroad from the United States today is dimensioned in SI units. Also, most computer-aided drafting packages available today allow the user to work in either the SI or the English Engineering system, while the machine does the conversion internally if required. The conversion to the SI system is continuing, but the pace has slowed considerably from what it was in the mid-1970s.

The English Engineering *FMLt* System

Force, mass, length, and time are all considered to be primary dimensions in the *FMLt* system. Thus this system necessarily falls into category 3 in Table 12-3; it is one of the very few that does. The *FMLt* system is most deeply embedded in the disciplines of engineering concerned with fluid flow, thermodynamics (the study of energy), and heat transfer (these subjects are sometimes grouped under the term *thermosciences*).

This system is one of the most confusing of all because of the use of the word "pound" as a measurement of both force and mass. Students should always speak carefully and with preciseness of meaning when using this word.

---

[7]The dimension/unit systems that define force to be one of the primary dimensions at one time had to rely on the gravitational attraction between the earth and a standard mass in order to define a standard force unit. Some authors often use the word *gravitational* as a part of the system name. Examples of such systems are the English Engineering system *FLt* and the metric mks system, the latter of which defines a kg_f as a primary force unit. Other systems, which define M as a primary dimension, such as the SI system, are sometimes referred as *absolute* systems. Since the English *FMLt* system chooses both F and M as primary, it is your choice to call it a gravitational or an absolute system.

Mass. The standard unit for the primary dimension of mass in the English Engineering *FMLt* system is the pound-mass, abbreviated as lb_{mass} or lb_m. Although there is an interesting history behind its evolution and adoption, the pound mass is now simply defined in terms of the standard kilogram-mass that was discussed under the SI system. This definition is

$$1 \ lb_m = 0.453 \ 592 \ 37 \ kg \qquad (12\text{-}7)$$

or the more familiar relationship

$$1 \ kg = 2.204 \ 622 \ 6 \ lb_m \qquad (12\text{-}8)$$

Distance. The standard and most often used unit for the dimension of length is the foot (ft), defined to be

$$1 \ ft = 0.3048 \ m \ (\text{exactly}) \qquad (12\text{-}9)$$

Other length units often used are the inch, the yard, the statute mile, and the nautical mile.

Time. The standard unit or scale for the dimension time is the second, defined the same way as in the SI system.

Force. Force is considered to be a primary (not derived) dimension in the English Engineering *FMLt* system. The standard unit for measuring the dimension of force in this system is the pound-force, abbreviated lb_{force} or lb_f. It is paramount that the reader know and understand that there is a fundamental difference between lb_m and lb_f, just as there is a fundamental difference between a kilogram (a mass) and a newton (a force) in the SI system.

We arbitrarily say that 1 lb_f is that force, when applied to a standard 1 lb_m (which is now well defined), which produces an acceleration of 32.1740 ft/s² (both foot and second are also now well defined). There are some details that need to be accommodated, such as how we measure the acceleration, but the experiment itself is well defined.

Now we have a unit that allows us to put meaningful numbers on forces just as we did in the SI system. (However, in the SI system, we did not arbitrarily choose the dimensions and units of force, but let Newton's law do it for us.) All that remains is to ensure that when we use these arbitrarily chosen units of F, M, L, and t in our physical laws of nature, we will get meaningful (i.e., correct) results. The discussions on k_n and k_g below show how this is done for Newton's second law and the universal law of gravitation.

k_n. As pointed out earlier, there must be some flexibility in the equations which relate measurable quantities so that arbitrary choices on the dimensions of quantities that appear in the equations do not overspecify the problem. In the English Engineering *FMLt* system, since F, L, M, and t are all arbitrarily defined, we cannot arbitrarily assign a value to k_n. If Newton's second law is to be satisfied for a 1 lb_m accelerating at 32.1740 ft/s² under the action of a force of magnitude of 1 lb_f, then we must have

$$1 \ lb_f = k_n \cdot 1 \ lb_m \cdot 32.1740 \ ft/s^2 \qquad (12\text{-}10)$$

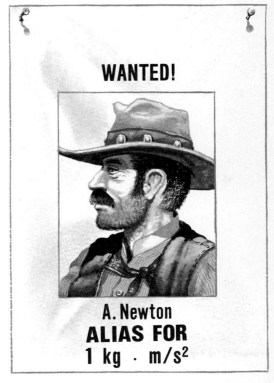

WANTED!

A. Newton
ALIAS FOR
1 kg · m/s²

[12-7]

You may wonder how the number 32.1740 ft/sec² was chosen to be a part of the definition of lb_f. Why wasn't 1 or 10 or perhaps, 136.3 ft/sec² chosen? In the evolution of the English system, people talked of taking the standard mass (1 lb_m) to a given place on earth (approximately sea level and 45° latitude) and measuring the gravitational force applied to that mass by the earth. This pull or force was called (calibrated as) 1 lb_f. In English units, the acceleration of an unrestrained object under just the influence of gravity at the subject location was 32.174 ft/s² (9.807 m/s²). An object weighing 1 lb_m at the subject location would have a mass of 1 lb_m. Thus the definition is appropriate.

The acceleration produced by only gravitational attraction acting on an unrestrained object is called the *local acceleration of gravity*. As was found by experiments conducted from the leaning tower of Pisa in Italy, such an acceleration is independent of the shape (neglecting air resistance) or mass of the object. However, this acceleration does depend on location with respect to the mutually attracting body. *At sea level on earth,* the local acceleration of gravity g, at any latitude ϕ, may be approximated from the following relationship:

$$g = 32.09(1 + 0.0053 \sin^2\phi) \ ft/s^2$$

or

$$g = 9.78(1 + 0.0053 \sin^2\phi) \ m/s^2$$

[12-8] These machine tools are calibrated in both SI and English units.

or

$$k_n = \frac{1}{32.1740} \frac{\text{lb}_f \cdot s^2}{\text{ft} \cdot \text{lb}_m} \qquad (12\text{-}11)$$

Notice that the dimensions of k_n are $F \cdot t^2/(L \cdot M)$, not t^2/L. Notice also that the units are not (s^2/ft). The lb_f and the lb_m do not cancel mathematically because they are not the same thing. A 1 lb_m, being a unit of mass, is independent of its environment. No matter where you take it, as long as you retain the original collection of matter that made up the 1 lb_m, it will remain a mass of 1 lb_m. At a location on earth where the local acceleration of gravity is 32.174 ft/s^2, the 1 lb_m will weigh, or be attracted by the earth, with a force of 1 lb_f. At any location where the local acceleration of gravity is different from 32.174 ft/s^2, the weight or force of attraction of a 1 lb_m will be different than 1 lb_f.

Forces can act on masses and cause the masses to accelerate, but forces are not masses. Mass is a property that a collection of matter possesses. Force is something that a collection of matter can experience. There is a very fundamental, but sometimes subtle, difference in the two.

It is common in engineering work that uses the $FMLt$ system, particularly the thermosciences, to use a term called g_c, which is simply $1/k_n$. In these cases Newton's law is written as

$$F = \frac{1}{g_c} Ma \qquad (12\text{-}12)$$

where for the $FMLt$ system, $g_c = 32.1740 \text{ ft} \cdot \text{lb}_m/(\text{lb}_f \cdot s^2)$. You will find the inclusion of g_c (or k_n, for that matter) to be somewhat controversial. Some engineering literature and many textbooks use it, some do not; some instructors require it, some do not.

Unfortunately, the tendency on the part of many students, when using Equation (12-12), is to associate g_c with the acceleration due to gravity. However, it should be clear by now that the term does not even possess the dimensions, not to mention the units, of an acceleration.

If g_c is used, it must be viewed as a fundamental constant inherent in the dimension/units system being used. Its value depends only on the dimension/units system chosen and not on the strength of the local gravitational field or other physical phenomena. In the English Engineering system being discussed here, the value is 32.1740 ft · $\text{lb}_m/(\text{lb}_f \cdot s^2)$; in the SI system, it is unity.

The personal preference of the authors is to write the basic equations without k_n or g_c when using any of the dimension/unit systems, including the English Engineering $FMLt$ system. Instead, we prefer a very careful units analysis of every equation used. Inconsistencies will appear during units reduction (the algebraic reduction of units involved in the equation) if g_c should have been included but was not. For example, if you are calculating a force and the units analysis produces a ft · $\text{lb}_m/(\text{lb}_f \cdot s^2)$ instead of a lb_f, you will know that something is wrong. A little experience will tell you that all you need do is divide by g_c. We reiterate this point in the section on units conversion later in this chapter.

With a workable Newton's second law in both the SI and English Engineering *FMLt* systems, we can now express 1 lb$_f$ in terms of the force unit in the SI system, the newton [remember, however, that the dimensions of force in the SI system are derived, whereas those in the English Engineering *FMLt* system are not]. The result is

$$1 \text{ lb}_f = 4.448\ 221\ 615\ 260\ 5 \text{ N} \qquad (12\text{-}13)$$

Example 12-6

Establish the foregoing relationship between lb$_f$ and newtons.

Solution

To determine force from Newton's second law, we need mass and acceleration. Therefore, let us assume that we are dealing with 1 kg of mass and that we wish to accelerate it at 1 m/s^2 (we may as well start off with some "nice" numbers). In the English system, the 1-kg mass is equivalent to 2.205 lb$_m$, while the acceleration of 1 m/s^2 is equivalent to

$$1\frac{\text{m}}{\text{s}^2} \cdot \frac{1 \text{ ft}}{0.3048 \text{ m}} = 3.281 \text{ ft/s}^2$$

In the metric system, a force of

$$F = 1 \text{ kg} \cdot 1 \text{ m/s}^2 = 1 \text{ N}$$

is required to produce the assumed acceleration. In the English system, a force of

$$F = 2.205 \text{ lb}_m \cdot 3.281 \text{ ft/s}^2 = 7.233 \text{ lb}_m \cdot \text{ft/s}^2$$

The units do not look familiar for a force. But since they involve lb$_m$, we should sense that we need g_c here (the constant we discussed above and not the local acceleration of gravity)

$$F = 7.234 \frac{\text{lb}_m \cdot \text{f}_t}{\text{s}^2} \cdot \frac{1}{32.174} \frac{\text{lb}_f \cdot \text{s}^2}{\text{ft} \cdot \text{lb}_m}$$

$$= 0.225 \text{ lb}_f$$

Thus 1 N = 0.225 lb$_f$, or 1 lb$_f$ = 4.45 N.

k_g. As in the SI system, all the quantities in the universal law of gravitation have now been determined in the Engineering *FMLt* system, except k_g. The dimensions and units of k_g are found from the form of the equation. The numerical value must come from experimental data. It is

$$k_g = 1.068\ 91 \times 10^{-9} \text{ lb}_f \cdot \text{ft}^2/\text{lb}_m^2 \qquad (12\text{-}14)$$

Example 12-7

Suppose that your bathroom scales indicated that your weight (the force with which the earth attracts you) was 150.0 lb$_f$ (if you are familiar with the accuracy of bathroom scales, you should question the resolution of this number). Knowing that the local acceleration of gravity was 32.00 ft/s^2, calculate your mass in kg.

Solution

Since metric units were requested, perhaps we should convert the given quantities (here your weight and the local acceleration of gravity) to metric values. Since we found earlier that 1 lb_f = 4.45 N, we easily calculate that

$$Wt = 150.0 \ lb_f \cdot 4.448 \ N/lb_f = 667.2 \ N$$

The local acceleration of gravity can also be converted since we know the conversion between ft and m:

$$g = 32.0 \ ft/s^2 \cdot 0.3048 \ m/ft = 9.75 \ m/s^2$$

Newton's law then tells us that

$$M = \frac{F}{a} = \frac{667.2 \ N}{9.754 \ m/s^2} = 68.40 \ \frac{N \cdot s^2}{m} = 68.40 \ kg$$

Before we leave this example, it may be constructive to convert this to mass in lb_m.

$$M = 68.40 \ kg \cdot 2.205 \ lb_m/kg = 150.8 \ lb_m$$

Note that the weight and mass in English units are not numerically equal, since the local acceleration of gravity differs slightly from the value of 32.174 ft/s^2 used to standardize the pound-mass.

The English Engineering *FLt* System

Force, length, and time are commonly said to be the primary dimensions in the *FLt* system. Thus this system necessarily falls into category 2 in Table 12-2. This system is used more in the engineering disciplines that deal primarily with mechanics (the study of the relation between applied forces, the motion of objects, and the stresses transmitted through the objects) than those that study thermoscience subjects.

This system is just as confusing as the *FMLt* system because the definitions within it seem to go in circles. This is more a result of the evolution of the system than anything else. If you enjoy the history of technology, you would find the study of this evolution quite interesting.

What we wish to stress here is that all the standards for the units used in this system, as well as those we discussed previously, are well established. Some, like the kg and its relative, the lb_m, require traceability to a single piece of matter, unique in the world. Others are based on physical phenomena that are highly reproducible, allowing anyone with the appropriate apparatus to duplicate the standard.

The word "pound" is used in the Engineering *FLt* system only in connection with forces. To avoid confusion, however, it is good practice to retain the additional word "force" (e.g., pound-force or lb_f) when specifying this unit.

Distance. The standard and most often used unit for the dimension of length is the foot (ft), just as in the *FMLt* system.

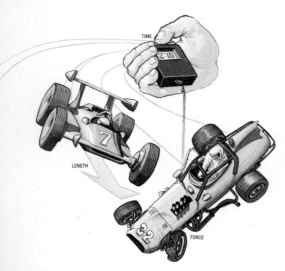

Time. The most common units of the dimension time are seconds, which are defined the same way as in both the English Engineering *FMLt* and SI systems.

Force. Force is considered to be a primary, not derived, dimension in the English Engineering *FLt* system, just as it is in the *FMLt* system. However, its definition in the *FLt* system seems to go in circles, because although mass is said to be a derived quantity, the definition of the standard force unit depends on having a defined mass.

The standard unit for force is the pound-force. Basically, it is defined the same way as the lb$_f$ is defined for the *FMLt* system. Note, however, that it depends on the definition of the lb$_m$. We have

$$1 \text{ lb}_f \bigg|_{FLt} = 1 \text{ lb}_f \bigg|_{FMLt} = 4.448\,221 \text{ } N \bigg|_{SI} \qquad (12\text{-}15)$$

k_n. k_n is arbitrarily defined to be unity.

Mass. To satisfy Newton's second law with arbitrarily chosen units on F, L, t, and k_n, we must now let the equation tell us the appropriate size or unit for mass. We have come nearly full circle; what we must do now is redefine the magnitude of the unit of mass which we will need for satisfying Newton's second law. That is, we have defined 1 lb$_m$ in order to define 1 lb$_f$. Now we are defining a new size or unit for mass so that we can use Newton's second law correctly with $k_n = 1$. Solving that equation for mass, we find

$$M = \frac{F}{k_n a} \stackrel{\mathrm{d}}{=} \frac{F}{1 \cdot L/t^2} \stackrel{\mathrm{d}}{=} \frac{Ft^2}{L} \qquad (12\text{-}16)$$

If we were to calculate the magnitude of the mass that would accelerate at 1 ft/s^2 under the action of a force of 1 lb$_f$, Newton's law tells us that the mass would be

$$M = \frac{1 \text{ lb}_f}{1 \text{ ft/s}^2} = 1 \frac{\text{lb}_f \cdot \text{s}^2}{\text{ft}} \qquad (12\text{-}17)$$

A mass of 1 lb$_f \cdot$ s^2/ft has traditionally been given the name *slug*. This is an alias in the same way that N is an alias for a force of 1 kg $\cdot$ m/s^2 [12-10]. Newton's second law in the English Engineering *FLt* system becomes

$$\mathbf{F} = M\mathbf{a} \qquad (12\text{-}18)$$

as long as forces are measured in lb$_f$, lengths in feet, times in seconds and masses in slugs. Note here for the *FLt* system that $k_n = 1/g_c = 1$, which is considerably different from k_n or g_c in the English Engineering *FMLt* system. It may be obvious now that g_c in the *FMLt* system, combined with the definition of the slug, reduces to

$$g_c = 32.1740 \text{ ft} \cdot \text{lb}_m/(\text{lb}_f \cdot \text{sec}^2)$$

$$= 32.1740 \text{ lb}_m/\text{slug} \qquad (12\text{-}19)$$

This means that g_c just plays the role of a mass conversion constant in

WANTED!

The Slug
ALIAS FOR
1 lb$_f$ $\cdot$ s^2/ft.

[12-10]

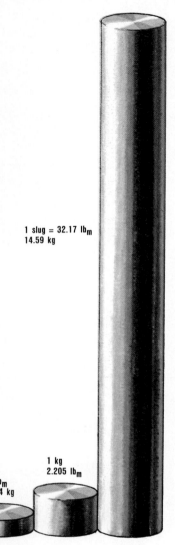

1 slug = 32.17 lb$_m$
14.59 kg

1 kg
2.205 lb$_m$

1 lb$_m$
.454 kg

[12-11] If these figures represent pieces of the same material, then the size of the figures denote the relative sizes of the lb$_m$, kg, and slug.

Newton's second law in the English Engineering *FMLt* system. In fact, this shows that

$$1 \text{ slug} = 32.1740 \text{ lb}_m = 14.593\ 879 \text{ kg} \qquad (12\text{-}20)$$

If size were proportional to mass (be careful here, for in the physical world, size and mass are not necessarily related), Figure [12-10] demonstrates the relative magnitudes of the lb$_m$, kg, and slug.

k$_g$. All the quantities in the universal law of gravitation have now been determined in the Engineering *FLt* system, except *k$_g$*. The dimensions and units can be determined from the form of the equation; the numerical value must come from experimental data. It is

$$k_g = 3.439\ 1 \times 10^{-8} \text{ ft}^4/(\text{lb}_f \cdot \text{s}^4)$$

or

$$k_g = 3.439\ 1 \times 10^{-8} \text{ ft}^3/(\text{slug} \cdot \text{s}^2) \qquad (12\text{-}21)$$

The latter set of units is obtained from the former by using the alias slug for the collection of terms, lb$_f \cdot$ s^2/ft.

Example 12-8
Solve for the mass (in lb$_m$) which is being accelerated at 6.14 ft/s^2 by a force of 196 lb$_f$.

Solution

$$F = Ma \quad \text{or} \quad M = \frac{F}{a}$$

$$M = \frac{196 \text{ lb}_f}{6.14 \text{ ft/s}^2} = 31.9 \frac{\text{lb}_f \cdot \text{s}^2}{\text{ft}}$$

Notice that the units do not involve lb$_m$. However, the collection of units that we do get is known by the alias, the slug. Hence

$$M = 31.9 \text{ slugs}$$

Since the conversion between slugs and lb$_m$ is 1 slug = 32.147 lb$_m$, then

$$M = 31.9 \text{ slugs}(32.174 \text{ lb}_m/\text{slug})$$

$$= 1.03 \times 10^3 \text{ lb}_m$$

The multiplicative factor 32.174 lb$_m$/slug is nothing more than unity since the numerator and the denominator are equal to one another. Therefore, use of it anywhere in an equation does not change the equality, but serves only to convert the units.

Problems

12-25. Change 100 N of force to lb$_f$.

12-26. In the English *FLt* system, what mass in slugs is necessary to produce a weight of 15.6 lb$_f$ at standard conditions?

12-27. In the English *FMLt* system, what mass in lb$_m$ is necessary to produce a weight of 195.3 lb$_f$ at standard conditions?

12-28. An interstellar explorer is accelerating uniformly at 58.6 ft/sec^2 in a spherical spaceship which has a total mass of 100,000 slugs. What is the force acting on the ship?

12-29. At a certain instant in time a space vehicle is being acted on by a vertically upward thrust of 497,000 lb$_f$. The mass of the space vehicle is 400,000 lb$_m$, and the acceleration of gravity is 32.1 ft/sec^2. Is the vehicle rising or descending? What is its acceleration? (Assume that "up" means radially outward from the center of the earth.)

12-30. Some interstellar adventurers land their spacecraft on a certain celestial body. Explain how they could calculate the acceleration of gravity at the point where they landed.

12-31. Using the relationship for g on page 395 and the fps gravitational system of units, determine the weight, at the latitude $0°$, of a stainless steel sphere whose mass is defined as 150 $lb_f \cdot sec^2/ft$.

12-32. The mass of solid propellant in a certain container is 5 kg. What is the weight of this material in newtons at a location in Greenland where the acceleration of gravity is 9.83 m/sec²? What is the weight in newtons?

12-33. If a gold sphere has a mass of 89.3 lb_m on earth, what would be its weight in lb_f on the moon, where the acceleration of gravity is 5.31 ft/sec²? What is the weight in SI units?

12-34. Assuming that the acceleration due to gravitation is 5.31 ft/sec² on the moon, what is the mass in slugs of 100 lb_m located on the moon? In SI units?

12-35. A silver bar weighs 382 lb_f at a point on the earth where the acceleration of gravity is measured to be 32.1 ft/sec². Calculate the mass of the bar in lb_m and slug units.

12-36. The acceleration of gravity can be approximated by the following relationship:

$$g = 980.6 - 3.086 \times 10^{-6}A$$

where g is expressed in cm/sec² and A is an altitude in centimeters. If a rocket weighs 10,370 lb_f at sea level and standard conditions, what will be its weight in dynes at an elevation of 50,000 ft? In SI units?

12-37. At a certain point on the moon the acceleration due to gravitation is 5.35 ft/sec². A rocket resting on the moon's surface at this point weighs 23,500 lb_f. What is its mass in slugs? In lb_m? In SI units?

12-38. If a 10-lb weight on the moon (where $g = 5.33$ ft/sec²) is returned to the earth and deposited at a latitude of $90°$ (see p. 281), how much would it weigh in the new location?

12-39. A 4.37-slug mass is taken from the earth to the moon and located at a point where $g = 5.33$ ft/sec². What is the magnitude of its mass in the new location?

12-40. The inertia force due to the acceleration of a rocket can be expressed as follows:

$$F = Ma$$

where F = unbalanced force

a = acceleration of the body

M = mass of the body

a. Given: $a = 439$ ft/s²; $M = 89.6$ $lb_f \cdot s^2/ft$
   Find: F in lb_f; in SI units.
b. Given: $F = 1500$ lb_f; $M = 26.4$ $lb_f \cdot s^2/ft$
   Find: a in ft/s²; in SI units
c. Given: $F = 49.3 \times 10^5$ lb_f; $a = 32.2$ ft/s²
   Find: M in $lb_f \cdot s^2/ft$; in SI units.
d. Given: $M = 9650$ $lb_f \cdot s^2/ft$; $a = 980$ cm/s²
   Find: F in lb_f; in SI units.

12-41. Choose k_g to be unity, and assume that F, t, and M are primary dimensions. Use the law of universal gravitation to define the secondary dimension of length. What are the dimensions of k_N?

ELECTRICAL, MAGNETIC, AND THERMAL QUANTITIES

Electrical and Magnetic Quantities

There are several observable or measurable quantities that are important in the fields of electricity and magnetism. In addition to length (L), force (F), work (W), power (P), and charge (Q), mentioned in in the first section of this chapter,[8] these include electrical current (I), electrostatic potential (ϕ), electric field [or voltage] (E), resistance (R), capacitance (C), inductance (L), magnetic flux [or magnetic induction] (B), auxiliary field [or magnetic field strength] (H), plus a few others. Each of these can be considered to be a dimension.

   From the list above, it appears that there are a large number of new and different dimensions with which we must learn to deal. However, due to the many fundamental laws that have been found to govern these fields, nearly all of the new dimensions are secondary

---

[8] You should not be surprised to see force and length (and their product, work) play intimate roles in electromagnetic theory. It is this interrelationship which makes, among other things, electromagnets attract ferrous metals and motor shafts turn.

ones. That is, these new dimensions we have listed can be expressed in terms of a much smaller subset of primary dimensions.

In discussing the lack of standards in dimensions and units encountered in mechanics (quantities involved in Newton's second law and the law of universal gravitation), we mentioned SI units and the two sets of commonly used English engineering units. Although the fields of electricity and magnetism do not embrace English engineering units, they do use several different sets of metric units. Among these are the rationalized mks system, the absolute esu system, and the absolute emu system.

The chief difference between the various dimension/unit systems used in electricity and magnetism is the set of primary dimensions that each system uses. Since the derived or secondary sets are chosen from the primary ones, the derived dimensions are also different. A more minor difference among the various systems is the actual units commonly used to quantify each dimension.

We will cover only one example of dimension/unit systems in electricity and magnetism because to address all of the different dimensions would require more space than we can devote here. We will cover this single example to reinforce the understanding of dimensions and units that you have gained from the earlier sections in this chapter. The important concept is that in establishing dimension/unit systems, we should let fundamental laws form as many of the relationships between quantities as we can. We should avoid overspecifying the quantities that are to be considered as primary.

The example shown below uses Coulomb's law, which relates the forces that two charged particles exert on one another. In equation form this law is expressed as

$$F = k_c \frac{q_1 q_2}{r^2} \tag{12-22}$$

where F is the magnitude of the force, q_1 and q_2 are the electrostatic charges on particles 1 and 2, respectively, r is the distance of separation between the two particles, and k_c is a constant of proportionality. You may remember this law being demonstrated in high school physics experiments, where it was shown that like charges repel and unlike charges attract each other [12-12].

Two of the quantities, force and distance, contained in Equation (12-22) were encountered earlier in our discussions on dimensions in mechanics. The other two, k_c and the q's, are new. Were we to establish arbitrarily that charge, q, and k_c are going to be primary dimensions, we would have to accept the fact that either force or length (or the combination of both) would be derived dimensions. This, of course, is going to lead to incompatibility with our concepts of force and length as being the same dimensions that we established earlier in this chapter. That would be both unfortunate and unwise.

A better alternative is to consider that force and length have already been established by our work in mechanics, and then to let Equation (12-22) establish either the dimensions of q or k_c (or the combination). The rationalized mks system and the absolute esu system do just that, although they differ on whether q or k_c is to be considered a primary dimension.

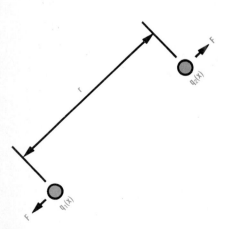

[12-12] Two like charges (i.e., both +, or both −) repelling each other.

The Rationalized MKS System

Dimensions. In the rationalized mks system, the dimension of charge (Q) is declared to be a primary dimension. The dimensions of force and length are taken from the SI system, where they have dimensions of ML/t^2 and L, respectively. With these choices, then, only the dimensions of k_c remain to be determined from Equation (12-22). Solving this equation for k_c yields

$$k_c = \frac{Fr^2}{q_1 q_2} \overset{\text{d}}{=} \frac{F \cdot L^2}{Q^2} \overset{\text{d}}{=} \frac{M \cdot L}{t^2} \frac{L^2}{Q^2} \overset{\text{d}}{=} \frac{M \cdot L^3}{t^2 \cdot Q^2} \qquad (12\text{-}23)$$

It is more traditional to write Equation (12-22) as

$$F = \frac{1}{4\pi\epsilon_0} \frac{q_1 q_2}{r^2} \qquad (12\text{-}24)$$

where $\epsilon_0 = 1/(4\pi k_c) \overset{\text{d}}{=} Q^2 \cdot t^2/(M \cdot L^3)$. The constant ϵ_0 is usually referred to as the dielectric constant or "permittivity" of free space.

Units. In the rationalized mks system, the common unit of mass is the kilogram, the unit of length is the meter, the unit of time is the second, and the unit of charge is the coulomb. One coulomb ($\mathcal{C}$) is the charge found on 6.24196×10^{18} electrons.

For a given pair of charged particles held a known distance apart, the force of attraction or repulsion is predetermined by nature and cannot be set arbitrarily. This means that for Equation (12-24) to give the correct answer, the value of ϵ_0 (or k_c) cannot be set arbitrarily. It must be determined experimentally. The common value used is $\epsilon_0 = 8.854 \times 10^{-12}\ \mathcal{C}^2 \cdot s^2/(kg \cdot m^3)$. Since a newton of force is $1\ kg \cdot m/s^2$, the collection of units on this ϵ_0 is identical to $\mathcal{C}^2/(N \cdot m^2)$. With this substitution, ϵ_0 is also equal to $8.854 \times 10^{-12}\ \mathcal{C}^2/(N \cdot m^2)$.

The Absolute ESU System

Dimensions. In the absolute esu system, k_c is arbitrarily taken as unity (a value of 1, no dimensions). The dimensions of force and length are assumed to be the same as the respective dimensions in the SI unit systems (but a different set of units is commonly used, as we will note in the next section). This leaves only the dimensions of charge to be determined. Solving Equation (12-24) for the product of the q's, we find

$$q_1 q_2 = \frac{F \cdot r^2}{k_c} \overset{\text{d}}{=} F \cdot L^2 \overset{\text{d}}{=} \frac{M \cdot L}{t^2} \cdot L^2 \overset{\text{d}}{=} \frac{M \cdot L^3}{t^2} \qquad (12\text{-}25)$$

Thus the dimensions of charge are $(M \cdot L^3/t^2)^{1/2}$ in the absolute esu system.

Units. The traditional units in the absolute esu system are the centimeter for length, the gram for mass, and the second for time. We have already noted that k_c has been chosen equal to 1. The (secondary) unit for force in Coulomb's law is the dyne, which is an alias for $1\ g \cdot cm/s^2$. A dyne is also the same as 10^{-5} N.

Two identical charged particles that repel each other with a force of 1 dyne (i.e., 1 g · cm/s²) when placed 1 cm apart must have a charge of 1 (dyne · cm²)$^{1/2}$ or 1 (g · cm/s²)$^{1/2}$, according to the absolute esu version of Coulomb's law. Since this looks like an appropriate place for an alias, we say that a charge of 1 (dyne · cm²)$^{1/2}$ will be called a *statcoulomb*. Thus the statcoulomb is an alias in the same way that the newton is an alias for 1 kg · m/s² of force. Now, in the following form of Coulomb's law,

$$F = \frac{q_1 q_2}{r^2} \qquad (12\text{-}26)$$

we will get meaningful results for forces (in dynes) when we use r's in cm and q's in statcoulombs.

Thermal Quantities

Dimensions. In the subjects of energy and thermodynamics, there is another set of dimensions, or measurable quantities, that become important. Fortunately, like the fields of electricity and magnetism, nearly all of these can be considered to be derived or secondary dimensions. Through basic laws and fundamental definitions, they can be related to the dimensions already discussed in this chapter (mass, length, time, force, etc.). Essentially the only exception to this is the fundamental quantity temperature (T). It must be considered as a primary dimension in problems involving the storage of energy within the atoms, molecules, and crystal structure that make up matter, the so-called internal energy.

Units. There are four sets of units in current use that allow us to put quantitative values or numbers on the primary dimension of temperature. Two of these four are said to be "absolute" temperature scales and two are said to be "relative" scales.

The absolute scales have their origin in the theory and principles of the science of thermodynamics. In this field we describe absolute zero as that temperature where nearly all molecular motion ceases.

The two relative scales are the Celsius (C) and the Fahrenheit (F) scales. The extent of the Celsius scale was defined by assigning the freezing temperature of water an arbitrary value of 0 "degrees," and the boiling temperature (at 1 atmosphere of pressure) an arbitrary value of 100 "degrees." The Fahrenheit scale was similarly established by assigning an arbitrary value of 32 degrees to the freezing point of water and a value of 212 degrees to the boiling point. Thus, the Celsius scale has 100 steps or "degrees" between the freezing and boiling points of water, while the Fahrenheit scale has 180 steps or degrees. From this we can see that a Celsius degree is not the same size as a Fahrenheit degree. The ratio of 100 to 180 is in the proportion of 5 to 9. This is the source of the $\frac{5}{9}$ and $\frac{9}{5}$ that appear in the conversion tables in Appendix I.

The two absolute scales are the Kelvin (K) scale and the Rankine (R) scale. Both start with 0 (zero) at absolute zero temperature, but each assigns different increments as the temperature increases. They are analogous to using two different measuring tapes, one calibrated in feet and one calibrated in meters, to measure distance from a specified point. Both measuring tapes might begin with their zero at

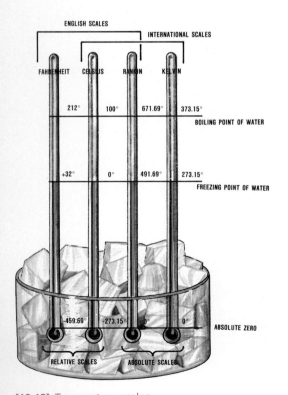

[12-13] Temperature scales.

the specified point but each progresses at different increments. When the tape marked in feet reads 10 ft, the tape marked in meters will show slightly over 3 m.

The size of the increments between temperatures is the same in the absolute Kelvin scale as in the relative Celsius scale. Similarly, the size of the increments between temperatures or the degrees is the same in the absolute Rankine scale as in the relative Fahrenheit scale. This means that the Kelvin degree is $\frac{9}{5}$ the size of a Rankine degree.

When you specify temperature it is imperative that you specify the scale that you are using. To say the temperature is 45 degrees or that the temperature has changed by 3 degrees is meaningless.

The Kelvin and Celsius scales are often referred to as the International scales since they, like the SI unit system, are almost uniformly adopted around the world. The Rankine and Fahrenheit scales, sometimes referred to as the English temperature scales, are now in use only in the United States.

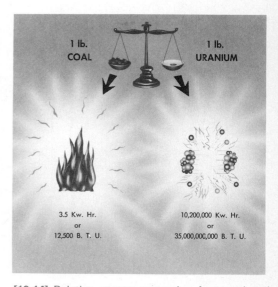

[12-14] Relative energy conversion from coal and nuclear fuels. In working with such energy sources, the engineer must be able to convert from one unit system to another.

UNITS CONVERSION

Conversion between units within the same dimension is not too difficult. That is, converting from one unit of mass to another unit of mass, or from one unit of length to another unit of length, requires the use of a simple conversion factor. A table of conversion factors is included in Appendix I to aid in this process.

Conversion between two or more different dimensions should not be attempted. It is wrong both conceptually and numerically. You can no more convert force units into mass units than you can length units into time units.

Engineers frequently work in several systems of units in the same calculation. In this case it is only necessary that each dimension be expressed in any valid set of units from any of the various unit systems. Numerical equality and unit homogeneity may be determined in any case by applying conversion factors to the individual terms of the expression.

Unfortunately, in much of the engineering literature, the equations used in a particular situation include one or more conversion factors. Therefore, considerable care must be exercised in using these expressions since they represent a "special case" rather than a "general condition." The engineer should form a habit of always checking the unit balance of all equations.

Example 12-9

Change a speed of 3000 miles per hour (mi/hr) to m/s.

Solution

Calling the speed V, we find

$$V = \left[3000 \ \frac{mi}{hr}\right]\underbrace{\left[\frac{5280 \ ft}{1 \ mi}\right]}_{\substack{\text{Conversion} \\ \text{factor} \\ 1}}\underbrace{\left[\frac{1 \ hr}{3600 \ s}\right]}_{\substack{\text{Conversion} \\ \text{factor} \\ 2}}\underbrace{\left[\frac{0.3048 \ m}{ft}\right]}_{\substack{\text{Conversion} \\ \text{factor} \\ 3}}$$

$$= 1341 \ m/s$$

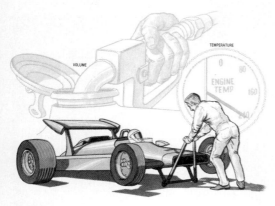

[12-15]

[12-16]

[12-17]

Here again, the three conversion factors are each equivalent to unity since the numerator and denominators are equivalent. These factors do not change the validity of the equation, only the units.

Example 12-10

Using the weight and mass from Example 12-7, determine the mass of the earth assuming that a distance of 6379 km separates you from the center of the earth.

Solution

Since this problem involves gravitational attraction, perhaps the law of universal gravitation, expressed in Equation (12-2), would be a good place to start. Since we know your mass and the force of attraction (your weight) and the distance of separation, we can calculate the mass of the earth from

$$M_e = \frac{Fr^2}{M_{you}k_g} = \frac{667.2 \text{ N} \cdot (6379 \times 10^3 \text{ m})^2}{68.40 \text{ kg} \cdot 6.673 \times 10^{-11} \text{ m}^3/(\text{kg} \cdot \text{s}^2)}$$

$$= 5.95 \times 10^{24} \text{ N} \cdot \text{s}^2/\text{m}$$

or, using the fact that the newton is an alias for a $\text{kg} \cdot \text{m/s}^2$,

$$M_e = 5.95 \times 10^{24} \frac{\text{kg} \cdot \text{m}}{\text{s}^2} \cdot \frac{\text{s}^2}{\text{m}} = 5.95 \times 10^{24} \text{ kg}$$

DERIVED DIMENSIONS/UNITS OF OTHER COMMONLY USED QUANTITIES

We end this chapter with a brief discussion of several common quantities defined and used by engineers. Conversion factors for various units are supplied in Appendix I.

Area and Volume

The dimensions of area and volume are L^2 and L^3, respectively. Commonly used units are:

Area: m^2 (SI preferred), cm^2, mm^2, in.^{2}, ft^2, yd^2, miles^2, acres, hectares.

Volume: m^3 (SI preferred), cm^3 (or cc), liter, in^3, ft^3, yd^3 (often called simply a yard for concrete or sand), gallons, Imperial gallons, barrels, bushels, board feet (for lumber), cords (for firewood).

Work

Work is the product of a force and a distance through which that force acts. Energy is the ability or capacity for doing work. Although the two quantities are conceptually different, they have the same dimensions and can, therefore, be expressed in the same units. The dimensions are $F \cdot L$. Commonly used units are kilojoules (SI preferred), joules, calories, kilocalories, electron-volts, in. $\cdot$ lb$_f$, ft $\cdot$ lb$_f$, therms, British thermal units (Btu), and horsepower-hours.

Power

Power is the time rate at which work is done. Hence its dimensions are $F \cdot L/t$. Common units are kilowatts (SI preferred), watts, calories/s, in. $\cdot$ lb$_f$/s, ft $\cdot$ lb$_f$/s, horsepower, and Btu/hr.

Pressure

Pressure is force per unit area, giving rise to dimensions of F/L^2. Common units are pascals (SI preferred), atmospheres, bars, lb$_f$/in.^{2}, lb$_f$/ft^2, millimeters or inches of Mercury, inches of water.

Mass Density

Mass density is mass per unit volume. Therefore, its dimensions are M/L^3. Common units are kg/m^3 (SI preferred), grams/liter, lb$_m$/ft^3, slugs/ft^3.

Specific Weight or Weight Density

Specific weight (or weight density) is a measure of the weight of a substance per unit volume. Its dimensions are F/L^3. Common units are N/m^3 (SI preferred), kg$_f$/m^3, lb$_f$/in.^{3}, lb$_f$/ft^3.

PROBLEMS

12-42. Referring to Example 12-10, calculate the mass of the sun if the earth (see Example 12-10) has an orbital diameter of 1.49×10^7 km and the force of attraction between the two celestial bodies is 1.44×10^{25} N.

12-43. Convert 76 N to dynes and lb$_f$.

12-44. Convert 2.67 in. to angstroms and miles.

12-45. Convert 26 knots to feet per second and meters per hour.

12-46. Convert 8.07×10^3 tons to newtons and ounces.

12-47. Convert 1.075 atmospheres to dynes per cm^2 and inches of mercury.

12-48. Convert 596 Btu to foot-pounds and joules.

12-49. Convert 26,059 watts to horsepower and ergs per second.

12-50. Convert 75 angstroms to feet.

12-51. Express 2903 ft^3 of sulfuric acid in gallons and cubic meters.

12-52. Change 1 Btu to horsepower-seconds.

12-53. A car is traveling 49 mi/hr. What is the speed in feet per second and meters per second?

12-54. A river has a flow of 3×10^6 gal per 24-hr day. Compute the flow in cubic feet per minute.

12-55. Convert 579 qt/sec to cubic feet per hour and cubic meters per second.

12-56. A copper wire is 0.0809 cm in diameter. What is the weight of 1000 m of the wire?

12-57. A cylindrical tank 2.96 ft high has a volume of 136 gal. What is its diameter?

12-58. A round iron rod is 0.125 in. in diameter. How long will a piece have to be to weigh 1 lb?

12-59. Find the weight of a common brick that is 2.6 in. by 4 in. by 8.75 in.

12-60. Convert 1 yd^2 to acres and square meters.

12-61. A container is 12 in. high, 10 in. in diameter at the top, and 6 in. in diameter at the bottom. What is the volume of this container in cubic inches? What is the weight of mercury that would fill this container?

12-62. How many gallons of water will be contained in a horizontal pipe 10 in. in internal diameter and 15 ft long, if the water is 6 in. deep in the pipe?

12-63. A hemispherical container 3 ft in diameter has half of its volume filled with lubricating oil. Neglecting the weight of the container, how much would the contents weigh if kerosene were added to fill the container to the brim?

12-64. What is the cross-sectional area of a railroad rail 33 m long that weighs 94 lb/yd?

12-65. A piece of cast iron has a very irregular shape and its volume is to be determined. It is submerged in water in a cylindrical tank having a diameter of 16 in. The water level is raised 3.4 in. above its original level. How many cubic feet are in the piece of cast iron? How much does it weigh?

12-66. A cylindrical tank is 22 ft in diameter and 8 ft high. How long will it take to fill the tank with water from a pipe which is flowing at 33.3 gal/min?

12-67. Two objects are made of the same material and have the same weights and diameters. One of the objects is a sphere 2 m in diameter. If the other object is a right cylinder, what is its length?

12-68. A hemisphere and cone are carved out of the same material and their weights are equal. The height of the cone is 3 ft. $10\frac{1}{2}$ in., while the radius of the hemisphere is 13 in. If a flat circular cover were to be made for the cone base, what would be its area in square inches?

12-69. An eight-sided wrought iron bar weighs 3.83 lb per linear foot. What will be its dimension across diagonally opposite corners?

12-70. The kinetic energy of a moving body in space can be expressed as follows:

$$KE = \frac{MV^2}{2}$$

where KE = kinetic energy of the moving body

   M = mass of the moving body

   V = velocity of the moving body

a. Given: $M = 539$ lb$_f \cdot$ s^2/ft; $V = 2900$ ft/s
   Find: KE in ft $\cdot$ lb$_f$; in SI units.
b. Given: $M = 42.6$ lb$_f \cdot$ s^2/ft; $KE = 1.20 \times 10^{11}$ ft $\cdot$ lb$_f$
   Find: V in ft/s; in SI units.
c. Given: $KE = 16,900$ in. $\cdot$ lb$_f$; $V = 3960$ in./min
   Find: M in slugs; in SI units.
d. Given: $M = 143$ g; $KE = 2690$ in. $\cdot$ lb$_f$
   Find: V in mi/hr; in SI units.

Chapter 13

Engineering Economy

In earlier chapters we learned that engineering is a profession that exists only to serve the needs of society. It is a people-oriented activity purposely directed toward the satisfaction of human wants and needs. These needs range from the physical, cultural, and economic to the spiritual. We also learn from our study of physics, and from the natural laws of thermodynamics in particular, that energy can neither be created nor destroyed and that *perfection is impossible* (from this it has been concluded that perpetual motion machines cannot be constructed). As an example, friction and heat losses are commonly found to be nemeses of perfection in the design of machines. These losses are the natural costs of operation. This means that every natural system in our universe operates in a manner such that its energy dissipates, and its behavior is irreversible. More simply, *there is a cost associated with the use of everything*.

Engineering economics is a study of various methods used to evaluate the worth of physical objects and services in relation to their cost. A mastery of these methods is particularly valuable, because they can be used to evaluate the costs before they are incurred. In this way the specific costs associated with the various design alternatives can be evaluated *before* investments of time and money are made. Engineering economy then becomes a very important tool of the engineer, because the same end result can often be attained by several different methods, each with its unique costs. However, this consideration is not unusual. Such an evaluation process occurs almost daily with most people.

There are a number of options available to a lending agency or person who possesses a sum of money. Some of these options are:

1. Hoard it in a secure place (i.e., bury it in the ground).

2. Loan it to someone for love or goodwill with expectancy of its return at a future date (i.e., loan it to your brother or to the local girl scouts organization).

3. Exchange it for products and services to enhance personal satisfaction (i.e., buy food, clothing, jewelry, or medical care).

4. Exchange it for potentially productive goods or properties (i.e., real estate, a taxi, or machine shop equipment).

5. Lend it on condition that the borrower will repay the principal sum with accrued interest at a future date (i.e., U.S. Treasury bonds, or a savings and loan company).

In any case, money is considered to be a valuable asset in all cultures, and the ability to use it wisely is a highly regarded attribute.

When a man says money can do anything, that settles it: he hasn't any.

Ed. Howe

Time is money.

Benjamin Franklin

Money is a stupid measurement of achievement but unfortunately it is the only universal measure we have.

Charles P. Steinmetz

[13-1]

The propensity to truck, barter, and exchange . . . is common to all men, and to be found in no other race of animals.

Adam Smith, 1723–1790
The Wealth of Nations

It is a socialist idea that making profits is a vice. I consider the real vice is making losses.

Winston Churchill

There are two things needed in these days; first, for rich men to find out how poor men live; and, second, for poor men to know how rich men work.

E. Atkinson

In modern business it is not the crook who is to be feared most, it is the honest man who doesn't know what he is doing.

Owen D. Young

. . . No business, no matter what its size, can be called safe until it has been forced to learn economy and to rigidly measure values of men and materials.

Harvey S. Firestone

The successful producer of an article sells it for more than it cost him to make, and that's his profit. But the customer buys it only because it is worth *more* to him than he pays for it, and that's his profit. No one can long make a profit *producing* anything unless the customer makes a profit *using* it.

Samuel B. Pettengill

I don't like to lose, and that isn't so much because it is just a football game, but because defeat means the failure to reach your objective. I don't want a football player who doesn't take defeat to heart, who laughs it off with the thought, "Oh, well, there's another Saturday." The trouble in American life today, in business as well as in sports, is that too many people are afraid of competition. The result is that in some circles people have come to sneer at success if it costs hard work and training and sacrifice.

Knute Rockne

TABLE 13-1   PREDICTED CASH FLOWS FOR THE FORD AND CHEVROLET

Year	Initial Cost Ford	Chevy	Operating Cost Ford	Chevy	Maintenance Cost Ford	Chevy	Salvage Cost Ford	Chevy
0	$5000	$4000	—	—	—	—	—	—
1	—	—	$1100	$1000	$200	$250	—	—
2	—	—	1100	1000	300	400	—	—
3	—	—	1100	1000	400	550	—	—
4	—	—	1100	1000	500	700	—	—
5	—	—	1100	1000	600	850	−$1000	−$500
Totals	$5000	$4000	$5500	$5000	$2000	$2750	−$1000	−$500

However, the options that are available concerning the strategies of using money are not well known by most people.

For example, let us examine Mary Brown's dilemma. She is considering purchasing either a Ford or a Chevrolet. Both are used cars, but the Chevrolet is a year older than the Ford. Using data supplied by friends who own similar models, Mary made an evaluation of the investment worth of the two automobiles. She assumed a five-year life for the two vehicles. Her estimated requirements for money to initiate and maintain these possible acquisitions over the years are shown in Table 13-1. Economists refer to such cash outflows or needs (and cash inflows) as *cash flows*. The salvage costs shown in year 5 are negative since they represent money coming back to Mary.

As shown in Table 13-2, Mary then tabulated the total cost for each automobile. On the basis of these calculations, Mary concluded that the Chevrolet was the better buy.

However, because of an omission on Mary's part, she should not infer from these calculations that it is more cost-effective to purchase the Chevrolet. She has ignored the cost of interest, often referred to as the time value of money. That is, in summing the cash flows over the projected years of life, she has implicitly assumed that $1 spent in year 1, for example, is the same as $1 spent in year 4.

This assumption is not valid. Remember, everything costs something . . . including the use of money. For example, Mary found that the dealership who offered the Ford for sale was willing to finance

TABLE 13-2   SUMMATION OF ESTIMATED COSTS

Cost	Ford	Chevrolet
Initial	$ 5000	$ 4000
Operating	5500	5000
Maintenance	2000	2750
Salvage	−1000	− 500
Total five-year cost	$11,500	$11,250

the purchase cost at 10 percent. For the Chevrolet she had to resort to a loan from her bank, where the interest rate was 15 percent.

The question arises as to how we should enter these interest rates into Mary's calculations. In order to explain this, let us review the basic methods of handling cash flows over time.

THE NATURE OF INTEREST

If we were to ask the question "would you rather have $1.00 now or $1.00 next year?", most people would choose to take $1.00 now. The principal reasons for this choice are (1) $1.00 now may be invested and earn interest so that it will be worth more than $1.00 after one year; and (2) due to inflation, $1.00 now will purchase more goods and services than $1.00 will a year from now. For the time being, we will ignore the effects of inflation and direct our attention toward the study of interest and its effects on cash flow.

In borrowing and lending situations, *interest* may be thought of as money paid (if you are borrowing) or earned (if you are lending) for the time use of money. It is the *time value of money*. The longer the money is held, the more the interest that is paid or earned. The magnitude of the interest rate is a function of the risk of loss of the borrowed sum, administrative expenses, and desired magnitude of profit or gain. The money being held by the borrower is called the *principal*. It will be held for one or more *periods of time* (weeks, months, quarters, years, etc.). The interest will be earned at a specified rate per period. This is the *interest rate*. For the most part, the rate of interest charged is a function of conditions experienced or specified by the lender. We will develop most of the concepts regarding interest from a borrower–lender viewpoint since this condition is familiar to most people. Later, we will modify the concept of interest so that it is more appropriate for engineering design decisions.

A loan transaction and the interest associated with it are viewed differently by the lender and the borrower. For example, the lender will consider the sum loaned as a negative cash flow and the payments received as positive cash flows. On the other hand, the borrower will consider the principal sum received as a positive cash flow and the repayments as negative cash flows.

To simplify understanding of the principles of engineering economy, there are several terms that need definition as follows:

i = interest rate for a given interest period (often given as a percent, but always used as a decimal fraction)

n = number of compounding interest periods

P = principal sum that exists at the beginning of an interest period at a time regarded as being the present

F = sum of money that exists in the future at the nth interest period measured from a time regarded as being the present

A = single amount in a series of n equal payments made at the end of each interest period

[13-2]

Money, which represents the prose of life, and which is hardly spoken of in parlors without an apology, is, in its effects and laws, as beautiful as roses.

Ralph Waldo Emerson
Nominalist and Realist, *1848*

Where profit is, loss is hidden nearby.

Japanese Proverb

Most men believe that it would benefit them if they could get a little from those who *have* more. How much more would it benefit them if they would learn a little from those who *know* more.

W. J. H. Boetcker

You're worth what you saved, not the million you made.

John Boyle O'Reilly
Rules of the Road

Money is the seed of money, and the first guinea is sometimes more difficult to acquire than the second millions.

Jean Jacques Rousseau
Discours sur l'origine et le fondement de l'inégalité parmi les hommes, 1754

90 percent of new products fail, in the sense that they are pulled off the market within four years of launch; in the more specialized areas, such as selling to the Original Equipment Market rather than to the consumer, possibly as many as two-thirds of the new products and processes lose money.

Design News
April 27, 1970

There is one rule for industrialists and that is: Make the best quality of goods possible at the lowest cost possible, paying the highest wages possible.

Henry Ford

Success is that old ABC—ability, breaks, and courage.

Charles Luckman
Quoted in the New York Mirror, *September 19, 1955*

Since there are a variety of ways in which loans can be repaid, we are often faced with the problem of comparing different alternatives that are possible. Thus we need a way to determine the equivalence of the alternatives. Two alternatives are said to be equivalent if they produce the same effect on the system. The concept of equivalence is very important in engineering economy studies. Computational techniques that provide methods of verifying the equivalence of alternatives are the backbone of engineering economic studies. They allow the analyst to evaluate the economic effect of a single project or proposal in many different ways which will result in the same conclusion. Equivalence also permits two or more alternatives to be compared on the basis of either present worth, annual worth, future worth, and so on, with the assurance that whichever basis is used for comparison, the same alternative will be superior. More will be mentioned about equivalent methods of comparing alternatives later in the chapter.

SIMPLE INTEREST

In simple-interest situations, the interest earned is directly proportional to the capital involved in the loan, and the interest rate is applied each period only to the principal amount.

Example 13-1

$100 is borrowed for five years at a simple interest rate of 10 percent per year. How much is owed at the end of five years?

Solution

In general, if a sum of money or principal (P) is borrowed for (n) time periods at a simple rate (i) per period (note that i is used as a decimal fraction), the amount owed (F), called a *future amount,* at the end of n periods is given by

$$F = P + nPi = P(1 + ni) \tag{13-1}$$

$$= \$100(1 + 5 \times 0.10)$$

$$= \$150$$

From the following tabulation we can see that this relationship is consistent throughout the lending period.

End of Year	*Interest During Year*	*Amount Owed at End of Year*
1	$10	$110
2	10	120
3	10	130
4	10	140
5	10	150

COMPOUND INTEREST

Most economic situations in business, government, and industry are governed by compound rather than simple interest. Therefore, for the remainder of this chapter all calculations will assume that the interest rate is compound rather than simple.

There are six common compound-interest factors (mathematical relationships) that we will use in comparing alternatives. These are shown in Table 13-3. To solve problems they are used in conjunction with cash flow quantities P, F, and A (see definition of terms, page 411) as follows:

$$\text{desired amount} = (\text{given amount}) \times \begin{pmatrix} \text{discrete compound-} \\ \text{interest factor} \end{pmatrix}$$

Examples:

$F = (P) \times$ (single-payment compound amount factor)

$A = (F) \times$ (sinking fund factor)

$P = (A) \times$ (uniform series present worth factor)

In the material that follows, the compound interest factors are explained and example problems are used to illustrate their application. In each case, the formulas used are for lump-sum cash flows and for interest that is compounded at the end of finite-length periods (such as a month or a year). To simplify the calculation process, tables of interest factors are included in Appendix III. Also, a type of shorthand code will be used to refer to these six types of problems. The code consists of four parts and is referred to as the interest *factor label*. In each case one part of the label will be unknown and the other three parts will be known. Placement of information within the label is invariable, and the parts are ordered as shown in Figure [13-3].

Examples of factor use:

1. Solve for F, the equivalent future worth. Given: P, i, n; use the label $(F/P, i, n)$.
2. Solve for P, the equivalent present worth. Given: A, i, n; use the label $(P/A, i, n)$.
3. Solve for the uniform payment, A. Given: F, i, n; use the label $(A/F, i, n)$.

When used as a part of a cash flow diagram it is convenient to use the circular form of the factor label in the manner shown in Figure [13-4]. Use of the factor label in cash flow diagrams is discussed in more detail in the following section.

CASH FLOW DIAGRAMS

In engineering economy studies, it is often helpful to draw diagrams that represent the flow of cash over time. These diagrams are similar to the free-body diagrams used in engineering mechanics. Vectors whose lengths are proportional to the magnitude of the cash flows

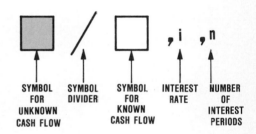

SYMBOL FOR UNKNOWN CASH FLOW   SYMBOL DIVIDER   SYMBOL FOR KNOWN CASH FLOW   INTEREST RATE   NUMBER OF INTEREST PERIODS

[13-3] Factor label.

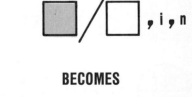

BECOMES

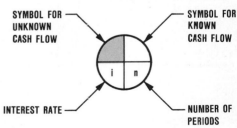

SYMBOL FOR UNKNOWN CASH FLOW   SYMBOL FOR KNOWN CASH FLOW

INTEREST RATE   NUMBER OF PERIODS

[13-4] Circular factor label.

There is this difference between those two temporal blessings, health and money: Money is the most envied, but the least enjoyed; Health is the most enjoyed, but the least envied; and this superiority of the latter is still more obvious when we reflect, that the poorest man would not part with health for money, but that the richest would gladly part with all their money for health.

Charles C. Colton

TABLE 13-3   **DISCRETE COMPOUND-INTEREST FACTORS AND LABELS**

Desired	Given	Factor Label	Factor[a]	Name of Factor
Single payment				
F	P	$F/P, i, n$	$(1 + i)^n$	Single-payment future worth
P	F	$P/F, i, n$	$\dfrac{1}{(1 + i)^n}$	Single-payment present worth
Equal payments				
F	A	$F/A, i, n$	$\dfrac{(1 + i)^n - 1}{i}$	Uniform series future worth
A	F	$A/F, i, n$	$\dfrac{i}{(1 + i)^n - 1}$	Sinking fund
P	A	$P/A, i, n$	$\dfrac{(1 + i)^n - 1}{i(1 + i)^n}$	Uniform series present worth
A	P	$A/P, i, n$	$\dfrac{i(1 + i)^n}{(1 + i)^n - 1}$	Capital recovery

[a] Note that three of these factors are reciprocals of the other three factors. Calculated values of the discrete compound interest factor are given in Appendix III for various i's and n's. *Symbols:*

i = Interest rate for a given interest period (often given as a percent, but always used as a decimal fraction).

n = Number of compounding interest periods.

P = A principal sum that exists at the beginning of an interest period at a time regarded as being the present.

F = A sum of money that exists in the future at the nth interest period measured from a time regarded as being the present.

A = A single amount in a series of n equal payments made at the end of each interest period.

are plotted proportionally in a vertical direction. Upward vectors represent positive cash flows or money earned. Downward vectors denote negative cash flows, or money paid. The horizontal axis represents time. Figure [13-5] illustrates the relationship between a present sum P and a future sum F using cash flows over time. Note that in Figure [13-5] the cash flow at time 0 is downward, whereas at time n, it is in an upward direction. This represents the situation as viewed by the lender. From the borrower's viewpoint, the directions of the vectors would be reversed. Without specifying the nature of the situation, the direction of the cash flow vectors is completely arbitrary.

COMPOUND INTEREST—SINGLE PAYMENTS

In compound-interest situations, the interest rate is applied each period to the principal amount plus all previous interest charges (or earnings). This means that for each time period the borrower pays

[13-5]

interest *on the interest previously paid* plus the interest on the principal that remains. This is advantageous to the lender.

Example 13-2

Suppose that the 10 percent per year interest rate of Example 13-1 is changed to a compound rate (i.e., $i = 10$ percent per year compounded yearly). How much is owed at the end of five years?

Solution

Let us first consider this problem from a year-by-year viewpoint. We could tabulate the interest owed as follows:

End of Year	Interest During Year	Amount Owed at End of Year
1	$10.00	$110.00
2	11.00	121.00
3	12.10	133.10
4	13.31	146.41
5	14.64	161.05

Note that the interest charged during any given year is the result of applying the interest rate (i) to the amount owed at the end of the preceding year. Table 13-4 generalizes the result for any principal (present value) (P), interest rate (i) per period, and accumulated amount (future value) (F).

We have seen that the relationship between a single amount P (present value) and future amount F is given by $F = P(1 + i)^n$. The factor $(1 + i)^n$ is called the *single payment future worth factor* (see Table 13-3) and its label is

$$(F/P, i, n) \quad \text{or}$$

The term F/P is read as "F given P." For example, this is the factor to be used to find F, given a value for $P = \$100$ (and also given the interest rate per period, 10 percent, and the number of periods, five).

TABLE 13-4 FUTURE VALUE, F, OF SINGLE AMOUNT—COMPOUND INTEREST

Period	Interest During Period	Accumulated Amount at the End of Period, F
1	Pi	$P + Pi = P(1 + i)$
2	$P(1 + i)i$	$P(1 + i) + P(1 + i)i = P(1 + i)^2$
3	$P(1 + i)^2 i$	$P(1 + i)^2 + P(1 + i)^2 i = P(1 + i)^3$
.	.	.
.	.	.
.	.	.
n	$P(1 + i)^{n-1} i$	$P(1 + i)^{n-1} + P(1 + i)^{n-1} i = P(1 + i)^n$

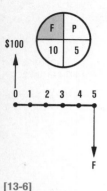

[13-6]

Thus:

$$F = P(F/P, i, n) \qquad (13\text{-}2)$$

$$= P(1 + i)^n$$

$$= \$100(1 + 0.10)^5$$

$$= \$100(1.6105) \qquad \text{(obtain this value from Appendix III)}$$

$$= \$161.05$$

Suppose that the future amount F is known and i and n have been specified, and it is desired to identify the present value P that will accumulate to F at interest rate i per period after n periods. The cash flow diagram for this situation is represented by Figure [13-6]. Equation (13-2) is solved for P. The result is

$$P = F\left[\frac{1}{(1 + i)^n}\right] = F(P/F, i, n) \qquad (13\text{-}3)$$

$$= 161.05\left[\frac{1}{(1 + 0.10)^5}\right]$$

$$= \$100$$

The factor $1/(1 + i)^n$ is called the *single-payment present worth factor* (see Table 13-3) and is abbreviated as $(P/F, i, n)$ or "P given F, i, n." Remember that

$$(P/F, i, n) = \frac{1}{(F/P, i, n)}$$

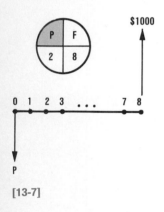

[13-7]

Example 13-3

How much money must be deposited in an account that earns 2 percent per quarter compounded quarterly so that $1000 accumulates after two years?

Solution

The first step in solving the problem is to draw a cash flow diagram that describes the transaction [13-7]. Use the label $(P/F, i, n)$ in Table 13-3.

$$P = F(P/F, i, n) = F\left[\frac{1}{(1 + i)^n}\right]$$

$$= \$1000\left[\frac{1}{(1 + .02)^8}\right] = \$1000\left[\frac{1}{1.1717}\right]$$

$$= \$1000(0.8535) = \$853.49$$

Notice that the value of n used in Example 13-3 is 8, since two years = eight quarters (a quarter is a three-month period). It is important to note that in all engineering economy calculations there must be agreement between i and n with respect to their labels. For example, if i is expressed as "percent per month compounded monthly," then n must also be expressed in months. The label associated with i is always "percent per time period compounded every time period." Usually, n is converted to time periods specified by i.

The justification of private profit is private risk.
Franklin D. Roosevelt

The highest use of capital is not to make more money, but to make money do more for the betterment of life.
Henry Ford

He who will not reason, is a bigot; he who cannot is a fool; and he who dares not, is a slave.
William Drummond

Tables for the *P/F*, *F/P*, and other factors soon to be developed are provided in Appendix III for selected values of *i* and a large range of values for *n*. The appropriate table to be used in a particular calculation is identified by the appropriate value of *i*. The value of *n* is found along the first column and then the numerical value of the factor is read below the appropriate factor symbols in the row associated with *n*.

For example, with *i* at 2 percent, the table headed "Decimal Interest Rate = .02" with *n* = 8 yields a *P/F* value of 0.8535 as calculated in Example 8-3. The 2 percent table is appropriate for all compounding periods where the interest rate is 2 percent per period, for example, 2 percent per month compounded monthly (in this case, *n* represents months), 2 percent per quarter, compounded quarterly (*n* represents quarters), 2 percent per year compounded yearly (*n* represents years), and so on.

COMPOUND INTEREST—SERIES OF EQUAL PAYMENTS

Suppose that you wish to borrow *P* dollars and plan to repay it at *i* percent compound interest in a series of equal payments of *A* dollars each over *n* periods. In considering such a loan you will want to know corresponding loan payments, *A*, for varying loan amounts (*P*'s). The cash flow is shown in Figure [13-8]. Note that the first *A* value will occur one time unit after the *P* has been received.

It can be shown that the value of *A* is given by

$$A = P\left[\frac{i(1 + i)^n}{(1 + i)^n - 1}\right] = P(A/P, i, n) \qquad (13\text{-}4)$$

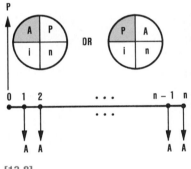

[13-8]

The factor (*A/P*, *i*, *n*) is known as the *capital recovery factor* (See Table 13-3).

If *A* is known and the value of *P* is desired, then

$$P = A\left[\frac{(1 + i)^n - 1}{i(1 + i)^n}\right] = A(P/A, i, n) \qquad (13\text{-}5)$$

The factor (*P/A*, *i*, *n*) is known as the *uniform series present worth factor* (see Table 13-3).

Example 13-4

A bank has offered to loan $4000 toward the purchase of a microcomputer. This loan will be financed at a rate of $1\frac{1}{2}$ percent per month compounded monthly over 48 months. What is the amount of each monthly payment?

Solution

First draw the cash flow diagram [13-9]. Using Appendix III, we have

$$A = \$4000(A/P, 1\tfrac{1}{2}, 48)$$

$$= 4000(0.0294)$$

$$= 117.60$$

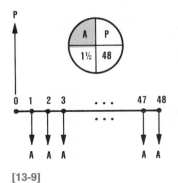

[13-9]

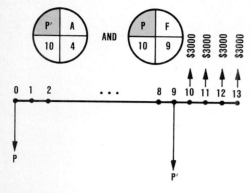

[13-10]

Example 13-5

How much money must Henry Jones deposit now which will earn a rate of interest of 10 percent per year compounded annually for him to be able to withdraw $3000 for each of four years beginning 10 years from now so that he can provide for his son's college education?

Solution

In this problem we are attempting to find the value of P at time 0 [13-10]. However, it is first necessary to convert the four $3000 withdrawals to an equivalent deposit P' at time 9. (Time 9 is one time unit before the first A value.) Then the P' value will be treated as a future amount (an F) and its value will be converted to an equivalent value at $t = 0$. Use the label P/A, i, n.

$$P' = \$3000(P/A, 10, 4)$$

$$= \$3000(3.1699) \qquad (Appendix\ III)$$

$$= \$9509.70$$

then use the label P/F, i, n.

$$P = \$9509.70(P/F, 10, 9)$$

$$= \$9509.70(0.4241)$$

$$= \$4033.06$$

Another situation occurs frequently in evaluating the worth of money. This is when it is desirable to determine the future value of a series of equal payments. The cash flow diagram for this situation is shown in Figure [13-11]. As we can see, equal payments of (A) are deposited every period at i percent interest per period (compounded every period) and the accumulated amount (including interest), F, results after n periods.

Note that in this case the accumulated value F occurs at the same time as the last deposit, A. This means that the last deposit earns no interest. The value of F is given by the relationship

$$F = A\left[\frac{(1 + i)^n - 1}{i}\right] = A(F/A, i, n) \qquad (13-6)$$

The factor (F/A, i, n) is called the *uniform series future worth factor* (see Table 13-3). Solving Equation (13-6) for A yields

$$A = F\left[\frac{i}{(1 + i)^n - 1}\right] = F(A/F, i, n) \qquad (13-7)$$

The factor (A/F, i, n) is called the *sinking fund factor* (see Table 13-3).

Let us examine several examples that illustrate the use of the (F/A, i, n) and (A/F, i, n) factors.

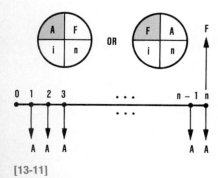

[13-11]

Example 13-6

$400 is deposited every six months into an account that earns 6 percent interest every six months, compounded semiannually. How much money has accumulated after five years?

Solution

The cash flow diagram is shown in Figure [13-12]. Use the label $(F/A, i, n)$.

$$F = \$400(F/A, 6, 10)$$

$$= \$400(13.1808)$$

$$= \$5232.72$$

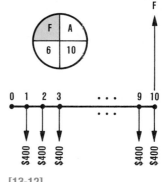

[13-12]

Example 13-7

How much money must be deposited annually into an account that earns 10 percent interest per year compounded annually so that $1500 is accumulated after five years?

Solution

The cash flow diagram is shown in Figure [13-13]. Use the label $(A/F, i, n)$.

$$A = \$1500(A/F, 10, 5)$$

$$= \$1500(0.1638)$$

$$= \$245.70$$

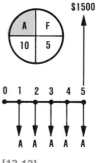

[13-13]

Example 13-8

$50 is deposited every month for two years. Beginning one year after the last deposit, the entire accumulated amount will be withdrawn in four equal monthly payments. What is the amount of these payments? Interest is at 2 percent per month compounded monthly.

Solution

The cash flow diagram is shown in Figure [13-14]. From the cash flow diagram we can see that there are three tasks to be performed. First, we must determine the future value F of the 24 equal payments of $50 each. Second, we must determine the value of F eleven months hence, X. Finally, we will disburse X in equal payments A over the next

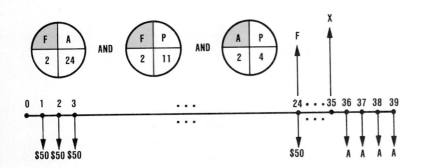

[13-14]

four-month period. The calculations for this transaction would be as follows:

Use the label $(F/A, i, n)$.

$$F = \$50(F/A, 2, 24)$$
$$= \$50(30.4218)$$
$$= \$1521.09$$

Hint: X is actually another future sum F.
Use the label $(F/P, i, n)$.

$$X = \$1521.09(F/P, 2, 11)$$
$$= \$1521.09(1.2434)$$
$$= \$1891.32$$

Use the label $(A/P, i, n)$.

$$A = \$1891.32(A/P, 2, 4)$$
$$= \$1891.32(0.2626)$$
$$= \$496.66$$

TABLE 13-5 SUMMARY OF THE BASIC INTEREST FACTORS

Cash Flow Diagram	To Find:	Given:	Use:
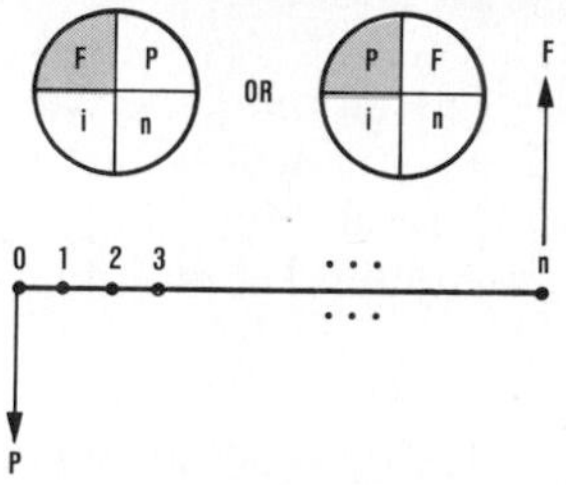	F	P	$F = P(F/P,i,n)$
	P	F	$P = F(P/F,i,n)$
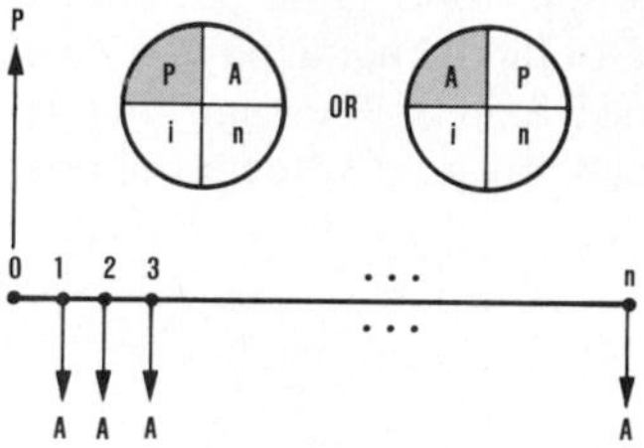	P	A	$P = A(P/A,i,n)$
	A	P	$A = P(A/P,i,n)$
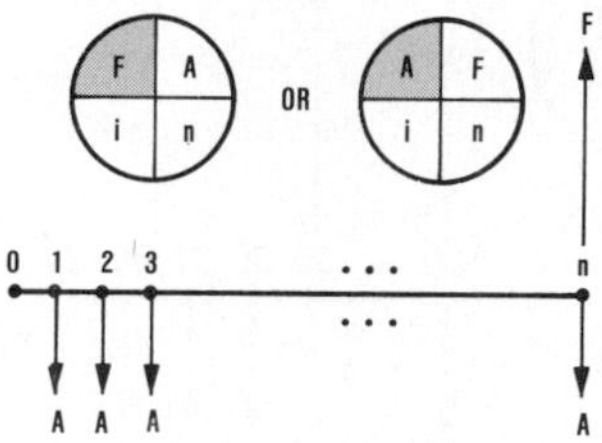	F	A	$F = A(F/A,i,n)$
	A	F	$A = F(A/F,i,n)$

Problems

13-1. $1000 is borrowed for a period of four years. How much must be paid by the borrower at the end of four years if:
  a.   15 percent simple interest is available?
  b.   Interest is at 15 percent per year compounded yearly?

13-2. What amount must you now deposit at 8 percent per year compounded yearly to accumulate $2000 after five years?

13-3. With interest at 7 percent per year compounded yearly, what equal annual amount would have to be deposited at the end of each of the next six years to accumulate $5000?

13-4. How much should you now deposit in an account that pays 10 percent per year compounded yearly so that you may withdraw $1000 each year for eight consecutive years beginning one year from now?

13-5. A wealthy relative has offered you $10,000 now or $4000 next year, $5000 after two years, and $2000 after three years. You conclude that you could earn 8 percent per year compounded yearly with any amounts you receive. Which of the two offers should you take? *Hint:* Compare the present worth of the three installments with $10,000.

13-6. Solve Problem 13-2 if interest is at $8\frac{1}{2}$ percent per year compounded yearly. *Hint:* Use the mathematical expression in Table 13-3 for the appropriate factor.

13-7. Solve Problem 13-3 if interest is at $7\frac{1}{4}$ percent per year compounded yearly. *Hint:* Use the mathematical expression in Table 13-3 for the appropriate factor.

13-8. Resolve Mary Brown's dilemma (see Table 13-1) by comparing the present worth of ownership of the Ford at 10 percent per year compounded yearly and the Chevrolet at 15 percent per year compounded yearly.

13-9. Convert the following cash flow to an equivalent uniform amount if the periods are years and the interest rate is 20 percent per year compounded yearly. *Hint:* First find a present worth (P) or future worth (F) value and then convert that value to a uniform amount (A).

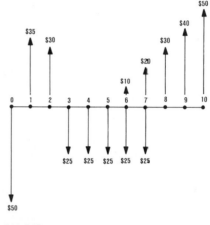

[13-P9]

THE NATURE OF INTEREST IN ENGINEERING ECONOMY STUDIES _____

Within an organization, there will be many projects that will compete for use of the organization's money. However, not all projects will be funded. Only those that make a compound return on the money invested in them (i.e., a rate of return) at least as great as the company's minimum attractive rate of return (MARR) have a chance of being funded. Among these, the likelihood of being funded depends on how large the estimated rate of return will be and how much of the company's money at this particular time is available for investment in engineering projects.

The company's MARR is usually based historically on what the company has been able to earn on previous engineering projects. Or it is based on the "cost of capital," which is the company's cost of borrowing money. Hence interest as we have used the term in the borrowing/lending situation developed in this chapter becomes "rate of return on investment."

METHODS OF ANALYSIS IN ENGINEERING ECONOMY STUDIES

The economic feasibility of launching engineering designs and other engineering projects are usually evaluated using one of the following methods:

1. Present worth (PW).
2. Annual worth (AW).
3. Future worth (FW).
4. Rate of return (RR).
5. Ratio of earnings to investment (RI).
6. Payback period (PP).

These methods are nothing more than applications of the concepts and factors which have already been discussed. The first five methods are equivalent treatments of the same cash flow. The result of any engineering economy study will be a recommendation; that is, when a single project is evaluated, the recommendation will be either to undertake the project or not to undertake the project. When several competing projects are compared, either a single project will be recommended or perhaps more than one will be recommended. If a project is recommended, it may or may not be undertaken by the company. An economic evaluation is only one of several considerations that must be made before a project is funded. Other considerations include environmental impact of the project, availability of funds for projects of this type, safety and health considerations, effect of the project on employees of the company, compliance with government regulations, and the judgment of the manager or managers within the company who must ultimately approve or disapprove such projects. In most instances, projects must be "sold" to the decision makers within the company, on the basis of both economic and noneconomic considerations. It has been said that a good idea (or project) will sell itself, but this belief does not seem to be valid in the competitive arenas of business, government, and industry.

It should be recognized that cash flows that will occur throughout the life of a design or other engineering project must be predicted during the developmental stages of the project in order to estimate the final total costs. Although the investment at time zero (the present time) is usually known with some certainty, most of the cash flows used in engineering economy evaluations are estimates of future returns, costs, or savings. The accuracy of the projected final costs will be as good or bad as the individual estimates that make up the total. For this reason, great care must be exercised in forecasting the future cash flows.

Each of the first five methods always seems to result in the same recommendation (i.e., do or do not undertake the project) when applied to the same cash flow situation. However, method 6, the payback period method will sometimes result in a different recommendation than the other methods. For each of the six methods listed above, incomes or savings are treated as a positive cash flow

An engineer is an unordinary person who can do for one dollar what any ordinary person can do for two dollars.

Certainly there are lots of things in life that money won't buy, but it's very funny—
Have you ever tried to buy them without money?
Ogden Nash
The Terrible People

and are given a "plus" sign. Investments, payments, and other costs are treated as negative cash flow and assigned a "minus" sign.

Present Worth Method

The term *present worth* (PW) method refers to a sum of money whose value at the present time (time zero) is equivalent to a series of cash receipts (+) and disbursements (−), each occurring at the end of a successive interest period following time zero. In determining the present worth of these inflows and outflows, their values must be discounted to time zero at an interest rate equal to MARR.

The algebraic sum of the inflow and outflow present worths is then computed. The resulting sum is called the *net present worth* (NPW). For a given project, if the net present worth is positive, it can be concluded that the project will earn more than the MARR. If the net present worth is negative, the project will probably earn less than the MARR. If the net present worth is zero, the project is predicted to earn at a rate exactly equal to the MARR.

In general, if a single project is being evaluated, the net present worth must be greater than zero in order for the project to be recommended. If two or more competing projects are being compared, the project having the largest *positive* net present worth will be recommended. However, there are cases where a project having a negative net present worth may be recommended. For example, if a problem must be solved and all projects directed at solving the problem result in a negative present worth, that project having the algebraically largest (although negative) net present worth would probably be selected.

Let us examine a situation involving net present worth.

Example 13-9

A small fabricating plant has purchased a metal-forming machine to perform several operations which previously had been done manually. The cost of the machine, including delivery and installation, is $30,000. The machine is expected to have a useful life of nine years and a salvage value of $3000 at that time. Estimated savings due to reduced labor costs and increased productivity are $6000 in year 1 with an increased savings of $500 for each successive year. The machine will need to be overhauled in years 3 and 6, with estimated costs of $2500 for each overhaul. The company's MARR is 15 percent. Compute the net present worth (NPW) of the investment.

Solution

The cash flow diagram is shown in Figure [13-15]. This problem will be solved in four steps.

Step 1: Convert the gradient series to a present worth [13-16].

$$PW_G = \$6000(P/A, 15, 9) + \$500(P/G, 15, 9)$$

$$= \$6000(4.7716) + \$500(14.7548)$$

$$= \$28,629.60 + \$7377.40$$

$$= \$36,007.00$$

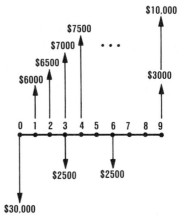

[13-15]

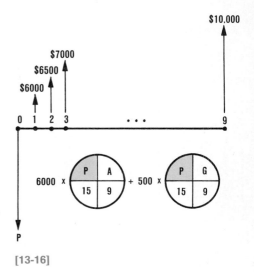

[13-16]

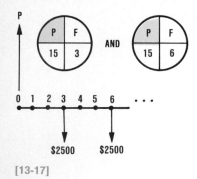

[13-17]

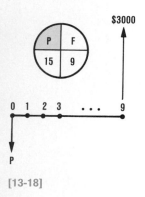

[13-18]

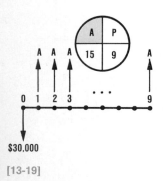

$30,000

[13-19]

Step 2: Convert each overhaul cost to a present worth [13-17].

$$PW_O = -\$2500(P/F, 15, 3) - \$2500(P/F, 15, 6)$$
$$= -\$2500(0.6575 + 0.4323)$$
$$= -\$2500(1.0898)$$
$$= -\$2724.50$$

Step 3: Convert the salvage value to a present worth [13-18].

$$PW_S = \$3000(P/F, 15, 9)$$
$$= \$3000(0.2843)$$
$$= \$852.90$$

Step 4: To obtain the net present worth (NPW), add all present worths, including the machine cost.

$$NPW = -\$30,000 + \$36,007 - \$2724.50 + 852.90$$
$$= \$4135.40$$

Since the NPW is positive, the investment in the new machine earns more than 15 percent. Thus the investment is recommended.

Annual Worth Method

The term *annual worth* (AW) refers to a uniform annual sum of money that is equivalent to a particular schedule of cash receipts and disbursements, minus an amount that represents the equivalent uniform annual cost of the capital invested.

In this method, the cash flows and capital recovery cost equivalency are converted to an equivalent net annual worth in a manner similar to the present worth method just discussed. If the AW is greater than zero, the project earns at a rate greater than MARR. If the AW is less than zero, then the project earns at a rate less than MARR. If the AW is zero, the project earns at a rate exactly equal to MARR. The comments made earlier concerning criteria and exceptions for recommending projects after a PW analysis are also valid for AW.

Example 13-10

Compute the AW for the cash flow of Example 13-9.

Solution:

This problem is also solved in five steps.

Step 1: Convert the machine cost to an annual amount [13-19].

$$AW_m = -\$30,000(A/P, 15, 9)$$
$$= -\$30,000(0.2096)$$
$$= -\$6288.00$$

Step 2: Convert the gradient series to an annual amount [13-20].

$$AW_G = \$6000 + \$500 (A/G, 15, a)$$

$$AW_G = \$6000 + \$500(3.0922)$$

$$= \$7546.10$$

Step 3: Convert the overhaul costs to a future amount [13-21] at year 9.

$$FW_O = -\$2500(F/P, 15, 6) + \$2500(F/P, 15, 3)$$

$$= -\$2500(2.3131 + 1.5209)$$

$$= -\$2500(3.8340)$$

$$= -\$9585.00$$

Step 4: Convert the salvage value and the FW_O value to an annual amount [13-22].

$$AW_{O+S} = (-\$9585 + 3000)(A/F, 15, 9)$$

$$= -\$6589(0.0596)$$

$$= -\$392.47$$

Step 5: Sum the annual worths of steps 1, 2, and 4.

$$AW = -\$6288.00 + \$7546.10 - \$392.47$$

$$= \$865.63$$

The net annual worth is positive and the project might be recommended. Note that the PW solution to Example 13-9 could have been used to obtain the AW. This computation would yield the following result:

$$AW = \$4135.40(A/P, 15, 9)$$

$$= \$4135.40(0.2096) = \$866.78$$

The $1.15 difference in the two solutions is due the rounding off of numbers in using the factors.

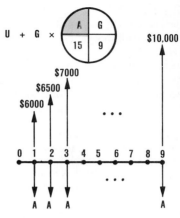

[13-20]

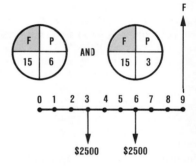

[13-21]

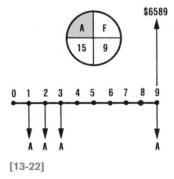

[13-22]

PROBLEMS

13-10. A company must decide between purchasing and leasing a large computing system. A five-year planning horizon will be used in the study. A new computing system will cost $600,000 and operating expenses are expected to be $25,000 per year. Maintenance charges will not be incurred by the company until the second year as the warranty extends over the first year. Maintenance costs in years 2 through 5 are expected to be $35,000. $40,000, $45,000, and $50,000, respectively. The system can be resold by the company for $150,000 after five years. On the other hand, a comparable system can be leased for $100,000 per year, with the payments being made at the *beginning* of each year. This is common practice in leasing arrangements. Operating expenses will be the same as those for ownership—$25,000 per year. In the case of a lease arrangement, the company will not have to pay any maintenance charges. The company's MARR is 20 percent per year compounded yearly.

Compare the two alternatives by calculating a present worth for each. *Hint:* To convert the beginning of the year lease costs to end of the year values, multiply by $(1 + 0.2)$. You now have five end-of-year costs in years 1 through 5 which can be converted to a present worth by multiplying by $(P/A, 20, 5)$.

13-11. The owner of an automatic car wash has been having difficulties in keeping the equipment operational. A decision must be made to either overhaul the present equipment or to replace it with new equipment. An overhaul will cost $10,000 and the equipment is expected to experience operating costs of $3000 per year over the next 10 years. Maintenance costs are expected to be $1000 in year 1 and will increase by $600 each year thereafter (i.e., $1600 in year 2, $2200 in year 3, etc.). If overhauled, the equipment will have no salvage value after 10 years.

On the other hand, new equipment costs $25,000 and is expected to have operating expenses of $1000 per year over the next 10 years. Maintenance charges are expected to be $500 in year 1 and will increase by $100 each year thereafter. This equipment can be resold after 10 years of use for $5000. Furthermore, if new equipment is purchased, the present equipment can be resold now for $5000.

Compare these two alternatives on the basis of annual worth (i.e., compute the net annual costs of each) if the MARR is 18 percent per year compounded yearly.

13-12. A bridge has a first cost of $75,000, annual maintenance costs of $5000, and must be repainted every five years at a cost of $8000. If $i = 10$ percent per year compounded yearly, what is the present worth?

13-13. A stadium has a first cost of $4,000,000 and annual maintenance costs of $300,000. Seats are replaced every 15 years at a cost of $400,000. Concrete supporting columns are replaced every 20 years at a cost of $1,000,000. What is the present worth if $i = 15$ percent per year compounded yearly?

13-14. Two possible types of road surface are being considered with cost estimates per mile as follows:

	Type I	Type II
First cost	$80,000	$120,000
Resurfacing period	12 years	18 years
Resurfacing cost	$50,000	$ 50,000
Average annual cost	$ 4,000	$ 3,000

Compare these two types on the basis of present worth using an interest rate of 8 percent per year compounded yearly.

13-15. Suppose that a product costs $1000 today. How much will be required to purchase this product three years from now if the rate of inflation is expected to be 8 percent each year? *Hint:* Simply multiply $1000 by the $(F/P, 8, 3)$ factor.

13-16. With respect to Problem 13-15, we will now consider the time value of money. With the same inflation rate of 8 percent, how much should be deposited now at a rate of return of 7 percent per year compounded yearly to have enough money to buy the product at the end of three years? *Hint:* Multiply the answer to Problem 13-15 by $(P/F, 7, 3)$.

13-17. Reconsider Problems 13-15 and 13-16. Suppose that the inflation rate is expected to be 8 percent in year 1, 9 percent in year 2, and 10 percent in year 3. The rate of return is still 7 percent per year compounded yearly. Under these conditions, how much should now be deposited to have enough money to buy the product at the end of three years?

Chapter 14

Engineering Design

DESIGN: THE ESSENCE OF ENGINEERING

In earlier chapters, engineers have been described primarily by the objects—the artifacts—which they have "produced." Indeed, it is through these engineering artifacts (artifacts which are loved and hated, often at the same time) that the image of the engineer has developed. The engineer is praised for designing the modern automobile and criticized for its polluting emissions; praised for the remarkable computer and criticized for its dehumanizing of the workplace; extolled for the number and variety of products that exist and criticized for the problems in disposing of them.

Although it is understandable and not unique to the engineering profession, this association with artifacts is unfortunate because it blocks from view the present role of engineers in society, the "tools" of their trade, and the process through which they work—all of which drastically affect the artifact produced. It is important that both the lay public and engineers understand and appreciate the engineer's environment so that they will better understand the artifacts of engineering.

The *process* the engineer uses to develop the ideas and the details for the artifacts used by mankind is commonly called the *engineering*

[14-1] The pyramids are the artifacts of early engineers.

427

[14-2] The modern automobile is the artifact by which the automotive engineer is known.

[14-3] Some artifacts of the aerospace engineer.

[14-4] The ever-shrinking microprocessor chip—an artifact of the computer engineer.

design process. It is this process that more adequately identifies engineering, and with it the engineer, than do the artifacts for which the process is ultimately responsible. Bridges, dams, and water delivery systems are among the artifacts by which the civil engineer is identified by the general public, while electric motors, TV sets, and computers are among the artifacts by which the electrical engineer is generally recognized. However, *it is the process—the engineering design process—by which the engineer brings these artifacts to fruition, that is the glue that binds together and truly identifies all engineers.*

It is important to point out that in our present technological structure, the physical act of building the artifact is *not* the direct culmination of the design process. Rather, it is the establishment of a set of specifications which, in and of themselves, completely *define* the artifact. Figure [14-5] is a block diagram that aids in defining the steps and responsibilities that are required in satisfying some technological need. The top block, where the need originates, represents the input of the client of the engineer.[1] Society, in general, nearly always has some self-serving input to the problem definition and this is shown as a separate beginning block. The intent of the latter input is to lessen the impact of the final solution on general segments of society, since, as John Muir the famous naturalist once said, "We can never do just one thing."

The large block in Figure [14-5] is intended to represent the role played by engineers. It is in this block where engineers decide how

---

[1]The word "client" is used here to represent the particular segment of society for which the engineer makes his services available. This could be governments, companies, individuals, and even engineers when they develop solutions to their own problems.

the need, as they understand it, would best[2] be met. Once the best solution has been determined, it is documented in a set of specifications that makes clear exactly what must be done to implement the chosen solution. This set of specifications dictates precisely what materials are to be used, the shapes into which these materials should be formed, and the manner by which these shapes are to be assembled together. We shall continually refer to this set of specifications as the DESIGN.[3] We have given this DESIGN its own block in Figure [14-5] for it takes on much significance in our technological world.

From this DESIGN, or set of specifications, skilled craftsmen—machinists, millrights, carpenters, welders, electricians, plumbers, steel erectors, and even factory workers—can determine *exactly* how to proceed in order to construct the solution. We have placed a block entitled "Implement the DESIGN" in Figure [14-5] to represent the role of these workers.

Finally, the bottom block of Figure [14-5] represents the emergence of the solution—the DESIGN embodied in an artifact—to the original need. As should be apparent from the various blocks in Figure [14-5], the success of the solution depends on *all* of the participants who have responsibilities in the various blocks or stages of this process, fulfilling those responsibilities.

The engineering design process, which takes place within the large block of Figure [14-5], is the subject of this chapter. Thus the engineering design process has as its goal or product the establishment of specifications which are sometimes referred to as, simply, *the DESIGN*.

In Chapter 9 we discuss the construction of documentation drawings, sometimes referred to as engineering drawings or blueprints. Documentation drawings generally form a very important part of the set of specifications which define the DESIGN, for without ambiguity, they help detail how the artifact is to be made. Perhaps it now becomes clearer why they are key items in establishing the engineer's responsibilities as opposed to the craftsman's responsibilities. The engineering design process must involve steps which will ensure that the artifact, if built according to the specifications, will perform its intended purpose. What is more, it must do so in a safe and environmentally compatible manner. It is the craftsman's responsibility to construct the artifact according to the engineer's specifications which establish the DESIGN.

Not all of the design work of an engineer may directly involve

[14-5] The steps and responsibilities in the creation of artifacts.

---

[2] We will return shortly to a discussion of the word "best" used to describe engineering DESIGNS.

[3] We will use the word DESIGN (in uppercase, letters in color) to signify the *product* of the design process as distinct from the process that produces the product. The word "design" (in lowercase, letters in black) will then be used to designate the *process* rather than the product of the process, and for any verbs (the act of designing) and modifying adjectives that are needed. At any point in the design process, the DESIGN consists of all of the specifications that describe the artifact or the findings at the current time. At the end of the process, the DESIGN is embodied in a formal set of specifications (reports, data, engineering drawings, etc.) that can be constructed or implemented by others.

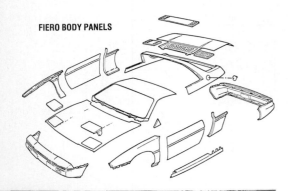

FIERO BODY PANELS

[14-6] The DESIGN is brought to fruition in the artifact shown in Figure [14-2].

. . . the process of design, the process of inventing physical things which display new physical order, organization, form, in response to function.
Christopher Alexander
Notes on the Synthesis of Form

[14-7] The consumer is often the judge of the successfulness of an artifact.

the design of an artifact. Designs can also describe recommendations arrived at through careful considerations, or technical processes which might be useful in some way. But nearly all of an engineer's work involves working within what we have called the design process. *Therefore, the essence of engineering is embodied in the process whereby DESIGNS are established. Engineers should be recognized as much by how they go about constructing their DESIGNS as by what they design.*

The processes leading to the establishment of a DESIGN, although not fully quantifiable, do involve many and varied steps. Often, one person cannot perform all of the necessary tasks alone. Hence it is common to find technical teams which, with members working in harmony, establish the definition of a DESIGN [14-9]. Typically, these teams consist of engineers and technologists (see Chapter 1). The role of the technologist on the team is to support the engineer in the establishment of the DESIGN: to compile data, to conduct tests, and to aid in the documentation efforts. Since technologists work at the direction of engineers, they generally are not directly responsible for ensuring the adequacy of the DESIGN in meeting its intended purpose. They are, however, essential in nurturing the DESIGN to completion.

It is the opportunity of defining and guiding the manner in which problems are to be solved that makes engineering and technology such exciting and gratifying professions. Whereas the craftsman's responsibility is to strictly follow a set of instructions (as stipulated in the DESIGN specifications), the engineer's role is to create those instructions. The technologist has the satisfaction of contributing to the creation of the DESIGN and of sharing with the engineer some of the satisfaction of seeing ideas come to fruition. Together with the ability to use creativity and knowledge, the engineer bears the important responsibility of developing DESIGNS that define an adequate and safe solution to some perceived need.

THE ENGINEERING DESIGN PROCESS: A DEFINITION

Before we investigate the intricacies of the engineering design process, it is helpful to summarize its intent [14-10]. The process itself has many aliases—design, design process, engineering design, engineering method, engineering—all of which refer to the typical way in which engineers practice their profession. For our immediate purposes, a definition of the process is as follows:

> *Engineering design process:* the strategy by which engineers use their unique set of "tools" in deciding how best to apply the available resources to adequately meet some existing need.

The process requires various amounts of creativity, exploration, knowledge, experience, and persistence from the engineer. Some engineers might compare it to the solving of a mystery, although it is critically different. The engineer (as a detective) sets out to collect the facts, analyze them, and draw conclusions, much as a police detective might do in trying to reconstruct a crime. However, engineers are not trying to reconstruct events that have already happened, but instead, they are attempting to build the future. They are trying to construct something that has not been, collecting uncertainties, not facts.

Unlike science, engineering design is not concerned with the pursuit of absolute truth and right. The *scientific method,* the procedure by which theories are compared to one another through experiment, is used by the scientist to decipher the facts and rules of the universe. In this way, scientists are much more like police detectives than are engineers, because they seek the facts which support something that already exists. Scientists continue their pursuit of truth regardless of whether it conflicts with society's perceptions of reality, just as Galileo pursued discovering the true center of the universe even when it conflicted with the official view of the church.

On the other hand, the engineer pursues a workable DESIGN that will satisfy a perceived need of society. As one engineering author states:[4] "In a society of cannibals, the engineer will try to design the most efficient kettle." That is perhaps too gruesome an idea, you say? An engineer would never do that! Remember, engineers are members of a profession that exists to serve the needs of the society in which they live. When society wants freeways, the engineer designs freeways. When society wants water purification plants, the engineer designs water purification plants. When society wants machines of war, the engineer designs machines of war. When society wants to have fun, the engineer designs recreational vehicles, reservoirs, amusement parks, camping equipment, and ski resorts. Whether society *should* want freeways, water purification plants, machines of war, or recreational aids are matters beyond the purview of the engineer as a professional, but not beyond his or her moral and citizenship responsibilities.

In the engineering design process, engineers continually compare workable systems that they have devised with other workable systems, using as a basis of comparison any of a number "measures of quality." These measures of quality may include lowest first cost, lowest cost to operate, easiest to manufacture, most reliable, lightest weight, most attractive, or easiest to operate. In these comparisons, there never seems to be one system that meets all of these desirable

---

[4] Billy V. Koen, *Definition of the Engineering Method,* (Washington, D.C.: American Society of Engineering Education, 1985).

[14-8] This horse was supposed to be a toy for my children, but the craftsman misread the drawing scale! Can it be used for some other purpose?

The most general survey shows us that the two foes of human happiness are pain and boredom.
Arthur Schopenhauer
Essays. Personality; or, What a Man is, *1853*

A scientist can discover a new star but he cannot make one. He would have to ask an engineer to do it for him.
Gordon L. Glegg

The scientist explores what is;
The engineer creates what has not been.
T. Von Karman

[14-9] Teams composed of technical people—engineers, technologists, technicians—often work together to establish a DESIGN.

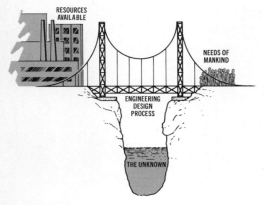

[14-10] The engineering design process is a bridge across the unknown, using available resources to provide an improved condition of life.

[14-11] Behold the turtle, he makes progress only when his neck is out.

[14-12] Engineering testing of recreational equipment.

qualities. Thus engineering trade-offs have to be made in order to pick what, in some sense, might be called the *best* DESIGN. To the engineer, *best* exists only in a relative, and often uncertain sense rather than in an absolute sense.

THE PERCEIVED NEED

The reason that an engineer begins the design process is that there exists an *unsatisfied perceived need*. This perceived need generally originates with the client (who could be society in general), or a small segment of society (e.g., citizens and their government agreeing that they need a better transportation system for their city), business enterprises (e.g., industrial investors saying that they need a new process in order to remain competitive in the marketplace), the company's management (e.g., the board of directors decreeing that the company will develop a new product), or with the engineer himself.

The need is referred to as a "perceived" one, because when the need is first recognized, it usually has not been studied extensively. It is typical of design-type problems that this perceived need is stated in inexact terms and parameters, lacking preciseness of definition—it is "fuzzy." For example, in the first of the examples in the preceding paragraph, the stated need for "a better transportation system" generally does not say that vehicles need to be built in a specified manner or that trolleys must operate along certain streets. Had it been possible to state the problem with precise definition, there would be no reason to seek a solution for it. Someone would have already solved it in order to provide the definition.

When someone says that a need exists, the statement is often nothing more than a signal to the engineer that *there may be* a need of some type, but it is recognized that the perceiver of the need may not totally understand the situation. As engineers assume the task of fulfilling the perceived need, they begin to study it and to understand it better. During this study period, it is not uncommon for the perceived need to change and to alter its appearance and for the real need to begin to emerge.

Engineers, by studying problems during the design process and by interacting with their clients, take the problems from ambiguous and uncertain statements of need to detailed and definitive methods of solution. But they must have a tolerance for the ambiguities and uncertainties encountered during the study period.

It is important that the DESIGN which the engineer defines in a set of specifications comes as close as possible to fulfilling the actual need, regardless of whether the actual need and the original perceived need are one and the same. If the true need is not established and satisfied, the DESIGN that the engineer defines will generally be less than satisfactory. The perceivers of the original need will generally continue to think that they have some unmet need as long as the real need is not met.

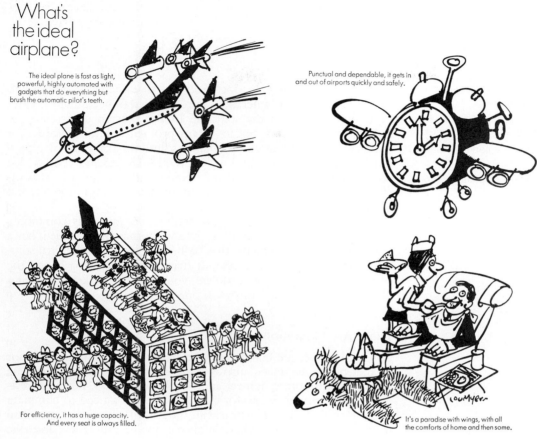

What's the ideal airplane?

The ideal plane is fast as light, powerful, highly automated with gadgets that do everything but brush the automatic pilot's teeth.

Punctual and dependable, it gets in and out of airports quickly and safely.

For efficiency, it has a huge capacity. And every seat is always filled.

It's a paradise with wings, with all the comforts of home and then some.

[14-13] Perceptions about the design of a new airplane may vary.

Design Objectives

Nearly all needs that engineers work to fulfill can be described using adjectives such as largest, smallest, maximum, minimum, most, lowest, highest, easiest, or least, as an inherent part of their solution requirements. Each of these terms implies comparisons between possible solutions, and each can be used to infer some sort of relative measure of quality for the chosen solution. Figure [14-14] lists example uses of a few of these terms in the statements of some typical problems.

Statements of both the need to be addressed and the inherent measures of quality to be used in judging any solution are referred to as the *design objectives*. The six examples given in Figure [14-14] are all statements of various design objectives.

Problems that contain these typical adjectives as a part of their statement are often referred to as *optimization problems*. That is, the solution being sought is to be, in some sense, an optimum or best one. We have already pointed out that *best* is a relative term to the engineer, for few, if any, solutions meet all of these desirable measures of quality in any absolute sense. Given unlimited problem-solving resources (time, finances, manpower), engineers might ultimately be able to define the optimum solution in meeting the design

[14-14] Examples of engineering problem statements.

Design a fertilizer-producing plant that requires the *smallest* initial cash outlay for constructing the plant.

Design a computer memory chip that has a *higher* reliability than the present memory chips.

Design a heating and cooling system for a residence that would require the *minimum* costs of operation over a 20-year lifetime.

Design a reservoir that would supply water to downstream users at the *smallest* possible costs.

Design an aircraft that *maximizes* the thrust-to-weight ratio in order to achieve high performance.

Design an automobile that is both functional and has *maximum* consumer appeal.

[14-15] Examples of design constraints.

The product must sell for less than $100.

The new lap-top computer must meet FAA regulations for emission of electromagnetic radiation on board aircraft.

This new internal combustion engine must not emit more than x ppm of CO or y ppm of NO_x.

The support system for the pipeline that is to carry the warm oil shall not allow the permafrost to be melted.

This product must meet FTC requirements for flammability.

This product must be built with the equipment now on hand.

objectives, but their resources are never unlimited. Therefore, they have to settle for the best they can do with what they have.

Constraints

It is typical of all engineering problems that there are design *constraints* or limitations inherent in the statement of need that the engineer must address in the solution. Indeed, design constraints are requirements that must be met by any problem solution in order for it to be considered as an acceptable solution. Examples of typical design constraints are found in Figure [14-15]. Included in the list are examples of constraints that originate with the client, with society in general (i.e., for the protection of the "common good"), and with consumers.

We have used the term "constraints" because it is commonly used in the field of optimization. Every optimization problem has a set of absolute requirements that "constrain" or affect the final solution. What this rightly implies is that if you were to change the constraints on a particular problem, you would very likely arrive at a different solution. *The constraints affect the solution.*

There Are Limitations

Obviously, there are many perceived needs of mankind that cannot be met by the engineer, just as there are many needs that can be satisfied without the engineer and his design process. For most residents of the planet Earth, for example, there is a need to eliminate the potential of thermonuclear war, indeed, of any type of war. The route of disarmament, whereby all nations give up their arsenals, involves political and societal solutions that are far outside the engineer's technical capabilities. The engineer's mastery of the design process is of little help here. On the other hand, the technological solution whereby each nation tries to outdo the offensive capabilities of the others on the pretense that no nation will make an offensive move against a stronger opponent is one that can and has been aided by the engineer. Such a solution partially explains the massive arsenals now in the possession of the world's superpowers. Constructing technological defensive systems that are capable of rendering ineffective any offensive strikes by a nation's opponents may be another solution that is amenable to the engineering design process.

We could, of course, continue to discuss the pros and cons of each solution that would decrease the potential of nuclear war, but that is beyond the scope of this text. What we do wish to point out is that there are problems which have technological solutions and there are others which do not. Others have possible solutions which fall into both categories. The engineer and others in society have personal responsibilities in the solving of any problem, and both have to know the extent of their responsibilities. The difficult thing about understanding these responsibilities is that they change depending on the problem and its requirements for solution. We will return to a discussion of these responsibilities in the summary.

[14-16] How might the design objectives of a commercial airliner differ for those of a military pursuit aircraft?

RESOURCES FOR FINDING A SOLUTION: AVAILABILITY AFFECTS THE DESIGN _____

Engineers always find that the resources available to them for pursuing a solution to an expressed need are limited in some way. These resources include technology, experience, knowledge, equipment, time, labor, and finances. The limitations are imposed by the client, by the current technological state of the art, by the time available, and by the engineer's education and experience.

The availability of problem-solving resources *always* affects the DESIGN in some manner; limited availability of different resources will yield different DESIGNS. Indeed, the limited availability of problem-solving resources is a constraint on the problem that affects the final solution in much the same way that the constraints inherent in the problem statement affect it.

For example, two different engineers or teams of engineers seeking to satisfy the same need would probably not end up with the same DESIGN or set of specifications because each has available a different set of limited resources. This explains, in part, why you seldom see two bridges which are identical, two airplanes which are clones of one another, or two computers which are the same—unless they were built from the same set of specifications.

Asked to estimate the number of gallons of water in an olympic-sized swimming pool, you would undoubtedly give a different answer if you had 1 minute to respond than if you had 1 day to respond. In this case, time would be a limited resource. With only $30,000 available for designing a race car, your set of specifications would define something different than if you had $1,000,000 available. Financial resources are always limited ("How do we get this project done within budget?").

Limited resources are a part of all engineering—a fact of life. Indeed, these limitations add interest and excitement to engineering because they add an extra challenge to any project.

[14-17] What statement of design objectives would lead to tests like this?

Which of you, intending to build a tower, sitteth not down first, and counteth the cost, whether he have sufficient to finish it?

Luke 14:28
The Holy Bible

THE ENGINEERING DESIGN PROCESS: THE BASIC INGREDIENTS AND THE ANATOMY _____

In this chapter we have briefly attempted to identify engineers by the manner in which they solve problems, rather than by the more traditional approach of defining them by the artifacts which they design. Having given this definition and discussed the differences between the methods of the engineer and those of the scientist, we will now devote the remainder of this chapter to examining the engineering design process.

The design process does not happen instantaneously. It takes place over a period of time, the length of which varies with the difficulty of the problem being solved and the (always limited) resources allocated to it. During this time period, the engineer accumulates information which, assuming the problem is solvable, eventually leads to an acceptable solution. We will refer to this

[14-18] What design objectives and constraints might apply to "manufacturability"?

"The horror of that moment," the king went on "I shall never, never forget!" "You will, though," the queen said, "if you don't make a memorandum of it."
Lewis Carroll, 1832–1898

BEWARE! Don't become victimized by habit.

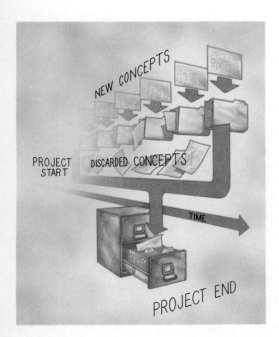

[14-19] Development of the data base.

information that the engineer accumulates as the *data base for the DESIGN*.

The word "data base" is a relatively new term in our vocabulary. Unfortunately, it has been given several different meanings during its short lifetime of usage. Here we will use the word in perhaps its most general sense: a collection of information that has been critiqued and cataloged in some manner. Various parts and pieces of the data base may exist, for example, in hard copy (in design notebooks, textbooks, and handbooks), on magnetic storage material (for computer retrieval and use), and (a must) in the brain of the designer.

Figure [14-19] schematically depicts the design data base at various times during the engineering design process. We have used a conventional office file folder as a symbolic representation of the data base. During the process some data are discarded (to some long-time storage medium, perhaps a file cabinet) and some are added.

Indeed, the data base (the file folder in Figure [14-19] *is the DESIGN as it is currently known at each point in time.* Thus we have shown it as a faint representation at the beginning of the process and as a well-defined symbol near the end. At the beginning, the DESIGN is not well known—it is "fuzzy" in the mind of the designer. It consists only of concepts and lacks many pieces of information. At that stage, it may not even contain the concept which may be used at the end.

When the engineering design process concludes, the data base contains all of the information that is needed to define the DESIGN. From this data base can come all of the information needed for the documentation drawings and/or the final report that constitute the information in the DESIGN block of Figure [14-5].

The size and content of the data base for a DESIGN will depend on several things. Typically, data bases for the design of new products or systems grow during the design process from something very small to something large enough to enable the DESIGN to be completed in a satisfactory manner. As the data base grows, it changes in the character of its detail. Initially, it acquires information of a very general nature with little detail. At some point, where the DESIGN is said to be frozen, it begins to take on a narrower scope but with greater detail.

In simple design tasks, the data base will probably remain small since the need for information is not great in developing a small and/or simple DESIGN. In design tasks of a repetitive nature (where a number of things are designed so as to be similar to an already designed item or process—for example, providing electrical service to many different commercial or residential buildings), the data base at the start of the project might be quite large in size and increase very little during the design. In situations such as the latter, the data base is used over and over. Also, it is generally not built from scratch each time a new DESIGN is needed.

THE FOUR INGREDIENTS _____

At various times while the DESIGN is maturing [14-19], the engineer will be performing different types of tasks. We will look at the anatomy of the process shortly in an attempt to identify these tasks. But first, we will discuss four basic ingredients—the basic building blocks—that are common to all of the tasks that can be identified within the process. Perhaps the most important, but least understood of these, is a source of ideas. Also important are the requirements of decision-making skills, the leveraging of information, and a set of engineering tools.

Basic Ingredient 1:
A Source of Ideas—
Oh, That It Be Plentiful!

Throughout the design process, the engineer must be the source of ideas. If he or she is unable to generate any ideas of how to satisfy the existing need, it will not be possible to satisfy that need. Generally, the more ideas the engineer or the technical team considers, the better the solution to the problem is likely to be. Thus it is imperative that there be sources of ideas for solving the problem. Indeed, in our study of the anatomy of the process, we will identify idea generation as one of the key tasks.

The need for ideas exists throughout the whole process, not just in the task where solutions to the need are generated. In the other tasks that we will identify, there will be a continual need for ideas on such things as:

How can we predict whether any particular solution will actually satisfy the need?

How can we compare the value of a solution with the value of another solution?

How can the risk involved in implementing the DESIGN be minimized?

How should the solution be documented so that it is clear to the craftsmen who will build it?

It is helpful to separate ideas into two categories: those that involve *adaptation* and those that involve *innovation* (see Topic 14-1). Adaptation utilizes the engineer's creative ability to generate solutions based on a knowledge of previous ideas and techniques. Although some are more skilled at it than others, all engineers and technologists can be successful at using adaptation for idea generation.

The innovation category contains all those ideas that involve a departure from what has been done before—ideas for new methods and/or devices. Although we cannot define when an idea ceases to be an adaptive one and begins to be recognized as an innovative one, the generation of innovative ideas is far less common than the generation of adaptive ideas.

New, innovative solutions always involve more uncertainty than do those based on known technology. Therefore, truly innovative

(text continues on page 441)

The person who is capable of producing a large number of ideas per unit of time, other things being equal, has a greater chance of having significant ideas.
J. P. Guilford

Everybody is ignorant, only in different subjects.
Will Rogers, 1879–1935

Originality is just a fresh pair of eyes.
W. Wilson

. . . every idea is the product of a single brain.
Bishop Richard Cumberland

If you want to kill an idea, assign it to a committee for study.

[14-20] I wonder . . .

Topic 14-1

Idea Generation

We do not have to teach people to be creative; we just have to quit interfering with their being creative.
Ross L. Mooney

Necessity may be the mother of invention, but imagination is its father.

They can have any color they want . . . just as long as it's black.
Henry Ford, 1863–1947

New things are made familiar, and familiar things are made new.
Samuel Johnson
Lives of the Poets, 1781

It's amazing what ordinary people can do if they set out without preconceived notions.
Charles F. Kettering
Forbes' Scrapbook of Thoughts in the Business of Life

It is obvious that invention or discovery, be it in mathematics or anywhere else, takes place by combining ideas.
Jacques Hadamard
An Essay on the Psychology of Invention in the Mathematical Fields

Important ideas are those that lie within the allowable scope of nature's laws.

Reason can answer questions, but imagination has to ask them.

U.S. space station engineers now have an in-depth data base on the last three Soviet space stations available for use in designing the U.S. station. NASA officials believe U.S. engineers will be able to learn much from the extensive Soviet experience in building and living on space stations. The report was compiled by B.J. Bluth, a sociologist working under a NASA study grant from a variety of sources, including translated Soviet technical journals.
Source: Aviation Week & Space Technology, Dec. 22, 1986.

Adaptation in Idea Generation

Ideas that apply devices, processes, and techniques already in existence to solving problems in the current design are called *adaptive* ideas. Indeed, the bulk of engineering design involves the applications of adaptive ideas. Indeed, engineers are perhaps best known for their skill at adapting old technology to new situations.

Awareness and interest are necessities in adapting previous ideas and techniques to the solution of new problems. Interest generated through education, experience, or other exposure (e.g., being well read and having wide interests) sharpens your awareness. So the more "things" of which you are aware, the more "things" you will have at your disposal for possible use in solving your problems. Let us consider some sources of adaptive ideas:

Something You Did Before.

If you have some previous experience in a similar situation, make use of it. What worked before may work again with only minor alteration. You may remove much of the risk associated with a solution to a problem if you can make no more than a small change in one of your previous DESIGNS to adapt it to the present design conditions.

Design in the consulting engineering field (electrical, structural, and mechanical consulting) and in much of industry is based largely on the use of successful ideas that worked in previous similar situations.

Something Someone Else Has Done.

Building on the experience of someone else may be the next best thing to having experience of your own in solving a problem. However, you must independently confirm that the previous solution was, in fact, a satisfactory solution and that there were no difficulties in its implementation. But be sure that you have all the facts regarding the apparent success of the previous idea.

Sources for borrowed ideas are:
Mother Nature.

Man has put to work the unique DESIGNS found in nature much less frequently than he has used the products of nature. In fact, man has often discovered the existence of some unique DESIGN in nature only after he himself had invented a similar device and learned to recognize its qualities.

Examples of DESIGNS in nature that man has learned from or may yet learn more about include the sonar of bats and porpoises, the snake's hypodermic fang, the propulsion of marine animals, the flight of birds, the bioluminescence of fireflies, and the generation of voltage by eels (see also 14-T1,T2,T3). Obviously, nature is filled with wondrous DESIGNS, DESIGNS that could be extremely useful to the engineer who is looking for new and better ways of solving problems. It was Aristotle who once wrote, "In the works of Nature *purpose,* not accident, is the main thing."
Other Members of Your Organization.

Most organizations consider the ideas of its employees to be the property of the organization. Using them, then, is nothing more than using ideas that the company already owns. There is nothing unethical or unprofessional in using such ideas under these circumstances. However, as a matter of courtesy, you should inform the originator of the idea that you will be using it.

Persons Outside Your Organization.

Once you go outside your organization for ideas, you raise ethical questions, if not legal questions, concerning the use of someone else's property. Society has long recognized the value of giving the originators of real and intellectual property (inventions and ideas) exclusive rights to that property in return for full disclosure. Accomplished through patent and copyright laws, this full disclosure stimulates further creative activity, while making the benefits of the inventions and ideas available to society through rights granted by the originator. It is your duty as a member of society to ensure that you have the right to use any ideas and materials.

Innovative Ideas

Ideas that involve new devices, processes, and techniques—technology that did not previously exist—are termed *innovative ideas*. It is these ideas that push our technological state of the art in quantum leaps since, once disclosed, they usually stimulate the flow of many adaptive ideas.

Truly innovative people often possess complex personalities and they are generally capable of seeing a problem in many different ways. They tend not to be overly conservative or parochial in their approach to solving problems and they are likely to be much more willing than the average person to take risks. They often possess a strong curiosity and a vivid imagination. Most also have a keen sense of humor and a high tolerance for ambiguity and uncertainty. They share with the very best adapters of ideas the characteristics of having wide interests, a good general and specialty education, and of being well read.

All of the attributes mentioned above work together to produce ideas that are both plentiful and diverse. The ability of the true innovator at *divergent thinking* (i.e., producing thoughts that are widely varied) is considerably better than that of the average person. Divergent thinking is especially important in the idea-generation task (discussed later in this chapter) of the design process.

[14-T1] Based on function and biological "manufacturing techniques," the egg is nature's perfect package. Its basic form has remained virtually unchanged for over 300 million years.

[14-T2] Honeybees build hexagonal cells with trihedral bases. This design is not only the strongest possible structure for a mass of adjacent cells, but it is also the one that requires the least amount of labor and wax.

[14-T3] Creatures in nature rely on visual contrast for self-preservation.

Topic 14-1

Idea Generation (continued)

Creativity is the art of taking a fresh look at old knowledge.

All men are born with a very definite potential for creative activity.

John E. Arnold

The mind is not a vessel to be filled but a fire to be kindled.

Plutarch

It takes courage to be creative. Just as soon as you have a new idea, you are a minority of one.

E. Paul Torrance

Disciplined thinking focuses inspiration rather than constricts it.

It is better to wear out than to rust out.

Bishop Richard Cumberland

Creativity is man's most challenging frontier!

A child is highly creative until he starts to school.

Stanley Czurles, Director of Art Education.
New York State College for Teachers

Imagination is more important than knowledge.

Albert Einstein, 1879–1955

An idea, in the highest sense of that word, cannot be conveyed but by a symbol.

Samuel Taylor Coleridge, 1772–1834
Biographia Literaria

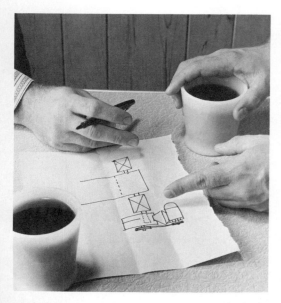

[14-T4] Not uncommonly, one team member's inspired idea will set off a chain reaction of ideas from other team members.

Both innovative thinkers and adaptive thinkers usually have fluency in generating new ideas from previous ideas as long as both the new and the old ideas are in the same "category." For example, if you were seeking ideas on possible uses of a common clay brick, both types of thinkers might be quick to respond and be quite productive in suggesting applications in a "building" category, since that is the traditional use for such an object. Typical responses might be: build a wall, build a patio, build a bookcase, build a fireplace, and so on. This ability to produce ideas within a given category is sometimes referred to as *ideational fluency*.

The innovative thinker, on the other hand, generally is better able to change categories than is the strictly adaptive thinker. That is, the innovative thinker might quickly jump to "nonbuilding" categories with such responses as: paperweight, boat anchor, displacer of water in a toilet tank, projectile for delivering notes, and so on. This ability to change categories is often referred to as *ideational flexibility*.

Idea Stimulation Techniques

There have been many techniques recommended for stimulating the generation of ideas. They include:

Brainstorming.

This technique uses an organized effort aimed at compiling all of the ideas that might be applicable for solving problems. This is accomplished (1) by releasing the imagination of the participants from restraints such as fear, conformity, and judgment, and (2) by providing a method to improve and combine ideas the moment an idea has been expressed. Although brainstorming is usually considered to be a group activity, there is evidence to suggest that the process can be as effective when done by a single individual.

The *Delphi technique* is a version of "storming" wherein the participants are separated from one another and asked to generate ideas. These ideas are then collected and distributed to all participants for evaluation.

No matter how storming is conducted it is imperative that the climate in which it takes place be noncritical. Any analysis or criticism of any of the ideas generated tends to thwart the process. After the storming session, there should be a "calming" period in which the ideas are allowed to germinate and take root. During this period, inappropriate ideas are culled out and appropriate ideas are improved.

Synectics—Use of Metaphors.

This technique is one whereby you try to think of analogies to your problem. It attempts to bring about one or more solutions to a problem by drawing seemingly unrelated ideas together and forcing them to complement each other. The synectics participant tries to imagine himself or herself as the "personality" of the inanimate object: "What would be my reaction if I were that gear (or drop of paint, or tank, or electron)?" Thus familiar objects take on strange appearances and actions, and strange concepts often become more comprehensible.

Visual Thinking.

Freehand drawing (see Chapter 7) often stimulates ideas because it forces the focusing of the mind on specific tasks.

solutions *may* involve more risk and, as a result, they do not always gain rapid acceptance by the people who ultimately approve the final DESIGN

Despite this admirable and understandable conservative bias in favor of adaptation, innovation moves our technology in giant leaps. We have already pointed out in Chapters 8 and 10 that the introduction of VisiCalc, the first computer spreadsheet program, was a significant innovation that is widely credited with spawning the microcomputer revolution. In fact, the history of technology is filled with examples where some truly significant idea or invention brought forth a flurry of activity that caused rapid advances in our technology—single ideas so significant that they stimulated many, many other people to build and improve on them with adaptive ideas.

In general, the engineering and technology professions are not particularly noted for their ability to generate truly innovative ideas, although some members of these professions are particularly adept at it. Indeed, those members who are highly proficient at true innovation are usually given their own generic label—inventor. Fortunately, everyone's creative ability can be stimulated and improved. Your creative talent is largely affected by your attitude, your education, your interests, and by the amount of mental exercise you do. Thus you can "invent" your own ability.

The generation of ideas is a necessary part of the engineering design process, but all of your ideas do not have to be truly innovative or new. The bulk of engineering involves making only small steps from what has been done before—the applications of adaptive ideas. There is a place and a need for a wide range of thinkers (idea generators) in the process.

No idea is so outlandish that it should not be considered with a searching but at the same time with a steady eye.
Winston Churchill, 1874–1965

A man must have a certain amount of intelligent ignorance to get anywhere.
Charles F. Kettering

Our doubts are traitors and make us lose the good we oft might win by fearing to attempt.
William Shakespeare, 1564–1616

Depend upon it, sir, when a man knows he is to be hanged in a fortnight, it concentrates his mind wonderfully.
Samuel Johnson
September 19, 1777

Those who dream by day are cognizant of many things which escape those who dream only by night.
Edgar Allen Poe, 1809–1849

More today than yesterday and more tomorrow than today, the survival of people and their institutions depends upon innovation.
Jack Morton, Innovation, 1969

The age is running mad after innovation. All the business of the world is to be done in a new way. Men are to be hanged in a new way.
Samuel Johnson, 1777

Whatever one man is capable of conceiving, other men will be able to achieve.
Jules Verne

Basic Ingredient 2:
An Ability to Make Decisions—
Deciding Among the Alternatives

Any experience with engineering design always demonstrates another requirement of all design activity: the requirement that decisions be made—decisions that must be based on the information at hand. Indeed, if any single activity best summarizes the engineering design process, it is *decision making*. The difficult part of decision making in engineering design is that the information at hand is usually either sparse or overwhelming in amount. It is feast or famine.

The need to make decisions implies that there will be alternatives that will emerge as the design process evolves. Indeed, the process can essentially be viewed as the selection of alternatives. The designer's creative abilities are important in generating the ideas that establish the alternatives. But the designer must be able to decide when an alternative is worthy of pursuit and when it should be put aside, if only temporarily. In Figure [14-17], deciding when a concept should be removed from the design data base and when it should be kept is often quite difficult when the DESIGN is "fuzzy."

Fortunately, all of us can improve our decision-making skills. In fact, it is often easier to correct an inability to make decisions than it is to correct the opposite tendency, that of making decisions too

[14-21] The mind can die from inactivity.

What good is electricity, Madam? What good is a baby?

Michael Faraday

An inventor is simply a fellow who doesn't take his education too seriously.

Charles F. Kettering, 1876–1958

"I can't believe that," said Alice. "Can't you?" the Queen said in a pitying tone. "Try again; draw a long breath and shut your eyes." Alice laughed. "There's no use trying," she said, "one can't believe impossible things."

"I daresay you haven't had much practice," said the Queen. "When I was younger, I always did it for half an hour a day. Why, sometimes I've believed as many as six impossible things before breakfast."

Lewis Carroll, 1832–1898
Through the Looking Glass

Isolation is the sum total of wretchedness to man.
Thomas Carlyle
Past and Present, 1795–1881

He that answereth a matter before he heareth it, it is folly and shame unto him.

Proverbs 18:13
The Holy Bible

[14-22]

quickly. Decisions made in a rigid fashion and with little thought often lock the designer into later difficulties that are hard to correct.

We devoted Chapter 7 to freehand drawing because it provides a quick means of developing your ideas while improving your visualization skills. Ideation drawing *forces you to make decisions* and resolve conflicts in your ideas—conflicts or uncertainties that can bounce around forever as long as they stay in your mind. Changes can be made in drawings by overdrawing earlier sketches or by the simple use of an eraser.

This same alteration privilege *may* hold for decisions made in other nondrawing parts of the design process. Indeed, early in the process, decisions should be made with the understanding that they may need to be reversed later. In the later stages of the process when the problem-solving resources begin to become scarce, it becomes harder and harder to reverse decisions once they are made. That is, as time goes on you become more and more committed to a particular DESIGN—in the words of the engineer, the DESIGN becomes *hardened*.

Basic Ingredient 3:
Ability to Leverage Information—
Expanding Your Design Data Base

In addition to idea generation and decision making, the design process is characterized by the "leveraging" of information. Leveraging information is the process of using the information you have to obtain additional *useful* information. It is an important part of all engineering design, because it allows the data base for the DESIGN to grow and the DESIGN to become better defined [14-19]. Experience, education, cleverness, and creativity of the designer are all important attributes in guiding the leveraging of information.

Physical means by which information is leveraged include library research (discovering what other people have written about some feature of your problem), discussing the DESIGN with other knowledgeable people (oral research instead of library research), applications of theoretical analyses (discovering what mathematics, science, and engineering science can tell about the DESIGN), and experimental testing (collecting your own information where analysis is thought to be inadequate).

In the early stages of the design process it is important to be able to convert small amounts of information (you usually have only a small amount of information in the early stages) into larger amounts—to make your information multiply [14-22]. In these early stages, it is usually undesirable to spend your problem-solving resources on efforts that convert large amounts of information into small returns. It is only later as the process is nearing a satisfactory conclusion that the expenditure of resources for conversion of large amounts of information into small returns is worthwhile and, often, necessary. These later stages of the process are where the DESIGN is becoming better defined or hardened.

When the design process seems to be going well and progress is being made on defining the DESIGN, leveraging of your information will seem easy. In reality, it will be the easy leveraging of information that will be making the design process appear to be going well. When the design process has slowed and little, if any, progress is being made in defining the DESIGN, leveraging of your information will seem very difficult. In reality, it will be the difficulty of leveraging information that will be thwarting the design process. It is important to be able to recognize when leveraging is becoming difficult, so that increased effort can be expanded to overcome the difficulty and get the design process back on track. We will return to this point in the later section "Pitfalls and Emotions of the Design Experience."

Basic Ingredient 4
Tools for the Engineer:
Heuristics—Tools that Work *Most* of the Time

The engineer's tool kit for design is stocked not only with information gained in an educational program and experience gained in the practice of the profession, but also with experience from all aspects of interacting with the environment. The tools in the tool kit come in a wide variety of sizes and shapes. At one extreme are those tools that are useful for coarse work, quick estimates, and "ballpark" numbers. At the other extreme are sophisticated tools that provide refined estimates and detailed information.

The tool kit of each engineer varies, since the tools are strongly tied to the education, culture, discipline, and experiences of its owner. However, all engineers have some common tools that identify them with various disciplines within their profession. For example, two civil engineers have more tools in common than do a civil engineer and an electrical engineer.

Koen[5] calls these tools of the engineer *heuristics*. "Heuristics" is a term commonly used in the field of artificial intelligence (often shortened to the acronym AI) to represent rules of thumb, relationships, and hints that are programmed into a computer in order to allow it to sort through alternatives, control the questioning, and arrive at *some* final conclusion, based on solicited responses from the operator. These heuristics generate the correct conclusion *most* of the time, but there is no guarantee that the conclusion will be correct *all* of the time.

Computer programs that allow the computer to play chess contain heuristics that define reasonable responses to opponent's moves rather than to analyze all the possible plays that are physically possible. The latter method, if it were ever programmed, would require huge computer resources. The heuristics, on the other hand, are based on suggestions and rules of thumb of good chess players; they represent what is thought to be good responses *most* of the time.

The heuristics in the engineer's tool kit serve the same purpose as those in AI. They are sets of rules and relationships that can be used to leverage information and provide aid and guidance in the

[14-23] Computerized information-retrieval systems give you rapid access to much more information than ever possible. Use them to leverage what you know.

[14-24] Chess-playing robots rely on heuristics.

[5] Billy V. Koen, *Definition of the Engineering Method*, (Washington D.C.: American Society for Engineering Education, 1985).

solution to a problem. One of the objects of a technical education is to deposit in your tool kit many of the heuristics you will need in order to begin functioning as a member of a technical team. Of these educationally deposited heuristics, some come from the field of physics, some from math, some from engineering science subject areas, and some from courses and electives in your major. Some will even come from your association with student professional societies.

After you begin your career, experience will add certain heuristics to your set. Participating in professional societies and in continuing education programs will help keep your set of heuristics up to date.

We have chosen to classify the heuristics used in the engineering profession into five categories (see Topic 14-2). These are:

1. Rules of thumb and orders of magnitude.
2. Rules for more refined analysis and experimentation.
3. Rules for reducing risk.
4. Rules that affect attitude.
5. Rules for allocating resources.

It is the unique set of engineering heuristics that distinguishes the engineering professional from other professionals, such as architects, graphics designers, and so on, who also do "design" work.

Topic 14-2

Engineering Heuristics

We have chosen to consider the heuristics of engineering to be divided into five categories. These are:

1. *Rules of thumb and order-of-magnitude analyses.* Rules of thumb are considered to be general rules that apply in some rough or crude way to a particular situation. They are useful in evaluating early concepts in the design process because they do not require large amounts of data in order to produce an answer. The answers they yield are not expected to be precise in any way, but they are useful in obtaining "ballpark" estimates. Order-of-magnitude analyses serve the same purpose, in that they yield answers that are usually good within a factor of 10 (said to be an "order of magnitude") or better.

In addition to being useful in evaluation of early concepts, these rules are also helpful when evaluating whether more "precise" methods are producing reasonable results. They give you some basis for comparing more "precise" answers.

2. *Rules for more refined analysis and experimentation.* The rules we include here are capable of giving more precise answers, when used carefully, than do the heuristics in category 1 above. We include here, among other principles, the *physical laws of nature* which were discussed at the beginning of Chapter 12. The applications of these laws in modeling and analysis work on the DESIGN, always requires that assumptions be made. The validity of the results depends on the validity of the assumptions made. Because assumptions are always necessary and because they are often difficult, if not impossible to evaluate, engineers always use scientific principles as heuristics—rules that work *most* of the time.

There are, of course, relationships which must be classed in this category in addition to the physical laws of nature. These include relationships between the properties of matter (often called "constitutive relationships"), various theoretical and experimental relationships and data that have been devised to model devices, and various design rules and principles.

We will not dwell on this category because one purpose of other courses in your educational curriculum is to bring these to your attention.

3. *Rules for reducing risk.* These rules, which should be self-explanatory in their purpose, have become very important in our post-industrialized society. Not only does the consumer of the DESIGN want lower risks associated with the use of the artifact designed, but so also is society very interested in protecting "the common good."

There are several ways in which engineers can ensure that their methodology will automatically reduce risk to all involved. These include the use of *factors of safety* to increase the safety and reliability of the DESIGN. Factors of safety are improvements in performance intended to more than offset any inaccuracy that may be inherent in the modeling of the artifact being designed. These inaccuracies arise because of the impossibility of evaluating all of the assumptions involved and/or the inability of the models being used to thoroughly represent the phenomena being modeled.

Another technique for controlling risk is to avoid making large changes in the technology being used in the DESIGN. Make small steps from what has been found to work. The lowest risk comes in using ideas that are adequate in similar situations. This is not necessarily meant to imply that innovative ideas, which often cause the technology to move in giant leaps, should be shunned. It simply means that more extensive testing and reviews must be applied to these innovations to ensure that they work.

Another risk-reducing rule is to always have in mind an option that you can fall back to if some unforeseen circumstance arises in pursuing your DESIGN. In the words of Koen, "always give yourself a chance to retreat."

4. *Rules that affect attitude.* This is a complex set of heuristics that are a part of every true engineer. This set alone distinguishes the actions of the engineer, for it instills the inquisitive nature and the penchant for problem solving that characterizes the profession. The set leads both to optimism in investigating solutions to new problems and to a disdain for the "unsolvable" problems of the world.

5. *Rules for allocating resources.* As with the other categories, there are many heuristics that are useful in allocating resources during the design process. As we did in the other categories, we will discuss only a few of these here. The ones we do discuss will be particularly useful in pursuing solutions to the problems in Chapter 15.

The first of these is to allocate enough of the design resources (time and/or money) to the most difficult and critical parts of the problem. Do not consume resources on trivial parts or on "nonproblems." If the difficult and critical features of the problem do not receive enough attention, a satisfactory DESIGN will probably never be established.

Another resource-allocating heuristic says that you should not consume more resources in trying to design around any unknown problem than it would take to solve that problem. In the words of Koen, if "the cost of *not* knowing exceeds the cost of finding out," then pursue "finding out." We will admit, however, that it is often difficult to estimate the price of "finding out."

As a final example, and one that will be extremely important in the

Development of a Project

1. Enthusiasm.
2. Disillusionment.
3. Panic.
4. Search for the guilty party.
5. Punishment of the innocent party.
6. Fame and honor for the non-participants.

OMNI, *July 1980*

It is important to point out that, in principle, your toolkit always bears a time stamp, that is, an invisible mark that tells the date when the tools were last updated. If you were to design an object five years from now, you would surely do it differently than you would do it today, even if you had the same resources available to you and the same constraints imposed. Personally, you will gain additional educational and practical experience during the intervening five years. During these years you will add some heuristics to your toolkit and you will deactivate others (indeed, you will be amazed at the number of heuristics you will add to your toolkit before graduation). In addition, the heuristics used in engineering will change during the next five years due to the experiences of engineers in general. If you keep current in your field, your own set of heuristics will change to reflect this.

THE DESIGN PROCESS: AN ANATOMY

We have discussed the role and purpose of the engineering design process as an activity and we described how the activity generally leads to an engineering design or set of specifications. Further, we have discussed some of the features that are common to all parts of the activity of producing an engineering design. That is, we have covered four ingredients that are required throughout the design process.

We now wish to analyze the activities of the engineer which permit the data base to grow with time and ultimately mature into the final DESIGN [14-5]. What is it that the engineer does, and when does he or she do it, in order to produce the DESIGN before the resources available for design are fully consumed?

It is appropriate to ask if there exists a standard procedure that, if followed, will guarantee the generation of an acceptable DESIGN. Has someone broken the process into its sequential steps that will enable you to know what to do next? That is, is the *morphology*—the structure or form—of the design process known?

Many practicing engineers have learned how to generate ideas for solving problems and how to take these ideas from conception to an acceptable DESIGN. But if you were to ask 100 different practic-

Topic 14-2

Engineering Hueristics (continued)

problems of Chapter 15, we point out that there will come a time in the design process when you *must* choose a concept to carry to completion. You will never have enough resources to carry all the early concepts to a final DESIGN. At various places in this chapter we have referred to this as "freezing" a concept or "freezing" the DESIGN. You must "freeze" a concept before you consume all of your resources.

Once you "freeze" the DESIGN, you then spend your remaining resources in adding detail to your data base and doing what is necessary to finalize the work. If unforeseen problems do arise, you should first try patching this chosen concept in some way rather than discarding the whole idea. Only as a last resort should you give it up and try something new (remember, "always give yourself a chance to retreat," but after the DESIGN is frozen, do so only when absolutely necessary).

ing engineers what steps they could identify in their approach to solving design problems, you would probably get a large number of different answers. These engineers would come closer to agreeing on a common set of tasks comprising the process (i.e., what they spend time doing) than they would agree on the order in which the tasks must be executed.

Those engineers who are involved in nearly repetitive design work where each DESIGN is very similar to a previous DESIGN may think of the process as a series of tasks that are sequentially executed. Do step 1, then step 2, followed by step 3, and so on, and you will arrive at the DESIGN. That, however, is an adopted heuristic which works for them. Like all heuristics, it has its limitations.

Engineers who are responsible for the design of new artifacts or who are involved in new projects would probably claim that they cannot assign any preferred order to an agreed-upon set of tasks. That is, the heuristic of the engineers discussed in the preceding paragraph does not work for them.

How, then, do you as a novice engineer or technologist, the new initiate to the design process, learn to function in this environment? One way, of course, is for you not to worry, but just to jump headlong into the process and learn to survive. Be careful! Such a technique is not very successful for learning to swim or to ride a bicycle, and experience tells us that it will not be successful here either.

A much more effective way is for you to become familiar with a general set of tasks and learn to adhere to some rules that govern where you should direct your attention while executing these tasks. Once you understand the tasks that can be identified and learn where to always direct your attention, you can gain valuable experience through examples and simple problems. Once you have a workable understanding, you may experiment and search for a better technique that works for you. We shall use the words "workable" and "search for better" again shortly.

Executing the engineering design process—knowing what to do next—is much like freehand drawing (see Chapter 7) in the sense that it is not an analytical, left-mode exercise. In its true form it is much more a holistic process; breaking it into steps and rules can never totally explain it. The instructional material of Chapter 7 stressed mental concentration on shape and form to avoid the common pitfalls, such as stereotypical substitution, while drawing. Similarly, we will try to focus your attention on the shape and form of the DESIGN at all times during the process.

Though this be madness, yet there is method in it.
Shakespeare, 1564–1616

The Tasks

In analyzing the anatomy of the engineering design process, it is convenient to consider the process as consisting of seven *tasks*. Seven is a large enough number to adequately represent the various activities that can be identified and small enough so that each activity can be distinguished from the others. The most natural order of these seven tasks is as follows.

Problem Definition. The object of this task is to determine the need to be satisfied, the constraints that are inherent in the need, and the resources available to satisfy it. The need is usually "fuzzy" in

The mere formulation of a problem is far more often essential than its solution, which may be merely a matter of mathematical or experimental skill. To raise new questions, new possibilities, to regard old problems from a new angle requires creative imagination and marks real advances in science.
Albert Einstein, 1879–1955

[14-25] In the design process one of the most difficult tasks is to define the problem accurately.

The engineer's first problem in any design situation is to discover what the problem really is.

One of our problems is trying to find out which way is up and which way is down.
John Young
Astronaut, Apollo Ten

Getting an idea should be like sitting down on a pin: it should make you jump up and do something.
E. L. Simpson

Society is never prepared to receive any invention. Every new thing is resisted, and it takes years for the inventor to get people to listen to him and years more before it can be introduced.
Thomas Alva Edison, 1847–1931

both its initial definition and in its accompanying limitations or constraints. Often, the resources are not well known. It is necessary to remove as much of the "fuzziness as possible" from the problem and to establish the resources (time, finances, etc.) available. During the design process, this definition should become further clarified, for the DESIGN should ideally present the engineer's response to his understanding of the problem.

Idea Generation and Germination. This task really consists of two sequential phases. The first is the generation phase, which has as its object the generation of ideas which might lead to a solution of the problem being defined. This is a creative task where both adaptive and innovative ideas are important. *Divergent thinking,* where you leverage ideas by letting ideas stimulate more ideas, is called for here. This is often done in "brainstorming" sessions, where the object is to generate and record ideas. It is important to postpone any evaluation of the ideas during these sessions since criticism stifles the imagination. Any technique for generating ideas is acceptable (see Topic 14-1).

The second part of this task is the idea-germination phase. This is a calming time when you let the idea seeds collected in the generation phase, germinate. As you go over and over the ideas in your mind, you will find some which have merit and some which are infeasible. This calming step begins the process of *convergent thinking,* in which you begin to converge on the better ideas by giving more consideration to some of the ideas and less to others.

Component Modeling. The ideas or concepts that survive the germination task above may consist of several parts or components. Usually, these components have some variables that can be controlled by the designer, but others which are then determined by the components' behavior. In order to be able to predict whether the components will work together and contribute to an acceptable DESIGN, the performances of each must be modeled.

The object of this task, then, is to collect or construct performance models of the components. Here the word "models" is meant to imply tabular data, graphical data, mathematical equations, computer programs, physical models whether full size or constructed to some scale, and so on—any type of model or simulation of the component which is helpful in evaluating performance. Many of the heuristics in an engineer's toolkit support this type of modeling.

System Modeling. If the DESIGN is to consist of several components or parts, some method must be used to determine if all of the parts, when used together, will produce an acceptable solution. This is the role of the System Modeling task. Like the models discussed under "Component Modeling" above, system models could be tabular data, graphical data, mathematical equations, computer programs, physical models, and so on. The component models play key roles in this system modeling.

Finding a "Workable" Design. The object of this task is to exercise the system models and study the present DESIGN (what is now represented by the data base) until *some* method of meeting the problem definition is found. This step may even require that you make major alterations in the present concept or even return to the idea-generation phase. Any solution that essentially meets the constraints of the problem is called the "workable" DESIGN. It may not meet all of the design objectives (e.g., it may not be the most efficient, or the most reliable), but it should meet essentially all of the technical requirements and the constraints on the solution. This step continues the process of "convergent" thinking started in the germination phase of the Idea Generation task.

The Search for "Better." Since the first "workable" DESIGN probably does not meet all of the design objectives, the object of this task is to use what has been learned from the earlier stages of the design process to generate solutions that come closer to meeting all of these objectives. In particular, some change in the DESIGN must be made in order to conduct this "search for better." It is important to realize that the "workable" DESIGN has to be compared to *another* DESIGN

The engineer must know the basis on which different DESIGNS are to be compared. That is, some "measure of quality" must be established in order to proceed with the Search for Better task. As we have discussed in the section on design objectives, the appropriate "measures of quality" are often established in the design objectives.

Documentation. The object of this task is to finish the documentation of the DESIGN, so that it can be understood and implemented by other people. This usually means writing reports, generating figures, finishing final communication drawings, completing documentation drawings, and presenting data in some useful form. An oral presentation may even be necessary. When the design process is concluded, the data base, which has grown during the design process, should contain all the necessary information for this task.

Since the design process is initiated with the perception that there is some unmet need, the *Problem Definition* has been placed at the beginning of the task list. *Documentation* is shown as the last task, because the set of specifications that defines the DESIGN is the desired product of the process. However, the order in which we have discussed these tasks should not necessarily imply any favored sequence for executing them. We will, instead, recommend that you go to the task that best serves your needs at the time, while always trying to visualize what your DESIGN will be.

Executing the Tasks in Homework Problems

After being assigned a textbook problem in a formal class, you have perhaps unknowingly executed a *few* of the tasks which we have just discussed. But which ones?

The difference between true design-type problems and most textbook-type problems is that the DESIGNS that you are dealing with in textbook problems are well known. They are no longer "fuzzy" and ambiguous. This also means that the details of all the

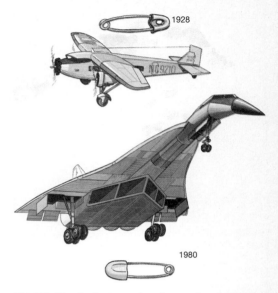

[14-26] Simple designs often have a longer life than complex designs.

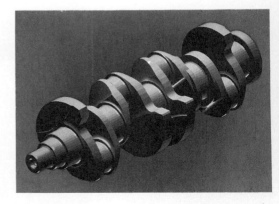

[14-27] Computer-generated solid model of a crankshaft.

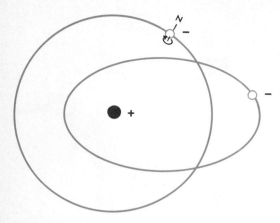

[14-28] Modified Bohr–Sommerfeldt model of the atom—a component model.

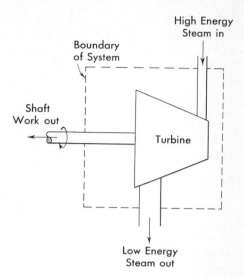

[14-29] Energy diagrams can be especially useful in understanding thermodynamic systems. They are component models.

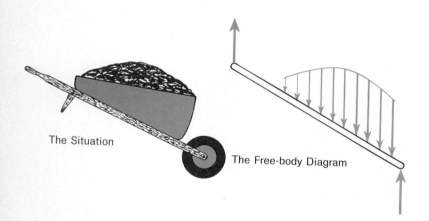

The Situation

The Free-body Diagram

[14-30] Free-body diagrams are simplistic models of physical systems.

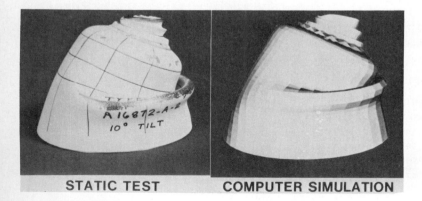

STATIC TEST   COMPUTER SIMULATION

[14-31] Component modeling. Computer simulation and actual test result made on a nose-section part of an aircraft.

[14-32] Full-scale system modeling of a proposed automobile. Does it meet the design constraint for passenger safety?

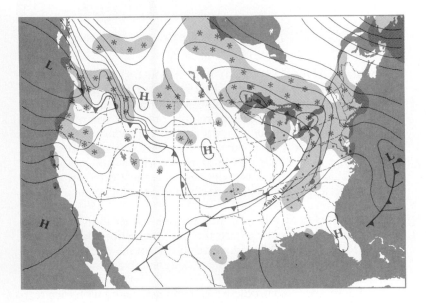

[14-33] A weather map is a system model.

[14-34] System model of New York City. Design studies of large projects frequently require the use of scale models, such as this one of New York City.

[14-35] Testing microprocessor circuitry for perform-ance—system modeling.

[14-36] The search for a better mouse trap continues.

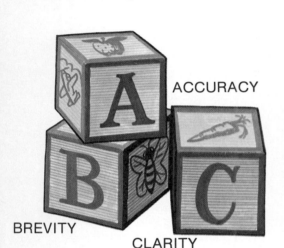

ACCURACY

BREVITY

CLARITY

A B Cs of COMMUNICATION

[14-37] Accuracy, brevity, and clarity are fundamen-tal to good documentation.

components are well known. It is as if you had been handed the symbolic file folder of Figure [14-19] late in the process when a DESIGN has almost, if not totally, been established.

Typically, textbook problems require only that you model components and/or the system to which the component belongs, in order to determine performance. Of course, you must document your findings in a problem write-up or report for submission to the instructor. Thus, most textbook problems are just exercises in modeling and performance predictions, followed by documentation (see Topic 14-3).

The intent—and it is an important one—is to allow you to become familiar with and exercise the new heuristics (the rules and procedures) which are being covered in the course that you are taking. That is, the class assignments are designed to add new heuristics to your engineering toolbox.

Although there is a problem statement associated with each textbook problem and ideas have to be generated about how to analyze it, these efforts are smaller in scale than what is implied in the Problem Definition and Idea Generation tasks in the design process.

In textbook problems, there is seldom a requirement that you verify that the system being studied will meet some design objective, or that you search for a better DESIGN

(text continues on page 455)

Topic 14-3

Design Problems Versus Textbook-Type Problems

It is useful at this point to compare real-world engineering design problems with textbook-type problems more commonly found in the engineering educational environment. Most textbook problems are analysis oriented. That is, they require calculations to be made—some component or system is to analyzed to predict its behavior.

The textbook-type problem corresponds to the activities in the *Component Modeling* and *System Modeling* tasks of the design process. However, in the textbook-type problem, the component or system being modeled is usually well defined in the problem statement. Also, the input variables are also given in the problem statement. All that remains is to exercise the model. As Einstein once said, the solution "may be merely a matter of mathematics or experimental skill."

In a design problem, the components and systems are not known until *you* decide what they will be. Even when you do decide what components are to be present in a DESIGN, it is not always straightforward to determine the values or range of the input variables that lead to a "workable" DESIGN, especially early in the design process. Since these inputs are necessary in predicting the output of the components and system, it is again difficult to proceed.

Component modeling and system modeling (analysis) play an important part in defining DESIGNS that have minimum risks associated with them. The good engineer must, therefore, have ample tools available for these tasks. To satisfy this requirement for modeling tools, engineering curricula emphasize the learning of rules and procedures of modeling or analysis through the use of the textbook-type problems much more than they do the philosophy and procedures associated with the establishment of a particular DESIGN.

To consider the difference between textbook- or analysis-type problems and design-type problems, let us consider the following examples:

Mechanics.

Given a block of known mass and shape (the DESIGN) and known imposed forces (the system inputs) [14-T5], determine the motion of the block as a function of time (predict the system performance). Notice that the inputs and the DESIGN are known and the performance is desired. However, in studying mechanics it becomes obvious that the inputs and the performance are related to the DESIGN through Newton's laws of motion. The performance is predicted through the proper application of these laws of motion.

Electrical Sciences.

Given a circuit of known resistors, capacitors, inductors, transistors, and so on (the DESIGN) and given imposed voltages on certain of these components (the inputs), determine the corresponding voltages at the other components (the performance) [14-T6]. Again, the inputs and the DESIGN are known and the performance is desired. Proper application of the principles of electrical circuit analysis allows the prediction of the desired performance.

Aerodynamics.

Given the specifications for a new airplane defining its shape, center of gravity, and so on (the DESIGN), predict its performance under known payload, flight conditions, and so on (the inputs). The inputs and DESIGN are again known, while the performance is desired. Proper application of aerodynamic principles and data can yield a prediction of the performance.

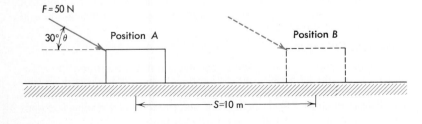

[14-T5] Typical mechanics problem.

Topic 14-3

Design Problems Versus Textbook-Type Problems (continued)

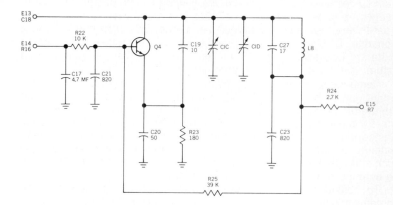

[14-T6] Circuit analysis courses will teach you to analyze the performance of electronic circuits.

Factory Planning.

Given a factory (a DESIGN) and its raw material inflow, hours of operation, and manpower availability (the inputs), predict the performance in producing goods. Proper application of data and principles studied in industrial engineering will yield the desired performance.

These textbook-type problems are said to be "well formulated." From the problem statement and the material just covered in the text, we often know how to proceed to a solution.

Contrast these problems now to real-world engineering problems, which are usually of the type in which the inputs and performance are known (i.e., the resources are known and a need is to be met) and the DESIGN is unknown. Here the engineer must devise a DESIGN that will satisfy the need in a satisfactory way. These real-world problems are much more difficult to solve than the analysis-type problems discussed above, since the principles of analysis learned in an educational program are extremely difficult to apply unless the DESIGNS are known. We could recast the four textbook-type examples above into design problems in the following way.

Mechanics.

Given the imposed forces (the inputs) and the desired motion of an object as a function of time (performance), define the object (the DESIGN) that will fulfill the need. Notice that the inputs and the performance are known and the DESIGN is desired. However, the principles embodied in Newton's laws of motion are most easily applied to known objects (i.e., bodies with known shapes and composition so that masses and moments of inertia can be determined). In this design setting, the laws of motion can be used to obtain the required mass and moments of inertia, but these quantities do not uniquely determine the shape and composition of the object.

Electrical Sciences.

Given certain available voltages (the inputs), determine the electronic circuit (the DESIGN) that will produce other desired voltages (the performance). Again, the inputs and the performance are known and the DESIGN is desired. The principles of electronic circuit analysis do not make the definition of the desired system unique or easy.

Executing the Tasks for Real-World Problems

In the typical true design problem, it is not as simple to work your way through the process as in the case of a textbook-type problem. First, for design problems, the initial problem definition is usually somewhat ambiguous and uncertain. Second, it is not as easy to generate ideas of how to solve the problem since there is no textbook with convenient equations and details that accompany the design problem statement.

Third, component modeling and system modeling are not straightforward in design problems, because the DESIGN is not well established, especially early in the process [14-19]. This leads to a fundamental difference between the typical textbook problem and true design problems because nearly all of the engineer's heuristics for modeling require definition of the system in order to proceed. In design problems, modeling is often needed at a time when the components that need to be modeled are not yet defined.

If the engineering design process were a sequential progression through the tasks we have discussed, the problem of modeling ill-defined components and systems would not arise. In any project, if you could completely define the problem and then conceive a solution that you know would work, the components which you were going to use would also be known. Under such conditions, the component modeling and the system modeling would be, in principle, quite possible. But in most design situations, the certainty that such a sequential progression makes possible is not there because of the inability of the designer to quickly establish the problem statement and then to quickly conceive an idea that will be lead to a *best* solution.

Rather than being a sequential series of steps, the engineering

Topic 14-3 ▰▰▰▰▰▰▰▰▰▰▰▰▰▰▰▰

Design Problems Versus Textbook-Type Problems (continued)

Aerodynamics.

Given a desired payload and specified flight conditions, and so on (the inputs), determine the DESIGN for a new airplane which will exhibit given performance characteristics (the performance). The inputs and performance are again known, while the DESIGN is desired. Proper application of aerodynamic principles will not yield a unique DESIGN

Factory Planning.

Given a raw material availability (an input), specify the factory, its hours of operation, and its manpower requirements (the DESIGN) in order to produce a desired output of finished goods (the performance). The applications of data and principles studied in industrial engineering will not yield a single, unique DESIGN here.

Notice that none of the design situations cited above has a single, unique DESIGN defined by the principles normally employed in modeling or analysis work. These problems are not well formulated; they are not accompanied with an obvious plan of attack. Their solution is not "merely a matter of mathematics or experimental skill."

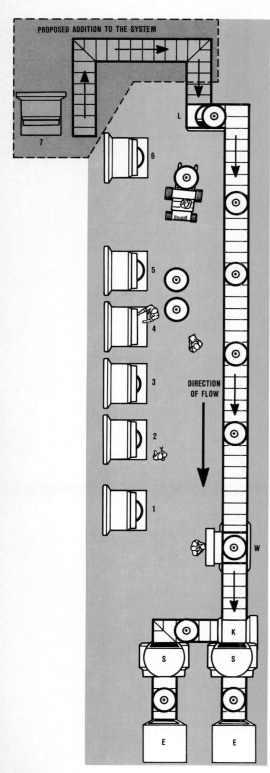

[14-T7] Conveyor system in paper mill. This design is well formulated.

design process should be thought of as one where, at any given time, you would work at the task that will most appropriately supply the immediate need for information. Go where the job can get done!

In this search for the DESIGN which will ultimately solve the problem, it is extremely important that you focus your attention on some measure of your progress. What better measure of your progress is there *than on the design data base*? Concentrate on its form and content, much like you concentrated on the shape and form of objects you drew in Chapter 7.[6] Keep trying to visualize and define what the final DESIGN would be, based on the ideas you are exploring. How will what you are currently doing affect the DESIGN? Is what you are doing relevant to the DESIGN? What is your DESIGN? Where is the data base missing data; where is it missing definition; where can it be strengthened; does it need a new idea; are the components in the DESIGN defined enough to enable some type of component modeling; is it time to freeze the DESIGN? Always keep "chipping away at the "fuzzy" statement of definition. Try out new ideas. Constantly question and assess your present concept of the DESIGN

Figure [14-38] demonstrates how the tasks we have defined earlier fit together. We have again used a drawing of an office file folder to symbolically represent the data base for the DESIGN. Its contents are small in the beginning and initially it contains only ideas or concepts that lack definition. At the completion of the process it contains much more information and completely defines the DESIGN Each of the tasks of *Problem Definition, Idea Generation, Component Modeling*, and *System Modeling* interact with the data base to produce changes in its contents. The task of *Finding a Workable Design* interacts with the output of the *System Modeling* task.

We pointed out much earlier that the design objective for a particular problem often provides clues as to what are the most important measures of quality for the solution. Thus the "workable" DESIGN has to be judged in some way against these measures of quality. This is done in the *Search for Better* task by changing the DESIGN in some way and comparing its quality against the first "workable" solution.

Let us now step through the engineering design process as depicted in Figure [14-38], in a manner that might be typical of its execution. As we have stated previously, in most design situations the perceived need usually makes the problem definition "fuzzy." This forces you to do some work on the *Problem Definition* task in order to better understand the problem. However, you will rarely do an exhaustive study of the problem before moving on to the *Idea Generation* task because the resource of time is always limited. An exhaustive study consumes time in producing information, the bulk of which may never be useful to you. Ideally, you would "define" the problem sufficiently to remove enough of the "fuzziness" to allow you to go on to the *Idea Generation* task, where you would search for

---

[6] Actually, the analogy between developing the DESIGN and developing a drawing is quite good. You begin your drawing on a blank piece of paper ("fuzzy" concepts and small data base) and eventually complete the work (you end with a well-defined DESIGN). You do not follow a sequential series of steps or tasks in the drawing process. You work at some parts of the drawing and then others; you bring some definition to a few parts and then to others.

some ideas that might be useful in solving it. However, the ideas generated may well force you to return to the first step to remove still more of the "fuzzy parts" of the problem definition. Some ideas will require more information regarding the problem definition before they can be evaluated and modeled.

This technique of returning to the *Problem Definition* stage only when there is a need to better understand the problem, or to remove some of its "fuzziness," is an important concept. Spend your time resources looking only for answers to questions you need answered. Remember, *leverage your information.* The ideas or concepts for solution that emerge from the *Idea Generation and Germination* task will give you direction in defining the items that should be addressed in the *Problem Definition* task. Always keep asking the question, "What is the DESIGN, and how does what I am doing affect it?"

New initiates to the design process seem to have a built-in resistance to doing anything until they thoroughly understand the problem. Generally, this is an induced behavior trait—an aversion to uncertainty and ambiguity—which can, and must, be overcome. The tendency is due largely to the designer's addiction to situations that leave little or no doubt as to what is to be done (these situations include textbook-type problems; directions from parents, teachers, and/or siblings; orders from the boss at work; and even "requests" from military or police officers), and to a lack of confidence in one's own abilities. Practice in the design environment lessens both difficulties.

Once you have generated some new ideas and they survive the *Idea Germination* stage, it is often appropriate to do some modeling of the ideas so that you might study their behavior in more detail. Thus you would proceed to the *Component Modeling* task. Specifically, you eventually want to be able to predict if one or more of the ideas will be useful in solving your problem. During this task there may well be a need to reduce more of the fuzziness of the problem statement by returning briefly to the *Problem Definition* task. Or, if the modeling quickly uncovers a flaw in an idea, you may need to return to the *Idea Generation* task to produce more ideas for evaluation.

Eventually, you want to arrive at the two tasks of *System Modeling* and *Finding a Workable Solution.* But at any time you may need to proceed to one of the other tasks in the process where you have already been.

In the *Finding a Workable Solution* task the emphasis is on extracting from your ideas and modeling some system that will actually solve the problem in some way. It may not, and probably will not, meet all of the design objectives imposed on the solution by the problem statement. However, in a technical sense, you must be convinced that it will work.

If you cannot find a "workable" DESIGN that satisfactorily solves the problem, you must revisit previous tasks to explore other DESIGNS. If you are successful in finding a satisfactory "workable" DESIGN, you will want to *Search for a Better* DESIGN. In this task you seek ways to improve on the "workable" system that was just defined so that it comes closer to meeting all of the conditions of the problem statement. For example, if the problem statement requires a DESIGN of high reliability, try to pinpoint what parts of the "work-

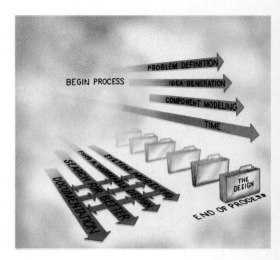

[14-38] The process.

able" DESIGN would most probably cause it to be unreliable. Can these parts be improved? If so, change the DESIGN and incorporate them.

For the adaptive thinker, it is usually easier to improve on an existing DESIGN (or something that works) than it is to conceive a whole new system. If you cannot determine ways to improve the DESIGN or if it is not satisfactory, you must return to an earlier task to pursue anew other good ideas.

With this brief introduction, perhaps the overlap of the tasks in Figure [14-36] now becomes clearer. The overlaps are used in leveraging information. If, while working in a given task, more information is necessary, you return or advance to the task that can supply the information. As more and more information is generated in each task, the need for information from other tasks is stimulated. In this way the data base grows—the problem becomes clearer and the best solution takes shape. Eventually, there comes a final trip to the *Documentation* phase when the DESIGN is fully documented; that is, the specifications for how to construct and implement the solution are complete.

It should now be clear that both the ability to generate ideas and skill in decision making play key roles in the design process. Without ideas, the solution never emerges. New ideas give you alternatives—they allow you to leverage your possibilities. But you must decide among the alternatives. Unless decisions are made, the process flounders even if ideas emerge.

A DESIGN EXAMPLE: DESIGNING A TRIP PLAN

To demonstrate some of the flavor of design, let us consider "designing" a travel plan. The example was chosen because it is devoid of any difficult analysis that might interfere with the understanding of the design process. It allows the philosophy of design to be demonstrated.

Imagine that you were asked to design a trip that would take a client from Lewisburg, Ohio, to Chicago, Illinois, in order to attend a meeting at McCormick Place, an exhibition hall on Lake Michigan. The plan is to include all of the necessary details so that the client could rely entirely on it for information. You may have been to Chicago, although probably not to McCormick Place, but it is doubtful that you have been to Lewisburg, Ohio. If this is true, you would find yourself lacking sufficient information to define the DESIGN immediately. However, you might have experience from previous traveling and trip planning and may not be too concerned with the unknowns with which you are confronted.

The statement of the problem may have already triggered in your mind an approach to constructing the trip plan. Without hesitation you may decide to find a road atlas of the central United States so that you could select what appear to be the major highways that connect the starting and ending locations. Or you may choose to find an airline guide listing all of the airline connections between the two locations.[7] But notice that nothing is contained in the original statement of the objective about the mode of transportation.

The original problem statement is, indeed, ill-defined or "fuzzy." The design objective contains no statements about cost, length of travel time available, or any other desirable goals that would serve as measures of quality for the DESIGN. Nor does it contain any constraints that must be met, such as that the trip must be made by public transportation, or that it must be made by private auto. Undoubtedly, the client does have some feelings about the trip, but this original problem statement does not make them clear.

It is instructive to stop and think about some of the uncertainties that will be encountered in designing the travel plan (the actual plan would constitute the DESIGN). Notice that we have said "uncertainties that will be encountered in *designing* the . . . plan," not "uncertainties that exist in the DESIGN or plan." The DESIGN or plan itself would not have any uncertainties in it[8]—it merely specifies exactly how the trip is to be made. It may, indeed, happen that the DESIGN which you devise is flawed (did you ever buy a product you thought was poorly designed?), and the client would have to alter the plan as he or she traveled. However, such flaws in the DESIGN are nearly always traceable to flaws, oversights, and lack of information during the design process. You might imagine that the client might have some choice unkind words for you if he or she arrived in the Corn Palace of Mitchell, South Dakota, instead of McCormick Place in Chicago.

Taking the advice of the preceding section about always directing your attention to the DESIGN, perhaps you should, at this point, consider what the final DESIGN might be. That is, you should look ahead and try to project what set of specifications would be required in order to specify the solution completely. This may give you direction in deciding what you need to do in the process.

The DESIGN might possibly be a written or recorded set of steps which detail what to do (what forms of transportation to take and/or highway routes to be used) and when to do it. Included would be information on where meals were to be obtained and where overnight lodging would be available. If tickets or reservations were needed for any part of the trip, the plan would specify how these are to be obtained. It would also contain information on the costs of executing the plan.

This initial look at what the DESIGN might consist of should certainly tell you that the transportation modes ultimately chosen will drastically affect the DESIGN. That is, the form of transportation chosen will dictate the financial resources required and the trip

---

[7] For this example, the relegation of the tasks to a travel agent or to the trip planning group of an automobile club is not an acceptable solution.

[8] The travel plan of the type desired here will obviously have uncertainties in it due to the fact that it must be based on assumptions concerning punctuality of public transportation, major accidents, weather, acts of war, and other such factors which are difficult, if not impossible, to anticipate. A good plan would have some assessment of these possibilities together with contingency plans for the more probable ones. Contingency plans specify alternatives that could be used if one of these possibilities does occur.

time, the latter of which will specify lodging requirements and number of meals on the "road." Since transportation seems to be a key ingredient of the plan, perhaps you should spend some time considering the transportation alternatives.

Transportation Alternatives. Since you need some ideas about the possible ways that this trip could be made, you need to work on the *Idea Generation* task [14-38]. Remember, *Idea Generation* is a *divergent thinking* task, and it is imperative that you not be critical during this phase—once you begin, just let the ideas flow until you seem to have an ample number. An example tabulation might include:

   Automobile or truck
      Company owned
      Rent
      Borrow
      Personal
   Airplane or helicopter
      Personal
      Private carrier
         Rent or charter
      Public carrier
         Charter
         Regularly scheduled flight
Bus

Train

Boat

Motorhome or camper

Motorcycle or motorscooter

Bicycle

Keep going—but do not be critical. The problem statement did not say what the client's real interests are.

Walk

Horseback
   Other beasts of burden
   Beast-drawn vehicles
Hitchhike

Roller skates
   Skateboard
Rocket

Hot-air balloon

Blimp

By mail

As freight

As a stowaway

Notice the *ideational fluency* shown in the first 11 major headings and the emergence of some *ideational flexibility* in the last seven (see Topic 14-1). Notice also that the quick look ahead to what the DESIGN might be has now allowed you to leverage your information. You have generated some ideas that have some potential for helping you solve your problems. You did not spend time collecting information on motels, hotels, and restaurants.

As you let these ideas germinate, you might realize that a road atlas would provide you with good information regarding *most* of your transportation ideas. It is perhaps the best *model* to use in considering your options. Thus once you have gained access to an atlas [14-39 and 14-40], you will already have begun to work on the Component Modeling task in the process.

A road atlas is a good modeling tool because it would make several things clear regarding the problem. First of all, you would find that there are no navigable waterways directly connecting the two endpoints of the trip, a fact that rules out the use of a water-dependent transportation mode, at least for a major part of the trip. Second, Lewisburg is not directly served by any major airport, which rules out airplane transportation as an exclusive solution. Using the single item of the road atlas, you have again leveraged your information and done something useful.

Lewisburg is about 25 miles from Dayton, Ohio, and about 60 miles from Cincinnati, Ohio, both of which are served by major airports (major airports are typically shown on road maps). Chicago is served by two major airports (O'Hare International and Midway) and one minor one (Meigs field along the lakefront). There seems to be the potential that a major part of the trip could be handled by public or private air carrier. Using the single item of the road atlas, you have again leveraged your information and done something useful.

Perhaps now that you understand the problem a little better, it is time to query the client about his or her desires—that is, return to the Problem Definition task. Suppose that the only additional information which you are able get is that the client wants a "fairly" quick trip that is not "too" expensive.

From the sense of distance obtained from the road atlas, you should be able to reason that several of the more esoteric modes of transportation which you have identified have just been ruled out due to the time that would be required (e.g., the bicycle, walking, beasts of burden, roller skates, hot-air balloon, and probably the blimp have essentially all been eliminated). Thus these concepts are removed from your possible concepts for the DESIGN.

Since railroad passenger service in the United States has been declining for some time, train transportation is probably not a reasonable solution. Certainly, it could be confirmed that no passenger trains pass through Lewisburg. So it seems reasonable to put this concept into the "holding" file.

If a regularly scheduled rocket service were available, it, no doubt, would be the fastest mode of transportation. However, since there is no such service in existence, the cost of implementing this solution would be extremely expensive. Thus this concept should also be put aside, as it is no longer a feasible one for your DESIGN.

Now the DESIGN which you are defining contains concepts

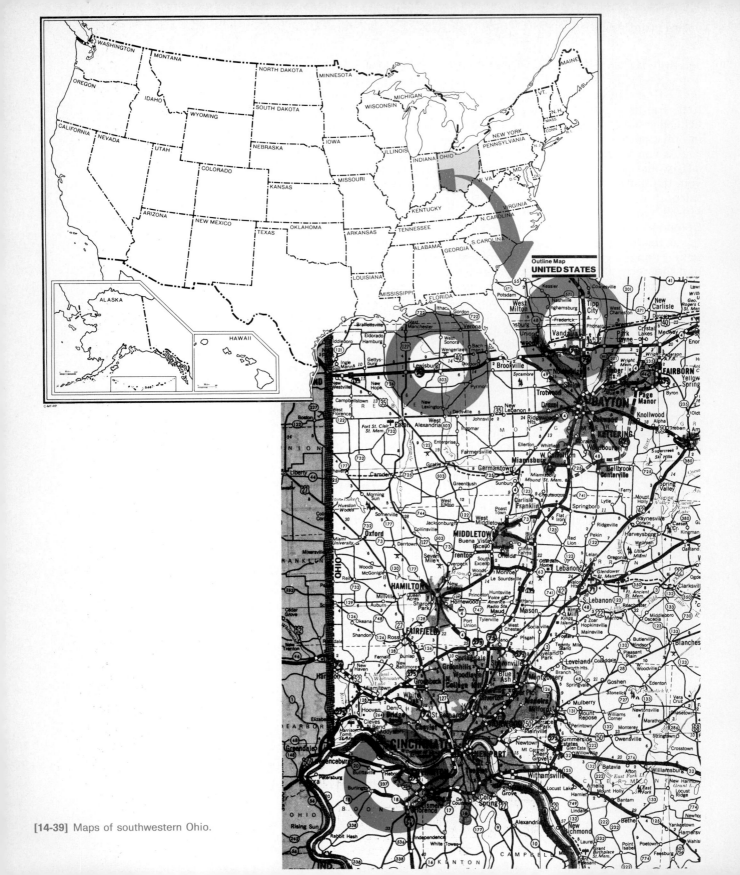

[14-39] Maps of southwestern Ohio.

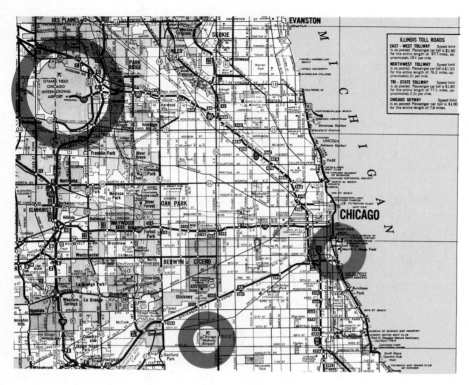

[14-40] Map of the Chicago vicinity.

which use the public highways, possibly in conjunction with regularly scheduled airlines. Your concepts are being narrowed in this *convergent thinking* stage. Those concepts that remain seem to have some reasonable chance of leading to an acceptable solution.

What does your DESIGN need now? Obviously, it needs to have further uncertainty removed regarding the specific modes of transportation and many other details defined. How would you decide on *the* transportation modes to be used in the final DESIGN? Obviously, you must make comparisons between the "quality" of the various ways of making the trip. Certainly, the concepts you are left with would lead to many possible ways of actually making the trip. You must have some way of assessing each (or at least several) of the possible ways.

What you need is a system model that will let you simulate each possible trip in order to assess its "quality." The system model would allow you to choose when you could leave and the model would tell you when you would arrive at your destination. Or solved in reverse, the system model would allow to specify when you had to arrive and it would tell you when you had to leave. In order to construct a system model, you must have component models of each segment of the trip.

As we pointed out earlier, the road atlas is, among other things, a component model of the highway system. The various methods of traveling the highways (auto, taxis, truck, motorhome, motorcycle) can be modeled (modeled with heuristics that are good most of the time) with average speeds that are reasonable for each method. Published schedules for ground-based public transportation modes (buses, airport limo, etc.) that use the public highways are nothing more than component models of the actual services offered. Once you define when you wish to start a trip and where you wish to go, these schedules will tell you when you will arrive. That is what component models, even those of the more usual engineering variety, do for you. If you supply the inputs, the model tells you the outputs.

Since the airline industry was deregulated a few years ago, the cost of airline travel has been quite low. In fact, you may have reason to believe that the cost may be low enough that the savings in time[9] are well worth any difference in cost, if indeed there is any difference.

An airline guide showing originations, destinations, and times of all regularly scheduled flights is a component model of commercial airline passenger services available [14-41]. Like other public transportation schedules, the airline guide relates the output (where you arrive and when) with the inputs (when and where you leave), which is exactly what component models are supposed to do.

This collection of component models can all be integrated into a system model that links up the time you leave Lewisburg to the time you arrive at McCormick Place in Chicago. This system model would tell you that if the client were to leave Lewisburg at 9 A.M. on Tuesday and proceed by auto on Interstate 70 to the Dayton airport, catch the 10:38 A.M. AA flight, arrive at O'Hare field at 10:45 (change of timezone), catch the 11:25 airport bus to the Hyatt Hotel in downtown Chicago, and catch a taxi from the hotel, he or she would very likely arrive at McCormick Place by 1:00 in the afternoon.

In addition, the system model can, and often must, tell you the estimated financial requirements for the trip. For the cost to be calculated, component cost models must be available for each component in the system. In Chapter 13 we discussed engineering economics and how alternatives having different economic conditions can be compared.

Each of the components of the Lewisburg-to-Chicago trip would have an associated cost model. Private-vehicle cost models often involve the estimated cost (the total of the operating cost—fuel and oil; maintenance costs—lubrication, tune-ups, repair; purchase price; and estimated lifetime) per mile and the mileage driven. The costs for airline and bus service are essentially the cost of the tickets (which depends on the class of service desired and available). The cost model of the taxi can be based on the taxi's published rates (often a city-approved rate structure expressed as a flat fee plus a mileage charge).

Once a good system model can be established, the *Finding a "Workable" DESIGN* task involves exercising the system model to find some specific trip that will meet the conditions of the problem. So far, the client has not placed any constraints on the problem other than the starting and ending locations. If the client were to now inform you that it is necessary to arrive at McCormick Place by 11 A.M.—noon on Wednesday, you should be able to use your system model to tell when he or she must leave Lewisburg. Not all possible trips will meet this condition.

[9] Estimated from the atlas, the flying time between Dayton and Chicago would probably be about 1 hour, while the shortest driving time might be around 8 hours.

[14-41] Airline schedules for flights to Chicago from a) Dayton, Ohio and b) Cincinnati, Ohio.

DAYTON, OHIO   **EST DAY**

D-DAY M-MGY

	8:00a	D	8:01a O	AA	241	FYBQM	72S	0
	8:55a	D	8:57a O	UA	257	FYBQH	72S	0
2	9:31a	D	10:45a O	AA	557	FYBQM	72S	1
X2	10:38a	D	10:45a O	AA	337	FYBQM	72S	0
	12:05P	D	12:08P O	UA	923	FYBQH	73S	0
	2:35P	D	2:35P O	UA	541	FYBQH	727	0
	5:06P	D	5:05P O	AA	294	FYBQM	72S	0
	5:55P	D	5:57P O	UA	911	FYBQM	72S	0
	8:10P	D	8:18P O	AA	121	FYBQM	72S	0
	8:46P	D	8:52P O	UA	581	FYBQH	72S	0

CONNECTIONS

X67	**7:05a**	D	7:35a CVG	DL★	3306	YBQM	EMB		0
	8:08a CVG	**8:05a** O	DL	400	FYBQM	72S	S	0
X7	**7:35a**	D	8:30a DTW	NW	961	YBQM	CVR	0
	9:25a DTW	**9:30a** O	NW	895	FYBQM	72S	0
X67	**9:10a**	D	9:50a IND	PI★	5043	YBHQK	J31	0
	10:20a IND	**10:20a** C	RU	480	Y	BE9	0

EX 28NOV

X67	**9:25a**	D	9:50a CVG	DL★	3127	YBQM	SF3	0
	12:00n CVG	**11:57a** O	DL	1116	FYBQM	72S	0
	10:45a	D	11:37a DTW	NW	963	YBQM	CVR	0
	12:30P DTW	**12:32P** O	NW	474	FYBQM	72S	0
X67	**1:40P**	D	2:20P IND	PI★	5024	YBHQK	J31	0
	2:40P IND	**2:40P** C	RU	482	Y	BE9	0
X67	**2:00P**	D	2:30P CVG	DL★	3341	YBQM	EMB	0
	3:08P CVG	**3:05P** O	DL	418	FYBQM	72S	0
	5:45P	D	6:40P DTW	NW	738	YBQM	CVR	0
	7:40P DTW	**7:44P** O	NW	709	FYBQM	D95	0
X6	**6:35P**	D	7:05P CVG	DL★	3129	YBQM	SF3	0
	7:43P CVG	**7:40P** O	DL	254	FYBQM	72S	0

CINCINNATI, OHIO   **EST CVG**

SEE ADVERTISEMENT THIS SECTION FOR: HILTON HOTELS

	8:00a	7:57a O	AA	933	FYBQM	72S	S	0
X7	8:08a	8:05a O	DL	400	FYBQM	72S	S	0
	8:12a	8:15a O	UA	459	FYBQH	72S	S	0
	11:43a	11:44a O	UA	367	FYBQH	72S		0
	12:00n	11:57a O	DL	1116	FYBQM	72S		0
	2:12P	2:09P O	AA	408	FYBQM	72S		0
	2:30P	2:34P O	UA	371	FYBQH	73S		0
	3:08P	3:05P O	DL	418	FYBQM	72S		0
	5:38P	5:44P O	UA	657	FYBQH	72S		0
X6	7:43P	7:40P O	DL	254	FYBQM	72S		0
	8:13P	8:15P O	AA	869	FnYnBQM	72S		0
	9:11P	9:15P O	UA	929	FnYnBQH	727		0

CONNECTIONS

X7	**6:50a**	7:45a DTW	NW	187	FYMBnQn	72S		0
	8:15a DTW	**8:21a** O	NW	991	FYBQM	72S	S	0
	10:45a	11:40a DTW	NW	996	FYBQM	D9S		0
	12:30P DTW	**12:32P** O	NW	474	FYBQM	72S		0
	2:35P	3:31P DTW	NW	728	FYBQM	D9S		0
	4:25P DTW	**4:31P** O	NW	1104	FYBQM	72S		0

Once established, the system model should, if exercised, be quite useful in finding a combination of transportation modes which would produce a trip that is "workable." Perhaps you will have to depart from Lewisburg by 6:45 A.M. to catch the 8 A.M. flight, but whatever it is, your system model will allow you to establish that it is indeed workable.

Having established a "workable" DESIGN, you now begin the *Search for Better* task. That is, you now make some change in the DESIGN you have just found to be workable, and exercise your system model to evaluate its potential for solving the problem. The change in the DESIGN you make could be either a major or a minor one. For example, an airline company flying out of the greater Cincinnati airport might have a flight to Chicago at a much cheaper fare than is available for flights out of Dayton. But this involves driving a greater distance. Your system model will tell when you would have to leave Lewisburg and what the economic cost would be.

If you do find that it is possible to complete the desired trip by traveling to Cincinnati, how do you make comparisons between this trip and the "workable" one through Dayton? That is where the measures of quality in the design objectives become useful. Does this new trip take less time? Does it cost less? You must get some sense of quality from these comparisons and use this sense of quality to choose one DESIGN over the other. If the trip through Dayton were both cheaper and quicker, there would be little question about which DESIGN would be better. However, if the flight through Cincinnati were cheaper but took twice as long to complete, an engineering trade-off would have to be made by you. That is, you would have to decide among these two nonoptimum solutions. Any other merits besides cost and time that one DESIGN would have over the other DESIGN might be useful in your trade-off decision.

You could, of course, continue the "Search for Better" forever—if you had available large problem-solving resources. Typically, you have to bring the search to an end by defining a "good" solution rather than the ultimate solution because you do not have large resources available.

Once you decide on the "best" solution, you must document it so that the client can make use of it. You have to itemize all aspects of the trip:

Specify when, where, and the cost to get necessary tickets.

Specify the financial resources required.

Specify the start time from Lewisburg.

Specify the mode of transportation between Lewisburg and Chicago.

Specify routes between Lewisburg and Chicago: roads, air routes, etc.

Specify arrival time.

Specify any lodging (where).

Specify any meals (when and where).

Specify any contingency plans that may be necessary.

etc.

We wish to point out that this exercise is truly a design-type problem in which both the resources and the need are known and the DESIGN that effectively applies the resources to meeting the need is to be determined.

If this were the typical textbook or analysis type of problem discussed in Topic 14-3, the DESIGN would be well known (the routes specified) and the inputs established (client leaves Lewisburg at known time). The time and place of arrival (the desired outputs) would then be sought. Finding the output "may be merely a matter of mathematics or experimental skill."

Of course, we have not discussed all of the details that this problem might entail. Hopefully, however, we have demonstrated a little of the nature and flavor of the design process:

Establish measures of quality.

Establish constraints.

Direct your attention on the DESIGN

Generate ideas.

Leverage your information.

Make decisions.

Use heuristics.

PITFALLS AND EMOTIONS OF THE DESIGN EXPERIENCE

It is one thing to read about the design process and another to actually participate in it. It will be helpful to return to selected parts of this chapter as you establish your solution to the problems at the end of both this chapter and Chapter 15.

If you are a "typical" novice in engineering or technology, there will be times when you will run out of ideas, when you do not know how to proceed, and when you think you are fruitlessly "spinning your wheels." However, in due time, most of you will be able to arrive at acceptable solutions to what, at the beginning of the projects, appeared to be very large, complicated, and fuzzy problems. Tables 14-1 and 14-2 are lists of the common pitfalls and emotions that novice (indeed even experienced) designers encounter. The pitfalls can appear before you at any stage in the design process and, if not overcome, they can siphon away your personal enthusiasm. When a poor emotional state is encountered, the process will come to a grinding halt.

Many of these pitfalls are caused by various needs which have been discussed earlier: a need to generate ideas, a need to make decisions, a need to leverage information, and a need to work within the limited resources. Other items which we have not discussed but which are very important include forcing a solution from a pet idea, applying convergent thinking too early in the process, attacking too large or too small a piece of the problem, and allowing your emotions to interfere.

Perhaps the most important advice that is contained in the list of

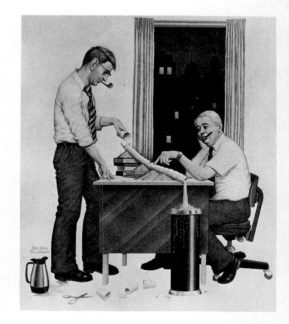

[14-42] Engineers are motivated to work at a particular task partly because of the exhilaration, thrill, special satisfaction, pride, and pleasure they get from completing a creative task. However, the path to success is not always easy, and on occasion frustration is rampant.

TABLE 14-1 COMMON PITFALLS AND POSSIBLE CURES

Pitfall or Problem	Possible Cure
Being overly rigid in defining the problem.	Redefine the problem in another way. Can you relax constraints?
Being too ambitious for available time and resources.	Ask "What practical thing can I do tomorrow, the next day, etc.?"
Having no plan or morphology.	Adopt one! Try the "starter set."
Following your plan too rigidly.	You may have to jump ahead or jump back over steps in your plan.
Not following good ideas through to final assessment.	Assessing good ideas sometimes takes time. It usually pays off.
Not knowing what to do next.	Review your work. Do anything relevant that you can do immediately.
Prematurely developing a fixed solution concept.	Abandon the solution idea and explore the problem further.
Solving an unimportant or trivial aspect of the problem.	Ensure that what you are doing is worthwhile. Talk to others about the problem.
Being unable to transform the mass of data collected during analysis into solution ideas.	Allow time for the problem to become defined in your mind. Write summaries of the data.
Collecting data without knowing how they can be used.	Do not collect data at random unless you have a "hunch" they may be useful somehow.
Lacking confidence to test ideas in real situations.	Courage is needed. Do it!
Getting disheartened when things are going badly.	Persevere. Remember, it is also important to find out which ideas will not work.
Being afraid to talk to others about the project.	Drag in everyone you can and excite their interest, or bore them to death with your enthusiasm. But observe any proprietary restrictions on the project!

Source: Adapted from *Design Project Guide Unit 12, T262 Man-Made Futures* (The Open University Press, 1975).

TABLE 14-2 EMOTIONS AND THEIR EFFECTS ON THE DESIGN PROCESS

Emotional State	Probable Consequence
Joy, enthusiasm	Productive
Frustration	Process stops
Fatigue, depression	Process stops
Watchful anticipation	Breakthroughs may occur

Table 14.1 is that when any pitfall becomes apparent, it is imperative that you do something to keep the design process moving. *Do anything relevant that you can do immediately.* This gives you the potential of discovering a new idea or a new piece of information, or of reconsidering some existing information that can be leveraged into a better idea to keep the process moving.

SUMMARY

Every practicing engineer is involved in some way in the engineering design process. No matter what the job title (e.g., research engineer, design engineer, development engineer, field engineer, etc.), the engineer uses some form of the process which we have just discussed. Technologists, in their support role for engineers, also participate in the process. In small organizations, the engineer and the technologist use the process as they participate in defining the overall artifact or solution to a problem. In larger organizations, they generally participate in specifying only a part of the overall artifact or solution, but they still use the process.

In nearly all cases the engineer and technologist work as members of a technical team. For example, in the design of a new airliner for a large airplane manufacturer, some members of the team will be responsible for the shape of the aerodynamic surfaces, some for the propulsion system, others for the structural integrity, and still others for the flight instrumentation. In such a situation, each group would individually be executing their own process as shown in Figure [14-19], defining their own particular part of the airplane. At the same time, all of the groups must be coordinating together and participating in one grand, overall design process so that proper integration of all the parts is successful. No matter what their involvement, the engineer and the technologist must understand how the process generally functions before they can fully understand and appreciate their own role.

The intent of this chapter has been to give an overview of the philosophy and methodology of engineering design and, as a comparison, to show how it differs from engineering modeling or analysis. The anatomy of the process that has been presented here is nothing more than a "starter set" of tasks that have recognition. This anatomy of the process has been found to be adequate and quite useful for engineering students, including freshman students who have very little in the way of engineering tools at their disposal.

Many professions, in addition to engineering, engage in "design." Most would agree that design is the generation of ideas and the making of decisions. Engineering design is set apart from other types of "design" by the origin of the need (the engineer serves the call of society rather that his own need for self-expression), the leveraging of information, and the unique set of heuristics which are used. Thus it is as much the methods that engineers use which identify their profession, as it is the needs they fulfill.

Engineers are basically servants of society; they attempt to fill society's needs *in those situations where it is within their capabilities to*

[14-43] This project could have used better coordination among the design team members.

do so. In problems that have purely technological solutions, societies of the past have generally granted to the engineer nearly full rights and privileges for constructing such solutions, perhaps placing only a limit on the financial resources available. From the early beginnings of engineering through the industrial revolution, this was the dominant relationship between society and engineers.

In the current postindustrial era, this relationship is changing. For several reasons, society is becoming more involved in the problem solutions. This involvement has manifested itself in many ways: public review of ideas during the design process, pollution standards, occupational safety and health requirements, consumer product safety legislation, and limits on natural resource consumption, to name only a few (see Chapter 6 for more). In more and more engineering problems, the solution must combine what society wants *and will allow* with what the engineer can produce.

In this environment of problem solving it is imperative that engineers understand society and that society understands engineers. However, the present level of each group's understanding of the other is not sufficient to support the type of interaction and cooperation that is needed. Some effort is being made to better educate engineers but there is no concerted effort to require society to better understand the engineer.

Engineers have two primary obligations. One is to keep the time stamp on their toolkit up to date so that their DESIGNS are arrived at using the set of heuristics which are considered to be "good" engineering practice at the time the DESIGN is finished. The other is for them to become responsible members of society at large and, in the words of Koen, to increase their "general sensitivity to the hopes and dreams of the human species."

PROBLEMS

14-1. a. In the next 3 minutes, estimate the number of whole concrete blocks (normal concrete blocks are 8 by 8 by 16 inches in size) that could simultaneously be placed inside your classroom.

b. Make a more accurate estimate between now and the next class period.

14-2. a. In the next 3 minutes, estimate the maximum number of inflated basketballs that could simultaneously be stuffed into your classroom.

b. Make a more accurate estimate between now and the next class period.

14-3. This exercise should ideally be done with a group of people. Within a 3-minute time limit, have each person make a written list of the names of all the colors that begin with the letter "b." Review all of the responses with the group at the end of the time period.

14-4. This exercise should ideally be done with a group of people. Within a 3-minute time limit, have each person make a written list of the names of all the birds of which they can think. Review all of the responses with the group at the end of the time period.

14-5. This exercise should ideally be done with a group of people. Within a 3-minute time limit, have each person make a written list of all the uses you can think of for a No. 2 lead pencil. Review all of the responses with the group at the end of the time period.

14-6. This forced-association exercise should ideally be done with a group of people. Within a 5-minute time limit, have each person make a written list of all the similarities between a cat and a baseball game that come to mind. Review all of the responses with the group at the end of the time period.

14-7. Formulate a design objective and set of constraints for an automobile that you would like to own.

14-8. Was your answer to Problem 14-7 affected more by what you know to be available on today's commercial automobiles or what you feel is needed but which may not yet be available?

14-9. Formulate a design objective and set of constraints for a military fighter aircraft.

14-10. For 10 minutes solo brainstorm the problem of disposal of home wastepaper. List your ideas for solution.

14-11. List five types of models that are routinely used by the average American citizen. Are these component models or system models?

14-12. Suggest several "highly desirable" alterations that would encourage personal travel by rail.

14-13. What are five ways in which you might accumulate a crowd of 100 people at the corner of Main Street and Central Avenue (or whatever major street corner there is in your town) at 6 A.M. on Saturday?

14-14. Starting with the words "As inevitable as," contrive six figures of speech similar to "As inevitable as night after day."

14-15. You have just been named president of the college or university that you now attend. List your first 10 official actions.

14-16. Describe the best original idea that you have ever had. Why has it (not) been adopted?

14-17. Discuss an idea that has been accepted within the past 10 years but which originally was ridiculed.

14-18. Describe what you would think the roles of the manufacturing engineer and the quality control engineer would be during the design process of a particular product.

14-19. List five familiar products whose design does not adequately consider human factors. Indicate why this is so.

14-20. At night you can hear a mouse gnawing wood inside your bedroom wall. Noise does not seem to encourage him to leave. Describe how you will get rid of him.

14-21. Describe an incident where a person or group abandoned a course of action because it was found that they were spending time working on the wrong problem.

14-22. Complete a trip plan (a DESIGN) that could be implemented by a functionally illiterate person (of which there are 20 million in the United States). The trip will take the person from Beckley, West Virginia, to Madison Square Garden in New York City, New York. The traveler must be able to implement the DESIGN without assistance. The traveler dislikes small confined spaces and enjoys beautiful scenery.

14-23. Complete a trip plan (a DESIGN) that could be implemented by a blind person. The trip will take the person from Pendleton, Oregon, to the Gateway Arch in St. Louis, Missouri. Travel time must not exceed 3 days and the trip should be "inexpensive."

14-24. Your company has available many pieces of lumber, but in only three cross-sectional sizes: 1 by 4 inches, 2 by 4 inches, and 4 by 4 inches. There is an assortment of lengths from 3 to 12 feet. Using visual thinking (freehand drawing), develop several concepts of how these might be used to construct loading docks that would allow semitruck trailers to be loaded without the need for a high-lift forklift. The height of the beds of the semitruck trailers are to be a distance of 3 feet off the ground. The load to be placed in the trailers is initially located at ground level.

14-25. Your company has available many pieces of thin-wall pipe in assorted lengths. Diameters between 1 and 10 cm are available in 1-cm increments (1, 2, 3, 4, etc.). Using visual thinking (freehand drawing), develop several concepts of how these pipes might be used to construct playground equipment for a children's playground.

14-26. Your company has available many pieces of angle iron in assorted lengths. These are available in 1- by 1-inch, 2- by 2-inch, and 3- by 3-inch cross-sectional dimensions. Using visual thinking (freehand drawing), develop several concepts of how these might be used to construct the frame for a utility trailer that would provide 16 cubic meters of carrying space.

14-27. Complete a DESIGN for a man's compact travel kit that can be carried in an inside coat pocket.

14-28. Complete a DESIGN for a device to aid federal or civil officers in the prevention or suppression of a crime.

14-29. The following problems are related. They are intentionally vague and ill-defined, like most real-life problems. Their purpose is to stimulate creativity and imaginative solutions, to permit students to find out for themselves, make assumptions, test them, compare ideas, build models, and prepare reports—written or oral—to convince a nontechnical audience. Give only as much aid or additional information as you believe to be absolutely essential. (Additional problems for this setting may suggest themselves.)

You are a Peace Corps volunteer (or a small team) about to be sent to a village of about 500 people in a primitive, underdeveloped country. The village lies 3000 feet below a steep escarpment in a valley through which a raging river flows. The river is about 80 feet wide, 4 to 8 feet deep, and too fast to wade or swim across. On your side of the river there is the village of mud huts in a clearing of the hardwood forest. The trees are no more than 40 feet tall. At the foot of the escarpment there is broken rock. Across the river there is another village, which cannot be reached except by a very long path and a difficult river crossing upstream. There are other villages on top of the escarpment. The people are small; few are over 5 feet 6 inches tall. They live mostly by hunting, gathering, and fishing, although they could benefit from trade with the people across the river and on the escarpment if communication were easier. Before you leave for your assignment, you should try to find solutions to one or more of the following problems:

a. How to improve communication, trade, and social contact between the two villages on each side of the river.

b. How to transport goods easily up and down the escarpment. There is a path up the escarpment, but it is steep, dangerous, and almost useless as a trade route.

c. Suggest a better way of hunting than with the bow and arrows now used. A crossbow has been suggested as being more powerful, easier to aim, and more accurate. Evaluate these claims and provide design criteria.

d. Provide for lighting of the huts. The villagers now use wicks dipped in open bowls of tallow. Can you improve their lamps so that they burn brighter, smoke less, and do not get blown out in the wind?

For each of these problems, select criteria for evaluating ideas. Choose several different solutions, check them against the criteria, and pick the best one; develop this idea by analysis and testing until you know how it will work. All the while, keep track of and test your assumptions whenever possible. Finally, prepare a way to convince the villagers of the value of your idea.

Chapter 15

Problems in Engineering Design

PURPOSE OF THE CHAPTER _____

The purpose of this chapter is simply to present an array of challenging design type problems which can be solved by beginning engineering and technology students who have acquired some skills in the use of the computer. Although the solutions may be programmed from "scratch" in a programming language, they are perhaps best solved on spreadsheets (covered in Chapter 10) or with equation solving packages such as TK!Solver or Formula/One.

The first three problems are not as difficult and involved as the last four. The first three thus serve as good precursors to any of the last four problems. The first three typically involve only modeling, decision making, and parameter analysis. Only a few hours of student time should be required for each.

Each of the last four problems involves many of the topics discussed in this text. Each involves sketching (Chapter 7), documentation drawing (Chapter 9), engineering economics (Chapter 13), units and dimensions (Chapter 12), computers (Chapter 8), and knowledge of the design process (Chapter 14). Thus they are capstone projects that tie nearly all of the material together.

For students carrying a normal course load, we have found a period of 5 weeks to be about the proper amount of time for working any of the last four problems. During this period students begin to understand the principles covered in this text. More important, it is during this time when they really begin to understand what engineering is all about.

PROBLEM 1: MANUFACTURING COST PROBLEM _____

Suppose that one of your concepts for packaging a liquid is to seal it in a cylindrical container. Suppose it has also been determined that the most marketable size would be a can that would enclose 250 cc of the liquid. Material of thickness 0.1 mm must be used to meet the structural constraints of the problem. The following component models have been established:

Cost of cylindrical container = material cost + welding cost

Material cost = volume of can material × cost/volume

472

Welding cost = length of welds × cost/length

Length of welds = 2 × circumference + height of can

Costs/volume are estimated to be in the range 0.01 to 0.02 dollar/cc of can material volume. Welding cost/length are estimated to be in the range of 0.001 and 0.002 dollars/mm of length. What would be the DESIGN of the cylindrical container you would propose? (Specify its diameter and height.)[1]

PROBLEM 2: PIPE AND PUMP DESIGN

The problem that you are concerned with here is the selection of pipe and a pump such that water can be pumped from point A to point B in the most economical fashion. Assume that the pressure needed at point A is given by

$$P = 9.6 \times 10^6 F^{1.94} L/D^{4.84}$$

where F is the flow rate (liters/second), L the distance (meters) between A and B, and D the actual (inside) diameter of the pipe (millimeters). Assume you have opted to supply this pressure with a pump connected to the pipe at point A.

Assume that the purchase and installation price of the pump is related to the problem variables by the equation

$$pump\$ = 118P^{0.95}D^{0.21}$$

and that the purchase and installation price of L meters of "standard"-size pipe of diameter D millimeters is given by

$$pipe\$ = 0.146D^{1.176}L$$

where the variables pump\$ and pipe\$ are in U.S. dollars and the other variables have the same meaning and units as they did in the first equation.

One important feature of this problem is that pipe is generally available in only certain "standard"-diameters, D. To obtain affordable pipe, you must choose one of the standard sizes available. That is, arbitrary diameters of pipe are not economical for you to use because you would have to have them specially made. This would be too expensive for use here. A list of standard pipe sizes (up to 12 in diameters) is given in Table 15-1. Your DESIGN must use one of these standard sizes.

The rate at which water is needed at point B is

$$F = 20 \text{ liters/second}$$

Point B is located

$$L = 240 \text{ meters}$$

from point A. Specify your DESIGN for requiring the minimum total (pipe and pump purchase and installation) costs for this project.

---

[1]This is a modified restatement of problem 5-4, from *Engineering Design*, G. Dieter, McGraw-Hill (1983), where a more sophisticated mathematical solution is requested than is expected here.

TABLE 15-1 STANDARD PIPE SIZES

Nominal Pipe ID (in.)	Actual Standard Pipe ID (in.)
$\frac{1}{8}$	0.269
$\frac{1}{4}$	0.364
$\frac{3}{8}$	0.439
$\frac{1}{2}$	0.622
$\frac{3}{4}$	0.824
1	1.049
$1\frac{1}{4}$	1.38
$1\frac{1}{2}$	1.61
2	2.067
$2\frac{1}{2}$	2.469
3	3.068
$3\frac{1}{2}$	3.548
4	4.026
5	5.047
6	6.065
8	8.071
10	10.02
12	12

PROBLEM 3: OPTIMUM DESIGN
FOR A SANDBOX _____

The object of this problem is to design a "sandbox" to be used in transporting 2,000,000 N (about 200 tons) of sand from the ocean to your backyard. The DESIGN consists of specifying the width, length, and the height of the box (these dimensions are all inside box dimensions). The box is to be made of 10-mm-thick steel plates which are welded together at the edges to form the box. The pickup truck that carries this box can carry a load of at most 20,000 newton. The cost of the steel plate is $0.50 per newton of weight; the transportation cost is $10.00 per trip. It is desirable to look for a DESIGN which has the least total cost, where total cost is the sum of the box cost and the transportation cost.

You have three design variables that need to be specified. You are to develop a program that will aid you in making your decisions. To keep the amount of work to a reasonable level you will be given a box height (see the end of this problem), and, thus you will only have to decide on the width and length of the box (the width and length are the dimensions of the bottom of the box).

To help you make your decision, you should do a parametric study of the problem and see how the cost varies as you change the width and/or length dimensions. To organize this study, you should generate a plot of cost versus box length for some fixed value of the box width. You then need to generate a series of such plots for a variety of box widths. By looking at these plots you can tell where the lowest cost is located, and you can also see what happens to the location of the lowest cost as you change the value of the width (i.e., determine the sensitivity of the DESIGN to the value of the width).

Your design documentation must include:

1. Any computer files used in solving the problems.
2. A hard copy of at least one of your plots.
3. A short report (one page maximum) discussing your DESIGN

Be sure to say what your final DESIGN is and why you selected it. Also discuss how sensitive your DESIGN is to the particular value of the width you have selected. Also, discuss how sensitive your DESIGN is to the plate thickness.

Additional Data. Your box height is to be 1000 mm (this may be modified by your instructor). The weight density of steel is 7.68×10^{-5} N/mm^3. The weight density of sand is 1.57×10^{-5} N/mm^3.

RULES AND REPORT REQUIREMENTS FOR
THE FOLLOWING DESIGN PROJECTS _____

The following general requirements apply to the Problems 4, 5, 6, and 7, that follow.

You must keep a design notebook, one with nonremovable pages, that documents your ideas and progress.

The design submittal (what you turn in to your instructor) is to include the final report, the design notebook, and files containing any computer related material used in this project.

You must submit the design sketches you developed in the geometry-related parts of your design. You must also submit final design drawings produced with a computer-aided drawing package to document the geometry related aspects of your design.

A final written report documenting your DESIGN is required. The report should include:

A table of contents.

An executive summary (a short description of the purpose and results of your studies for the busy reader).

An introduction.

A description of the variable in the problem.

A description of your DESIGN.

A description of the sensitivities of your DESIGN to variables in the problem.

A summary of your findings and some ideas on what needs further work.

An appendix documenting your DESIGN with appropriate:
Graphs.
Tables.
Design sketches and final drawings.
Other pertinent documenting evidence.

The narrative should refer, by number, to your tables and/or graphs, and so on. You may use discipline-dependent units in your report, but you should include the number and units for the equivalent SI (metric) units in parentheses alongside the discipline-dependent units. For example, if you state a volume in 3 ac-ft, which are common units in hydrology, include $(1.2 \times 10^3 \text{ m}^4)$ next to it, as in

$$3 \text{ ac-ft } (1.2 \times 10^3 \text{ m}^4)$$

Remember, the reader of your report is a busy person. From your report, he or she needs to quickly grasp what you have done and gain confidence that you did a reasonable job. But do not write as if your reader knows, ahead of time, the answer to the problem you have solved. Avoid such phrases as "the results were as expected."

It is suggested that you write (and word process) as you go along. When you get the project finished you can then collect all of the parts and edit them into your final report.

Careful inclusion of graphic material can do much to clarify your design concepts, but be careful to avoid redundant or unnecessary "filler" material. Any graphic material included in your final report should function to improve communication.

PROBLEM 4: SOLAR PHOTOVOLTAIC POWER PLANT

Purpose. The purpose of this project is to design a solar power plant that uses solar cells to convert sunlight to electrical energy. For purposes of this design, solar cells are similar to batteries in that they produce a dc (for direct-current) electrical voltage and current, except that solar cells do it only when sunlight falls on their surface. Since solar cells must be exposed to the sun, they cannot be placed on top of one another; they must be placed side by side in a sunny spot.

Goal. The desirable goal in this project is to design the plant to produce a given voltage and current while minimizing the land area that is covered by the solar cells and their supporting structures.

More Information. An analysis supplement on performance modeling as it relates to this problem is given in the section "Performance Modeling." There are many ways to use this material and to proceed to an acceptable result.

The geometry-related aspects of your DESIGN are discussed in the section "Geometry-Related Design." The geometry aspects include the layout of the cells on their supporting structure, together with a full description of the supporting structure, and a site plan.

Refer to the previous section entitled "Rules and Report Requirements For the Following Design Projects" on page 474, for further information on rules and requirements.

Performance Modeling

Modeling the Cells. The solar cells which you have at your disposal are 7.5 cm in diameter and have a negligible thickness. If you wish, you may cut the cells into two equal halves or into four equal quarters. Schematically, the cells can be represented as shown in Figure [15-1].

The relationship between the current and voltage of the solar cells available to you for this project can be modeled adequately by the following equation:

$$I = I_{sc} \frac{e^{V/V_T} - e^{V_{oc}/V_T}}{1 - e^{V_{oc}/V_T}} \qquad \text{amperes} \qquad (15\text{-}1)$$

where V is the voltage (volts) across the two electrical leads emanating from the cell and I is the electrical current (amperes) in those leads. You are only interested in solutions of this equation for which

$$V \geq 0 \qquad \text{and} \qquad I \geq 0$$

V_T is a temperature-dependent variable given by

$$V_T = 8.625 \times 10^{-5} TK \qquad \text{volts}$$

where TK is the temperature of the cell in kelvin.[2]

In Equation (15-1) above, V_{oc} is called the open-circuit voltage since it corresponds to the voltage produced when the current is zero

---

[2]The following formula allows you to convert from degrees Fahrenheit (TF) to degrees Kelvin (TK):

$$TK = \frac{5}{9}(TF - 32) + 273$$

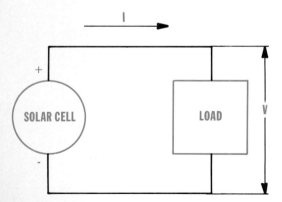

I

+

SOLAR CELL   LOAD   V

-

[15-1] Schematic of a solar cell connected to a load.

(when the leads are open). You are to treat V_{oc} as a parameter that is independent of your DESIGN, including the number of parts into which you divide the cell. For your final DESIGN assign V_{oc} a value of 0.6 V. Note that the largest voltage, V, that a single cell can produce for $I \geq 0$ occurs at $I = 0$ and is V_{oc}.

I_{sc} is called the short-circuit current; that is, it is the current when the leads are shorted directly together (not through the load). I_{sc} is proportional to the area of the cell and to the intensity of the sunlight reaching the cell. The short circuit current (I_{sc}) for the cells you are to use is given by

$$I_{sc} = SIF \times A \qquad \text{amperes}$$

where SIF is the solar intensity fraction of normal sunlight[3] $(0 \leq SIF \leq 1)$ falling on the cells, and A is equal to

1.00   if you have *not* physically split the 7.5-cm-diameter cells.

0.50   for each individual part if you have split the cells into two equal parts.

0.25   for each individual part if you have split the cells into four equal parts.

Note that I_{sc}, for a fixed cell size and fixed sunlight, is the largest current that can be produced by a single cell; it occurs at a voltage of $V = 0$.

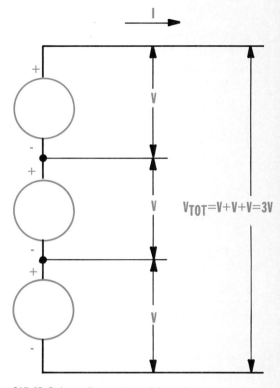

[15-2] Solar cells connected in series.

Connecting the Cells Together in a Circuit. You will find that neither the maximum voltage nor the maximum current that a single cell will produce can meet the design requirements for your problem. Therefore, you will have to wire or connect the solar cells into circuits in order to get the job done. This is identical to placing flashlight batteries in series (as in a two-cell flashlight) or in parallel arrangements to meet the voltage and/or current requirements of some electrical gadget.

Series connections. Wiring the cells in series is demonstrated in Figure [15-2]. In series strings, the voltages V of each cell add to give the total voltage V_{tot} across the string. The current I through each of the cells is the same as the total current I_{tot} in the string.

Parallel connections. Wiring the cells in parallel is demonstrated in Figure [15-3]. In parallel strings, the currents I through each cell add to give the total currents I_{tot} across the string. The voltage V across each of the cells is the same as the total voltage V_{tot} across the circuit.

Series-parallel connections. Wiring the cells in series-parallel combinations is demonstrated in Figure [15-4]. (For your application, you will want to keep the number of solar cells in each series string the same.) The voltage V_{tot} is the voltage across

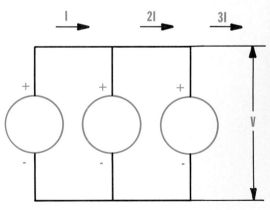

[15-3] Solar cell connected in parallel.

[3] Here normal sunlight is considered to be that which falls on the cells at noon on the day for which you are designing. At SIF = 1, the intensity is 1 kilowatt of sunlight per square meter.

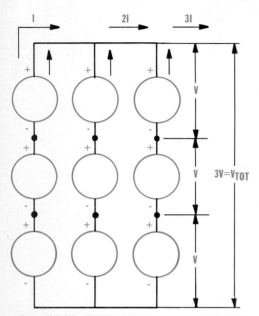

$V_{TOT} = N_S V$; NS = NUMBER OF CELLS IN EACH PARALLEL STRING

$I_{TOT} = N_p I$; N_p = NUMBER OF PARALLEL STRINGS

[15-4] Series strings of solar cells connected in parallel.

each series string in the circuit. V_{tot} is also equal to the sum of the voltages across each cell in any series string. The total current I_{tot} is equal to the sum of the currents in each series string. The current I through each cell is the same as the current through the series string of which the cell is a part.

Design Point for the Array. The load that your power plant has to support requires a voltage in the range of 100 to 1000 volts. The current will be in the range of 500 to 5000 amperes. Your instructor will assign specific values for these quantities after your design efforts are underway.

The nature of the device that makes up the load is such that it forces the voltage on your solar cell circuit to be held at the voltage specified. The load has to be met when the cell temperature TF is 90° F and the solar intensity fraction is $SIF = 1$.

Geometry-Related Design

Of course, it is impractical to just lay solar cells out on the ground, wire them together, and sit back and collect electrical energy. The cells need to be mounted on some surface and this surface needs to be supported by some structure. The term "array" is usually used to describe the cells mounted on a surface. These arrays then have to be supported by a structure.

Array Configuration. Your DESIGN should specify the geometry for positioning the cells on an appropriate surface (you also must specify the material from which the surface is to be made). That is, the array configuration or geometry should be determined by you.

To prevent electrical arcing between cells, you should allow 5 mm separation between any two cells (or parts of cells if you have chosen to split individual cells), regardless of how you configure the arrays.

The glue or actual mounting technique for holding the cells on this surface need not concern you, for another designer within your organization specializes in this problem. Similarly, weatherizing the cells should be of no concern to you.

Support Structure. Your DESIGN should specify the support structure that will hold the arrays. You may use wood or metal structural members, but these members should be made of standard-size materials (e.g., 2 × 4 in. wood members, 2 × 2 in. angle iron, etc.). Your instructor should be able to help you with these choices.

This support structure must orient the cells so that they are tilted up from the horizontal at an angle of 35° and generally point toward the south. Figure [15-5] should clarify this.

The land on which you are to build this power plant is flat and level. Land preparation is not a concern for you.

Your company owns equipment that will produce a foundation that consists of four concrete pads each having one 30 mm diameter anchor bolt. The four bolts with each foundation should be used to

EAST ELEVATION   **SOUTH ELEVATION**

secure the structure. The bolts are located at the corners of an imaginary rectangle with sides of 1 m by 2 m. Each bolt is located in the center of a 0.25- $\times$ 0.25-m pad. The tops of all the pads are flat and in the same horizontal plane. The top of each pad is located 0.5 m above the ground. These foundations can be replicated over and over to hold as many arrays as you need.

Remember, though, the object of the project is to minimize the use of land. This means that you want to use as few arrays as you can and as few solar cells as you can.

PROBLEM 5: SOLVENT SEPARATION PLANT _____

Purpose.   The purpose of this project is to design a system that will separate two solvents which have been mixed together during a manufacturing process. Together, the solvents have no value, but separated and returned to their original pure states, they are valuable since they could be used over again in the manufacturing process. The solvent mixture to be separated has previously been disposed of by discharging it into dry wells. However, this practice has recently been found to be contaminating underground water supplies. Since one of the solvents has been shown to be a carcinogen, another method of disposal or, better still, a method of recycling is sought. Unfortunately, this system must be placed in a conspicuous location. Thus, in addition to being functional, your DESIGN must be as esthetically pleasing as possible. (Note, a solvent that has been of recent environmental concern is TCE which is short for trichloroethylene.)

Goal.   The goal of this project is to design the desired solvent recovery or separation plant so that it produces the separated solvents at the lowest possible cost and is as esthetically pleasing as possible.

More Information.   Information on this problem is given in the section entitled "Solvent Separation."

An analysis supplement on performance modeling as it relates to this problem is given in the section "Performance Modeling." There are many ways to use this material in order to proceed to an acceptable result.

The geometry-related aspects of your DESIGN are discussed briefly in the section "The Geometry-Related Design."

Refer to the previous section "Rules and Report Requirements For The Following Design Projects" on page 474, for further information on rules and requirements.

Solvent Separation. The proposed solvent recovery plant will take advantage of the fact that one of the solvents will have a lower boiling temperature than the other. The solvent which has the lower boiling temperature is said to be the one of higher volatility. The liquid of higher volatility can be boiled off from the mixture and recondensed back to a liquid at a place removed from the mixture. Thus, an essentially pure, liquid solvent (the more volatile one) may be collected at one point, while the other solvent (the less volatile one) collects as a fairly pure liquid at some other location in the device.

Figure [15-6] is a schematic of a system commonly used to separate mixtures of liquids of differing volatilities. The liquid mixture of

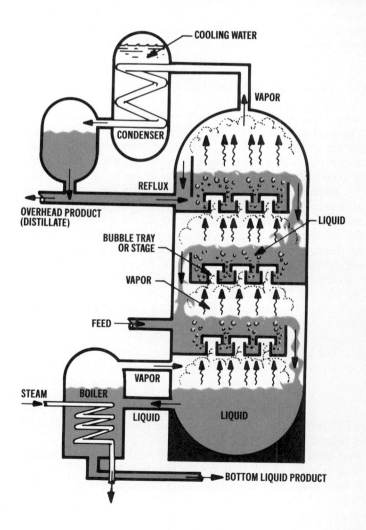

[15-6] Schematic of the Solvent Separation System.

solvents enters the *tower* at the point labeled *feed*. This liquid, combined with liquid from the top, flows to the bottom of the device where it eventually enters the *boiler* through the *liquid* line near the bottom of the tower. The device labeled as a *boiler* is where part of the liquid is boiled off from the mixture. The *steam* provides the necessary energy for boiling in the *boiler*. The vapor generated in the boiler is returned to the tower. The liquid leaving the boiler along the line marked *bottom liquid product* is very pure, consisting of essentially only the solvent with the lower volatility.

As the vapor progresses upward through the *tower*, it comes in contact with liquid in the *bubble trays* or *stages*. The composition of the liquid and of the vapor changes along the height of the tower. If the tower has a sufficient number of bubble trays, the vapor that reaches the top of the tower consists essentially of only the higher volatility solvent. The vapor at the top of the tower enters a condenser whose purpose it is to cool and recondense the vapor back into a liquid (this liquid will be referred to as the *condensate*). Thus, if there are enough trays, this condensate is then very pure; it consists of essentially only the solvent with the higher volatility.

If the tower has enough trays, you could, in principle, remove and collect the condensate that emerges from the condenser, for this is one of the products you seek. In practice, however, some of the condensate must be returned to the tower for it provides the liquid which flows from tray to tray. Without this reinjection of some of the product back into the tower, the device will not function. Figure [15-6] shows this reinjected fluid as the *reflux*. The line labeled *distillate* is where the higher volatility product is removed.

The physics and chemistry of how the system works are not important here; you need only to realize that some of the product emerging from the condenser must be returned to the tower. All of the condensate that emerges from the condenser may not be extracted as *distillate*. If you accept this fact, then you can complete your part of the DESIGN without understanding the detailed physics or chemistry.

Modeling

Performance Modeling:

   R: The most important parameter governing the design of this solvent recovery system is the reflux ratio, R, defined as:

$$R = \frac{\textit{Reflux (i.e., the amount of condensate returned to the tower)}}{\textit{Distillate collected and removed}}$$

A value of $R = 0$ means that all of the condensate from the condenser is collected and removed; none is returned to the tower. As stated previously, the system does not work well at this value of R. A value of $R = 100$ means that 100 times more condensate is being returned to the tower than is being removed and collected. Since R is a ratio of two amounts having the same dimensions, R is dimensionless.

   It can be shown that the amount of vapor rising in the column and the amount of solvent passing through either the condenser or the boiler are all proportional to $(1 + R)$. This means that the tower cross-sectional area, the amount of cooling water required for the condenser, and the amount of steam required for the boiler are all proportional to $(1 + R)$. Since the cross-sectional area of the tower is proportional to $(1 + R)$, the diameter of the column is proportional to $(1 + R)^{0.5}$. Physically, this means that as R increases, the diameter of the tower must become larger.

N: There is an important relationship between the reflux ratio, R, and the number of bubble trays, N, necessary for proper operation of the system. If too few trays are present, then the product emerging from the top of the tower will not be sufficiently pure. If too many trays are present, the excess trays only add to the expense of the apparatus without improving the performance.

Experience shows that when large reflux ratios are being used (R large), fewer trays are needed (N small). When small reflux ratios are used (R small), more trays are required (N large). The relationship between R and a satisfactory number of trays, N, has experimentally been found to be given by:

$$\Upsilon = \frac{N - N_{\min}}{N + 1} = 1 - \exp\left[\left\{\frac{1 + 54.4X}{11 + 117.2X} \cdot \frac{X - 1}{X^{0.5}}\right\}\right]$$

where

$$X = \frac{R - R_{\min}}{R + 1}$$

and $R_{\min}$ and $N_{\min}$ are determined by the particular solvents being separated. This relationship between R and N has been found to apply over a large range of variables that might be thought to affect the problem. The relationship is more than adequate for the design specifications being requested of you.

$N_{\min}$ is expected to be somewhere between 5 and 10 for the solvents of interest here, while $R_{\min}$ is expected to be in the range of .5 to 2. Values for the solvents you will design for will be supplied to you by your instructor as these data become available.

V: The amount of the volatile solvent being processed also impacts the cross-sectional area of the tower and the amount of cooling water and steam required. It can be shown that the cross-sectional area of the tower, the amount of cooling water, and the steam requirements are proportional to the flowrate of the distillate removed from the device (this flowrate will be referred to as V and will be considered to have units of liters/second). Since the cross-sectional area of the tower is proportional to V, the diameter of the column is proportional to $V^{0.5}$.

The appropriate value for V is still being assessed. However, it is expected to be in the range of 0.2 to 2 liters/second. The actual value will be supplied through your instructor.

Important Parameters: Parameters which will be important in determining your DESIGN, along with their relationship to R, N and/or V, are:

TABLE 15-2

Parameters	Relationship to R, N, or V	Units[4]
Diameter of the tower	$C_D[V(R+1)]^{1/2}$	m
Height of the tower	$C_H(N+1)$	m
Cooling capacity of the condenser	$C_{cool}V(R+1)$	$\dfrac{kJ}{s}$
Heating capacity of the boiler	$C_{heat}V(R+1)$	$\dfrac{kJ}{s}$

In Table 15-2, C_D, C_H, C_{cool}, and C_{heat} are coefficients whose values are being determined by another group within your organization. The values will be relayed to you by your instructor as they become available. At this time, it appears that these coefficients will fall into the ranges given in Table 15-3.

Economic Modeling. Your economic model should allow you to pick the design that best meets the design objectives. It should account for all costs expected to be incurred during the lifetime of the system.

Capital Costs. Capital costs that will be incurred are those related to the construction of the tower, the associated piping, the construction of the trays, and the construction or purchase of the condenser and boiler. Many of these costs may be estimated using a 0.6 power relationship to extrapolate known costs to new situations. This is explained in the following details.

Tower and Associated Piping Costs. These costs (here called T&PCosts) can be adequately estimated from the following relationship:

$$T\&PCost = A\_known\_cost \cdot (A/A\_known)^{0.6}$$

where

T&PCost is the installed cost of the tower and piping,

TABLE 15-3

Coefficient	Probable Range	Units[5]
C_D	.5 to 3	$m(s/1)^{1/2}$
C_H	.2 to 2	m
C_{cool}	200 to 1000	kJ/l
C_{heat}	200 to 1000	kJ/l

---

[4]kJ is the standard abbreviation for 1000 joules; m is the standard abbreviation for meters.

[5]l is the standard abbreviation for liters.

A is the cylindrical surface area of the cylindrical tower, *A_known_cost* is the known cost of building a tower of known cylindrical surface area *A_known*.

It is known that on a previous project, it costs $10,000 for a cylindrical tank of surface area of 50 m². This $10,000 included installation and much associated piping.

Tray Costs: These costs are largely dependent on the area covered by the tray. You may estimate the costs of single trays from the following formula:

$$TrayCost = T\_known\_cost \cdot (T/T\_known)^{0.6}$$

where

TrayCost is the installed cost of the trays,

T is the area covered by a single tray,

T_known_cost is the known cost of building a tray of known area *T_known*.

You may use *T_known_cost* = $2,000 for a tray of *T_known* = 3 m². You may also use 80% of the cross-sectional area of the tower as the area covered by a single tray.

Combined Tower, Piping, Tray Costs: If the tower gets quite tall, the costs of construction usually increase because of additional complications in construction and installation. You are to account for this effect in the following way:

Actual_TPT_cost =
$$(T\&PCost + N \cdot TrayCost) \cdot (Height/H\_known)^{Hexp}$$

where:

Actual_TPT_cost is the actual installed tower, piping, and tray cost,

T&PCost is the tower and piping cost from above,

TrayCost is the tray cost from above,

N is the number of trays,

Height is the height of the tower,

H_known is the height of tower used to establish *T&PCosts*,

Hexp is an exponent used to adjust costs for height.

H_known is 8 meters, and *Hexp* is thought to be near 1.3.

Condenser Cost: You may estimate the installed costs of the condenser from the following formula:

$$CondCost = C\_known\_cost \cdot (C\_cap/C\_known)^{0.6}$$

where

CondCost is the installed cost of the condenser,

C_cap is the required cooling capacity of the condenser,

C_known_cost is the known installed cost of a condenser of cooling capacity *C_known*.

You may use *C_known_cost* = $9,000 for a condenser of known cooling capacity *C_known* = 900 kJ/sec.

Boiler Cost. You may estimate the installed costs of the boiler from following formula:

$$Boilcost = B\_known\_cost \cdot (B\_cap/B\_known)^{0.6}$$

where

Boilcost is the installed cost of the boiler,

B_cap is the required heating capacity for the boiler,

B_known_cost is the known installed cost of a boiler of known heating capacity B_known.

You may use $B\_known\_cost =$ \$12,000 for a boiler of known heating capacity of $B_known = $ 1,000 kJ/sec.

Operating and Maintenance Costs   *Cooling Water Costs:* The cost of cooling water for operating the condenser can be computed from the following equation:

$$Oper\$\_Cool = Water\_Cost \cdot \_conv \cdot C\_cap$$

where

$Oper\$\_Cool$ is the cost in \$/yr,

$Water\_cost$ is the water cost in \$/ft^3.

C_conv is a conversion factor dependent on the high volatility solvent,

C_cap is the required cooling capacity of the condenser.

Water costs are estimated to be in the range of \$0.001 to \$0.002 per ft^3. For the high volatility solvent to be used here, C_conv is expected to be near 26,700 s $\cdot$ ft^3/(kJ $\cdot$ yr).

Steam Supply Costs. The costs of the steam for operating the boiler can be computed from the following equation:

$$Oper\$\_Boil = Steam\_cost \cdot B\_conv \cdot B\_cap$$

where:

$Oper\$\_Boil$ is the cost in \$/year,

$Steam\_cost$ is the steam cost in \$/lb$_m$,

B_conv is a conversion factor dependent on the high volatility solvent,

B_cap is the required heating capacity of the boiler.

Steam-costs are estimated to be in the range of \$0.0025 to \$0.01 per lb$_m$. For the high volatility solvent to be used here, B_conv is expected to be near 6,515 lb$_m$ $\cdot$ s/(kJ $\cdot$ yr).

Maintenance Costs. You should make some estimate of the dollar value and frequency of maintenance to the facility for use in your economic modeling. Maintenance may depend on tower height, so you might consider increasing the maintenance costs as $Height^{Hexp}$.

Geometry-Related Design

Graphics requirements include sketches and final drawings showing the geometry and size of the solvent recovery system, including a site plan. The graphics materials should also address the structure for supporting the system. Maintenance access to the tower (e.g., for cleaning) is to be specifically included. The site plan must show your creative approach in dealing with the visual impact of the separation plant.

The Design Criteria for the Solvent Separating System.   Your goal is to choose the solvent recovery system that is the most economically attractive (within the constraints given above under "Economic Modeling") in meeting all of the project requirements and is as esthetically pleasing as possible.

PROBLEM 6: WIND TURBINE FARM

Purpose.   The purpose of this project is to design a wind turbine generator (WTG) "farm" that will supply a given fraction of an existing electrical load. In the last 10 years, wind powered generators, or windmills, have become serious contenders in the search for renewable energy sources. These devices convert the kinetic energy inherent in the wind into mechanical motion (usually the rotation of a shaft). When this mechanical motion is used to turn the armature of an electrical generator, electrical energy can be produced.

Goal.   The desirable goal in this project is to design the most economical windmill/generator system that will provide a given fraction of the electrical energy currently being consumed on Orkney, a group of islands off the northern tip of Scotland.

More Information.   Information on Orkney, its terrain, and its current electrical consumption is given in the section "Orkney."
An analysis supplement on performance modeling as it relates to this problem is given in the section "Performance Modeling." There are many ways to use this material in order to proceed to an acceptable result. The blade/shaft/generator unit that you will use must be selected from an existing inventory of designs. This inventory is described in the section "Inventory of WTG Designs."
The geometry-related aspects of your DESIGN are discussed briefly in the section "Geometry-Related Design." The geometry aspects include the site plan (where the windmills, access roads, power lines, etc., are to be placed on the chosen site) as well as a full description of the structure you have chosen for supporting the blade/shaft/generator unit you select.
Refer to the previous section entitled "Rules and Report Requirements For the Following Design Projects" on page 474, for further information on rules and requirements.

Orkney

The archipelago known as Orkney forms the northern most county of Scotland. Of the approximately 70 separate islands that constitute Orkney, the largest is known as Mainland. A map of that island is given in Figure [15-7].

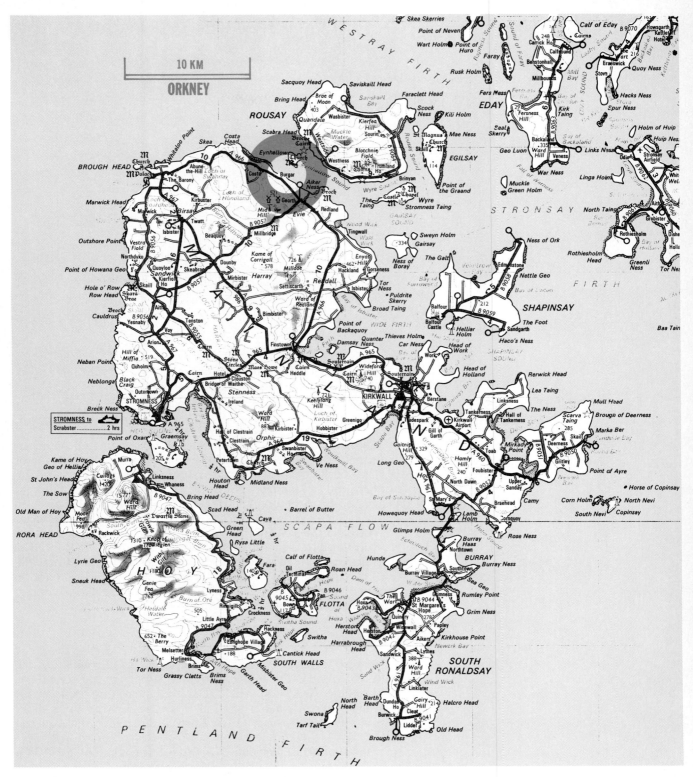

[15-7] Orkney.

The Burgar Hill area of Orkney.

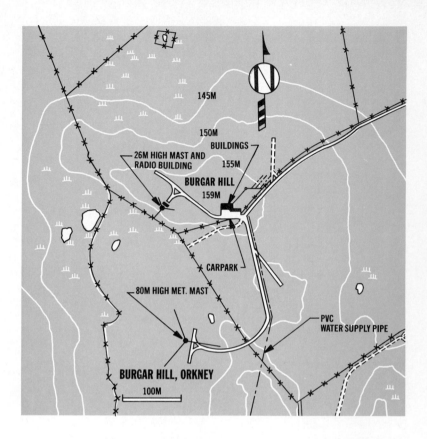

Orkney is located in a region having very good wind conditions and seems ideal for WTGs. One of the better sites for installation of WTGs is on Burger Hill just southeast of the town of Burger on the northern end of Mainland. A map of the local area can be found in Figure [15-8]. A photograph of the typical terrain is also given in Figure [15-9].

A total of 74.5×10^6 kilowatt-hours (kWh) of electrical energy was consumed by the 8400 residents of the islands in 1981. Electricity is generated by diesel-engine-driven generators located near the town of Kirkwall. These engines burn heavy fuel oil, which must be imported. In 1981 the fuel cost to operate these diesel engines amounted to 2.8 pence per kilowatt-hour.

Modeling

This project involves WTGs that have their axes of rotation of the blades horizontal. These are referred to simply as horizontal axis wind turbines (HAWTs). There is another interesting group which have their axes of rotation vertical. They are referred to as either vertical axis wind turbines (VAWTs) or as Darrieus wind turbines.

Modeling the Blade/Shaft/Generator. It is common for WTGs to be designed to produce a given electrical power when operating in a given wind speed. The power produced under these conditions is

called the "rated power" (P_{rated}) and the wind speed (at the hub or axis of the machine) at which the rated power is produced is called the "rated speed" (W_{rated}). Unfortunately, the wind does not always blow at the rated speed, so it is important to understand how much power is produced at other wind speeds.

Figure [15-10] gives the typical performance of a wind turbine as a function of wind speed. At very low speeds, the turbine, if it turns at all, does not produce enough electrical power to overcome its own losses. However, at some "cut-in" wind speed (W_{ci}) a useful amount of power begins to be produced. This power increases until it reaches the P_{rated} at W_{rated}. Beyond W_{rated}, the power is constant until the "cut-out" wind speed (W_{co}) is reached. At W_{co} and beyond the blades are feathered in order to keep the blades from breaking. That is, beyond W_{co} no electrical power is produced.

The shaft speed of the WTG is basically controlled at a constant value by the electrical grid to which the machine is connected. In a 60-cycle alternating-current (ac) grid the shafts on most generators connected to the grid are all locked in unison at 1800 revolutions per minute (rpm). However, these details should be of no concern to you in your analysis.

Modeling the Wind. You will need information on winds on Orkney to couple with the blade/shaft/generator model in order to calculate yearly electrical energy production from your WTG farm.

A histogram giving the number of hours per year of occurrence of various wind speeds is available for the north shore of Mainland Orkney. A tabulated presentation of this is given in Table 15-4. Each line on Table 15-4 has a mean bin wind speed (in meters per second) followed by the number of hours per year of occurrence. For example, a mean bin wind speed of 5 m/s represents wind speeds greater

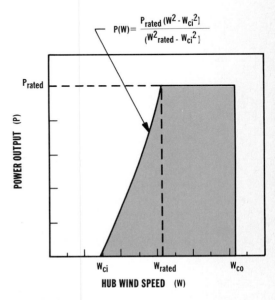

$$P(W) = \frac{P_{\text{rated}} (W^2 - W_{\text{ci}}^2)}{(W_{\text{rated}}^2 - W_{\text{ci}}^2)}$$

[15-10] The variation of output power with wind speed for a typical horizontal axis wind turbine.

TABLE 15-4 WIND SPEED DATA

Wind Speed Median Bin Value (m/s)	Hours/Year at Speed (hrs)
1	88
2	169
3	235
4	309
5	360
6	406
7	486
8	527
9	494
10	545
11	468
12	477
13	429
14	408
15	377
16	355
17	336
18	294
19	247
20	212
21	170
22	147
23	110
24	88
25	75
26	60
27	50
28	38
29	26
30	19
31	17
32	11
33	9
34	6
35	4
36	3
37	2
38	1
39	1
40	1

(continued)

than 4.5 m/s but less than or equal to 5.5 m/s. The wind information for which you are given data were collected at a height above ground of 3 m.

Economic Modeling. WTGs are just in their infancy. As a result, accurate costs for production line models are not well established. However, there is evidence that they can be produced for $1000 (U.S.)/kW$_{rated}$ or less; kW$_{rated}$ here stands for "rated kilowatt of power output." Use this value in estimating that part of your system's cost.

This project concerns itself with WTG farms that provide a relatively small part of the total electrical needs. In these situations it can be shown that the number of diesel-powered generators cannot be reduced just because there are wind turbines available. This is due to the fact that the wind is usually not blowing when the consumers request the maximum amount of power. Therefore, there must be enough conventional equipment available to meet the maximum needs. The existing diesel generators are simply operated fewer hours per year. *As a result of this, WTGs only produce a savings in fuel.* They do not decrease the amount of money needed for equipment, operations, maintenance, and so on, of the conventional generation equipment.

Thus the WTG needs to be paid for out of fuel savings that accrue over the lifetime of the WTG. You may assume that each of the blade/shaft/generator units in the design inventory has a lifetime of 20 years. You will need to assume an interest rate for these calculations.

Geometry-Related Design

One of your many tasks is to design the support structure for the blade/shaft/generator you choose to use. The blade/shaft/generator choice must come from an existing inventory of designs which is described briefly in the next section. This support structure will take the place of the large question mark (?) on the sketch shown in Figure [15-11] of a typical WTG. The choice of height (*H* on the sketch) will be yours. In general, wind speeds increase with height but so do construction costs.

The blade/shaft/generator designs in the available inventory all may be considered to have similar mounting pads that include a turntable that allows rotation of the unit around a vertical axis through 360° in order to keep the blades always headed into the wind. The control system which monitors the direction of the wind and adjusts the rotation of the blade/shaft/generator is a part of each design in the inventory. Do not concern yourself with the details of this control feature, but do remember that the blade/shaft/generator unit must turn to face the wind.

The shaft/transmission/generator in each design is housed in a cylindrical metal enclosure called the nacelle. The diameter of the nacelle for each design in the inventory is 7.5 percent of the blade diameter. The length of each nacelle is equal to 6 of its diameters.

The base of the turntable of each unit is circular with a diameter equal to the diameter of the nacelle which it supports. It is this base of the turntable that will mount to the top of your structure.

The weight of each unit in the design inventory can be calculated from the fact that each weighs $400\ N/P_{rated}$, where P_{rated} is in kW. That is, a 2-kW_{rated} unit would weigh 800 N.

Maintenance is a major problem with WTGs. To reduce maintenance problems, your DESIGN for the structure must address this issue. It should be clear how reasonable access to the WTG is to be gained.

The geometry aspects of your DESIGN must include a site plan showing where your wind turbines will be located. As a "rule of thumb" you should separate the WTGs horizontally in your field by at least 4 blade diameters. This will avoid downwind machines being affected by wind disturbances created by upwind machines. The site plan should also show access roads and power lines. All WTGs in the field should be wired to a central collection building where the voltage is stepped up to the value used on the electrical grid that distributes electrical energy to the residents. The transforming of the voltage is not a concern of yours, but the location of the building in which it takes place is.

Inventory of WTG Designs

Your task in this project is *not* to design a new set of blades, a new shaft/transmission, a new generator, or a new nacelle. This task is being handled by another group within your organization. You are, however, responsible for choosing a blade/transmission/generator design from the existing inventory of designs that were devised by this other group.

A list of the pertinent properties (P_{rated}, W_{rated}, W_{ci}, W_{co}) of a number of WTGs from which you must select one for your final DESIGN WTG farm is available in Table 15-5.

Design Criteria for the WTG Farm

Your DESIGN must supply 10 percent (your instructor may furnish you a different number) of the electrical energy consumed on Orkney. Because you cannot install a fraction of a WTG, you will not be able to supply exactly this amount. You should exceed this amount but only by the smallest difference that is physically possible.

Your design goal is to choose the system that is the most economically attractive (within the constraints given above in the section "Economic Modeling"). You are not being asked to estimate the costs for building your structures, but by comparing your blade/shaft/generator costs to fuel savings, you must estimate how much money you can afford to spend on your parts of the DESIGN.

PROBLEM 7: RESERVOIR AND DAM DESIGN _____

Purpose. The purpose of this project is to design a water-retaining reservoir that would tap the runoff from a watershed area and provide downstream water users with a continuous water flow,

TABLE 15-4   WIND SPEED DATA (cont.)

Wind Speed Median Bin Value (m/s)	Hours/Year at Speed (hrs)
41	0
42	0
43	0
44	0
45	0
46	0
47	0
48	0
49	0
50	0

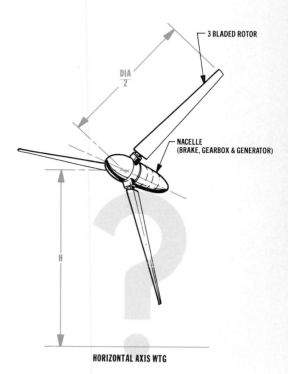

HORIZONTAL AXIS WTG

[15-11]

TABLE 15-5 TURBINE CHARACTERISTICS

Turbine	Rated Power (kW)	Blade Diameter (m)	Rated Wind Speed (m/s)	Cut-In Wind Speed (m/s)	Cut-Out Wind Speed (m/s)
A	100	18	14	6	30
B	65	15	16	5	25
C	300	22	13	5	25
D	60	16	15	4	25
E	65	16	15	4	27
F	65	15	15	3	22
G	75	22	9	4	22
H	65	15	18	4	30
I	40	13	13	4	22
J	75	10	18	5	27
K	95	11	18	5	27
L	400	24	16	5	27
M	50	17	10	5	20
N	100	17	13	5	20
O	340	31	13	5	25
P	75	16	18	5	25
Q	80	16	17	5	25
R	250	24	16	5	25
S	40	12	13	5	27
T	250	20	17	8	27
U	3000	60	17	8	27

independent of yearly fluctuations of rainfall. Most of the better sites for large reservoirs in the United States now have such facilities built on them. However, there are many smaller sites which have potential for such use. Construction of reservoirs on many of these potential sites offers promises of steadier sources of water for domestic, irrigation, and recreational uses, as well as reduced damage due to flooding.

Goal. The goal of this project is to minimize the economic value[6] of the water that would be supplied to downstream water users were this project to be built. The establishment of this value would allow an economic assessment of the overall project to be made. Recreational uses of the reservoir are to be maintained.

More Information. Information on the chosen reservoir site and requirements of this project are given in the section "Proposed Dam and Reservoir."

---

[6] Economic value as used here means the minimum price that you must receive for the water (dollar/amount) in order to yield a zero net present value of all income and expenditures associated with this project over its lifetime.

An analysis supplement on modeling as it relates to this problem is given in the section "Modeling." There are many ways to use this material in order to proceed to an acceptable result.

The geometry-related aspects of your DESIGN are discussed briefly in the section "The Geometry-Related Design." The geometry aspects include the site plan (where the reservoir, access roads, and a boat ramp are to be placed on the chosen site) as well as a description of the dam (with spillway) you have chosen.

Refer to the previous section entitled "Rules and Report Requirements For the Following Design Projects" on page 474, for further information on rules and requirements.

Proposed Dam and Reservoir

A topographic map of the site to be studied in this project is shown in Figure [15-12]. The terrain essentially dictates that the dam be placed at the site marked "Dam." The topography also makes it clear

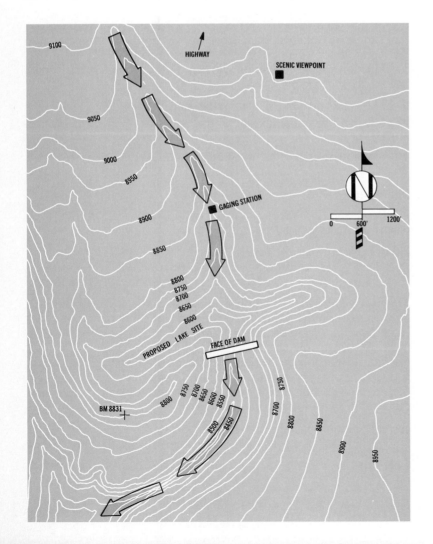

[15-12] The dam site.

that the elevation of the top of the dam should not exceed approximately 8750 ft.

The intended recreational use for the reservoir is trout fishing. Thus, in addition to an access road to the dam for maintenance purposes, there must be a boat ramp and concomitant access road. The access roads, either individually or together, must tie into a major highway shown just north of the proposed reservoir site on the map.

To avoid occasional killing off of the fish population that would be caused by removing all of the water from the reservoir during extended dry spells, a minimum amount of water (and thus a minimum surface elevation) must be maintained at all times. The minimum requirement is to be specified and accounted for in the final DESIGN.

The dam or reservoir must include a spillway for accommodating the wet years when it may not be possible to impound all of the water that flows into the reservoir. The spillway protects the integrity of the dam by keeping the water from flowing over the top of the dam and eroding it.

In dry years, when water cannot be taken from the reservoir, water must be obtained from other sources by the dam operator so that commitments to downstream users are met.

The following information will be provided to you over the next few weeks as it is developed:

■   The water demand that must be met.

■   The cost of construction of the dam (as a function of dam height).

■   The cost incurred in obtaining makeup water.

Modeling

Performance Modeling.   An appropriate model for the reservoir is given, in words, as

inflows over some time period
— outflows over some time period
= change in storage during the time period

This is a general conservation principle known as conservation of mass or "continuity." Example "inflows" are the streamflows that enter the reservoir. Examples of "outflows" are the water removed to meet user demands, any water that cannot be retained when the reservoir is full (it must be sent over the dam spillway), evaporation, and seepage into the ground.[7] The current amount of water stored in the reservoir can be found from the amount available at an earlier time by the (word) equation.

current amount = amount at an earlier time
+ change in storage during the time period

where the last item in the equation is obtained from the previous

---

[7] You may neglect this outflow for this study.

word equation. "Changes in storage during the time period" can be either positive (the amount of water in the reservoir has increased) or negative (the amount of water in the reservoir has decreased).

Converting these equations to algebraic relationships and applying them repeatedly to the reservoir (which starts empty) in sequential yearly intervals over the years allows you to track the amount of water in the reservoir at the end of each year. However, you must not allow the elevation of the surface of your reservoir to exceed the spillway height or to go below the minimum possible height specified by recreational demands.[8]

It should be clear that for a given dam site and given topography of the reservoir bottom, there should be a relationship between the amount of water impounded and the surface elevation. From the topographic map it has been determined that a suitable relationship between the volume of water in the reservoir (*Vol* in ac-ft/yr[9]) and the surface elevation of the reservoir (*El* in feet) is given by

$$El = 8554 + 0.032008Vol - 0.0000013167Vol^2$$

for $0 < Vol \leq 11,000$ ac-ft.

There is also a relationship between the surface area of the reservoir (*area* in acres) and the elevation of the surface (*El* in feet). It is given to sufficient accuracy by

$$area = 176,202 - 41.47El + 0.00244El^2$$

Reservoir Modeling: A Deterministic Approach.   In anticipation of the future need for data, the U.S. Bureau of Reclamation has been collecting streamflow information at the point marked "Gaging Station" on Figure [15-12]. Table 15-6 includes yearly streamflow data for the years 1925–1981. The units for the flows are acre-feet/year.

These historical data can be used for the "inflows over some time period" in your reservoir model. They will "drive" your model and tell you what the reservoir would have done in the past.

Economic Modeling.   Your economic model should allow you to establish the value of the water which you will supply to the downstream users. It is important to realize that your contractual obligation to the water users will be to supply a specified amount of water each year, regardless of where it will come from.

Your economic model would establish an absolute minimum that you would have to charge your customers in order to recover the costs of capital improvements (here the costs of building the dam), estimated maintenance, and water you need to supply (from some other source) when your reservoir surface is at its minimum

---

[8] Actually, yearly time intervals do not give strictly accurate results when spillage and minimum reservoir conditions are encountered, but they suffice for your purposes here. If you can model this problem using yearly intervals, you can certainly model it using monthly, weekly, daily, or even hourly intervals.

[9] An acre-foot is a volume unit commonly used by hydrologists; it is equivalent to 43,560 ft^3.

TABLE 15-6 YEARLY AVERAGE STREAMFLOW INFORMATION FOR GAGING STATION NEAR PROPOSED RESERVOIR SITE

Year	Flow (ac-ft/yr)	Year	Flow (ac-ft/yr)
1925	1834	1954	1887
1926	3425	1955	1387
1927	3315	1956	1432
1928	1666	1957	1967
1929	1996	1958	3648
1930	2435	1959	1454
1931	2660	1960	3347
1932	4872	1961	1209
1933	2272	1962	3495
1934	1488	1963	2063
1935	3660	1964	1695
1936	3227	1965	3490
1937	3816	1966	4419
1938	1921	1967	1650
1939	1951	1968	3739
1940	1754	1969	2493
1941	7476	1970	1722
1942	2814	1971	1334
1943	2606	1972	2190
1944	1741	1973	6830
1945	2887	1974	1324
1946	1733	1975	3010
1947	1727	1976	1918
1948	2560	1977	1382
1949	3484	1978	4091
1950	1342	1979	6863
1951	1194	1980	5759
1952	4920	1981	1757
1953	1470		

elevation. That is, your economic model is to determine the "value" of the water to be supplied to the users.

Another group within your organization is charged with estimating construction costs of the dam. When your dam DESIGN has "jelled" somewhat, that group will use past experience with dams of the size and structure you have chosen, to estimate your dam's cost. Your instructor will supply you with information on the costs of the dam as they become available. You should make some estimate of the dollar value and frequency of maintenance to the facility for use in your economic modeling.

When you need to supply water but the reservoir surface is at its minimum elevation, you must supply water from another source (e.g., purchase it elsewhere and truck it in). As stated above, your economic model should include these costs for "makeup" water; your instructor will supply you with information on the costs of such make-up water.

Geometry-Related Design

It will be important to you and to the recreational users of your reservoir to establish where the high, low, and average water lines will be for your reservoir. Even a histogram of surface elevation might be useful. It is also important to establish where on the reservoir your facilities (your boat ramp) will be located and how access from the highway to the dam and the facilities is to be made. These are all part of the geometry-related design that needs to be specified.

The dam you will propose also needs to be specified. The geometrical aspects of its specification include type of construction, height, location of spillway, location of the standpipe for water withdrawal, and so on. Details of how it is to be tied into the canyon walls and calculations regarding its structural integrity are assigned to another group within your organization; these particular features are of no major concern to you.

Design Criteria for the Reservoir

Your DESIGN must supply the water demand of downstream water users as specified by your instructor while meeting the minimum lake elevation requirements for recreational purposes.

Your goal is to choose the dam/reservoir that is the most economically attractive (within the constraints given in the section "Economic Modeling"), in meeting all of the water requirements.

Appendix I

Conversion Tables

Example: Convert 87.5 cm to miles (see Table 1).

$$1 \text{ cm} = 6.214 \times 10^{-6} \text{ mile}$$
$$87.5 \text{ cm} = 87.5(6.214 \times 10^{-6}) \text{ mile}$$
$$= 5.44 \times 10^{-4} \text{ mile}$$

Example: Convert 14.7 lb/in.^{2} to pascals (see Table 7).

$$1 \text{ lb/in.}^2 = 6.895 \times 10^3 \text{ pascals}$$
$$14.7 \text{ lb/in.}^2 = 14.7(6.895 \times 10^3) \text{ pascals}$$
$$= 1.014 \times 10^5 \text{ pascals}$$

In the following tables, the E, minus or plus an exponent, following a value indicates the power of 10 by which this value should be multiplied. Thus $6.214 \text{ E} - 06$ means 6.214×10^{-6}.

1. LENGTH EQUIVALENTS

cm	in.	ft	m	mi[a]	
1	3.937 E − 01	3.281 E − 02	1.0 E − 02	6.214 E − 06	cm
2.540	1	8.333 E − 02	2.54 E − 02	1.578 E − 05	in.
3.048 E + 01	1.2 E + 01	1	3.048 E − 01	1.894 E − 04	ft
1.0 E + 02	3.937 E + 01	3.281	1	6.214 E − 04	m
1.609 E + 05	6.336 E + 04	5.280 E + 03	1.609 E + 05	1	mi

[a] Mile.

Additional measures

Metric: 1 km = 10^3 m
       1 mm = 10^{-3} m
       1 μm = 10^{-6} m (micrometer or micron)
       1 Å = 10^{-10} m (angstrom)

English: 1 mil = 10^{-3} in.
       1 yd = 3.0 ft
       1 rod = 5.5 yd = 16.5 ft
       1 furlong = 40 rod = 660 ft

2. AREA EQUIVALENTS

m^2	$in.^2$	ft^2	acres	mi^2	
1	1.55 E + 03	1.076 E + 01	2.471 E − 04	3.861 E − 07	m^2
6.452 E − 04	1	6.944 E − 03	1.594 E − 07	2.491 E − 10	$in.^2$
9.290 E − 02	1.44 E + 02	1	2.296 E − 05	3.587 E − 08	ft^2
4.047 E + 03	6.273 E + 06	4.356 E + 04	1	1.562 E − 03	acres
2.590 E + 06	4.018 E + 09	2.788 E + 07	6.40 E + 02	1	mi^2

Additional measures

1 hectare = 10^4 m^2

3. VOLUME EQUIVALENTS

cm^3	$in.^3$	ft^3	gal (U.S.)	
1	6.102 E − 02	3.532 E − 05	2.642 E − 04	cm^3
1.639 E + 01	1	5.787 E − 04	4.329 E − 03	$in.^3$
2.832 E + 04	1.728 E + 03	1	7.481	ft^3
3.785 E + 03	2.31 E + 02	1.337 E − 01	1	gal (U.S.)

Additional measures

Metric: 1 liter = 10^3 cm^3
          1 m^3 = 10^6 cm^3

English: 1 quart = 0.250 gal (U.S.)
          1 bushel = 9.309 gal (U.S.)
          1 barrel = 42 gal (U.S.)
             (petroleum measure only)
          1 imperial gal = 1.20 gal (U.S.) approx.
          1 board foot (wood) = 144 $in.^3$
          1 cord (wood) = 128 ft^3

4. MASS EQUIVALENTS

kg	slug	lb_m[a]	g	
1	6.85 E − 02	2.205	1.0 E + 03	kg
1.46 E + 01	1	3.22 E + 01	1.46 E + 04	slug
4.54 E − 01	3.11 E − 02	1	4.54 E + 02	lb_m
1.0 E − 03	6.85 E − 05	2.205 E − 03	1	g

[a] Not recommended.

5. FORCE EQUIVALENTS

N[a]	lb_f[b]	dyn[c]	kg_f[d]	g_f[d]	Poundal[d]	
1	2.248 E − 01	1.0 E + 05	1.019 E − 01	1.019 E + 02	7.234	N
4.448	1	4.448 E + 05	4.54 E − 01	4.54 E + 02	3.217 E + 01	lb_f
1.0 E − 05	2.248 E − 06	1	1.02 E − 06	1.02 E − 03	7.233 E − 05	dyn
9.807	2.205	9.807 E + 05	1	1.0	7.093 E + 01	kg_f
9.087 E − 03	2.205 E − 03	9.087 E + 02	1.0 E − 03	1	7.093 E − 02	g_f
1.382 E − 01	3.108 E − 02	1.383 E + 04	1.410 E − 02	1.410 E + 01	1	poundal

Additional measures

1 metric ton = 10^3 kg$_f$ = 2.205 × 10^3 lb$_f$

1 pound troy = 0.8229 lb$_f$

1 oz^b = 6.25 × 10^{-2} lb$_f$

1 oz troy = 6.857 × 10^{-2} lb$_f$

1 grain = 0.0001429 lb$_f$

[a] Newton.
[b] Avoirdupois.
[c] Dyne.
[d] Not recommended.

6. VELOCITY AND ACCELERATION EQUIVALENTS

a. Velocity

cm/s	ft/sec	mi/hr (mph)	km/h	
1	3.281 E − 02	2.237 E − 02	3.60 E − 02	cm/s
3.048 E + 01	1	6.818 E − 01	1.097	ft/sec
4.470 E + 01	1.467	1	1.609	mi/hr
2.778 E + 01	9.113 E − 01	6.214 E − 01	1	km/h

b. Acceleration

cm/s^2	ft/sec^2	$\bar{g}^a$	
1	3.281 E − 02	1.019 E − 03	cm/s^2
3.048 E + 01	1	3.109 E − 03	ft/sec^2
9.807 E + 02	3.217 E + 01	1	$\bar{g}$

Additional measures

1 knot = 1.152 mi/hr

[a] Standard acceleration of gravity.

7. PRESSURE EQUIVALENTS

dyn/cm^2	N/m^2 (pascal)	lb/in.^{2} (psi)	lb$_f$/ft^2 (psf)	atm^a	Head^b in. (Hg)	ft (H$_2$O)	
1	1.0 E − 01	1.45 E − 05	2.089 E − 03	9.869 E − 07	2.953 E − 05	3.349 E − 05	dyn/cm^2
1.0 E + 01	1	1.45 E − 04	2.089 E − 02	9.869 E − 06	2.953 E − 04	3.349 E − 04	N/m^2
6.895 E − 04	6.895 E + 03	1	1.44 E + 02	6.805 E − 02	2.036	2.309	lb$_f$/in.^{2}
4.788 E + 02	4.788 E + 01	6.944 E − 03	1	4.725 E − 04	1.414 E − 02	1.603 E − 02	lb/ft^2
1.013 E + 06	1.013 E + 05	1.47 E + 01	2.116 E + 03	1	2.992 E + 01	3.393 E + 01	atm
3.336 E + 04	3.386 E + 03	4.912 E − 01	7.073 E + 01	3.342 E − 02	1	1.134	in. (Hg)
2.986 E + 04	2.986 E + 03	4.331 E − 01	6.237 E + 01	2.947 E − 02	8.819 E − 01	1	ft (H$_2$O)

Additional measure

1 bar = 1 dyne/cm^2

[a] Standard atmospheric pressure.
[b] At standard gravity and 0°C for Hg, 15°C for H$_2$O.

8. WORK AND ENERGY EQUIVALENTS

J^a	ft-lb$_f$	W-h	Btu^b	kcal^c	kg-m	
1	7.376 E − 01	2.778 E − 04	9.478 E − 04	2.388 E − 04	1.020 E − 01	J
1.356	1	3.766 E − 04	1.285 E − 03	3.238 E − 04	1.383 E − 01	ft-lb$_f$
3.60 E + 03	2.655 E + 03	1	3.412	8.599 E − 01	3.671 E + 02	W-h
1.055 E + 03	7.782 E + 02	2.931 E − 01	1	2.520 E − 01	1.076 E + 02	Btu
4.187 E + 03	3.088 E + 03	1.163	3.968	1	4.269 E + 02	kcal
9.807	7.233	2.724 E − 03	9.295 E − 03	2.342 E − 03	1	kg-m

Additional measures

1 newton-meter = 1 J   1 therm = 10^{-5} Btu

1 erg = 1 dyne-cm = 10^{-7} J   1 million electron volts (MeV) = 1.602×10^{-13} J

1 cal = 10^{-3} kcal

[a] Joule.
[b] British thermal unit.
[c] Kilocalorie.

9. POWER EQUIVALENTS

J/s	ft-lb$_f$/s	hp^a	kW	Btu/h	
1	7.376 E − 01	1.341 E − 03	1.0 E − 03	3.412	J/s
1.356	1	1.818 E − 03	1.356 E − 03	4.626	ft-lb$_f$/s
7.457 E + 02	5.50 E + 02	1	7.457 − 01	2.545 E + 03	hp
1.0 E + 03	7.376 E + 02	1.341	1	3.412 E + 03	kW
2.931 E − 01	2.162 E − 01	3.930 E − 04	2.931 E − 04	1	Btu/h

Additional Measures

1 W = 10^{-3} kW   1 poncelet = 100 kg-m/s = 0.9807 kW

1 cal/s = 14.29 Btu/h   1 ton of refrigeration = 1.2×10^4 Btu/h

[a] Horsepower.

Time

1 week	7 days	168 hours	10,080 minutes	604,800 seconds
1 mean solar day		24 hours	1440 minutes	86,400 seconds
1 calendar year	365 days	8760 hours	5.256 E + 05 minutes	3.1536 E + 07 seconds
1 tropical mean solar year	365.2422 days (basis of modern calendar)			

Temperature

$\Delta 1°$ Celsius (C) = $\Delta 1°$ Kelvin (K) = $\Delta 1.8°$ Fahrenheit (F) = $1.8°$ Rankine (R)

$0°C = 273.15°K = 32°F = 491.67°R$ $°C = \frac{5}{9}(°F - 32)$

$0°K = -273.15°C = -459.67°F = 0°R$ $°F = \frac{9}{5}(°C + 32)$

Electrical Quantities

1 coulomb	1.036×10^5 faradays	0.1 abcoulomb	2.998×10^9 statcoulombs
1 ampere		0.1 abampere	2.998×10^9 statcoulombs
1 volt	10^3 millivolts	10^8 abvolts	3.335×10^{-3} statvolt
1 ohm	10^6 megohms	10^9 abohms	1.112×10^{-12} statohm
1 farad	10^6 microfarads	10^{-9} abfarad	8.987×10^{11} statfarads

Special Tables and Formulas

SPECIFIC GRAVITIES AND SPECIFIC WEIGHTS

Material	Average specific gravity	Average specific weight, lb_f/ft^3	Material	Average specific gravity	Average specific weight lb_f/ft^3
Acid, sulfuric	1.80	112	Granite, solid	2.70	172
Air, S.T.P.	0.001293	0.0806	Graphite	1.67	135
Alcohol, ethyl	0.790	49	Gravel, loose, wet	1.68	105
Aluminum, cast	2.65	165			
Asbestos	2.5	153	Hickory	0.77	48
Ash, white	0.67	42			
Ashes, cinders	0.68	44	Ice	0.91	57
Asphaltum	1.3	81	Iron, gray cast	7.10	450
			Iron, wrought	7.75	480
Babbitt metal, soft	10.25	625			
Basalt, granite	1.50	96	Kerosene	0.80	50
Brass, cast-rolled	8.50	534			
Brick, common	1.90	119	Lead	11.34	710
Bronze	8.1	509	Leather	0.94	59
			Limestone, solid	2.70	168
Cedar, white, red	0.35	22	Limestone, crushed	1.50	95
Cement, portland, bags	1.44	90			
Chalk	2.25	140	Mahogany	0.70	44
Clay, dry	1.00	63	Manganese	7.42	475
Clay, loose, wet	1.75	110	Marble	2.70	166
Coal, anthracite, solid	1.60	95	Mercury	13.56	845
Coal, bituminous, solid	1.35	85	Monel metal, rolled	8.97	555
Concrete, gravel, sand	2.3	142			
Copper, cast, rolled	8.90	556	Nickel	8.90	558
Cork	0.24	15			
Cotton, flax, hemp	1.48	93	Oak, white	0.77	48
Copper ore	4.2	262	Oil, lubricating	0.91	57
Earth	1.75	105	Paper	0.92	58
			Paraffin	0.90	56
Fir, Douglas	0.50	32	Petroleum, crude	0.88	55
Flour, loose	0.45	28	Pine, white	0.43	27
			Platinum	21.5	1330
Gasoline	0.70	44			
Glass, crown	2.60	161	Redwood, California	0.42	26
Glass, flint	3.30	205	Rubber	1.25	78
Glycerine	1.25	78			
Gold, cast-hammered	19.3	1205	Sand, loose, wet	1.90	120

(Continued on next page)

Material	Average specific gravity	Average specific weight, lb_f/ft^3	Material	Average specific gravity	Average specific weight lb_f/ft^3
Sandstone, solid	2.30	144	Tungsten	19.22	1200
Seawater	1.03	64	Turpentine	0.865	54
Silver	10.5	655			
Steel, structural	7.90	490	Water, 4°C	1.00	62.4[a]
Sulfur	2.00	125	Water, snow, fresh fallen	0.125	8.0
Teak, African	0.99	62			
Tin	7.30	456	Zinc	7.14	445

[a]The value for the specific weight of water which is usually used in problem solutions is 62.4 lb_f/ft^3 or 8.34 lb_f/gal.

Appendix III

Interest Factors

DECIMAL INTEREST RATE =   .005

PERIOD	F/P	P/F	F/A	A/F	P/A	A/P	P/G	A/G
1.0000	1.0050	0.9950	1.0000	1.0000	0.9950	1.0050	-0.0002	-0.0002
2.0000	1.0100	0.9901	2.0050	0.4988	1.9851	0.5038	0.9911	0.4992
3.0000	1.0151	0.9851	3.0150	0.3317	2.9702	0.3367	2.9589	0.9962
4.0000	1.0202	0.9802	4.0301	0.2481	3.9505	0.2531	5.9027	1.4942
5.0000	1.0253	0.9754	5.0503	0.1980	4.9259	0.2030	9.8033	1.9902
6.0000	1.0304	0.9705	6.0755	0.1646	5.8964	0.1696	14.6597	2.4862
7.0000	1.0355	0.9657	7.1059	0.1407	6.8621	0.1457	20.4485	2.9799
8.0000	1.0407	0.9609	8.1414	0.1228	7.8230	0.1278	27.1781	3.4741
9.0000	1.0459	0.9561	9.1821	0.1089	8.7791	0.1139	34.8251	3.9668
10.0000	1.0511	0.9513	10.2280	0.0978	9.7304	0.1028	43.3886	4.4591
11.0000	1.0564	0.9466	11.2792	0.0887	10.6770	0.0937	52.8547	4.9503
12.0000	1.0617	0.9419	12.3356	0.0811	11.6190	0.0861	63.2181	5.4409
13.0000	1.0670	0.9372	13.3973	0.0746	12.5562	0.0796	74.4646	5.9305
14.0000	1.0723	0.9326	14.4643	0.0691	13.4887	0.0741	86.5894	6.4194
15.0000	1.0777	0.9279	15.5365	0.0644	14.4166	0.0694	99.5741	6.9069
16.0000	1.0831	0.9233	16.6142	0.0602	15.3399	0.0652	113.4268	7.3942
17.0000	1.0885	0.9187	17.6973	0.0565	16.2586	0.0615	128.1250	7.8804
18.0000	1.0939	0.9141	18.7858	0.0532	17.1728	0.0582	143.6676	8.3660
19.0000	1.0994	0.9096	19.8797	0.0503	18.0824	0.0553	160.0369	8.8504
20.0000	1.1049	0.9051	20.9791	0.0477	18.9874	0.0527	177.2373	9.3344
21.0000	1.1104	0.9006	22.0840	0.0453	19.8880	0.0503	195.2454	9.8173
22.0000	1.1160	0.8961	23.1945	0.0431	20.7841	0.0481	214.0701	10.2997
23.0000	1.1216	0.8916	24.3104	0.0411	21.6757	0.0461	233.6800	10.7808
24.0000	1.1272	0.8872	25.4320	0.0393	22.5629	0.0443	254.0886	11.2613
25.0000	1.1328	0.8828	26.5591	0.0377	23.4457	0.0427	275.2728	11.7409
26.0000	1.1385	0.8784	27.6919	0.0361	24.3240	0.0411	297.2325	12.2197
27.0000	1.1442	0.8740	28.8304	0.0347	25.1980	0.0397	319.9551	12.6976
28.0000	1.1499	0.8697	29.9746	0.0334	26.0677	0.0384	343.4387	13.1749
29.0000	1.1556	0.8653	31.1245	0.0321	26.9331	0.0371	367.6720	13.6513
30.0000	1.1614	0.8610	32.2801	0.0310	27.7941	0.0360	392.6403	14.1268
31.0000	1.1672	0.8567	33.4414	0.0299	28.6508	0.0349	418.3365	14.6012
32.0000	1.1730	0.8525	34.6087	0.0289	29.5033	0.0339	444.7663	15.0751
33.0000	1.1789	0.8482	35.7817	0.0279	30.3515	0.0329	471.9098	15.5481
34.0000	1.1848	0.8440	36.9606	0.0271	31.1956	0.0321	499.7631	16.0203
35.0000	1.1907	0.8398	38.1454	0.0262	32.0354	0.0312	528.3165	16.4916
36.0000	1.1967	0.8356	39.3361	0.0254	32.8710	0.0304	557.5638	16.9622
42.0000	1.2330	0.8110	46.6066	0.0215	37.7983	0.0265	747.1963	19.7680
48.0000	1.2705	0.7871	54.0979	0.0185	42.5804	0.0235	959.9282	22.5439
54.0000	1.3091	0.7639	61.8168	0.0162	47.2214	0.0212	1194.2380	25.2902
60.0000	1.3489	0.7414	69.7701	0.0143	51.7256	0.0193	1448.6560	28.0065
66.0000	1.3898	0.7195	77.9650	0.0128	56.0970	0.0178	1721.8060	30.6934
72.0000	1.4320	0.6983	86.4089	0.0116	60.3396	0.0166	2012.3570	33.3506
78.0000	1.4755	0.6777	95.1093	0.0105	64.4570	0.0155	2319.0510	35.9783
84.0000	1.5204	0.6577	104.0740	0.0096	68.4531	0.0146	2640.6770	38.5764
90.0000	1.5666	0.6383	113.3110	0.0088	72.3313	0.0138	2976.0870	41.1452
96.0000	1.6141	0.6195	122.8287	0.0081	76.0953	0.0131	3324.1980	43.6847
102.0000	1.6632	0.6013	132.6354	0.0075	79.7482	0.0125	3683.9640	46.1949
108.0000	1.7137	0.5835	142.7400	0.0070	83.2935	0.0120	4054.3900	48.6760
114.0000	1.7658	0.5663	153.1517	0.0065	86.7342	0.0115	4434.5430	51.1280
120.0000	1.8194	0.5496	163.8795	0.0061	90.0735	0.0111	4823.5200	53.5510

DECIMAL INTEREST RATE = .01

PERIOD	F/P	P/F	F/A	A/F	P/A	A/P	P/G	A/G
1.0000	1.0100	0.9901	1.0000	1.0000	0.9901	1.0100	-0.0001	-0.0001
2.0000	1.0201	0.9803	2.0100	0.4975	1.9704	0.5075	0.9803	0.4975
3.0000	1.0303	0.9706	3.0301	0.3300	2.9410	0.3400	2.9212	0.9933
4.0000	1.0406	0.9610	4.0604	0.2463	3.9020	0.2563	5.8043	1.4875
5.0000	1.0510	0.9515	5.1010	0.1960	4.8534	0.2060	9.6099	1.9800
6.0000	1.0615	0.9420	6.1520	0.1625	5.7955	0.1725	14.3200	2.4709
7.0000	1.0721	0.9327	7.2135	0.1386	6.7282	0.1486	19.9166	2.9602
8.0000	1.0829	0.9235	8.2857	0.1207	7.6517	0.1307	26.3807	3.4477
9.0000	1.0937	0.9143	9.3685	0.1067	8.5660	0.1167	33.6948	3.9335
10.0000	1.1046	0.9053	10.4622	0.0956	9.4713	0.1056	41.8435	4.4179
11.0000	1.1157	0.8963	11.5668	0.0865	10.3676	0.0965	50.8063	4.9005
12.0000	1.1268	0.8874	12.6825	0.0788	11.2551	0.0888	60.5682	5.3814
13.0000	1.1381	0.8787	13.8093	0.0724	12.1337	0.0824	71.1123	5.8607
14.0000	1.1495	0.8700	14.9474	0.0669	13.0037	0.0769	82.4215	6.3383
15.0000	1.1610	0.8613	16.0969	0.0621	13.8650	0.0721	94.4805	6.8143
16.0000	1.1726	0.8528	17.2579	0.0579	14.7179	0.0679	107.2728	7.2886
17.0000	1.1843	0.8444	18.4304	0.0543	15.5622	0.0643	120.7828	7.7613
18.0000	1.1961	0.8360	19.6147	0.0510	16.3983	0.0610	134.9955	8.2323
19.0000	1.2081	0.8277	20.8109	0.0481	17.2260	0.0581	149.8946	8.7017
20.0000	1.2202	0.8195	22.0190	0.0454	18.0455	0.0554	165.4654	9.1693
21.0000	1.2324	0.8114	23.2392	0.0430	18.8570	0.0530	181.6940	9.6354
22.0000	1.2447	0.8034	24.4716	0.0409	19.6604	0.0509	198.5649	10.0998
23.0000	1.2572	0.7954	25.7163	0.0389	20.4558	0.0489	216.0654	10.5625
24.0000	1.2697	0.7876	26.9735	0.0371	21.2434	0.0471	234.1789	11.0236
25.0000	1.2824	0.7798	28.2432	0.0354	22.0231	0.0454	252.8923	11.4830
26.0000	1.2953	0.7720	29.5256	0.0339	22.7952	0.0439	272.1953	11.9409
27.0000	1.3082	0.7644	30.8209	0.0324	23.5596	0.0424	292.0689	12.3970
28.0000	1.3213	0.7568	32.1291	0.0311	24.3164	0.0411	312.5038	12.8515
29.0000	1.3345	0.7493	33.4504	0.0299	25.0658	0.0399	333.4856	13.3044
30.0000	1.3478	0.7419	34.7849	0.0287	25.8077	0.0387	355.0013	13.7556
31.0000	1.3613	0.7346	36.1327	0.0277	26.5423	0.0377	377.0386	14.2052
32.0000	1.3749	0.7273	37.4941	0.0267	27.2696	0.0367	399.5846	14.6531
33.0000	1.3887	0.7201	38.8690	0.0257	27.9897	0.0357	422.6277	15.0994
34.0000	1.4026	0.7130	40.2577	0.0248	28.7027	0.0348	446.1562	15.5441
35.0000	1.4166	0.7059	41.6603	0.0240	29.4086	0.0340	470.1575	15.9871
36.0000	1.4308	0.6989	43.0769	0.0232	30.1075	0.0332	494.6198	16.4285
42.0000	1.5188	0.6584	51.8790	0.0193	34.1581	0.0293	650.4510	19.0424
48.0000	1.6122	0.6203	61.2226	0.0163	37.9740	0.0263	820.1443	21.5976
54.0000	1.7114	0.5843	71.1410	0.0141	41.5687	0.0241	1001.5720	24.0944
60.0000	1.8167	0.5504	81.6696	0.0122	44.9550	0.0222	1192.8040	26.5333
66.0000	1.9285	0.5185	92.8460	0.0108	48.1451	0.0208	1392.0940	28.9145
72.0000	2.0471	0.4885	104.7099	0.0096	51.1504	0.0196	1597.8650	31.2386
78.0000	2.1730	0.4602	117.3037	0.0085	53.9814	0.0185	1808.6980	33.5059
84.0000	2.3067	0.4335	130.6722	0.0077	56.6484	0.0177	2023.3130	35.7170
90.0000	2.4486	0.4084	144.8632	0.0069	59.1609	0.0169	2240.5660	37.8724
96.0000	2.5993	0.3847	159.9272	0.0063	61.5277	0.0163	2459.4270	39.9727
102.0000	2.7592	0.3624	175.9180	0.0057	63.7573	0.0157	2678.9840	42.0184
108.0000	2.9289	0.3414	192.8925	0.0052	65.8578	0.0152	2898.4190	44.0103
114.0000	3.1091	0.3216	210.9113	0.0047	67.8365	0.0147	3117.0080	45.9488
120.0000	3.3004	0.3030	230.0386	0.0043	69.7005	0.0143	3334.1120	47.8348

DECIMAL INTEREST RATE = .015

PERIOD	F/P	P/F	F/A	A/F	P/A	A/P	P/G	A/G
1.0000	1.0150	0.9852	1.0000	1.0000	0.9852	1.0150	-0.0001	-0.0001
2.0000	1.0302	0.9707	2.0150	0.4963	1.9559	0.5113	0.9703	0.4961
3.0000	1.0457	0.9563	3.0452	0.3284	2.9122	0.3434	2.8828	0.9899
4.0000	1.0614	0.9422	4.0909	0.2444	3.8544	0.2594	5.7090	1.4812
5.0000	1.0773	0.9283	5.1522	0.1941	4.7826	0.2091	9.4218	1.9700
6.0000	1.0934	0.9145	6.2295	0.1605	5.6972	0.1755	13.9942	2.4564
7.0000	1.1098	0.9010	7.3230	0.1366	6.5982	0.1516	19.4004	2.9403
8.0000	1.1265	0.8877	8.4328	0.1186	7.4859	0.1336	25.6140	3.4216
9.0000	1.1434	0.8746	9.5593	0.1046	8.3605	0.1196	32.6105	3.9006
10.0000	1.1605	0.8617	10.7027	0.0934	9.2222	0.1084	40.3652	4.3770
11.0000	1.1779	0.8489	11.8632	0.0843	10.0711	0.0993	48.8546	4.8510
12.0000	1.1956	0.8364	13.0412	0.0767	10.9075	0.0917	58.0544	5.3224
13.0000	1.2136	0.8240	14.2368	0.0702	11.7315	0.0852	67.9425	5.7915
14.0000	1.2318	0.8118	15.4503	0.0647	12.5433	0.0797	78.4962	6.2580
15.0000	1.2502	0.7999	16.6821	0.0599	13.3432	0.0749	89.6943	6.7221
16.0000	1.2690	0.7880	17.9323	0.0558	14.1312	0.0708	101.5145	7.1837
17.0000	1.2880	0.7764	19.2013	0.0521	14.9076	0.0671	113.9365	7.6428
18.0000	1.3073	0.7649	20.4893	0.0488	15.6725	0.0638	126.9399	8.0995
19.0000	1.3269	0.7536	21.7966	0.0459	16.4261	0.0609	140.5047	8.5537
20.0000	1.3469	0.7425	23.1236	0.0432	17.1686	0.0582	154.6113	9.0055
21.0000	1.3671	0.7315	24.4704	0.0409	17.9001	0.0559	169.2408	9.4548
22.0000	1.3876	0.7207	25.8375	0.0387	18.6208	0.0537	184.3750	9.9016
23.0000	1.4084	0.7100	27.2250	0.0367	19.3308	0.0517	199.9961	10.3460
24.0000	1.4295	0.6995	28.6334	0.0349	20.0304	0.0499	216.0852	10.7879
25.0000	1.4509	0.6892	30.0629	0.0333	20.7196	0.0483	232.6259	11.2274
26.0000	1.4727	0.6790	31.5138	0.0317	21.3986	0.0467	249.6013	11.6644
27.0000	1.4948	0.6690	32.9866	0.0303	22.0676	0.0453	266.9950	12.0990
28.0000	1.5172	0.6591	34.4813	0.0290	22.7267	0.0440	284.7899	12.5311
29.0000	1.5400	0.6494	35.9986	0.0278	23.3760	0.0428	302.9718	12.9608
30.0000	1.5631	0.6398	37.5385	0.0266	24.0158	0.0416	321.5247	13.3881
31.0000	1.5865	0.6303	39.1016	0.0256	24.6461	0.0406	340.4340	13.8129
32.0000	1.6103	0.6210	40.6881	0.0246	25.2671	0.0396	359.6845	14.2353
33.0000	1.6345	0.6118	42.2984	0.0236	25.8789	0.0386	379.2626	14.6553
34.0000	1.6590	0.6028	43.9329	0.0228	26.4817	0.0378	399.1538	15.0728
35.0000	1.6839	0.5939	45.5919	0.0219	27.0755	0.0369	419.3452	15.4880
36.0000	1.7091	0.5851	47.2758	0.0212	27.6606	0.0362	439.8232	15.9007
42.0000	1.8688	0.5351	57.9229	0.0173	30.9940	0.0323	568.0120	18.3265
48.0000	2.0435	0.4894	69.5649	0.0144	34.0425	0.0294	703.5374	20.6665
54.0000	2.2344	0.4475	82.2948	0.0122	36.8305	0.0272	844.2087	22.9215
60.0000	2.4432	0.4093	96.2142	0.0104	39.3802	0.0254	988.1572	25.0928
66.0000	2.6715	0.3743	111.4342	0.0090	41.7120	0.0240	1133.7950	27.1815
72.0000	2.9211	0.3423	128.0765	0.0078	43.8446	0.0228	1279.7830	29.1891
78.0000	3.1941	0.3131	146.2739	0.0068	45.7949	0.0218	1424.9960	31.1169
84.0000	3.4926	0.2863	166.1717	0.0060	47.5786	0.0210	1568.5020	32.9666
90.0000	3.8189	0.2619	187.9288	0.0053	49.2098	0.0203	1709.5320	34.7397
96.0000	4.1758	0.2395	211.7189	0.0047	50.7016	0.0197	1847.4600	36.4379
102.0000	4.5660	0.2190	237.7321	0.0042	52.0659	0.0192	1981.7880	38.0630
108.0000	4.9926	0.2003	266.1760	0.0038	53.3137	0.0188	2112.1220	39.6169
114.0000	5.4592	0.1832	297.2778	0.0034	54.4548	0.0184	2238.1660	41.1014
120.0000	5.9693	0.1675	331.2858	0.0030	55.4984	0.0180	2359.6990	42.5183

DECIMAL INTEREST RATE =   .0175

PERIOD	F/P	P/F	F/A	A/F	P/A	A/P	P/G	A/G
1.0000	1.0175	0.9828	1.0000	1.0000	0.9828	1.0175	0.0001	0.0001
2.0000	1.0353	0.9659	2.0175	0.4957	1.9487	0.5132	0.9662	0.4958
3.0000	1.0534	0.9493	3.0528	0.3276	2.8980	0.3451	2.8649	0.9886
4.0000	1.0719	0.9330	4.1062	0.2435	3.8310	0.2610	5.6640	1.4785
5.0000	1.0906	0.9169	5.1781	0.1931	4.7479	0.2106	9.3318	1.9655
6.0000	1.1097	0.9011	6.2687	0.1595	5.6490	0.1770	13.8374	2.4495
7.0000	1.1291	0.8856	7.3784	0.1355	6.5347	0.1530	19.1516	2.9308
8.0000	1.1489	0.8704	8.5076	0.1175	7.4051	0.1350	25.2448	3.4091
9.0000	1.1690	0.8554	9.6564	0.1036	8.2605	0.1211	32.0884	3.8846
10.0000	1.1894	0.8407	10.8254	0.0924	9.1012	0.1099	39.6550	4.3571
11.0000	1.2103	0.8263	12.0149	0.0832	9.9275	0.1007	47.9179	4.8268
12.0000	1.2314	0.8121	13.2251	0.0756	10.7396	0.0931	56.8508	5.2936
13.0000	1.2530	0.7981	14.4566	0.0692	11.5377	0.0867	66.4282	5.7575
14.0000	1.2749	0.7844	15.7096	0.0637	12.3220	0.0812	76.6247	6.2185
15.0000	1.2972	0.7709	16.9845	0.0589	13.0929	0.0764	87.4172	6.6767
16.0000	1.3199	0.7576	18.2817	0.0547	13.8505	0.0722	98.7817	7.1320
17.0000	1.3430	0.7446	19.6017	0.0510	14.5951	0.0685	110.6953	7.5844
18.0000	1.3665	0.7318	20.9447	0.0477	15.3269	0.0652	123.1357	8.0340
19.0000	1.3904	0.7192	22.3112	0.0448	16.0461	0.0623	136.0811	8.4806
20.0000	1.4148	0.7068	23.7017	0.0422	16.7529	0.0597	149.5112	8.9245
21.0000	1.4395	0.6947	25.1165	0.0398	17.4476	0.0573	163.4046	9.3655
22.0000	1.4647	0.6827	26.5560	0.0377	18.1303	0.0552	177.7417	9.8036
23.0000	1.4904	0.6710	28.0208	0.0357	18.8013	0.0532	192.5034	10.2388
24.0000	1.5164	0.6594	29.5111	0.0339	19.4607	0.0514	207.6708	10.6713
25.0000	1.5430	0.6481	31.0276	0.0322	20.1088	0.0497	223.2252	11.1009
26.0000	1.5700	0.6369	32.5706	0.0307	20.7458	0.0482	239.1489	11.5276
27.0000	1.5975	0.6260	34.1405	0.0293	21.3718	0.0468	255.4251	11.9515
28.0000	1.6254	0.6152	35.7380	0.0280	21.9870	0.0455	272.0363	12.3726
29.0000	1.6539	0.6046	37.3634	0.0268	22.5917	0.0443	288.9667	12.7909
30.0000	1.6828	0.5942	39.0173	0.0256	23.1859	0.0431	306.1998	13.2063
31.0000	1.7123	0.5840	40.7001	0.0246	23.7699	0.0421	323.7208	13.6189
32.0000	1.7422	0.5740	42.4124	0.0236	24.3439	0.0411	341.5145	14.0287
33.0000	1.7727	0.5641	44.1546	0.0226	24.9080	0.0401	359.5662	14.4358
34.0000	1.8037	0.5544	45.9273	0.0218	25.4624	0.0393	377.8617	14.8400
35.0000	1.8353	0.5449	47.7310	0.0210	26.0073	0.0385	396.3875	15.2414
36.0000	1.8674	0.5355	49.5663	0.0202	26.5428	0.0377	415.1302	15.6400
42.0000	2.0723	0.4826	61.2726	0.0163	29.5679	0.0338	531.4420	17.9736
48.0000	2.2996	0.4349	74.2631	0.0135	32.2939	0.0310	652.6118	20.2085
54.0000	2.5519	0.3919	88.6787	0.0113	34.7504	0.0288	776.5419	22.3463
60.0000	2.8318	0.3531	104.6757	0.0096	36.9640	0.0271	901.5028	24.3886
66.0000	3.1425	0.3182	122.4276	0.0082	38.9589	0.0257	1026.0790	26.3375
72.0000	3.4872	0.2868	142.1270	0.0070	40.7565	0.0245	1149.1260	28.1949
78.0000	3.8698	0.2584	163.9874	0.0061	42.3764	0.0236	1269.7280	29.9631
84.0000	4.2943	0.2329	188.2461	0.0053	43.8362	0.0228	1387.1670	31.6443
90.0000	4.7654	0.2098	215.1659	0.0046	45.1517	0.0221	1500.8880	33.2410
96.0000	5.2882	0.1891	245.0389	0.0041	46.3371	0.0216	1610.4800	34.7557
102.0000	5.8683	0.1704	278.1890	0.0036	47.4053	0.0211	1715.6470	36.1910
108.0000	6.5121	0.1536	314.9758	0.0032	48.3680	0.0207	1816.1930	37.5495
114.0000	7.2265	0.1384	355.7983	0.0028	49.2354	0.0203	1912.0050	38.8339
120.0000	8.0192	0.1247	401.0990	0.0025	50.0171	0.0200	2003.0340	40.0470

DECIMAL INTEREST RATE = .02

PERIOD	F/P	P/F	F/A	A/F	P/A	A/P	P/G	A/G
1.0000	1.0200	0.9804	1.0000	1.0000	0.9804	1.0200	-0.0000	-0.0000
2.0000	1.0404	0.9612	2.0200	0.4951	1.9416	0.5151	0.9609	0.4949
3.0000	1.0612	0.9423	3.0604	0.3268	2.8839	0.3468	2.8456	0.9867
4.0000	1.0824	0.9238	4.1216	0.2426	3.8077	0.2626	5.6168	1.4751
5.0000	1.1041	0.9057	5.2040	0.1922	4.7134	0.2122	9.2397	1.9603
6.0000	1.1262	0.8880	6.3081	0.1585	5.6014	0.1785	13.6793	2.4421
7.0000	1.1487	0.8706	7.4343	0.1345	6.4720	0.1545	18.9025	2.9207
8.0000	1.1717	0.8535	8.5829	0.1165	7.3255	0.1365	24.8769	3.3959
9.0000	1.1951	0.8368	9.7546	0.1025	8.1622	0.1225	31.5707	3.8679
10.0000	1.2190	0.8203	10.9497	0.0913	8.9826	0.1113	38.9538	4.3366
11.0000	1.2434	0.8043	12.1687	0.0822	9.7868	0.1022	46.9964	4.8020
12.0000	1.2682	0.7885	13.4120	0.0746	10.5753	0.0946	55.6695	5.2641
13.0000	1.2936	0.7730	14.6803	0.0681	11.3483	0.0881	64.9460	5.7229
14.0000	1.3195	0.7579	15.9739	0.0626	12.1062	0.0826	74.7982	6.1785
15.0000	1.3459	0.7430	17.2934	0.0578	12.8492	0.0778	85.2000	6.6307
16.0000	1.3728	0.7284	18.6392	0.0537	13.5777	0.0737	96.1267	7.0798
17.0000	1.4002	0.7142	20.0120	0.0500	14.2918	0.0700	107.5533	7.5255
18.0000	1.4282	0.7002	21.4122	0.0467	14.9920	0.0667	119.4558	7.9680
19.0000	1.4568	0.6864	22.8405	0.0438	15.6784	0.0638	131.8116	8.4072
20.0000	1.4859	0.6730	24.2973	0.0412	16.3514	0.0612	144.5976	8.8431
21.0000	1.5157	0.6598	25.7832	0.0388	17.0112	0.0588	157.7933	9.2759
22.0000	1.5460	0.6468	27.2989	0.0366	17.6580	0.0566	171.3768	9.7053
23.0000	1.5769	0.6342	28.8449	0.0347	18.2922	0.0547	185.3280	10.1316
24.0000	1.6084	0.6217	30.4218	0.0329	18.9139	0.0529	199.6275	10.5546
25.0000	1.6406	0.6095	32.0302	0.0312	19.5234	0.0512	214.2560	10.9743
26.0000	1.6734	0.5976	33.6708	0.0297	20.1210	0.0497	229.1956	11.3909
27.0000	1.7069	0.5859	35.3442	0.0283	20.7069	0.0483	244.4279	11.8042
28.0000	1.7410	0.5744	37.0511	0.0270	21.2812	0.0470	259.9358	12.2143
29.0000	1.7758	0.5631	38.7921	0.0258	21.8443	0.0458	275.7031	12.6213
30.0000	1.8114	0.5521	40.5679	0.0247	22.3964	0.0447	291.7130	13.0250
31.0000	1.8476	0.5412	42.3793	0.0236	22.9377	0.0436	307.9500	13.4255
32.0000	1.8845	0.5306	44.2269	0.0226	23.4683	0.0426	324.3996	13.8229
33.0000	1.9222	0.5202	46.1114	0.0217	23.9885	0.0417	341.0468	14.2171
34.0000	1.9607	0.5100	48.0336	0.0208	24.4985	0.0408	357.8778	14.6081
35.0000	1.9999	0.5000	49.9943	0.0200	24.9986	0.0400	374.8786	14.9960
36.0000	2.0399	0.4902	51.9942	0.0192	25.4888	0.0392	392.0361	15.3807
42.0000	2.2972	0.4353	64.8620	0.0154	28.2347	0.0354	497.5964	17.6235
48.0000	2.5871	0.3865	79.3532	0.0126	30.6731	0.0326	605.9608	19.7555
54.0000	2.9135	0.3432	95.6726	0.0105	32.8382	0.0305	715.1760	21.7788
60.0000	3.2810	0.3048	114.0510	0.0088	34.7608	0.0288	823.6918	23.6960
66.0000	3.6950	0.2706	134.7480	0.0074	36.4681	0.0274	930.2944	25.5098
72.0000	4.1611	0.2403	158.0562	0.0063	37.9840	0.0263	1034.0500	27.2233
78.0000	4.6861	0.2134	184.3050	0.0054	39.3301	0.0254	1134.2590	28.8394
84.0000	5.2773	0.1895	213.8654	0.0047	40.5255	0.0247	1230.4130	30.3615
90.0000	5.9431	0.1683	247.1552	0.0040	41.5869	0.0240	1322.1640	31.7928
96.0000	6.6929	0.1494	284.6449	0.0035	42.5294	0.0235	1409.2920	33.1369
102.0000	7.5373	0.1327	326.8643	0.0031	43.3663	0.0231	1491.6800	34.3972
108.0000	8.4882	0.1178	374.4104	0.0027	44.1095	0.0227	1569.2970	35.5773
114.0000	9.5591	0.1046	427.9548	0.0023	44.7694	0.0223	1642.1780	36.6808
120.0000	10.7651	0.0929	488.2546	0.0020	45.3554	0.0220	1710.4110	37.7113

DECIMAL INTEREST RATE = .025

PERIOD	F/P	P/F	F/A	A/F	P/A	A/P	P/G	A/G
1.0000	1.0250	0.9756	1.0000	1.0000	0.9756	1.0250	-0.0000	-0.0000
2.0000	1.0506	0.9518	2.0250	0.4938	1.9274	0.5188	0.9518	0.4938
3.0000	1.0769	0.9286	3.0756	0.3251	2.8560	0.3501	2.8090	0.9835
4.0000	1.1038	0.9060	4.1525	0.2408	3.7620	0.2658	5.5268	1.4691
5.0000	1.1314	0.8839	5.2563	0.1902	4.6458	0.2152	9.0621	1.9506
6.0000	1.1597	0.8623	6.3877	0.1566	5.5081	0.1816	13.3735	2.4280
7.0000	1.1887	0.8413	7.5474	0.1325	6.3494	0.1575	18.4212	2.9013
8.0000	1.2184	0.8207	8.7361	0.1145	7.1701	0.1395	24.1664	3.3704
9.0000	1.2489	0.8007	9.9545	0.1005	7.9709	0.1255	30.5721	3.8355
10.0000	1.2801	0.7812	11.2034	0.0893	8.7521	0.1143	37.6030	4.2965
11.0000	1.3121	0.7621	12.4835	0.0801	9.5142	0.1051	45.2244	4.7534
12.0000	1.3449	0.7436	13.7955	0.0725	10.2578	0.0975	53.4035	5.2062
13.0000	1.3785	0.7254	15.1404	0.0660	10.9832	0.0910	62.1085	5.6549
14.0000	1.4130	0.7077	16.5189	0.0605	11.6909	0.0855	71.3088	6.0995
15.0000	1.4483	0.6905	17.9319	0.0558	12.3814	0.0808	80.9755	6.5401
16.0000	1.4845	0.6736	19.3802	0.0516	13.0550	0.0766	91.0797	6.9766
17.0000	1.5216	0.6572	20.8647	0.0479	13.7122	0.0729	101.5948	7.4091
18.0000	1.5597	0.6412	22.3863	0.0447	14.3534	0.0697	112.4945	7.8375
19.0000	1.5986	0.6255	23.9460	0.0418	14.9789	0.0668	123.7540	8.2619
20.0000	1.6386	0.6103	25.5446	0.0391	15.5892	0.0641	135.3492	8.6823
21.0000	1.6796	0.5954	27.1832	0.0368	16.1845	0.0618	147.2569	9.0986
22.0000	1.7216	0.5809	28.8628	0.0346	16.7654	0.0596	159.4550	9.5110
23.0000	1.7646	0.5667	30.5844	0.0327	17.3321	0.0577	171.9225	9.9193
24.0000	1.8087	0.5529	32.3490	0.0309	17.8850	0.0559	184.6384	10.3237
25.0000	1.8539	0.5394	34.1577	0.0293	18.4244	0.0543	197.5838	10.7241
26.0000	1.9003	0.5262	36.0117	0.0278	18.9506	0.0528	210.7398	11.1205
27.0000	1.9478	0.5134	37.9120	0.0264	19.4640	0.0514	224.0881	11.5130
28.0000	1.9965	0.5009	39.8598	0.0251	19.9649	0.0501	237.6117	11.9015
29.0000	2.0464	0.4887	41.8563	0.0239	20.4535	0.0489	251.2942	12.2861
30.0000	2.0976	0.4767	43.9027	0.0228	20.9303	0.0478	265.1197	12.6668
31.0000	2.1500	0.4651	46.0002	0.0217	21.3954	0.0467	279.0731	13.0436
32.0000	2.2038	0.4538	48.1502	0.0208	21.8492	0.0458	293.1400	13.4165
33.0000	2.2588	0.4427	50.3540	0.0199	22.2919	0.0449	307.3064	13.7856
34.0000	2.3153	0.4319	52.6128	0.0190	22.7238	0.0440	321.5593	14.1508
35.0000	2.3732	0.4214	54.9282	0.0182	23.1451	0.0432	335.8861	14.5122
36.0000	2.4325	0.4111	57.3014	0.0175	23.5562	0.0425	350.2743	14.8697
42.0000	2.8210	0.3545	72.8397	0.0137	25.8206	0.0387	437.2888	16.9357
48.0000	3.2715	0.3057	90.8595	0.0110	27.7731	0.0360	524.0365	18.8685
54.0000	3.7939	0.2636	111.7568	0.0089	29.4568	0.0339	608.9409	20.6723
60.0000	4.3998	0.2273	135.9914	0.0074	30.9086	0.0324	690.8646	22.3518
66.0000	5.1024	0.1960	164.0960	0.0061	32.1606	0.0311	769.0186	23.9119
72.0000	5.9172	0.1690	196.6888	0.0051	33.2401	0.0301	842.8876	25.3576
78.0000	6.8622	0.1457	234.4864	0.0043	34.1709	0.0293	912.1698	26.6943
84.0000	7.9580	0.1257	278.3201	0.0036	34.9736	0.0286	976.7278	27.9276
90.0000	9.2288	0.1084	329.1537	0.0030	35.6658	0.0280	1036.5490	29.0629
96.0000	10.7026	0.0934	388.1050	0.0026	36.2626	0.0276	1091.7140	30.1058
102.0000	12.4118	0.0806	456.4705	0.0022	36.7773	0.0272	1142.3700	31.0619
108.0000	14.3938	0.0695	535.7535	0.0019	37.2210	0.0269	1188.7130	31.9366
114.0000	16.6924	0.0599	627.6975	0.0016	37.6037	0.0266	1230.9710	32.7354
120.0000	19.3581	0.0517	734.3243	0.0014	37.9337	0.0264	1269.3890	33.4634

DECIMAL INTEREST RATE = .03

PERIOD	F/P	P/F	F/A	A/F	P/A	A/P	P/G	A/G
1.0000	1.0300	0.9709	1.0000	1.0000	0.9709	1.0300	-0.0000	-0.0000
2.0000	1.0609	0.9426	2.0300	0.4926	1.9135	0.5226	0.9426	0.4926
3.0000	1.0927	0.9151	3.0909	0.3235	2.8286	0.3535	2.7728	0.9803
4.0000	1.1255	0.8885	4.1836	0.2390	3.7171	0.2690	5.4383	1.4631
5.0000	1.1593	0.8626	5.3091	0.1884	4.5797	0.2184	8.8887	1.9409
6.0000	1.1941	0.8375	6.4684	0.1546	5.4172	0.1846	13.0761	2.4138
7.0000	1.2299	0.8131	7.6625	0.1305	6.2303	0.1605	17.9547	2.8818
8.0000	1.2668	0.7894	8.8923	0.1125	7.0197	0.1425	23.4806	3.3450
9.0000	1.3048	0.7664	10.1591	0.0984	7.7861	0.1284	29.6119	3.8032
10.0000	1.3439	0.7441	11.4639	0.0872	8.5302	0.1172	36.3087	4.2565
11.0000	1.3842	0.7224	12.8078	0.0781	9.2526	0.1081	43.5329	4.7049
12.0000	1.4258	0.7014	14.1920	0.0705	9.9540	0.1005	51.2481	5.1485
13.0000	1.4685	0.6810	15.6178	0.0640	10.6350	0.0940	59.4195	5.5872
14.0000	1.5126	0.6611	17.0863	0.0585	11.2961	0.0885	68.0141	6.0210
15.0000	1.5580	0.6419	18.5989	0.0538	11.9379	0.0838	77.0001	6.4500
16.0000	1.6047	0.6232	20.1569	0.0496	12.5611	0.0796	86.3476	6.8742
17.0000	1.6528	0.6050	21.7616	0.0460	13.1661	0.0760	96.0279	7.2936
18.0000	1.7024	0.5874	23.4144	0.0427	13.7535	0.0727	106.0136	7.7081
19.0000	1.7535	0.5703	25.1169	0.0398	14.3238	0.0698	116.2787	8.1179
20.0000	1.8061	0.5537	26.8704	0.0372	14.8775	0.0672	126.7985	8.5229
21.0000	1.8603	0.5375	28.6765	0.0349	15.4150	0.0649	137.5495	8.9231
22.0000	1.9161	0.5219	30.5368	0.0327	15.9369	0.0627	148.5093	9.3186
23.0000	1.9736	0.5067	32.4529	0.0308	16.4436	0.0608	159.6565	9.7093
24.0000	2.0328	0.4919	34.4265	0.0290	16.9355	0.0590	170.9709	10.0954
25.0000	2.0938	0.4776	36.4593	0.0274	17.4131	0.0574	182.4335	10.4768
26.0000	2.1566	0.4637	38.5530	0.0259	17.8768	0.0559	194.0259	10.8535
27.0000	2.2213	0.4502	40.7096	0.0246	18.3270	0.0546	205.7306	11.2255
28.0000	2.2879	0.4371	42.9309	0.0233	18.7641	0.0533	217.5318	11.5930
29.0000	2.3566	0.4243	45.2188	0.0221	19.1885	0.0521	229.4135	11.9558
30.0000	2.4273	0.4120	47.5754	0.0210	19.6004	0.0510	241.3612	12.3141
31.0000	2.5001	0.4000	50.0027	0.0200	20.0004	0.0500	253.3607	12.6678
32.0000	2.5751	0.3883	52.5027	0.0190	20.3888	0.0490	265.3992	13.0169
33.0000	2.6523	0.3770	55.0778	0.0182	20.7658	0.0482	277.4640	13.3616
34.0000	2.7319	0.3660	57.7302	0.0173	21.1318	0.0473	289.5435	13.7018
35.0000	2.8139	0.3554	60.4621	0.0165	21.4872	0.0465	301.6265	14.0375
36.0000	2.8983	0.3450	63.2759	0.0158	21.8323	0.0458	313.7027	14.3688
42.0000	3.4607	0.2890	82.0232	0.0122	23.7014	0.0422	385.5022	16.2650
48.0000	4.1323	0.2420	104.4084	0.0096	25.2667	0.0396	455.0253	18.0089
54.0000	4.9341	0.2027	131.1374	0.0076	26.5777	0.0376	521.1155	19.6073
60.0000	5.8916	0.1697	163.0534	0.0061	27.6756	0.0361	583.0524	21.0674
66.0000	7.0349	0.1421	201.1626	0.0050	28.5950	0.0350	640.4405	22.3969
72.0000	8.4000	0.1190	246.6671	0.0041	29.3651	0.0341	693.1224	23.6036
78.0000	10.0301	0.0997	301.0018	0.0033	30.0100	0.0333	741.1120	24.6955
84.0000	11.9764	0.0835	365.8804	0.0027	30.5501	0.0327	784.5432	25.6806
90.0000	14.3005	0.0699	443.3487	0.0023	31.0024	0.0323	823.6300	26.5667
96.0000	17.0755	0.0586	535.8500	0.0019	31.3812	0.0319	858.6376	27.3615
102.0000	20.3890	0.0491	646.3011	0.0015	31.6985	0.0315	889.8592	28.0726
108.0000	24.3456	0.0411	778.1858	0.0013	31.9642	0.0313	917.6011	28.7072
114.0000	29.0699	0.0344	935.6629	0.0011	32.1867	0.0311	942.1697	29.2720
120.0000	34.7110	0.0288	1123.6990	0.0009	32.3730	0.0309	963.8634	29.7737

DECIMAL INTEREST RATE = .04

PERIOD	F/P	P/F	F/A	A/F	P/A	A/P	P/G	A/G
1.0000	1.0400	0.9615	1.0000	1.0000	0.9615	1.0400	-0.0000	-0.0000
2.0000	1.0816	0.9246	2.0400	0.4902	1.8861	0.5302	0.9245	0.4902
3.0000	1.1249	0.8890	3.1216	0.3203	2.7751	0.3603	2.7025	0.9738
4.0000	1.1699	0.8548	4.2465	0.2355	3.6299	0.2755	5.2669	1.4510
5.0000	1.2167	0.8219	5.4163	0.1846	4.4518	0.2246	8.5546	1.9216
6.0000	1.2653	0.7903	6.6330	0.1508	5.2421	0.1908	12.5062	2.3857
7.0000	1.3159	0.7599	7.8983	0.1266	6.0021	0.1666	17.0657	2.8433
8.0000	1.3686	0.7307	9.2142	0.1085	6.7327	0.1485	22.1805	3.2944
9.0000	1.4233	0.7026	10.5828	0.0945	7.4353	0.1345	27.8011	3.7391
10.0000	1.4802	0.6756	12.0061	0.0833	8.1109	0.1233	33.8812	4.1772
11.0000	1.5395	0.6496	13.4863	0.0741	8.7605	0.1141	40.3770	4.6090
12.0000	1.6010	0.6246	15.0258	0.0666	9.3851	0.1066	47.2476	5.0343
13.0000	1.6651	0.6006	16.6268	0.0601	9.9856	0.1001	54.4544	5.4533
14.0000	1.7317	0.5775	18.2919	0.0547	10.5631	0.0947	61.9616	5.8658
15.0000	1.8009	0.5553	20.0236	0.0499	11.1184	0.0899	69.7353	6.2721
16.0000	1.8730	0.5339	21.8245	0.0458	11.6523	0.0858	77.7439	6.6720
17.0000	1.9479	0.5134	23.6975	0.0422	12.1657	0.0822	85.9579	7.0656
18.0000	2.0258	0.4936	25.6454	0.0390	12.6593	0.0790	94.3495	7.4530
19.0000	2.1068	0.4746	27.6712	0.0361	13.1339	0.0761	102.8930	7.8341
20.0000	2.1911	0.4564	29.7781	0.0336	13.5903	0.0736	111.5645	8.2091
21.0000	2.2788	0.4388	31.9692	0.0313	14.0292	0.0713	120.3411	8.5779
22.0000	2.3699	0.4220	34.2479	0.0292	14.4511	0.0692	129.2022	8.9406
23.0000	2.4647	0.4057	36.6179	0.0273	14.8568	0.0673	138.1281	9.2973
24.0000	2.5633	0.3901	39.0826	0.0256	15.2470	0.0656	147.1009	9.6479
25.0000	2.6658	0.3751	41.6459	0.0240	15.6221	0.0640	156.1037	9.9925
26.0000	2.7725	0.3607	44.3117	0.0226	15.9828	0.0626	165.1209	10.3312
27.0000	2.8834	0.3468	47.0842	0.0212	16.3296	0.0612	174.1381	10.6640
28.0000	2.9987	0.3335	49.9675	0.0200	16.6631	0.0600	183.1421	10.9909
29.0000	3.1186	0.3207	52.9662	0.0189	16.9837	0.0589	192.1202	11.3120
30.0000	3.2434	0.3083	56.0849	0.0178	17.2920	0.0578	201.0615	11.6274
31.0000	3.3731	0.2965	59.3283	0.0169	17.5885	0.0569	209.9553	11.9371
32.0000	3.5081	0.2851	62.7014	0.0159	17.8735	0.0559	218.7921	12.2411
33.0000	3.6484	0.2741	66.2095	0.0151	18.1476	0.0551	227.5631	12.5395
34.0000	3.7943	0.2636	69.8578	0.0143	18.4112	0.0543	236.2604	12.8324
35.0000	3.9461	0.2534	73.6521	0.0136	18.6646	0.0536	244.8765	13.1198
36.0000	4.1039	0.2437	77.5982	0.0129	18.9083	0.0529	253.4049	13.4018
42.0000	5.1928	0.1926	104.8195	0.0095	20.1856	0.0495	302.4367	14.9828
48.0000	6.5705	0.1522	139.2630	0.0072	21.1951	0.0472	347.2443	16.3832
54.0000	8.3138	0.1203	182.8451	0.0055	21.9930	0.0455	387.4433	17.6167
60.0000	10.5196	0.0951	237.9903	0.0042	22.6235	0.0442	422.9964	18.6972
66.0000	13.3107	0.0751	307.7666	0.0032	23.1218	0.0432	454.0844	19.6388
72.0000	16.8422	0.0594	396.0558	0.0025	23.5156	0.0425	481.0167	20.4552
78.0000	21.3108	0.0469	507.7699	0.0020	23.8269	0.0420	504.1692	21.1597
84.0000	26.9649	0.0371	649.1237	0.0015	24.0729	0.0415	523.9430	21.7649
90.0000	34.1193	0.0293	827.9814	0.0012	24.2673	0.0412	540.7368	22.2825
96.0000	43.1717	0.0232	1054.2940	0.0009	24.4209	0.0409	554.9310	22.7236
102.0000	54.6260	0.0183	1340.6500	0.0007	24.5423	0.0407	566.8775	23.0979
108.0000	69.1193	0.0145	1702.9830	0.0006	24.6383	0.0406	576.8948	23.4145
114.0000	87.4580	0.0114	2161.4500	0.0005	24.7142	0.0405	585.2667	23.6814
120.0000	110.6622	0.0090	2741.5560	0.0004	24.7741	0.0404	592.2427	23.9057

DECIMAL INTEREST RATE = .05

PERIOD	F/P	P/F	F/A	A/F	P/A	A/P	P/G	A/G
1.0000	1.0500	0.9524	1.0000	1.0000	0.9524	1.0500	-0.0000	-0.0000
2.0000	1.1025	0.9070	2.0500	0.4878	1.8594	0.5378	0.9070	0.4878
3.0000	1.1576	0.8638	3.1525	0.3172	2.7232	0.3672	2.6346	0.9675
4.0000	1.2155	0.8227	4.3101	0.2320	3.5459	0.2820	5.1027	1.4390
5.0000	1.2763	0.7835	5.5256	0.1810	4.3295	0.2310	8.2368	1.9025
6.0000	1.3401	0.7462	6.8019	0.1470	5.0757	0.1970	11.9678	2.3579
7.0000	1.4071	0.7107	8.1420	0.1228	5.7864	0.1728	16.2319	2.8052
8.0000	1.4775	0.6768	9.5491	0.1047	6.4632	0.1547	20.9697	3.2445
9.0000	1.5513	0.6446	11.0265	0.0907	7.1078	0.1407	26.1265	3.6758
10.0000	1.6289	0.6139	12.5779	0.0795	7.7217	0.1295	31.6517	4.0990
11.0000	1.7103	0.5847	14.2068	0.0704	8.3064	0.1204	37.4985	4.5144
12.0000	1.7959	0.5568	15.9171	0.0628	8.8632	0.1128	43.6237	4.9219
13.0000	1.8856	0.5303	17.7129	0.0565	9.3936	0.1065	49.9876	5.3215
14.0000	1.9799	0.5051	19.5986	0.0510	9.8986	0.1010	56.5534	5.7133
15.0000	2.0789	0.4810	21.5785	0.0463	10.3796	0.0963	63.2876	6.0973
16.0000	2.1829	0.4581	23.6574	0.0423	10.8378	0.0923	70.1592	6.4736
17.0000	2.2920	0.4363	25.8403	0.0387	11.2741	0.0887	77.1400	6.8423
18.0000	2.4066	0.4155	28.1323	0.0355	11.6896	0.0855	84.2038	7.2033
19.0000	2.5269	0.3957	30.5389	0.0327	12.0853	0.0827	91.3270	7.5569
20.0000	2.6533	0.3769	33.0659	0.0302	12.4622	0.0802	98.4879	7.9029
21.0000	2.7860	0.3589	35.7192	0.0280	12.8211	0.0780	105.6667	8.2416
22.0000	2.9253	0.3419	38.5051	0.0260	13.1630	0.0760	112.8455	8.5729
23.0000	3.0715	0.3256	41.4304	0.0241	13.4886	0.0741	120.0081	8.8970
24.0000	3.2251	0.3101	44.5019	0.0225	13.7986	0.0725	127.1397	9.2139
25.0000	3.3863	0.2953	47.7270	0.0210	14.0939	0.0710	134.2270	9.5237
26.0000	3.5557	0.2812	51.1133	0.0196	14.3752	0.0696	141.2579	9.8265
27.0000	3.7334	0.2678	54.6690	0.0183	14.6430	0.0683	148.2220	10.1224
28.0000	3.9201	0.2551	58.4024	0.0171	14.8981	0.0671	155.1095	10.4114
29.0000	4.1161	0.2429	62.3225	0.0160	15.1411	0.0660	161.9120	10.6936
30.0000	4.3219	0.2314	66.4386	0.0151	15.3724	0.0651	168.6220	10.9691
31.0000	4.5380	0.2204	70.7606	0.0141	15.5928	0.0641	175.2327	11.2381
32.0000	4.7649	0.2099	75.2986	0.0133	15.8027	0.0633	181.7386	11.5005
33.0000	5.0032	0.1999	80.0635	0.0125	16.0025	0.0625	188.1345	11.7565
34.0000	5.2533	0.1904	85.0667	0.0118	16.1929	0.0618	194.4162	12.0063
35.0000	5.5160	0.1813	90.3200	0.0111	16.3742	0.0611	200.5801	12.2498
36.0000	5.7918	0.1727	95.8360	0.0104	16.5468	0.0604	206.6231	12.4872
37.0000	6.0814	0.1644	101.6278	0.0098	16.7113	0.0598	212.5428	12.7185
38.0000	6.3855	0.1566	107.7091	0.0093	16.8679	0.0593	218.3372	12.9440
39.0000	6.7047	0.1491	114.0946	0.0088	17.0170	0.0588	224.0048	13.1636
40.0000	7.0400	0.1420	120.7993	0.0083	17.1591	0.0583	229.5446	13.3774

DECIMAL INTEREST RATE = .06

PERIOD	F/P	P/F	F/A	A/F	P/A	A/P	P/G	A/G
1.0000	1.0600	0.9434	1.0000	1.0000	0.9434	1.0600	-0.0000	-0.0000
2.0000	1.1236	0.8900	2.0600	0.4854	1.8334	0.5454	0.8900	0.4854
3.0000	1.1910	0.8396	3.1836	0.3141	2.6730	0.3741	2.5692	0.9612
4.0000	1.2625	0.7921	4.3746	0.2286	3.4651	0.2886	4.9455	1.4272
5.0000	1.3382	0.7473	5.6371	0.1774	4.2124	0.2374	7.9345	1.8836
6.0000	1.4185	0.7050	6.9753	0.1434	4.9173	0.2034	11.4593	2.3304
7.0000	1.5036	0.6651	8.3938	0.1191	5.5824	0.1791	15.4496	2.7676
8.0000	1.5938	0.6274	9.8975	0.1010	6.2098	0.1610	19.8415	3.1952
9.0000	1.6895	0.5919	11.4913	0.0870	6.8017	0.1470	24.5766	3.6133
10.0000	1.7908	0.5584	13.1808	0.0759	7.3601	0.1359	29.6022	4.0220
11.0000	1.8983	0.5268	14.9716	0.0668	7.8869	0.1268	34.8701	4.4213
12.0000	2.0122	0.4970	16.8699	0.0593	8.3838	0.1193	40.3367	4.8112
13.0000	2.1329	0.4688	18.8821	0.0530	8.8527	0.1130	45.9628	5.1920
14.0000	2.2609	0.4423	21.0150	0.0476	9.2950	0.1076	51.7127	5.5635
15.0000	2.3966	0.4173	23.2759	0.0430	9.7122	0.1030	57.5544	5.9260
16.0000	2.5403	0.3936	25.6725	0.0390	10.1059	0.0990	63.4591	6.2794
17.0000	2.6928	0.3714	28.2128	0.0354	10.4773	0.0954	69.4009	6.6240
18.0000	2.8543	0.3503	30.9056	0.0324	10.8276	0.0924	75.3567	6.9597
19.0000	3.0256	0.3305	33.7599	0.0296	11.1581	0.0896	81.3060	7.2867
20.0000	3.2071	0.3118	36.7855	0.0272	11.4699	0.0872	87.2302	7.6051
21.0000	3.3996	0.2942	39.9927	0.0250	11.7641	0.0850	93.1133	7.9151
22.0000	3.6035	0.2775	43.3922	0.0230	12.0416	0.0830	98.9410	8.2166
23.0000	3.8197	0.2618	46.9957	0.0213	12.3034	0.0813	104.7005	8.5099
24.0000	4.0489	0.2470	50.8155	0.0197	12.5504	0.0797	110.3810	8.7951
25.0000	4.2919	0.2330	54.8644	0.0182	12.7834	0.0782	115.9730	9.0722
26.0000	4.5494	0.2198	59.1563	0.0169	13.0032	0.0769	121.4682	9.3414
27.0000	4.8223	0.2074	63.7057	0.0157	13.2105	0.0757	126.8598	9.6029
28.0000	5.1117	0.1956	68.5280	0.0146	13.4062	0.0746	132.1418	9.8568
29.0000	5.4184	0.1846	73.6397	0.0136	13.5907	0.0736	137.3094	10.1032
30.0000	5.7435	0.1741	79.0580	0.0126	13.7648	0.0726	142.3586	10.3422
31.0000	6.0881	0.1643	84.8015	0.0118	13.9291	0.0718	147.2862	10.5740
32.0000	6.4534	0.1550	90.8896	0.0110	14.0840	0.0710	152.0899	10.7987
33.0000	6.8406	0.1462	97.3430	0.0103	14.2302	0.0703	156.7679	11.0165
34.0000	7.2510	0.1379	104.1835	0.0096	14.3681	0.0696	161.3189	11.2275
35.0000	7.6861	0.1301	111.4345	0.0090	14.4982	0.0690	165.7425	11.4319
36.0000	8.1472	0.1227	119.1206	0.0084	14.6210	0.0684	170.0385	11.6298
37.0000	8.6361	0.1158	127.2678	0.0079	14.7368	0.0679	174.2070	11.8212
38.0000	9.1542	0.1092	135.9039	0.0074	14.8460	0.0674	178.2489	12.0065
39.0000	9.7035	0.1031	145.0581	0.0069	14.9491	0.0669	182.1650	12.1857
40.0000	10.2857	0.0972	154.7616	0.0065	15.0463	0.0665	185.9566	12.3590

DECIMAL INTEREST RATE = .07

PERIOD	F/P	P/F	F/A	A/F	P/A	A/P	P/G	A/G
1.0000	1.0700	0.9346	1.0000	1.0000	0.9346	1.0700	0.0000	0.0000
2.0000	1.1449	0.8734	2.0700	0.4831	1.8080	0.5531	0.8735	0.4831
3.0000	1.2250	0.8163	3.2149	0.3111	2.6243	0.3811	2.5061	0.9549
4.0000	1.3108	0.7629	4.4399	0.2252	3.3872	0.2952	4.7948	1.4155
5.0000	1.4026	0.7130	5.7507	0.1739	4.1002	0.2439	7.6467	1.8650
6.0000	1.5007	0.6663	7.1533	0.1398	4.7665	0.2098	10.9784	2.3032
7.0000	1.6058	0.6227	8.6540	0.1156	5.3893	0.1856	14.7149	2.7304
8.0000	1.7182	0.5820	10.2598	0.0975	5.9713	0.1675	18.7890	3.1466
9.0000	1.8385	0.5439	11.9780	0.0835	6.5152	0.1535	23.1405	3.5517
10.0000	1.9672	0.5083	13.8165	0.0724	7.0236	0.1424	27.7156	3.9461
11.0000	2.1049	0.4751	15.7836	0.0634	7.4987	0.1334	32.4666	4.3296
12.0000	2.2522	0.4440	17.8885	0.0559	7.9427	0.1259	37.3507	4.7025
13.0000	2.4098	0.4150	20.1407	0.0497	8.3577	0.1197	42.3303	5.0649
14.0000	2.5785	0.3878	22.5505	0.0443	8.7455	0.1143	47.3719	5.4167
15.0000	2.7590	0.3624	25.1291	0.0398	9.1079	0.1098	52.4462	5.7583
16.0000	2.9522	0.3387	27.8881	0.0359	9.4467	0.1059	57.5272	6.0897
17.0000	3.1588	0.3166	30.8403	0.0324	9.7632	0.1024	62.5924	6.4110
18.0000	3.3799	0.2959	33.9991	0.0294	10.0591	0.0994	67.6221	6.7225
19.0000	3.6165	0.2765	37.3790	0.0268	10.3356	0.0968	72.5992	7.0242
20.0000	3.8697	0.2584	40.9955	0.0244	10.5940	0.0944	77.5092	7.3163
21.0000	4.1406	0.2415	44.8652	0.0223	10.8355	0.0923	82.3395	7.5990
22.0000	4.4304	0.2257	49.0058	0.0204	11.0612	0.0904	87.0794	7.8725
23.0000	4.7405	0.2109	53.4362	0.0187	11.2722	0.0887	91.7203	8.1369
24.0000	5.0724	0.1971	58.1768	0.0172	11.4693	0.0872	96.2546	8.3923
25.0000	5.4274	0.1842	63.2491	0.0158	11.6536	0.0858	100.6766	8.6391
26.0000	5.8074	0.1722	68.6766	0.0146	11.8258	0.0846	104.9815	8.8773
27.0000	6.2139	0.1609	74.4840	0.0134	11.9867	0.0834	109.1657	9.1072
28.0000	6.6488	0.1504	80.6978	0.0124	12.1371	0.0824	113.2266	9.3290
29.0000	7.1143	0.1406	87.3467	0.0114	12.2777	0.0814	117.1623	9.5427
30.0000	7.6123	0.1314	94.4609	0.0106	12.4090	0.0806	120.9720	9.7487
31.0000	8.1451	0.1228	102.0732	0.0098	12.5318	0.0798	124.6551	9.9471
32.0000	8.7153	0.1147	110.2184	0.0091	12.6466	0.0791	128.2121	10.1381
33.0000	9.3254	0.1072	118.9336	0.0084	12.7538	0.0784	131.6436	10.3219
34.0000	9.9781	0.1002	128.2590	0.0078	12.8540	0.0778	134.9509	10.4987
35.0000	10.6766	0.0937	138.2371	0.0072	12.9477	0.0772	138.1354	10.6687
36.0000	11.4240	0.0875	148.9137	0.0067	13.0352	0.0767	141.1991	10.8321
37.0000	12.2236	0.0818	160.3377	0.0062	13.1170	0.0762	144.1442	10.9891
38.0000	13.0793	0.0765	172.5614	0.0058	13.1935	0.0758	146.9731	11.1398
39.0000	13.9948	0.0715	185.6407	0.0054	13.2649	0.0754	149.6884	11.2845
40.0000	14.9745	0.0668	199.6355	0.0050	13.3317	0.0750	152.2929	11.4234

DECIMAL INTEREST RATE = .08

PERIOD	F/P	P/F	F/A	A/F	P/A	A/P	P/G	A/G
1.0000	1.0800	0.9259	1.0000	1.0000	0.9259	1.0800	0.0000	0.0000
2.0000	1.1664	0.8573	2.0800	0.4808	1.7833	0.5608	0.8573	0.4808
3.0000	1.2597	0.7938	3.2464	0.3080	2.5771	0.3880	2.4450	0.9487
4.0000	1.3605	0.7350	4.5061	0.2219	3.3121	0.3019	4.6501	1.4040
5.0000	1.4693	0.6806	5.8666	0.1705	3.9927	0.2505	7.3725	1.8465
6.0000	1.5869	0.6302	7.3359	0.1363	4.6229	0.2163	10.5233	2.2764
7.0000	1.7138	0.5835	8.9228	0.1121	5.2064	0.1921	14.0242	2.6937
8.0000	1.8509	0.5403	10.6366	0.0940	5.7466	0.1740	17.8061	3.0985
9.0000	1.9990	0.5002	12.4876	0.0801	6.2469	0.1601	21.8081	3.4910
10.0000	2.1589	0.4632	14.4866	0.0690	6.7101	0.1490	25.9769	3.8713
11.0000	2.3316	0.4289	16.6455	0.0601	7.1390	0.1401	30.2657	4.2395
12.0000	2.5182	0.3971	18.9771	0.0527	7.5361	0.1327	34.6340	4.5958
13.0000	2.7196	0.3677	21.4953	0.0465	7.9038	0.1265	39.0463	4.9402
14.0000	2.9372	0.3405	24.2149	0.0413	8.2442	0.1213	43.4723	5.2731
15.0000	3.1722	0.3152	27.1521	0.0368	8.5595	0.1168	47.8857	5.5945
16.0000	3.4259	0.2919	30.3243	0.0330	8.8514	0.1130	52.2641	5.9046
17.0000	3.7000	0.2703	33.7503	0.0296	9.1216	0.1096	56.5884	6.2038
18.0000	3.9960	0.2502	37.4503	0.0267	9.3719	0.1067	60.8426	6.4920
19.0000	4.3157	0.2317	41.4463	0.0241	9.6036	0.1041	65.0134	6.7697
20.0000	4.6610	0.2145	45.7620	0.0219	9.8181	0.1019	69.0899	7.0370
21.0000	5.0338	0.1987	50.4230	0.0198	10.0168	0.0998	73.0630	7.2940
22.0000	5.4365	0.1839	55.4568	0.0180	10.2007	0.0980	76.9257	7.5412
23.0000	5.8715	0.1703	60.8933	0.0164	10.3711	0.0964	80.6726	7.7786
24.0000	6.3412	0.1577	66.7648	0.0150	10.5288	0.0950	84.2998	8.0066
25.0000	6.8485	0.1460	73.1060	0.0137	10.6748	0.0937	87.8042	8.2254
26.0000	7.3964	0.1352	79.9545	0.0125	10.8100	0.0925	91.1842	8.4352
27.0000	7.9881	0.1252	87.3509	0.0114	10.9352	0.0914	94.4391	8.6363
28.0000	8.6271	0.1159	95.3389	0.0105	11.0511	0.0905	97.5687	8.8289
29.0000	9.3173	0.1073	103.9660	0.0096	11.1584	0.0896	100.5739	9.0133
30.0000	10.0627	0.0994	113.2833	0.0088	11.2578	0.0888	103.4558	9.1897
31.0000	10.8677	0.0920	123.3460	0.0081	11.3498	0.0881	106.2163	9.3584
32.0000	11.7371	0.0852	134.2137	0.0075	11.4350	0.0875	108.8575	9.5197
33.0000	12.6761	0.0789	145.9508	0.0069	11.5139	0.0869	111.3820	9.6737
34.0000	13.6901	0.0730	158.6269	0.0063	11.5869	0.0863	113.7925	9.8208
35.0000	14.7854	0.0676	172.3170	0.0058	11.6546	0.0858	116.0920	9.9611
36.0000	15.9682	0.0626	187.1024	0.0053	11.7172	0.0853	118.2839	10.0949
37.0000	17.2456	0.0580	203.0706	0.0049	11.7752	0.0849	120.3714	10.2225
38.0000	18.6253	0.0537	220.3162	0.0045	11.8289	0.0845	122.3579	10.3440
39.0000	20.1153	0.0497	238.9415	0.0042	11.8786	0.0842	124.2470	10.4598
40.0000	21.7245	0.0460	259.0569	0.0039	11.9246	0.0839	126.0422	10.5699

DECIMAL INTEREST RATE = .09

PERIOD	F/P	P/F	F/A	A/F	P/A	A/P	P/G	A/G
1.0000	1.0900	0.9174	1.0000	1.0000	0.9174	1.0900	0.0000	0.0000
2.0000	1.1881	0.8417	2.0900	0.4785	1.7591	0.5685	0.8417	0.4785
3.0000	1.2950	0.7722	3.2781	0.3051	2.5313	0.3951	2.3861	0.9426
4.0000	1.4116	0.7084	4.5731	0.2187	3.2397	0.3087	4.5113	1.3925
5.0000	1.5386	0.6499	5.9847	0.1671	3.8897	0.2571	7.1111	1.8282
6.0000	1.6771	0.5963	7.5233	0.1329	4.4859	0.2229	10.0924	2.2498
7.0000	1.8280	0.5470	9.2004	0.1087	5.0330	0.1987	13.3746	2.6574
8.0000	1.9926	0.5019	11.0285	0.0907	5.5348	0.1807	16.8877	3.0512
9.0000	2.1719	0.4604	13.0210	0.0768	5.9952	0.1668	20.5711	3.4312
10.0000	2.3674	0.4224	15.1929	0.0658	6.4177	0.1558	24.3728	3.7978
11.0000	2.5804	0.3875	17.5603	0.0569	6.8052	0.1469	28.2481	4.1510
12.0000	2.8127	0.3555	20.1407	0.0497	7.1607	0.1397	32.1590	4.4910
13.0000	3.0658	0.3262	22.9534	0.0436	7.4869	0.1336	36.0732	4.8182
14.0000	3.3417	0.2992	26.0192	0.0384	7.7862	0.1284	39.9634	5.1326
15.0000	3.6425	0.2745	29.3609	0.0341	8.0607	0.1241	43.8069	5.4346
16.0000	3.9703	0.2519	33.0034	0.0303	8.3126	0.1203	47.5850	5.7245
17.0000	4.3276	0.2311	36.9737	0.0270	8.5436	0.1170	51.2821	6.0024
18.0000	4.7171	0.2120	41.3014	0.0242	8.7556	0.1142	54.8860	6.2687
19.0000	5.1417	0.1945	46.0185	0.0217	8.9501	0.1117	58.3868	6.5236
20.0000	5.6044	0.1784	51.1602	0.0195	9.1285	0.1095	61.7770	6.7675
21.0000	6.1088	0.1637	56.7646	0.0176	9.2922	0.1076	65.0510	7.0006
22.0000	6.6586	0.1502	62.8734	0.0159	9.4424	0.1059	68.2048	7.2232
23.0000	7.2579	0.1378	69.5320	0.0144	9.5802	0.1044	71.2360	7.4357
24.0000	7.9111	0.1264	76.7899	0.0130	9.7066	0.1030	74.1433	7.6384
25.0000	8.6231	0.1160	84.7010	0.0118	9.8226	0.1018	76.9265	7.8316
26.0000	9.3992	0.1064	93.3241	0.0107	9.9290	0.1007	79.5863	8.0156
27.0000	10.2451	0.0976	102.7233	0.0097	10.0266	0.0997	82.1241	8.1906
28.0000	11.1672	0.0895	112.9684	0.0089	10.1161	0.0989	84.5420	8.3571
29.0000	12.1722	0.0822	124.1355	0.0081	10.1983	0.0981	86.8423	8.5154
30.0000	13.2677	0.0754	136.3077	0.0073	10.2737	0.0973	89.0280	8.6657
31.0000	14.4618	0.0691	149.5754	0.0067	10.3428	0.0967	91.1025	8.8083
32.0000	15.7634	0.0634	164.0372	0.0061	10.4062	0.0961	93.0691	8.9436
33.0000	17.1821	0.0582	179.8006	0.0056	10.4644	0.0956	94.9315	9.0718
34.0000	18.7284	0.0534	196.9827	0.0051	10.5178	0.0951	96.6935	9.1933
35.0000	20.4140	0.0490	215.7111	0.0046	10.5668	0.0946	98.3590	9.3083
36.0000	22.2513	0.0449	236.1251	0.0042	10.6118	0.0942	99.9320	9.4171
37.0000	24.2539	0.0412	258.3764	0.0039	10.6530	0.0939	101.4163	9.5200
38.0000	26.4367	0.0378	282.6303	0.0035	10.6908	0.0935	102.8158	9.6172
39.0000	28.8160	0.0347	309.0670	0.0032	10.7255	0.0932	104.1345	9.7090
40.0000	31.4095	0.0318	337.8831	0.0030	10.7574	0.0930	105.3762	9.7957

DECIMAL INTEREST RATE = .1

PERIOD	F/P	P/F	F/A	A/F	P/A	A/P	P/G	A/G
1.0000	1.1000	0.9091	1.0000	1.0000	0.9091	1.1000	0.0000	0.0000
2.0000	1.2100	0.8264	2.1000	0.4762	1.7355	0.5762	0.8264	0.4762
3.0000	1.3310	0.7513	3.3100	0.3021	2.4869	0.4021	2.3291	0.9366
4.0000	1.4641	0.6830	4.6410	0.2155	3.1699	0.3155	4.3781	1.3812
5.0000	1.6105	0.6209	6.1051	0.1638	3.7908	0.2638	6.8618	1.8101
6.0000	1.7716	0.5645	7.7156	0.1296	4.3553	0.2296	9.6842	2.2236
7.0000	1.9487	0.5132	9.4872	0.1054	4.8684	0.2054	12.7631	2.6216
8.0000	2.1436	0.4665	11.4359	0.0874	5.3349	0.1874	16.0287	3.0045
9.0000	2.3579	0.4241	13.5795	0.0736	5.7590	0.1736	19.4215	3.3724
10.0000	2.5937	0.3855	15.9374	0.0627	6.1446	0.1627	22.8914	3.7255
11.0000	2.8531	0.3505	18.5312	0.0540	6.4951	0.1540	26.3963	4.0641
12.0000	3.1384	0.3186	21.3843	0.0468	6.8137	0.1468	29.9012	4.3884
13.0000	3.4523	0.2897	24.5227	0.0408	7.1034	0.1408	33.3772	4.6988
14.0000	3.7975	0.2633	27.9750	0.0357	7.3667	0.1357	36.8005	4.9955
15.0000	4.1772	0.2394	31.7725	0.0315	7.6061	0.1315	40.1520	5.2789
16.0000	4.5950	0.2176	35.9497	0.0278	7.8237	0.1278	43.4164	5.5493
17.0000	5.0545	0.1978	40.5447	0.0247	8.0216	0.1247	46.5820	5.8071
18.0000	5.5599	0.1799	45.5992	0.0219	8.2014	0.1219	49.6396	6.0526
19.0000	6.1159	0.1635	51.1591	0.0195	8.3649	0.1195	52.5827	6.2861
20.0000	6.7275	0.1486	57.2750	0.0175	8.5136	0.1175	55.4069	6.5081
21.0000	7.4003	0.1351	64.0025	0.0156	8.6487	0.1156	58.1095	6.7189
22.0000	8.1403	0.1228	71.4028	0.0140	8.7715	0.1140	60.6893	6.9189
23.0000	8.9543	0.1117	79.5431	0.0126	8.8832	0.1126	63.1462	7.1085
24.0000	9.8497	0.1015	88.4974	0.0113	8.9847	0.1113	65.4813	7.2881
25.0000	10.8347	0.0923	98.3471	0.0102	9.0770	0.1102	67.6964	7.4580
26.0000	11.9182	0.0839	109.1818	0.0092	9.1609	0.1092	69.7940	7.6187
27.0000	13.1100	0.0763	121.1000	0.0083	9.2372	0.1083	71.7773	7.7704
28.0000	14.4210	0.0693	134.2100	0.0075	9.3066	0.1075	73.6495	7.9137
29.0000	15.8631	0.0630	148.6310	0.0067	9.3696	0.1067	75.4146	8.0489
30.0000	17.4494	0.0573	164.4941	0.0061	9.4269	0.1061	77.0766	8.1762
31.0000	19.1944	0.0521	181.9435	0.0055	9.4790	0.1055	78.6396	8.2962
32.0000	21.1138	0.0474	201.1379	0.0050	9.5264	0.1050	80.1078	8.4091
33.0000	23.2252	0.0431	222.2517	0.0045	9.5694	0.1045	81.4856	8.5152
34.0000	25.5477	0.0391	245.4768	0.0041	9.6086	0.1041	82.7773	8.6149
35.0000	28.1025	0.0356	271.0245	0.0037	9.6442	0.1037	83.9872	8.7086
36.0000	30.9127	0.0323	299.1270	0.0033	9.6765	0.1033	85.1194	8.7965
37.0000	34.0040	0.0294	330.0397	0.0030	9.7059	0.1030	86.1781	8.8789
38.0000	37.4044	0.0267	364.0437	0.0027	9.7327	0.1027	87.1673	8.9562
39.0000	41.1448	0.0243	401.4480	0.0025	9.7570	0.1025	88.0908	9.0285
40.0000	45.2593	0.0221	442.5928	0.0023	9.7791	0.1023	88.9525	9.0962

DECIMAL INTEREST RATE =   .12

PERIOD	F/P	P/F	F/A	A/F	P/A	A/P	P/G	A/G
1.0000	1.1200	0.8929	1.0000	1.0000	0.8929	1.1200	0.0000	0.0000
2.0000	1.2544	0.7972	2.1200	0.4717	1.6901	0.5917	0.7972	0.4717
3.0000	1.4049	0.7118	3.3744	0.2963	2.4018	0.4163	2.2208	0.9246
4.0000	1.5735	0.6355	4.7793	0.2092	3.0373	0.3292	4.1273	1.3589
5.0000	1.7623	0.5674	6.3528	0.1574	3.6048	0.2774	6.3970	1.7746
6.0000	1.9738	0.5066	8.1152	0.1232	4.1114	0.2432	8.9302	2.1720
7.0000	2.2107	0.4523	10.0890	0.0991	4.5638	0.2191	11.6443	2.5515
8.0000	2.4760	0.4039	12.2997	0.0813	4.9676	0.2013	14.4715	2.9131
9.0000	2.7731	0.3606	14.7757	0.0677	5.3283	0.1877	17.3563	3.2574
10.0000	3.1058	0.3220	17.5487	0.0570	5.6502	0.1770	20.2541	3.5847
11.0000	3.4785	0.2875	20.6546	0.0484	5.9377	0.1684	23.1288	3.8953
12.0000	3.8960	0.2567	24.1331	0.0414	6.1944	0.1614	25.9523	4.1897
13.0000	4.3635	0.2292	28.0291	0.0357	6.4235	0.1557	28.7024	4.4683
14.0000	4.8871	0.2046	32.3926	0.0309	6.6282	0.1509	31.3624	4.7317
15.0000	5.4736	0.1827	37.2797	0.0268	6.8109	0.1468	33.9202	4.9803
16.0000	6.1304	0.1631	42.7533	0.0234	6.9740	0.1434	36.3670	5.2147
17.0000	6.8660	0.1456	48.8837	0.0205	7.1196	0.1405	38.6973	5.4353
18.0000	7.6900	0.1300	55.7497	0.0179	7.2497	0.1379	40.9080	5.6427
19.0000	8.6128	0.1161	63.4397	0.0158	7.3658	0.1358	42.9979	5.8375
20.0000	9.6463	0.1037	72.0524	0.0139	7.4694	0.1339	44.9676	6.0202
21.0000	10.8038	0.0926	81.6987	0.0122	7.5620	0.1322	46.8188	6.1913
22.0000	12.1003	0.0826	92.5026	0.0108	7.6446	0.1308	48.5543	6.3514
23.0000	13.5523	0.0738	104.6029	0.0096	7.7184	0.1296	50.1776	6.5010
24.0000	15.1786	0.0659	118.1552	0.0085	7.7843	0.1285	51.6929	6.6406
25.0000	17.0001	0.0588	133.3339	0.0075	7.8431	0.1275	53.1046	6.7708
26.0000	19.0401	0.0525	150.3339	0.0067	7.8957	0.1267	54.4177	6.8921
27.0000	21.3249	0.0469	169.3740	0.0059	7.9426	0.1259	55.6369	7.0049
28.0000	23.8839	0.0419	190.6989	0.0052	7.9844	0.1252	56.7674	7.1098
29.0000	26.7499	0.0374	214.5828	0.0047	8.0218	0.1247	57.8141	7.2071
30.0000	29.9599	0.0334	241.3327	0.0041	8.0552	0.1241	58.7821	7.2974
31.0000	33.5551	0.0298	271.2926	0.0037	8.0850	0.1237	59.6761	7.3811
32.0000	37.5817	0.0266	304.8477	0.0033	8.1116	0.1233	60.5010	7.4586
33.0000	42.0915	0.0238	342.4295	0.0029	8.1354	0.1229	61.2612	7.5302
34.0000	47.1425	0.0212	384.5210	0.0026	8.1566	0.1226	61.9612	7.5965
35.0000	52.7996	0.0189	431.6635	0.0023	8.1755	0.1223	62.6052	7.6577
36.0000	59.1356	0.0169	484.4632	0.0021	8.1924	0.1221	63.1970	7.7141
37.0000	66.2318	0.0151	543.5987	0.0018	8.2075	0.1218	63.7406	7.7661
38.0000	74.1797	0.0135	609.8306	0.0016	8.2210	0.1216	64.2394	7.8141
39.0000	83.0812	0.0120	684.0102	0.0015	8.2330	0.1215	64.6968	7.8582
40.0000	93.0510	0.0107	767.0914	0.0013	8.2438	0.1213	65.1159	7.8988

DECIMAL INTEREST RATE = .15

PERIOD	F/P	P/F	F/A	A/F	P/A	A/P	P/G	A/G
1.0000	1.1500	0.8696	1.0000	1.0000	0.8696	1.1500	-0.0000	-0.0000
2.0000	1.3225	0.7561	2.1500	0.4651	1.6257	0.6151	0.7561	0.4651
3.0000	1.5209	0.6575	3.4725	0.2880	2.2832	0.4380	2.0712	0.9071
4.0000	1.7490	0.5718	4.9934	0.2003	2.8550	0.3503	3.7864	1.3263
5.0000	2.0114	0.4972	6.7424	0.1483	3.3522	0.2983	5.7751	1.7228
6.0000	2.3131	0.4323	8.7537	0.1142	3.7845	0.2642	7.9368	2.0972
7.0000	2.6600	0.3759	11.0668	0.0904	4.1604	0.2404	10.1924	2.4498
8.0000	3.0590	0.3269	13.7268	0.0729	4.4873	0.2229	12.4807	2.7813
9.0000	3.5179	0.2843	16.7858	0.0596	4.7716	0.2096	14.7548	3.0922
10.0000	4.0456	0.2472	20.3037	0.0493	5.0188	0.1993	16.9795	3.3832
11.0000	4.6524	0.2149	24.3493	0.0411	5.2337	0.1911	19.1289	3.6549
12.0000	5.3503	0.1869	29.0017	0.0345	5.4206	0.1845	21.1849	3.9082
13.0000	6.1528	0.1625	34.3519	0.0291	5.5831	0.1791	23.1352	4.1438
14.0000	7.0757	0.1413	40.5047	0.0247	5.7245	0.1747	24.9725	4.3624
15.0000	8.1371	0.1229	47.5804	0.0210	5.8474	0.1710	26.6930	4.5650
16.0000	9.3576	0.1069	55.7175	0.0179	5.9542	0.1679	28.2960	4.7522
17.0000	10.7613	0.0929	65.0751	0.0154	6.0472	0.1654	29.7828	4.9251
18.0000	12.3755	0.0808	75.8364	0.0132	6.1280	0.1632	31.1565	5.0843
19.0000	14.2318	0.0703	88.2118	0.0113	6.1982	0.1613	32.4213	5.2307
20.0000	16.3665	0.0611	102.4436	0.0098	6.2593	0.1598	33.5822	5.3651
21.0000	18.8215	0.0531	118.8101	0.0084	6.3125	0.1584	34.6448	5.4883
22.0000	21.6447	0.0462	137.6316	0.0073	6.3587	0.1573	35.6150	5.6010
23.0000	24.8915	0.0402	159.2764	0.0063	6.3988	0.1563	36.4988	5.7040
24.0000	28.6252	0.0349	184.1679	0.0054	6.4338	0.1554	37.3023	5.7979
25.0000	32.9190	0.0304	212.7930	0.0047	6.4641	0.1547	38.0314	5.8834
26.0000	37.8568	0.0264	245.7120	0.0041	6.4906	0.1541	38.6918	5.9612
27.0000	43.5353	0.0230	283.5688	0.0035	6.5135	0.1535	39.2890	6.0319
28.0000	50.0656	0.0200	327.1041	0.0031	6.5335	0.1531	39.8283	6.0960
29.0000	57.5755	0.0174	377.1697	0.0027	6.5509	0.1527	40.3146	6.1541
30.0000	66.2118	0.0151	434.7452	0.0023	6.5660	0.1523	40.7526	6.2066
31.0000	76.1436	0.0131	500.9570	0.0020	6.5791	0.1520	41.1466	6.2541
32.0000	87.5651	0.0114	577.1005	0.0017	6.5905	0.1517	41.5006	6.2970
33.0000	100.6998	0.0099	664.6655	0.0015	6.6005	0.1515	41.8184	6.3357
34.0000	115.8048	0.0086	765.3653	0.0013	6.6091	0.1513	42.1033	6.3705
35.0000	133.1755	0.0075	881.1701	0.0011	6.6166	0.1511	42.3586	6.4019
36.0000	153.1519	0.0065	1014.3460	0.0010	6.6231	0.1510	42.5872	6.4301
37.0000	176.1246	0.0057	1167.4970	0.0009	6.6288	0.1509	42.7916	6.4554
38.0000	202.5433	0.0049	1343.6220	0.0007	6.6338	0.1507	42.9742	6.4781
39.0000	232.9248	0.0043	1546.1660	0.0006	6.6380	0.1506	43.1374	6.4985
40.0000	267.8636	0.0037	1779.0900	0.0006	6.6418	0.1506	43.2830	6.5168

DECIMAL INTEREST RATE = .18

PERIOD	F/P	P/F	F/A	A/F	P/A	A/P	P/G	A/G
1.0000	1.1800	0.8475	1.0000	1.0000	0.8475	1.1800	0.0000	0.0000
2.0000	1.3924	0.7182	2.1800	0.4587	1.5656	0.6387	0.7182	0.4587
3.0000	1.6430	0.6086	3.5724	0.2799	2.1743	0.4599	1.9355	0.8902
4.0000	1.9388	0.5158	5.2154	0.1917	2.6901	0.3717	3.4828	1.2947
5.0000	2.2878	0.4371	7.1542	0.1398	3.1272	0.3198	5.2313	1.6728
6.0000	2.6996	0.3704	9.4420	0.1059	3.4976	0.2859	7.0834	2.0252
7.0000	3.1855	0.3139	12.1415	0.0824	3.8115	0.2624	8.9670	2.3526
8.0000	3.7589	0.2660	15.3270	0.0652	4.0776	0.2452	10.8292	2.6558
9.0000	4.4355	0.2255	19.0859	0.0524	4.3030	0.2324	12.6329	2.9358
10.0000	5.2338	0.1911	23.5213	0.0425	4.4941	0.2225	14.3525	3.1936
11.0000	6.1759	0.1619	28.7552	0.0348	4.6560	0.2148	15.9717	3.4303
12.0000	7.2876	0.1372	34.9311	0.0286	4.7932	0.2086	17.4811	3.6470
13.0000	8.5994	0.1163	42.2187	0.0237	4.9095	0.2037	18.8765	3.8449
14.0000	10.1473	0.0985	50.8181	0.0197	5.0081	0.1997	20.1577	4.0250
15.0000	11.9738	0.0835	60.9653	0.0164	5.0916	0.1964	21.3269	4.1887
16.0000	14.1290	0.0708	72.9391	0.0137	5.1624	0.1937	22.3885	4.3369
17.0000	16.6723	0.0600	87.0681	0.0115	5.2223	0.1915	23.3482	4.4708
18.0000	19.6733	0.0508	103.7404	0.0096	5.2732	0.1896	24.2123	4.5916
19.0000	23.2145	0.0431	123.4137	0.0081	5.3162	0.1881	24.9877	4.7003
20.0000	27.3931	0.0365	146.6281	0.0068	5.3527	0.1868	25.6813	4.7978
21.0000	32.3238	0.0309	174.0212	0.0057	5.3837	0.1857	26.3000	4.8851
22.0000	38.1421	0.0262	206.3450	0.0048	5.4099	0.1848	26.8506	4.9632
23.0000	45.0077	0.0222	244.4872	0.0041	5.4321	0.1841	27.3394	5.0329
24.0000	53.1091	0.0188	289.4949	0.0035	5.4509	0.1835	27.7725	5.0950
25.0000	62.6687	0.0160	342.6039	0.0029	5.4669	0.1829	28.1555	5.1502
26.0000	73.9491	0.0135	405.2727	0.0025	5.4804	0.1825	28.4935	5.1991
27.0000	87.2599	0.0115	479.2219	0.0021	5.4919	0.1821	28.7915	5.2425
28.0000	102.9667	0.0097	566.4817	0.0018	5.5016	0.1818	29.0537	5.2810
29.0000	121.5007	0.0082	669.4485	0.0015	5.5098	0.1815	29.2842	5.3149
30.0000	143.3709	0.0070	790.9493	0.0013	5.5168	0.1813	29.4864	5.3448
31.0000	169.1776	0.0059	934.3202	0.0011	5.5227	0.1811	29.6638	5.3712
32.0000	199.6296	0.0050	1103.4980	0.0009	5.5277	0.1809	29.8190	5.3945
33.0000	235.5630	0.0042	1303.1280	0.0008	5.5320	0.1808	29.9549	5.4149
34.0000	277.9643	0.0036	1538.6910	0.0006	5.5356	0.1806	30.0736	5.4328
35.0000	327.9979	0.0030	1816.6550	0.0006	5.5386	0.1806	30.1773	5.4485
36.0000	387.0375	0.0026	2144.6530	0.0005	5.5412	0.1805	30.2677	5.4623
37.0000	456.7044	0.0022	2531.6910	0.0004	5.5434	0.1804	30.3465	5.4744
38.0000	538.9111	0.0019	2988.3950	0.0003	5.5452	0.1803	30.4152	5.4849
39.0000	635.9152	0.0016	3527.3070	0.0003	5.5468	0.1803	30.4749	5.4941
40.0000	750.3800	0.0013	4163.2220	0.0002	5.5482	0.1802	30.5269	5.5022

DECIMAL INTEREST RATE = .2

PERIOD	F/P	P/F	F/A	A/F	P/A	A/P	P/G	A/G
1.0000	1.2000	0.8333	1.0000	1.0000	0.8333	1.2000	0.0000	0.0000
2.0000	1.4400	0.6944	2.2000	0.4545	1.5278	0.6545	0.6944	0.4545
3.0000	1.7280	0.5787	3.6400	0.2747	2.1065	0.4747	1.8519	0.8791
4.0000	2.0736	0.4823	5.3680	0.1863	2.5887	0.3863	3.2986	1.2742
5.0000	2.4883	0.4019	7.4416	0.1344	2.9906	0.3344	4.9061	1.6405
6.0000	2.9860	0.3349	9.9299	0.1007	3.3255	0.3007	6.5806	1.9788
7.0000	3.5832	0.2791	12.9159	0.0774	3.6046	0.2774	8.2551	2.2902
8.0000	4.2998	0.2326	16.4991	0.0606	3.8372	0.2606	9.8831	2.5756
9.0000	5.1598	0.1938	20.7989	0.0481	4.0310	0.2481	11.4335	2.8364
10.0000	6.1917	0.1615	25.9587	0.0385	4.1925	0.2385	12.8871	3.0739
11.0000	7.4301	0.1346	32.1504	0.0311	4.3271	0.2311	14.2330	3.2893
12.0000	8.9161	0.1122	39.5805	0.0253	4.4392	0.2253	15.4667	3.4841
13.0000	10.6993	0.0935	48.4966	0.0206	4.5327	0.2206	16.5883	3.6597
14.0000	12.8392	0.0779	59.1959	0.0169	4.6106	0.2169	17.6008	3.8175
15.0000	15.4070	0.0649	72.0351	0.0139	4.6755	0.2139	18.5095	3.9588
16.0000	18.4884	0.0541	87.4421	0.0114	4.7296	0.2114	19.3208	4.0851
17.0000	22.1861	0.0451	105.9306	0.0094	4.7746	0.2094	20.0419	4.1976
18.0000	26.6233	0.0376	128.1167	0.0078	4.8122	0.2078	20.6805	4.2975
19.0000	31.9480	0.0313	154.7400	0.0065	4.8435	0.2065	21.2439	4.3861
20.0000	38.3376	0.0261	186.6880	0.0054	4.8696	0.2054	21.7395	4.4643
21.0000	46.0051	0.0217	225.0256	0.0044	4.8913	0.2044	22.1742	4.5334
22.0000	55.2061	0.0181	271.0307	0.0037	4.9094	0.2037	22.5546	4.5941
23.0000	66.2474	0.0151	326.2369	0.0031	4.9245	0.2031	22.8867	4.6475
24.0000	79.4969	0.0126	392.4843	0.0025	4.9371	0.2025	23.1760	4.6943
25.0000	95.3962	0.0105	471.9811	0.0021	4.9476	0.2021	23.4276	4.7352
26.0000	114.4755	0.0087	567.3773	0.0018	4.9563	0.2018	23.6460	4.7709
27.0000	137.3706	0.0073	681.8529	0.0015	4.9636	0.2015	23.8353	4.8020
28.0000	164.8447	0.0061	819.2233	0.0012	4.9697	0.2012	23.9991	4.8291
29.0000	197.8136	0.0051	984.0681	0.0010	4.9747	0.2010	24.1406	4.8527
30.0000	237.3764	0.0042	1181.8820	0.0008	4.9789	0.2008	24.2628	4.8731
31.0000	284.8516	0.0035	1419.2580	0.0007	4.9824	0.2007	24.3681	4.8908
32.0000	341.8219	0.0029	1704.1100	0.0006	4.9854	0.2006	24.4588	4.9061
33.0000	410.1863	0.0024	2045.9310	0.0005	4.9878	0.2005	24.5368	4.9194
34.0000	492.2236	0.0020	2456.1180	0.0004	4.9898	0.2004	24.6038	4.9308
35.0000	590.6683	0.0017	2948.3410	0.0003	4.9915	0.2003	24.6614	4.9406
36.0000	708.8019	0.0014	3539.0090	0.0003	4.9929	0.2003	24.7108	4.9491
37.0000	850.5623	0.0012	4247.8120	0.0002	4.9941	0.2002	24.7531	4.9564
38.0000	1020.6750	0.0010	5098.3740	0.0002	4.9951	0.2002	24.7894	4.9627
39.0000	1224.8100	0.0008	6119.0490	0.0002	4.9959	0.2002	24.8204	4.9681
40.0000	1469.7720	0.0007	7343.8580	0.0001	4.9966	0.2001	24.8469	4.9728

Index

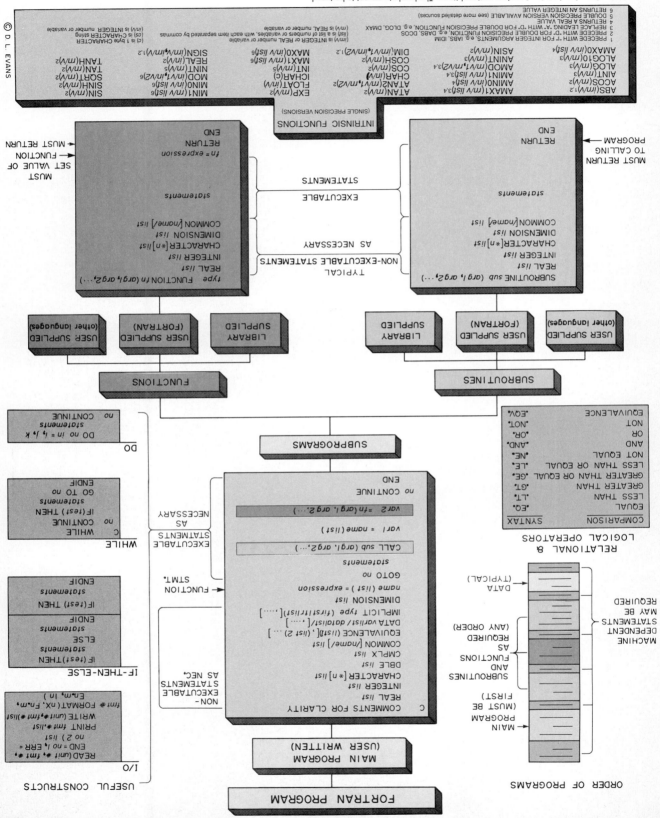